CHIROPTICAL SPECTROSCOPY

CHIROPTICAL SPECTROSCOPY

Fundamentals and Applications

Prasad L. Polavarapu

Vanderbilt University, Nashville, Tennessee, USA

CRC Press is an imprint of the
Taylor & Francis Group, an **informa** business

CRC Press
Taylor & Francis Group
6000 Broken Sound Parkway NW, Suite 300
Boca Raton, FL 33487-2742

First issued in paperback 2019

ISBN-13: 978-1-4200-9246-2 (hbk)
ISBN-13: 978-0-367-86456-9 (pbk)

Library of Congress Cataloging-in-Publication Data

Names: Polavarapu, Prasad L.
Title: Chiroptical spectroscopy : fundamentals and applications / Prasad L. Polavarapu.
Description: Boca Raton : Taylor & Francis, 2017. | "A CRC title." | Includes
bibliographical references and index.
Identifiers: LCCN 2016009687 | ISBN 9781420092462 (alk. paper)
Subjects: LCSH: Optical spectroscopy. | Spectrum analysis. | Vibrational spectra.
Classification: LCC QC454.O66 P65 2017 | DDC 543/.5--dc23
LC record available at https://lccn.loc.gov/2016009687

Visit the Taylor & Francis Web site at
http://www.taylorandfrancis.com

and the CRC Press Web site at
http://www.crcpress.com

To Bharathi, Sruthi, and Aaseesh

Contents

Appendixes

Preface

Optical activity is a phenomenon that is well known to scientists (chemists, biochemists, and physicists, in particular). For some, the familiarity with optical activity comes from the rotation of plane polarized visible light, often referred to as optical rotation (OR). This property when normalized for concentration and path length is called specific rotation (SR), and when displayed as a wavelength resolved plot is called optical rotatory dispersion (ORD). In recent years, SR is often referred to as specific optical rotation (SOR). Another familiar form of optical activity is the differential absorption of left and right circularly polarized visible light, often referred to as circular dichroism (CD) and more specifically as electronic circular dichroism (ECD). The scientific research in ORD and ECD originated in the nineteenth century and has been widespread. For aged research areas such as these, one may wonder if there is anything new that is left to be discovered. It would not be surprising for people with an open mind that such complacency can be out of place in scientific research. Not only did significant new developments take place in these areas but also new variants of optical activity were discovered in the late twentieth century. The new variants include CD in the infrared region, known as infrared vibrational circular dichroism (VCD), and differential vibrational Raman scattering of left and right circularly polarized visible light, known as vibrational Raman optical activity (VROA). These two new areas are referred together as vibrational optical activity (VOA). New developments in the methods for measurement and interpretation of OR and ECD have also taken place at about the same time. The four methods mentioned above (ORD, ECD, VCD, and VROA) represent four different branches of chiroptical spectroscopy. An increasing number of research groups have started utilizing these methods in the quest for advancing science. Since most of the research in individual research laboratories is carried out by graduate and undergraduate students, an influx of new young scientists has occurred. The excellent books (Barron 2004; Charney 1985; Nafie 2011; Polavarapu 1998; Stephens et al. 2012) available in the market on this subject have focused either on one or two of these areas or are written for people with an advanced background. As a result, a starting young scientist is often burdened with collecting pieces of information from different sources. My desire for writing a book that explains chiroptical spectroscopy starting from an elementary level, and progressing to the level of helping to adopt this research area, has emerged in this context with beginning researchers in mind.

My goal for this book is to provide a convenient and comfortable reference for those who are novice to chiroptical spectroscopy. At the same time, I wanted to lead them to a level where they can feel comfortable to pursue the advanced level of inquiry, if they so desire. At the first instinct, it may sound impossible to achieve these goals in one monograph. Nevertheless, I tried to achieve these goals by providing the main text at an elementary level, keeping novice students in mind, in an easily readable language. However, when it comes to describing the quantum mechanical underpinnings, there is no escape from jumping into involved mathematical formulations. To make this a smooth transition, I have presented the equations in as much detail as possible, so a starting researcher does not have to get bogged down in deriving them.

It is my hope that this book will serve as a guiding source for both entry level and seasoned practitioners of these research areas, as much of the needed information for both groups of researchers has been provided. In addition, this book is intended to serve as a textbook for an introductory course on chiroptical spectroscopy.

Greater benefit to society can be realized if an author can help the novice understand the material easily and thoroughly. Therefore, it is my sincere hope that this book will attract and help a new generation of researchers in chiroptical spectroscopy to become proficient in this scientific area.

Finally, I am privileged to be associated with excellent teachers, research advisers, collaborators, postdoctoral research associates, graduate students, and undergraduate students, who conducted research with me in the last four decades. With immense pleasure, I acknowledge my productive association with them and express my gratitude for facilitating the never-ending learning process.

Prasad L. Polavarapu
Vanderbilt University

References

Barron, L. D. 2004. *Molecular Light Scattering and Optical Activity*, 2nd ed. Cambridge, UK: Cambridge University Press.

Charney, E. 1985. *The Molecular Basis of Optical Activity*. Malabar, FL: Krieger Publishing Co.

Nafie, L. A. 2011. *Vibrational Optical Activity: Principles and Applications*. New York, NY: John Wiley and Sons.

Polavarapu, P. L. 1998. *Vibrational Spectra: Principles and Applications with Emphasis on Optical Activity*. New York, NY: Elsevier.

Stephens, P. J., F. Devlin, and J. R. Cheeseman. 2012. *VCD Spectroscopy for Organic Chemists*. London, UK: CRC Press.

Author

Prasad L. Polavarapu, PhD, was born and raised in Gudlavalleru, a small village in the state of Andhra Pradesh, India. His parents were farmers by profession and did not have a formal educational background. Nevertheless, his father provided never-ending support for him to pursue higher education.

Dr. Polavarapu joined the PhD program at the Indian Institute of Technology, Madras (now renamed as Chennai), after earning his master's degree from Birla Institute of Technology, Pilani. He earned his doctoral degree under the supervision of the late Professor Surjit Singh at the Indian Institute of Technology, where his research focused on the interpretation of vibrational infrared absorption intensities. On completion of his doctoral degree, he moved to the laboratory of Professor Duane F. Burrow at the University of Toledo, Toledo, Ohio, in 1977 for postdoctoral research on vibrational Raman optical activity. In the subsequent year, he joined the laboratory of Professor Laurence A. Nafie at Syracuse University, Syracuse, New York, where he carried out research work on vibrational Raman optical activity and vibrational circular dichroism. Dr. Polavarapu joined Vanderbilt University, Nashville, Tennessee, in 1980 as an assistant professor of chemistry. He was promoted to associate professor in 1987 and then to professor in 1994.

Dr. Polavarapu has made pioneering contributions in the research areas of optical rotation, vibrational Raman optical activity, vibrational circular dichroism, and electronic circular dichroism. He has published approximately 260 peer-reviewed research papers in leading scientific journals.

Dr. Polavarapu edited the book *Polarization Division Interferometry*, which was published in 1997 by John Wiley & Sons. One of the contributors of that book, Dr. John Mather, won the 2006 Nobel Prize in Physics. Dr. Polavarapu wrote the book *Vibrational Spectroscopy: Principles and Applications with Emphasis on Optical Activity*, which was published by Elsevier in 1998. More recently, he coedited two volumes on *Comprehensive Chiroptical Spectroscopy*, which were published by John Wiley & Sons in 2012.

Dr. Polavarapu was awarded the Jeffrey Nordhaus Award in 2010 for excellence in undergraduate teaching at Vanderbilt University. He was elected as a fellow of the American Association for Advancement of Science in 2012 for distinguished contributions to the field of chiral molecular structure determination.

Dr. Polavarapu continues his research work in chiroptical spectroscopy while teaching undergraduate and graduate courses at Vanderbilt University.

Author

1

Polarized Light

Interaction between electromagnetic radiation and matter is the basis of a broad research area that is generally categorized as spectroscopy. When the matter under consideration is made up of chiral molecules (see Chapter 2), a special branch of spectroscopy, labeled as chiroptical spectroscopy, becomes pertinent. The words "chiroptical spectroscopy" can be construed to mean the study of chiral systems using optical spectroscopy methods. To understand the principles of chiroptical spectroscopy, it is necessary to gain insight into the properties of electromagnetic radiation, chiral molecules, and the interaction between them. These concepts form the basis for the starting chapters in this book.

1.1 Introduction to Light

Visible light represents electromagnetic radiation that can be detected with the human eye. This statement automatically raises thoughts about "invisible" light. To distinguish between visible and invisible light, we need to consider the wave properties of electromagnetic radiation. A wave (see Figure 1.1a) is associated with periodic oscillatory variation of its height (referred to as amplitude) in time. Two identical points in a wave constitute one cycle. For example, two consecutive crest points, two consecutive trough points, or three consecutive zero points constitute one cycle or one wave. If an observer stands at one point and watches the waves pass by in time, the number of cycles passed in a given second is labeled as frequency (ν). The unit of frequency is cycles/second (or simply s^{-1}), which is designated as hertz (Hz; 1 Hz = s^{-1}). A cosine wave with unit amplitude is represented by the function $\cos 2\pi\nu t$, where t stands for time in seconds. This cosine wave is shown in Figure 1.1a, where the definition of a cycle is depicted. The alternate representations of frequency of oscillation are wavelength and wavenumber. If a stationary wave is held along the z axis (see Figure 1.1b) and the distance between three consecutive zero points (or between two crests or between two troughs) of that wave is measured, then that distance is labeled as wavelength and represented by symbol λ [expressed as meters/cycle or simply as meters (m)]. This definition of wavelength is shown in Figure 1.1b, where a

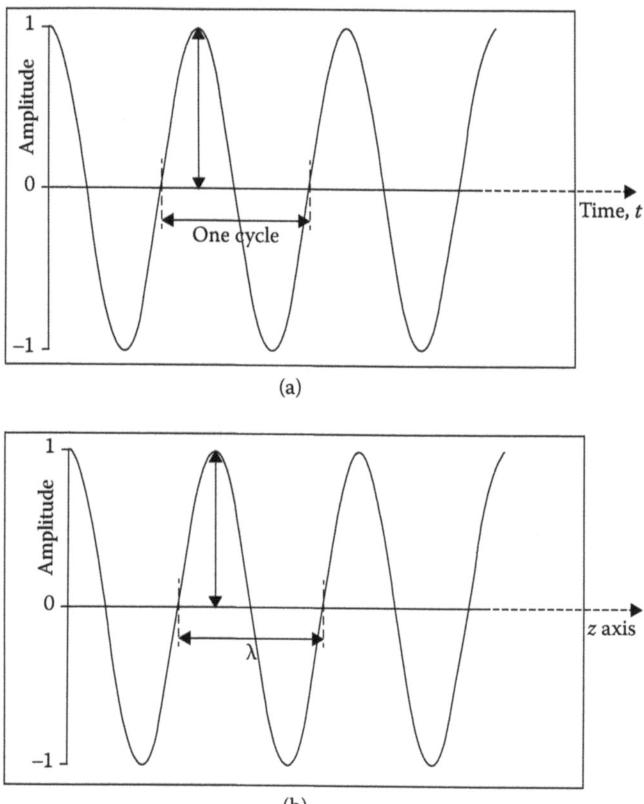

FIGURE 1.1
(a) Depiction of the amplitude and cycle of an oscillating cosine wave, cos $2\pi\nu t$, with unit amplitude as a function of time. (b) Depiction of the amplitude and wavelength λ of an oscillating cosine wave, cos $(2\pi z/\lambda)$, propagating along the z axis with unit amplitude.

wave with unit amplitude, represented by the cosine function $\cos 2\pi z/\lambda$, is depicted. The product of wavelength and frequency of a wave is equal to the speed of that wave c, which is expressed as $\text{m}\cdot\text{s}^{-1}$:

$$c = \lambda\nu = \frac{\text{m}}{\text{cycle}} \times \frac{\text{cycle}}{\text{s}} = \frac{\text{m}}{\text{s}} = \text{m}\,\text{s}^{-1} \tag{1.1}$$

If an electromagnetic wave passes through a medium with refractive index n_m, then the speed of electromagnetic wave changes as $c_m = c/n_m$. Then, Equation 1.1 modifies as $c_m = \lambda_m \nu$, where λ_m is the wavelength of the wave in that medium (Levine 2009). For the current purposes, consider the waves propagating in vacuum, where $n_m = 1$. Just as the number of cycles per second is labeled as frequency, the number of waves per unit length is labeled as wavenumber ($\bar{\nu}$). Since one wave occupies a length of λ, wavenumber is equal to the reciprocal of wavelength. Therefore, $\bar{\nu} = 1/\lambda$, and its units are expressed in m^{-1}. Thus, an alternate form of Equation 1.1 is as follows:

$$\nu = \bar{\nu}c = m^{-1} \times \frac{m}{s} = s^{-1} \tag{1.2}$$

The wavelength λ (or equivalently frequency ν or wavenumber $\bar{\nu}$) of a wave determines the properties of that wave. Human eyes respond to the electromagnetic waves of wavelength between ~400 and 800 nm (1 nm = 10^{-9} m), and this portion of electromagnetic radiation is referred to as the visible light. The different colors that human eyes can recognize (violet, indigo, blue, green, yellow, orange, and red, abbreviated as VIBGYOR) are the components of the visible light. The portions of electromagnetic radiation with wavelengths greater or less than the wavelengths in the visible light range are invisible to the human eye.

From here onward, the word "light" will be used to represent electromagnetic radiation (regardless of whether it is visible or invisible to the human eyes). The speed of light in vacuum is a constant, 2.99792458×10^8 m·s^{-1}. Wavelength and frequency of a light wave are related through Equations 1.1 and 1.2. The different wavelength ranges of electromagnetic radiation constitute different portions of the electromagnetic "spectrum." Selected portions of electromagnetic spectrum and their names are summarized in Table 1.1.

Wavelength is expressed in nanometers (nm) for the ultraviolet–visible region and micrometers (μm) for the infrared region. Wavenumber is usually expressed in cm^{-1}. For example, a light wave with 200-nm wavelength represents ultraviolet light and corresponds to a wavenumber of 50,000 cm^{-1} $\left(\frac{1}{200 \times 10^{-9}\,m \times 100\,\frac{cm}{m}} = 50,000\,cm^{-1}\right)$ and a frequency of 1.5×10^{15} Hz $\left(50,000\,cm^{-1} \times 2.99792458 \times 10^{10}\,\frac{cm}{s} = 1.5 \times 10^{15}\,s^{-1}\right)$.

TABLE 1.1

Selected Regions of Electromagnetic Radiation[a,b]

Region	Wavelength, λ	Wavenumber, $\bar{\nu}$	Frequency, ν (Hz)	Energy, E (kJ/mol)
Ultraviolet	200–400 nm	50,000–25,000 cm^{-1}	1.5×10^{15}–0.75×10^{15}	598–299
Visible	400–800 nm	25,000–12,500 cm^{-1}	0.75×10^{15}–0.37×10^{15}	299–150
Near-infrared	0.8–3 μm	12,500–3,333 cm^{-1}	0.37×10^{15}–0.1×10^{15}	150–39.9
Mid-infrared	3–25 μm	3,333–400 cm^{-1}	1.0×10^{14}–0.1×10^{14}	39.9–4.78
Far-infrared	25–100 μm	400–10 cm^{-1}	12×10^{12}–3×10^{12}	4.78–1.20

[a] 1 nm = 10^{-9}m; 1 μm = 10^{-6} m; $\bar{\nu} = 1/\lambda$; $\nu = c/\lambda$; speed of light, $c = 2.99792458 \times 10^{10}$ cm/s; energy per photon is $E = hc/\lambda$; Planck's constant, $h = 6.62606896 \times 10^{-34}$ J·s^{-1}; kJ = 1,000 J; cal = 4.184 J; energy per mole of photons is $E = Nhc/\lambda$, where N is the Avogadro's number; $N = 6.02214179 \times 10^{23}$ mol^{-1}.

[b] Wavelengths are expressed in nanometer for the ultraviolet–visible region and micrometer for the infrared region. Wavenumbers are expressed in cm^{-1}.

1.2 Polarization States

An electromagnetic wave is associated with electric and magnetic fields oscillating perpendicular to each other. Waves are represented in three mutually perpendicular Cartesian axes x, y, and z. In a right-handed coordinate system, the thumb, index, and middle fingers of right hand represent, respectively, the x, y, and z axes (see Figure 1.2). If an electromagnetic wave is represented to propagate along the z axis, then electric and magnetic fields will be oscillating in the two perpendicular xz and yz planes containing this z axis. For the sake of simplicity, we will restrict our discussion, in this chapter, to the electric field alone. If light is represented to propagate along z axis, with its electric field oscillating in the xz plane, then light is said to be x-polarized or linearly polarized along the x axis or plane polarized in the xz plane. If light is represented to propagate along the z axis, with its electric field oscillating in the yz plane, then light is said to be y-polarized or linearly polarized along the y axis or plane polarized in the yz plane. If the electric field is represented to oscillate arbitrarily in the planes containing the z axis, then that light is said to be unpolarized. Optical devices, known as polarizers, are used to select a desired polarization. Also the optical devices known as polarization rotators are used to rotate the linear polarization to a desired orientation.

Two or more waves that have the same frequency and the same relationship between corresponding points at all times are referred to as coherent waves. Coherent light waves are normally realized by taking a parent light wave from a given light source and resolving it into component waves. More details on how to realize the x- and y-polarized coherent light components are given later. By combining the x-polarized and y-polarized coherent light wave components, one can generate a variety of polarization states. First, let us consider an x-polarized wave component propagating in time t, along

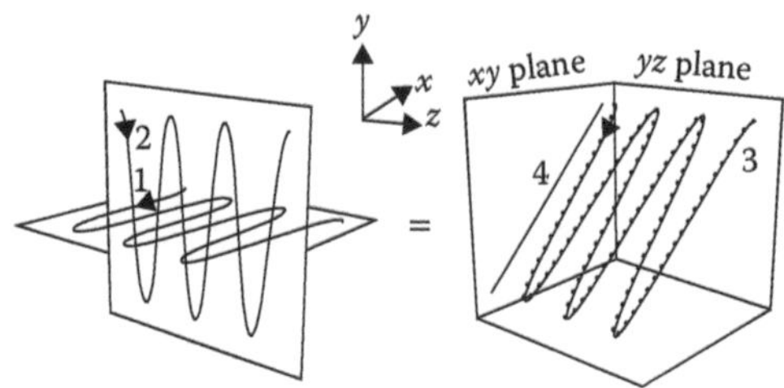

FIGURE 1.2
The x-polarized cosine wave with unit amplitude, $\cos\Theta_t$ (curve 1 in the xz plane), and the y-polarized cosine wave with unit amplitude, $\cos\Theta_t$ (curve 2 in the yz plane), both propagating along the z axis, result in a linearly polarized wave (curve 3) with its plane bisecting the $+x$ and $+y$ axes. The projection of this linearly polarized wave in the xy plane (shown as curve 4) gives a straight line at 45° from the $+x$ and $+y$ axes.

z axis, in the xz plane, with maximum amplitude F_{x0}. This wave F_x is represented by Equation 1.3:

$$F_x = F_{x0}\cos 2\pi\nu t = F_{x0}\cos\Theta_t \tag{1.3}$$

where the abbreviation $\Theta_t = 2\pi\nu t$ is introduced. Similarly, a y-polarized wave component propagating in the yz plane is represented by, $F_y = F_{y0}\cos\Theta_t$, where F_{y0} is its maximum amplitude. We can choose the amplitudes to be equal $F_{x0} = F_{y0} = F_0$. For convenience in depicting, we will use unit amplitudes, that is, $F_0 = 1$. As shown in Figure 1.2, when the observer looks in to the incoming waves (into the z axis in right-handed coordinate system), their combination appears as a light wave whose electric vector oscillates at 45° between the $+x$ and $+y$ axes, that is, in the xy plane bisecting the $+x$ and $+y$ axes. The projection of the resulting wave on to the xy plane appears as a diagonal line in the xy plane. The resulting linearly polarized light is represented by Equation 1.4:

$$\mathbf{F} = F_0(\mathbf{u}\cos\Theta_t + \mathbf{v}\cos\Theta_t) \tag{1.4}$$

where **u** and **v** are unit vectors along the x and y axes, respectively. When both coherent waves in the xz and yz planes are represented by the same function, $\cos\Theta_t$, these waves are said to have zero phase difference. Figure 1.2 demonstrates that two coherent plane-polarized waves, propagating in mutually perpendicular planes, without any phase difference between them, combine to result in a wave whose polarization axis is in between those of parent waves. This observation can be used in reverse to resolve a plane-polarized electric vector of a given frequency into two mutually perpendicular plane-polarized electric vectors of the same frequency whose planes are at +45° and –45° to that of the parent electric vector. This concept is used to generate two mutually perpendicular coherent plane-polarized electric vectors from a linearly polarized parent electric vector of a light source. This concept is used, for example, in polarization division interferometry (Polavarapu 1997).

1.2.1 Dependence on Phase Difference

If the wave in the xz plane is represented by $F_0\cos\Theta_t$ and that in the yz plane by $F_0\cos(\Theta_t + \delta)$, then these waves are said to have a phase difference of δ. Variation of this phase difference between 0 and 2π will lead to changes in the polarization state of the resulting wave. Since 2π corresponds to 360°, a phase difference between 0 and 2π corresponds to that between 0° and 360°. For the current discussion, we will introduce δ into the y-polarized wave and write Equation 1.4 as

$$\mathbf{F} = F_0(\mathbf{u}\cos\Theta_t + \mathbf{v}\cos(\Theta_t + \delta)) \tag{1.5}$$

and evaluate the polarization state of the resulting wave. The polarization states obtained by varying δ in increments of $\pi/6$ are shown in Figure 1.3. In this figure, the resulting wave and its projection onto the xy plane are

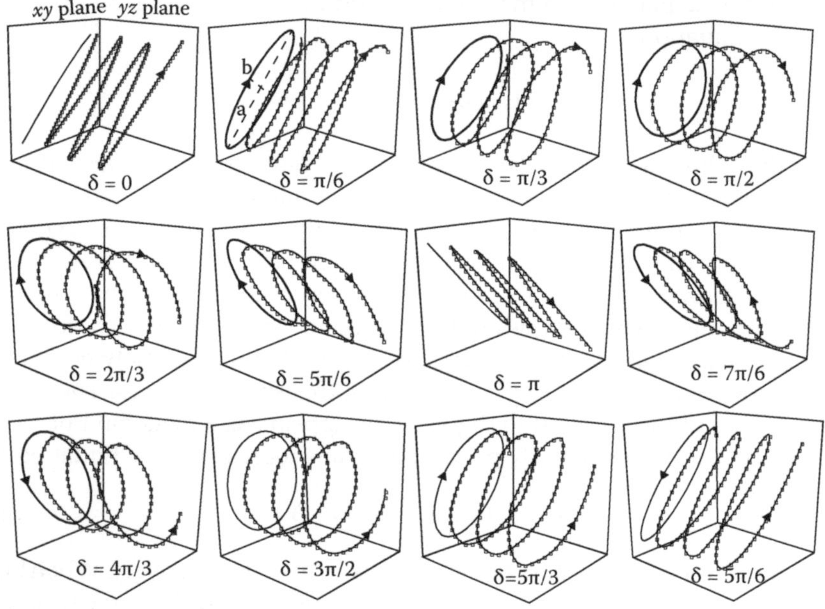

FIGURE 1.3
Depiction of the changes in the polarization state as a function of phase difference between the x-polarized and y-polarized cosine waves propagating along the z direction. The phase shift, δ, introduced into the y-polarized cosine wave is indicated on the figures.

also shown as a function of δ. As already mentioned earlier, the polarization of the resulting wave at $\delta = 0$ is linear with its polarization axis at 45° from $+x$ and $+y$ axes. It can be seen from Figure 1.3 that for a phase difference in the range of $\pi/2 > \delta > 0$, the projection of the resulting wave acquires an elliptical shape (which is referred to as the polarization ellipse), and the electric vector rotates in a clockwise direction. This is shown in Figure 1.3 for two representative situations with $\delta = \pi/6$ and $\delta = 2\pi/6 = \pi/3$. The long axis of the polarization ellipse is referred to as the major axis and the one perpendicular to it as the minor axis. Designating the length of major axis of the polarization ellipse as "a" and that of minor axis as "b," the ratio b/a can be seen to increase (from zero at $\delta = 0$) toward 1 as δ approaches $\pi/2$. This process is akin to watching a collapsed rubber balloon being breathed into, where the balloon takes an elliptical shape with initial breath of air and progresses toward a circle as more air is blown in. An ellipse collapses into a line along its major axis when the length of the minor axis becomes zero. Therefore, for $\delta = 0$, as shown in Figure 1.3, the projected shape (which we called earlier as a straight line) can be viewed as an ellipse with $b = 0$. The angle between the major axis of the polarization ellipse and $+x$ axis is called the azimuth and designated as θ. This angle is 45° for all three cases, $\delta = 0$, $\pi/6$, and $\pi/3$, discussed earlier.

At $\delta = 3\pi/6 = \pi/2$, as shown in Figure 1.3, the projection of the resulting wave onto the xy plane indicates that its shape becomes circular (lengths of major and minor axes are equal; i.e., $a = b$) with the electric vector rotating in clockwise direction. This polarization state is referred to as right circular polarization (RCP), and the wave is referred to as right circularly polarized wave. This right circularly polarized wave is represented by the expression:

$$\mathbf{F}_{RCP} = F_0 \left(\mathbf{u} \cos\Theta_t + \mathbf{v} \cos(\Theta_t + \pi/2)\right) = F_0 \left(\mathbf{u} \cos\Theta_t - \mathbf{v} \sin\Theta_t\right) \quad (1.6)$$

The individual wave components with a phase difference of $\pi/2$, along with the resulting RCP wave and its projection onto the xy plane, are shown in Figure 1.4. The two individual components that lead to this RCP light wave are said to have a phase difference of $\pi/2$. Since one full wave corresponds to 2π, a phase difference of $\pi/2$ corresponds to one-quarter wavelength or $\lambda/4$. In other words, if the y-polarized wave in the yz plane is advanced in time, by $\lambda/4$ relative to the x-polarized wave in the xz plane, then the resulting wave is right circularly polarized.

For phase differences in the range of $\pi > \delta > \pi/2$, as shown in Figure 1.3, the projected shape of the resulting wave distorts from circular (at $\delta = \pi/2$) into elliptical, with the electric vector still rotating in a clockwise direction. However, note that the major axis of the polarization ellipse is now oriented at 135° from $+x$ axis (i.e., between the positive y axis and negative x axis). This orientation is orthogonal to that for $\pi/2 > \delta > 0$ (where it was at 45° to the $+x$ axis, i.e., oriented between the positive x axis and positive y axis). Two representative examples with $\delta = 2\pi/3$ and $5\pi/6$ are shown in Figure 1.3. The ratio of the length of the minor axis to the major axis of the polarization ellipse decreases from 1 (at a phase difference of $\pi/2$) toward zero as the phase difference approaches π.

At a phase difference of $\delta = \pi$, as shown in Figure 1.3, the projected shape of the resulting wave becomes a straight line (equivalent to an ellipse with $b = 0$). That means the resulting wave at a phase difference of π is linearly polarized. However, note that the plane of polarization resulting from $\delta = \pi$ is at 135° from the $+x$ axis, that is, between the $-xz$ and $+yz$ planes (while the plane of polarization at $\delta = 0$ was at 45° from $+x$ axis, i.e., between the $+xz$ and $+yz$

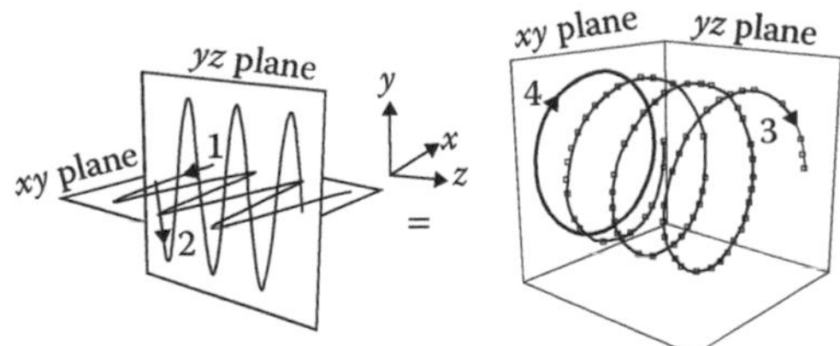

FIGURE 1.4
The x-polarized cosine wave with unit amplitude, $\cos\Theta_t$ (curve 1 in the xz plane), and phase-shifted ($\delta = \pi/2$) y-polarized cosine wave with unit amplitude, $\cos(\Theta_t + \pi/2)$ (curve 2 in the yz plane), both propagating along the z axis, result in a right circularly polarized wave (curve 3). The projection of this right circularly polarized wave in the xy plane is shown as curve 4.

planes). Thus, a phase difference of π between the coherent waves propagating in the *xz* and *yz* planes rotates the plane of polarization, around the propagating *z* axis, by 90°, from $\theta = 45°$ to 135°. Since one full wave corresponds to 2π, a phase difference of π corresponds to one-half wavelength or $\lambda/2$. In other words, if a *y*-polarized wave in the *yz* plane is advanced or shifted forward in time, by $\lambda/2$ relative to the *x*-polarized wave in the *xz* plane, then the plane of polarization of the resulting wave is rotated around the *z* axis by 90°. This linearly polarized wave is represented by the following expression:

$$\mathbf{F} = F_0\,(\mathbf{u}\cos\Theta_t + \mathbf{v}\cos(\Theta_t + \pi)) = F_0\,(\mathbf{u}\cos\Theta_t - \mathbf{v}\cos\Theta_t) \tag{1.7}$$

The individual wave components with a phase difference of π, along with the resulting electric vector and its projected shape, are shown in Figure 1.5. Note that a phase difference of $-\pi$ will give the same result as that for a phase difference of $+\pi$, so the advancing (shifting forward) of the *y*-polarized wave or retardation (shifting in reverse) of the *y*-polarized wave (relative to the *x*-polarized wave) by one-half wavelength has the same effect.

For phase differences in the range of $3\pi/2 > \delta > \pi$, as shown in Figure 1.3, the projected shape of the resulting electric vector acquires an elliptical shape again, but the electric vector of the resulting wave now rotates in a counterclockwise direction. Note that this sense of rotation is opposite to that for phase differences in the range $\pi > \delta > 0$. Two representative examples with phase difference of $7\pi/6$ and $4\pi/3$ are shown in Figure 1.3. The ratio of the length of the minor axis to the major axis of the polarization ellipse starts out at zero (at $\delta = \pi$) and approaches 1 as the phase difference approaches $3\pi/2$.

At a phase difference of $3\pi/2$, the projected shape of the resulting electric vector is a circle with its electric vector rotating in a counterclockwise direction. This polarization state is referred to as left circular polarization (LCP). This left circularly polarized wave is represented by the following expression:

$$\mathbf{F}_{LCP} = F_0\,(\mathbf{u}\cos\Theta_t + \mathbf{v}\cos(\Theta_t + 3\pi/2)) = F_0\,(\mathbf{u}\cos\Theta_t + \mathbf{v}\sin\Theta_t) \tag{1.8}$$

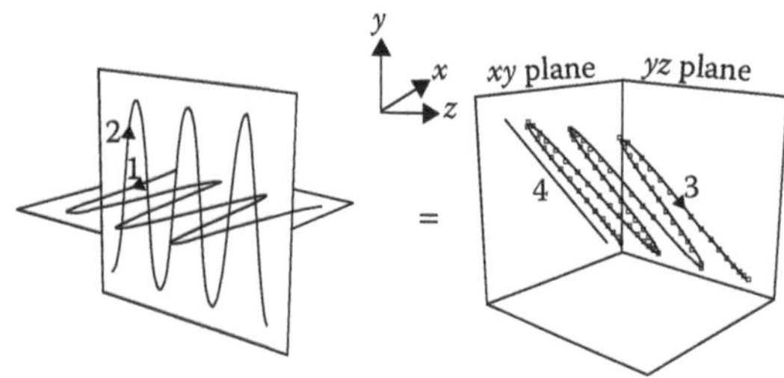

FIGURE 1.5
The *x*-polarized cosine wave with unit amplitude, $\cos\Theta_t$ (curve 1 in the *xz* plane), and the phase-shifted ($\delta = \pi$) *y*-polarized cosine wave with unit amplitude, $\cos(\Theta_t + \pi)$ (curve 2 in the *yz* plane), both propagating along the *z* axis, result in a linearly polarized wave (curve 3) with its plane bisecting the $-x$ and $+y$ axes. The projection of this linearly polarized wave in the *xy* plane is shown as curve 4.

The individual wave components with $3\pi/2$ phase difference, along with the resulting LCP wave and its projected shape in the *xy* plane, are shown in Figure 1.6. The two individual components that lead to this LCP light wave are said to have a phase difference of $3\pi/2$. Since one full wave corresponds to 2π, a phase difference of $3\pi/2$ corresponds to three quarters of a wavelength or $3\lambda/4$. In other words, if a *y*-polarized wave in the *yz* plane is advanced or shifted forward in time along the *z* axis, by $3\lambda/4$ relative to the *x*-polarized wave in the *xz* plane, then the resulting wave is left circularly polarized. An alternate way to generate an LCP wave is to slow down the *y*-polarized wave by $-\lambda/4$ (i.e., introduce a phase difference of $\delta = -\pi/2$), instead of advancing it by $3\lambda/4$, relative to the *x*-polarized wave. This is because $\mathbf{F} = F_0\,(\mathbf{u}\cos\Theta_t + \mathbf{v}\cos(\Theta_t - \pi/2)) = F_0\,(\mathbf{u}\cos\Theta_t + \mathbf{v}\sin\Theta_t)$, which is same as Equation 1.8.

For phase differences in the range of $2\pi > \delta > 3\pi/2$, as shown in Figure 1.3, the projected shape of the resulting electric vector distorts from a circle (at $\delta = 3\pi/2$) to an elliptical shape, with the electric vector of the resulting wave rotating in a counterclockwise direction (as it was between π and $3\pi/2$), but the azimuth of the polarization ellipse is back to 45° (as it was for the range, $\pi/2 < \delta > 0$). Two representative examples with a phase difference of $5\pi/3$ and $11\pi/6$ are shown in Figure 1.3. The ratio of the length of the minor axis to the major axis of the polarization ellipse decreases from 1 (at $\delta = 3\pi/2$) and approaches zero as the phase difference approaches 2π. Note that a phase difference of 2π is the same as 0 and that beyond 2π the polarization cycle that we discussed between 0 and 2π repeats itself.

The observations made so far are summarized in Table 1.2 and are also shown in Figure 1.7 in the form of a "polarization wheel." This "wheel" has two circles and two spokes: one vertical spoke that bisects the inner circle and one horizontal spoke that bisects the outer circle. Opposite ends of the vertical spoke are capped with linear polarizations, and the opposite ends of the horizontal spoke are caped with circular polarizations. The vertical spoke separates the light waves that have a clockwise sense of rotation ($\pi > \delta > 0$) from those that have a counterclockwise sense of rotation ($2\pi > \delta > \pi$). The horizontal spoke separates the light waves that have an azimuth $\theta = 45°$ ($\pi/2 > \delta \geq 0$; and $2\pi \geq \delta > 3\pi/2$) from those that have an azimuth $\theta = 135°$ ($3\pi/2 > \delta > \pi/2$). It is useful to

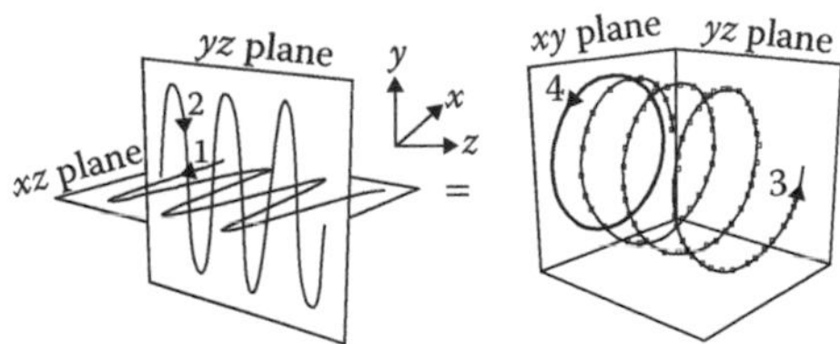

FIGURE 1.6
The *x*-polarized cosine wave with unit amplitude, $\cos\Theta_t$ (curve 1 in the *xz* plane), and the phase-shifted ($\delta = 3\pi/2$) *y*-polarized cosine wave with unit amplitude $\cos(\Theta_t + 3\pi/2)$ (curve 2 in the *yz* plane), both propagating along the *z* axis, result in a left circularly polarized wave (curve 3). The projection of this left circularly polarized wave in the *xy* plane is shown as curve 4.

TABLE 1.2

Dependence of Resulting Polarization on the Differences in Amplitudes and Phases of x-Polarized and y-Polarized Coherent Wave Components Propagating in z Direction

Amplitudes	Phase Difference[a]	Polarization State	Sense of Rotation	θ[b]
$E_{xo} = E_{yo}$	$\delta = 0$	Linear		45°
	$\pi/2 > \delta > 0$	Elliptical	Clockwise	45°
	$\delta = \pi/2$	Right circular	Clockwise	
	$\delta = \pi > \delta > \pi/2$	Elliptical	Clockwise	135°
	$\delta = \pi$	Linear		135°
	$\delta = 3\pi/2 > \delta > \pi$	Elliptical	Counterclockwise	135°
	$\delta = 3\pi/2$	Left circular	Counterclockwise	
	$\delta = 2\pi > \delta > 3\pi/2$	Linear	Counterclockwise	45°
$E_{xo} > E_{yo}$; $E_{yo} \neq 0$	$\delta = 0$	Linear		$0 > \theta < 45°$
$E_{xo} < E_{yo}$; $E_{xo} \neq 0$	$\delta = 0$	Linear		$90° > \theta > 45°$
$E_{xo} > E_{yo}$; $E_{yo} \neq 0$	$\delta = \pi/2$	Elliptical	Clockwise	0°
$E_{xo} < E_{yo}$; $E_{xo} \neq 0$	$\delta = \pi/2$	Elliptical	Clockwise	90°

[a] Phase δ is introduced into the y-polarized component.

[b] θ, referred to as the azimuth, is between the $+x$ axis and the major axis of the polarization ellipse. Linear polarization can be viewed as a special case of elliptical polarization by collapsing the ellipse so that its minor axis vanishes and its major axis becomes the linear polarization axis.

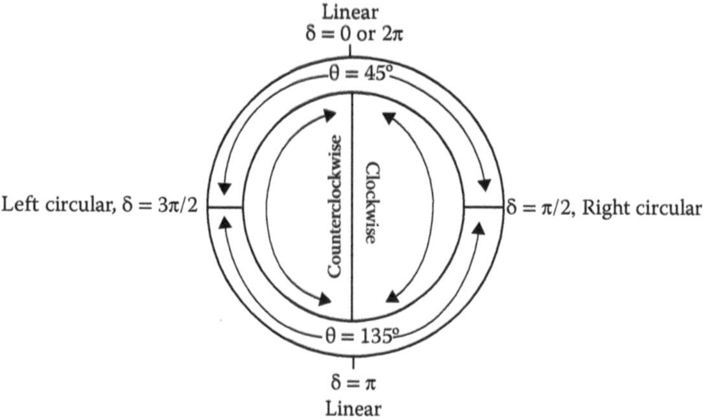

FIGURE 1.7
A "polarization wheel" summarizing the polarization states generated by two orthogonal plane-polarized coherent cosine waves of equal amplitudes as a function of the phase difference between them.

remember that the sense of rotation for the polarization vector changes whenever the polarization state goes through a linear polarization, while the azimuth changes whenever the polarization state goes through a circular polarization.

From the earlier paragraphs, it can be seen that right and left circularly polarized light waves can be generated from two plane-polarized coherent light waves, propagating in mutually orthogonal planes, by modulating the phase shift from $+\lambda/4$ to $-\lambda/4$ (or to $+3\lambda/4$). This is the principle behind the optical devices that modulate the polarization of light between the LCP and RCP states. Similarly, the polarization of light can be modulated between two orthogonal plane polarizations by modulating the phase shift from 0 to $\pm\lambda/2$.

A linearly polarized wave can also be regarded as a superposition of right and left circularly polarized waves. For example, if we combine the right circularly polarized wave, represented by Equation 1.6, with the left circularly polarized wave, represented by Equation 1.8, the result is a plane-polarized wave that oscillates in the *xz* plane.

1.2.2 Dependence on Amplitude Difference

In the discussion so far, we focused on the influence of phase difference on the polarization state by keeping the amplitudes of participating coherent waves equal and unity. In these cases (see Figure 1.3), the polarization ellipse is seen to be oriented at either 45° or 135° from the *x* axis. However, the orientation of the polarization ellipse also depends on the difference in the amplitudes of two participating coherent wave components. For the case of linearly polarized light generated from two orthogonal coherent waves with equal amplitudes, and $\delta = 0$ (see Figure 1.2), we have seen that the resultant polarization vector is at 45° to the $+x$ axis. Now let us modify the amplitude of the *y*-polarized wave component to be one-half of that of the *x*-polarized wave component; that is, $F_{x0} = F_0$ and $F_{y0} = F_0/2$. These individual components, resulting polarization vector and its projected shape in the *xy* plane, are shown in Figure 1.8. The projected shape of the resulting polarization vector is a straight line (as it was when $\delta = 0$ and $F_{x0} = F_{y0}$), but this straight line is not oriented at 45°, unlike when $F_{x0} = F_{y0}$ and $\delta = 0$; see Figure 1.2) and is at less than 45° from the $+x$ axis.

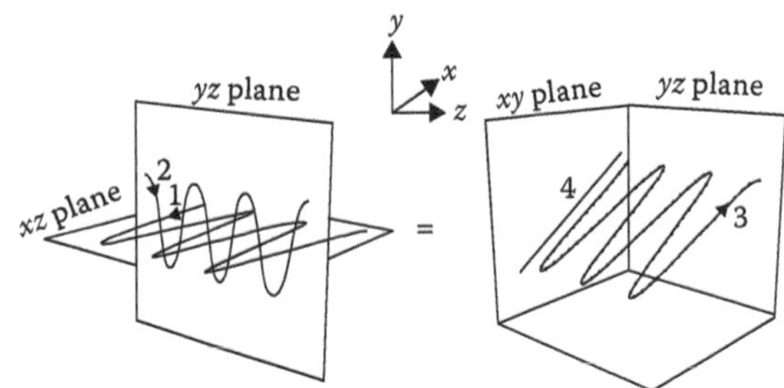

FIGURE 1.8
The *x*-polarized cosine wave with unit amplitude, $\cos\Theta_t$ (curve 1 in the *xz* plane), and the *y*-polarized cosine wave with an amplitude of 0.5, ½ $\cos\Theta_t$ (curve 2 in the *yz* plane), both propagating along the *z* axis, result in a linearly polarized wave (curve 3). The projection of this linearly polarized wave in the *xy* plane is shown as curve 4. The angle between the polarization vector and *x* axis is not 45°, unlike the analogous case with equal amplitudes and $\delta = 0$ (see Figure 1.2).

This observation indicates that a difference in the amplitudes of participating coherent waves also influences the orientation of the resulting polarization ellipse. Additional examples involving the influence of amplitude difference at a nonzero phase difference are discussed in the next section.

1.2.3 Dependence on Both Phase and Amplitude Differences

Although we have considered the influence of differences in phase and amplitude of the participating coherent waves separately in Figures 1.2 and 1.8, for a general situation, one has to consider the differences in both phase and amplitude, of participating coherent wave components, together. One such situation is presented in Figure 1.9, where the *y*-polarized component is shifted by $\pi/3$, relative to the *x*-polarized component, and the amplitude of the *y*-polarized component is one-half of the *x*-polarized component; that is, $F_{x0} = F_0$ and $F_{y0} = F_0/2$. In the absence of amplitude difference, one would have obtained a polarization ellipse with clockwise rotation and its major axis at 45° from the *x* axis (see Figure 1.3). With the incorporation of amplitude difference the major axis of this polarization ellipse is shifted from 45° to the *x* axis, although a clockwise sense of rotation is maintained. Therefore, the orientation of the polarization ellipse is influenced by the difference in amplitudes of the participating coherent waves. An explanation for this changing orientation of polarization ellipse is given as advanced material in Appendix 1.

Special situations arise when the phase difference is $\pi/2$ or $3\pi/2$. To consider these situations, let us consider the right circularly polarized wave (see Figure 1.4) and multiply the *y*-polarized wave propagating in the *yz* plane with 0.5. That is, the *y*-polarized wave in the *yz* plane now is represented by $\frac{1}{2}\cos(\Theta_t + \pi/2)$ and the *x*-polarized wave propagating in the *xz* plane, is represented by $\cos\Theta_t$. If the amplitudes were equal, the polarization state of the resulting wave should have been RCP, as in Figure 1.4. However, as shown in Figure 1.10, the projection of the resulting wave appears as an ellipse with its electric vector

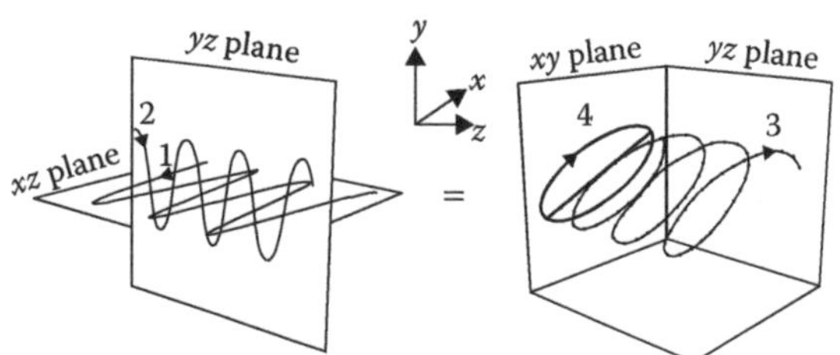

FIGURE 1.9
The *x*-polarized cosine wave with unit amplitude, $\cos\Theta_t$ (curve 1 in the *xz* plane), and phase-shifted ($\delta = \pi/3$) *y*-polarized cosine wave with an amplitude of 0.5, $\frac{1}{2}\cos(\Theta_t + \pi/3)$ (curve 2 in the *yz* plane), both propagating along the *z* axis, result in an elliptically polarized wave (curve 3) rotating in the clockwise direction. The projection of this elliptically polarized wave in the *xy* plane is shown as curve 4. The angle between the major axis of the ellipse (dotted line) and *x* axis is different from 45°, unlike in Figure 1.3, where $F_{xo} = F_{yo}$ and $\delta = \pi/3$.

oscillating in the clockwise direction. Note that the major axis of the polarization ellipse is seen to be along the x axis, which is also the axis of the x-polarized electric vector that has larger amplitude. Now let us exchange amplitudes of the x- and y-polarized waves. That is, the amplitude of the x-polarized wave is multiplied by 0.5, so our x-polarized wave in the xz plane now is ½ $\cos\Theta_t$ and the y-polarized wave propagating in the yz plane is $\cos(\Theta_t + \pi/2)$. As shown in Figure 1.11, the projection of the resulting wave appears as an ellipse with its electric vector oscillating in a clockwise direction. Note however, that the major axis of polarization ellipse is now seen to be along the y axis, which is also the axis of the y-polarized electric vector that has larger amplitude.

These observations, which are also summarized in Table 1.2, indicate that if circularly polarized light is disturbed by changing the amplitudes of participating coherent waves, then the projected shape is elongated, from a circle into an ellipse, along the polarization axis of a component that has larger amplitude.

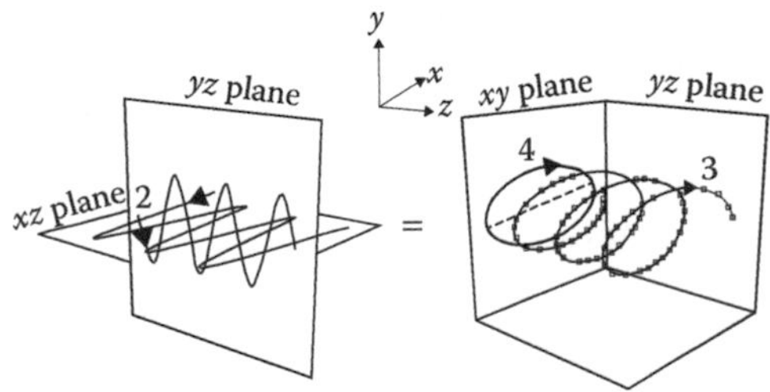

FIGURE 1.10
The x-polarized cosine wave with unit amplitude, $\cos\Theta_t$ (curve 1 in the xz plane), and phase-shifted ($\delta = \pi/2$) y-polarized cosine wave with an amplitude of 0.5, ½ $\cos(\Theta_t + \pi/2)$ (curve 2 in the yz plane), both propagating along the z axis, result in an elliptically polarized wave (curve 3) rotating in the clockwise direction. The projection of this elliptically polarized wave in the xy plane is shown as curve 4. The angle between the major axis of ellipse (dotted line) and the x axis is 0°.

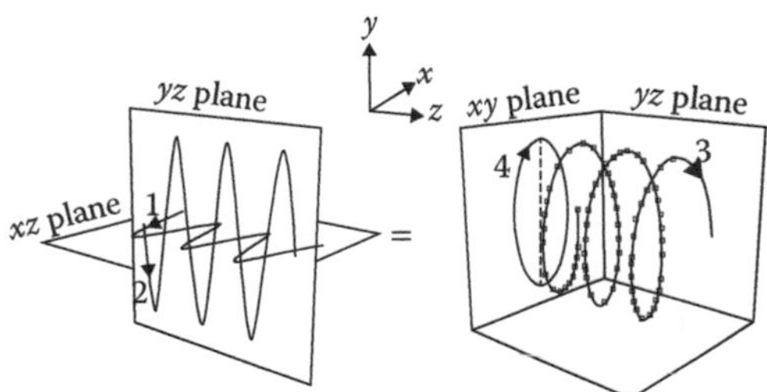

FIGURE 1.11
The x-polarized cosine wave with an amplitude of 0.5, ½ $\cos\Theta_t$ (curve 1 in the xz plane), and the phase-shifted ($\delta = \pi/2$) y-polarized cosine wave with an unit amplitude, $\cos(\Theta_t + \pi/2)$ (curve 2 in the yz plane), both propagating along the z axis, result in an elliptically polarized wave (curve 3) rotating in the clockwise direction. The projection of this elliptically polarized wave in the xy plane is shown as curve 4. The angle between the major axis of the ellipse (dotted line) and the x axis is 90°.

1.2.4 Generalization

Generalizing the observations presented earlier and summarized in Table 1.2, when a light wave is represented as a superposition of two coherent waves in orthogonal planes,

$$\mathbf{F} = \mathbf{u}\, F_x + \mathbf{v}\, F_y = \mathbf{u}\, F_{x0} \cos(\Theta_t + \delta_1) + \mathbf{v}\, F_{y0} \cos(\Theta_t + \delta_2) \tag{1.9}$$

one can achieve perfect circular polarization only when $F_{x0} = F_{y0}$ and $\delta_2 - \delta_1 = \pm\pi/2, \pm 3\pi/2$. Here F_{x0} and F_{y0} are the maximum amplitudes of the waves in the xz and yz planes, respectively; δ_1 and δ_2 are, respectively, the phase shifts for the waves in the xz and yz planes. Similarly, linear polarization is obtained when $\delta_2 - \delta_1 = 0, \pm\pi$. When F_{x0} and F_{y0} are equal and $\delta_2 - \delta_1 = 0, \pm\pi$, the resulting linear polarization vector will be at 45° or 135° from the x axis. When F_{x0} and F_{y0} are not equal and $\delta_2 - \delta_1 = 0, \pm\pi$, the resulting linear polarization vector will not be at these angles. For values of $\delta_2 - \delta_1$ other than $0, \pm\pi/2, \pm\pi$, and $\pm 3\pi/2$, the projected shape of the resulting light wave will be elliptical with its major axis oriented at an angle to the x axis, and this angle depends on the differences in both amplitudes and phases of participating coherent waves.

1.2.5 Advanced Representations

The elementary level description given in the earlier sections for polarized light is sufficient for progressing through this book. However, advanced material would be needed to understand the intricate details (Barron 2004). Readers interested in such advanced material can refer to Appendix 1.

1.3 Energies

Even though light is described as a wave in the previous section, it can also be considered as consisting of a stream of packets of energy, called photons. The energy associated with a photon is $E = h\nu$, where h is Planck's constant and ν is the frequency of the light wave. These packets may be viewed as particles. Then, light is considered to exhibit dual behavior as waves or as particles (which is referred to as wave-particle duality). A photon of higher frequency (or shorter wavelength) will have higher energy and that of lower frequency (or longer wavelength) will have lower energy. The energies associated with one Avogadro number of photons (i.e., one mol photons) in different spectral regions are summarized in Table 1.1. As an example, energy associated with a photon of 400-nm wavelength is

$$E = \frac{hc}{\lambda} = \frac{6.62606896 \times 10^{-34}\,\text{J s} \times 2.99792458 \times 10^{10}\,\text{cm s}^{-1}}{400 \times 10^{-7}\,\text{cm}} = 4.966 \times 10^{-19}\,\text{J}$$

For a mol of these photons, energy is 4.966×10^{-19} J $\times$ 6.0221415×10^{23} mol^{-1} = 29.9×10^4 J/mol = 299 kJ/mol. Using 1 cal = 4.184 J, this energy corresponds to ~71 kcal/mol. Energies associated with photons in the ultraviolet and visible regions are sufficient to excite (or dislocate) the electrons in the chemical bonds of a molecule, and therefore, ultraviolet–visible radiation is used to study the electronic transitions in a molecule. Energies associated with photons in the near infrared region are sufficient to excite loosely held electrons of a molecule (such as those in inorganic complexes), and therefore, near infrared radiation is used to study low-lying electronic transitions. The interaction with circularly polarized ultraviolet, visible, or near infrared light during these electronic transitions in chiral molecules leads to electronic optical rotation (OR) and circular dichroism (CD) phenomena. OR represents the rotation of linearly polarized light and results from circular birefringence (CB). CB is the difference in refractive indices of left and right circularly polarized light. CD is the difference in absorption of left and right circularly polarized light.

The exchange of energies associated with photons in the mid-infrared radiation by molecules leads to fundamental vibrational transitions. Near-infrared radiation is also suitable for studying certain molecular vibrational transitions known as overtone transitions. The absorption of circularly polarized infrared light during these vibrational transitions in chiral molecules leads to vibrational circular dichroism.

The scattering of monochromatic circularly polarized visible light leads to a novel phenomenon known as circular difference Raman scattering, which is referred to as Raman optical activity.

Thus, the interaction of circularly polarized light resulting in electronic and vibrational transitions of chiral molecules forms the basis of chiroptical spectroscopy as detailed in later chapters.

References

Barron, L. D. 2004. *Molecular Light Scattering and Optical Activity*, 2nd ed. Cambridge, UK: Cambridge University Press.

Levine, I. N. 2009. *Physical Chemistry*, 6th ed. New York, NY: McGraw Hill.

Polavarapu, P. 1997. *Principles and Applications of Polarization-Division Interferometry*. Chichester, UK: John Wiley & Sons.

2

Chiral Molecules

2.1 Classification of Molecules

For the purpose of developing concepts needed for chiroptical spectroscopy, molecules can be categorized into two different groups. The criterion needed to distinguish these two groups of molecules is to ask whether a given molecule possesses certain symmetry properties. The symmetry properties to consider are (a) reflection symmetry (also referred to as plane of symmetry), (b) rotation-reflection symmetry (also referred to as improper rotation axis), and (c) inversion symmetry (or a point of inversion). If all atoms in a given molecule reside on a plane, then that molecule is said to have a plane of symmetry. Alternately, if identical atoms in a given molecule reflect on to each other through a plane, then that molecule is also said to have a plane of symmetry. If a rotation of the molecule about an axis followed by a reflection perpendicular to that rotation axis brings the identical atoms in that molecule into each other, then the molecule is said to have an improper rotation axis (or rotation-reflection symmetry). A point of inversion symmetry is a point through which the identical atoms invert into each other (vide infra). The planes, axes, and point of symmetry are referred to as the symmetry elements. Although we are only concerned here with the three symmetry elements mentioned above, additional symmetry elements can also be present in a molecule. These additional symmetry elements and associated material can be found in books on group theory (Cotton 1971), but they are not necessary for the present discussion.

To elaborate on the three symmetry elements mentioned above, let us consider methane (CH_4) molecule. The four surrounding H atoms connected to the central C atom yield a tetrahedral arrangement around C, which can be visualized by placing the central atom at the center of a cube and the surrounding H atoms at four different corners of that cube. Such an arrangement of the atoms of CH_4 molecule in a cube is shown in Figure 2.1. Unless one can visualize this structure in three dimensions, it can be difficult sometimes to grasp this drawing. Therefore an equivalent drawing in a trigonal pyramidal arrangement of atoms is also shown in Figure 2.1, where three of the H atoms are at the base of a pyramid and the fourth H atom at the apex of that pyramid. Another equivalent arrangement, also shown in Figure 2.1, is to place two H atoms and one C

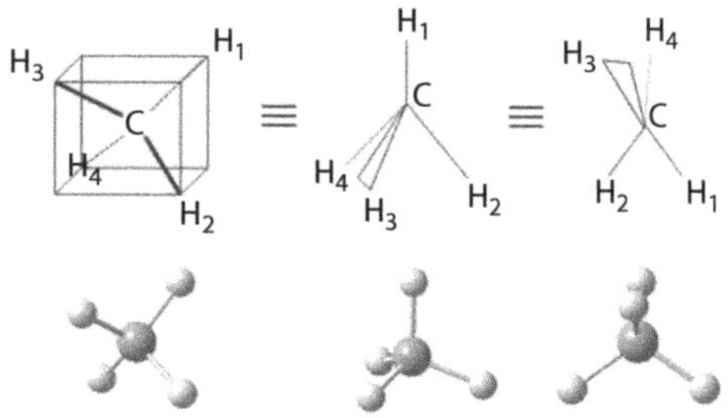

FIGURE 2.1
Three different, but equivalent, depictions of tetrahedral arrangement of H atoms in a methane molecule. In the leftmost figure, H atoms are arranged at four different corners of a cube. In the middle figure, three H atoms are at the base of a pyramid and the fourth H atom is at the apex. In the rightmost figure, two H atoms and the C atom are in the plane of the paper, and the remaining two H atoms in a perpendicular plane. For a better visualization, the top three drawings are shown as ball-and-stick representations below them.

atom in one plane and the remaining two H atoms in a perpendicular plane. Using these three different visualizations, one can quickly notice that the CH_4 molecule has reflection symmetry in the plane containing H_1, C, and H_2 atoms (the H_3 and H_4 atoms reflect onto each other). Note that the subscripts 1–4 on the H atoms are for accounting purposes only, and these subscripts do not make the H atoms different from each other. Although it is sufficient to note that the CH_4 molecule has plane(s) of symmetry, a curious mind will recognize that there are in fact a total of six different planes of symmetry in the CH_4 molecule, namely the planes containing H_1CH_2, H_1CH_3, H_1CH_4, H_2CH_3, H_2CH_4, and H_3CH_4.

The CH_4 molecule can also be used to demonstrate the presence of rotation-reflection symmetry (or an improper rotation axis). For this purpose, we need to view the CH_4 molecule in relation to the *x*, *y*, *z*, Cartesian axes that go through the centers of three mutually perpendicular faces of the cube (see Figure 2.2). The *y*-axis directed behind the plane of the paper is not shown in Figure 2.2. Now consider the effect of rotation and reflection on the CH_4 molecule in two different steps, by holding the frame of the cube fixed and rotating or reflecting the atoms inside the cube. In step (a), the molecule is rotated by 90° around the *z*-axis, which makes the four H atoms switch to the corners of the cube that were previously unoccupied. Note that the structure obtained after rotation by 90° around the *z*-axis is not equivalent to the structure that we started with. In step (b), we consider the structure obtained after rotation in step (a) and reflect the atoms through the *xy*-plane, which is perpendicular to the *z*-axis. Now the H atoms are brought to those corners of the cube that were previously occupied [before rotation in step (a)]. Thus the two structures (before and after the combined rotation-reflection) are equivalent, and the CH_4 molecule is said to have rotation-reflection symmetry.

Inversion symmetry is defined relative to the central point within the molecule. The (*x*, *y*, *z*) Cartesian coordinates of this central point are defined as

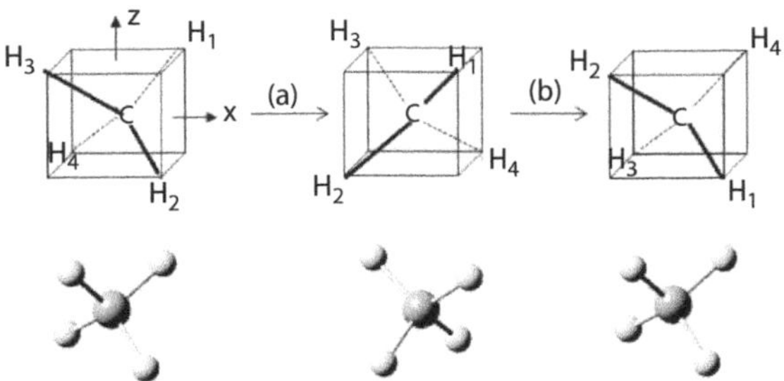

FIGURE 2.2
Demonstration of rotation-reflection symmetry in methane. In step (a), a 90° rotation around the vertical z-axis is performed. In step (b), reflection in the horizontal xy-plane, which is perpendicular to the z-axis, is carried out. The structure on the left is indistinguishable from that on the right. For a better visualization, the top three drawings are shown as ball-and-stick representations below them.

(0, 0, 0). If a molecule has inversion symmetry, then for every atom with (x, y, z) coordinates given as (a, b, c), there should be an equivalent atom at $(-a, -b, -c)$ coordinates. The CH_4 molecule does not have inversion symmetry, so we need to consider a different molecule to demonstrate this symmetry. Consider a planar *trans*-A_2B_2 molecule as shown in Figure 2.3. Let us define the positive x-axis as directed from the center of the A–A bond toward the A_1 atom, and the positive y-axis as directed from the center of the A–A bond toward the top of the page and perpendicular to the x-axis. The z-axis is perpendicular to the plane of the paper. Then the x-, y-, and z-coordinates for the point of inversion are (0, 0, 0), and those for the four atoms are $A_1(+x_{A1}, 0, 0)$, $A_2(-x_{A1}, 0, 0)$, $B_1(+x_{B1}, -y_{B1}, 0)$, and $B_2(-x_{B1}, +y_{B1}, 0)$. Since the atoms B_1 and B_2 invert into each other, as are the atoms A_1 and A_2, the planar *trans*-A_2B_2 molecule is said to have inversion symmetry. Note that the planar *trans*-A_2B_2 molecule also has a plane of symmetry, since all atoms are in one plane.

Molecules that possess one or more of the three symmetry elements that we just discussed (namely, plane of symmetry, improper axis, and point of inversion) are referred to as *achiral* molecules. A chemical sample made of achiral molecules is referred to as an achiral sample. Both the CH_4 and the planar *trans*-A_2B_2 molecules are achiral, so the CH_4 and *trans*-hydrogen peroxide (*trans*-H_2O_2) chemical samples are achiral samples. The mirror image of an achiral molecule is superimposable on the parent molecule. This can be seen in Figure 2.4, where the parent CH_4 molecule (a) and its mirror image (b) are shown. When the mirror image (b) is rotated by 90° around the vertical z-axis, it becomes (c), which is superimposable on the parent molecule (a).

Molecules that do not possess the three symmetry elements that we discussed earlier are referred to as *chiral* molecules. It is useful to note that chiral molecules belong to the point groups designated as C_n, D_n, T, O, and I. More details on point groups can be found in the books focused on that topic (Cotton 1971). A chemical sample made up of chiral molecules is referred to as

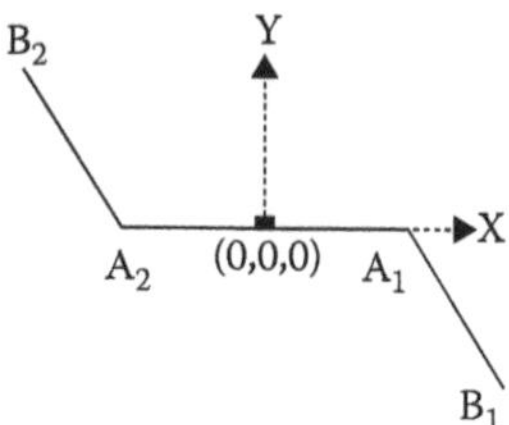

FIGURE 2.3
Demonstration of the inversion symmetry in a planar *trans*-A_2B_2 molecule. See text.

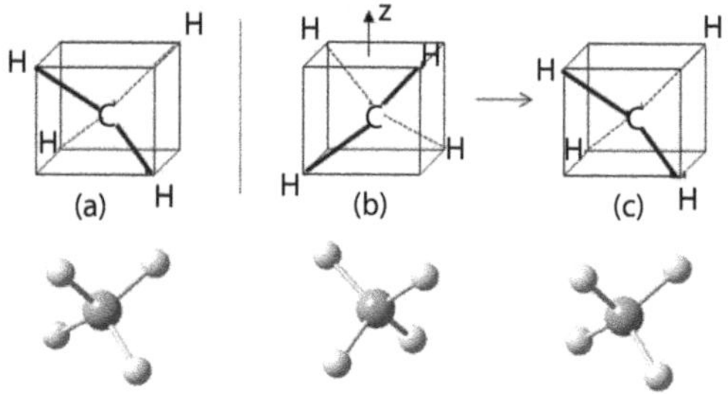

FIGURE 2.4
Mirror images of achiral molecules are superimposable. (a) Parent methane molecule, (b) mirror image of (a), and (c) rotation of the mirror image around the vertical *z*-axis by 90° yielding identical figure to (a). The figures (a) and (c) are superimposable. For a better visualization, the top three drawings are shown as ball-and-stick representations below them.

a chiral sample. The mirror image of a chiral molecule is not superimposable on its parent molecule. As an example, consider a substituted CH_4 molecule where three H atoms of CH_4 are replaced by F, Cl, and Br. The resulting molecule, bromochlorofluoromethane, CHFClBr, does not have a plane of symmetry, improper rotation axis, or point of inversion and is therefore chiral. The parent and its mirror image of CHFClBr are shown in Figure 2.5 in three different perspectives. In the top three drawings (a)–(c), the terminal atoms are placed at the corners of a cube. In the middle three drawings (d)–(f), the terminal atoms are placed at the corners of a pyramid. The mirror image of (a) is (b) and that of (d) is (e.) Note that since C–H, C–F, C–Cl, and C–Br bond lengths are all different, as are bond angles, representation with a cube or a pyramid is not strictly correct, but this is useful for illustrative purposes. The ball-and-stick representations, shown underneath pyramidal representations, provide a more realistic perspective. Rotation of (b) around the vertical *z*-axis by 90° yields the structure (c). Rotation of (e) around the *z*-axis by 120° yields the structure (f). From these rotated structures, it can be quickly noticed that, unlike in the case of CH_4, the parent CHFClBr molecule and its mirror image do not superimpose, regardless of whichever way we rotate the mirror image.

The parent chiral molecule and its nonsuperimposable mirror image are referred to as the *enantiomers* or as the enantiomeric pair. Sometimes

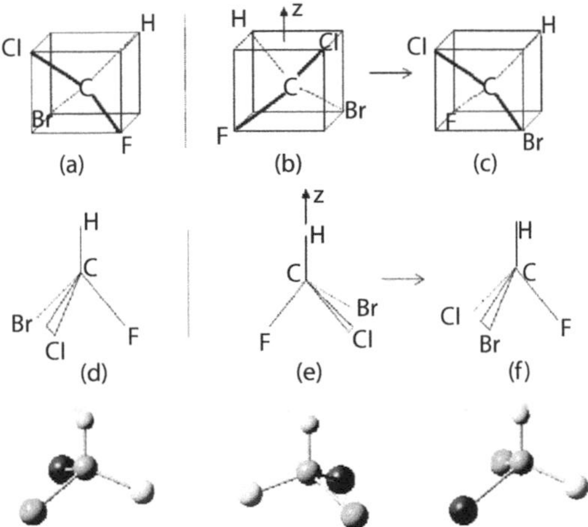

FIGURE 2.5
The mirror image of a chiral molecule is not superimposable on the parent molecule. A bromochlorofluoromethane (CHFClBr) molecule is shown, in three different perspectives, with the terminal atoms at the corners of the cube (top) and at the corners of the pyramid (middle). Figures (a) and (d) represent a parent CHFClBr molecule while (b) and (e) represent mirror images of (a) and (d); (c) is obtained by rotating the mirror image (b) around the vertical z-axis by 90°. (f) is obtained by rotating the mirror image (e) around the vertical z-axis by 120°. Figure (a) is not superimposable on (b) or (c). Figure (d) is not superimposable on (e) or (f). For a better visualization, drawings (d)–(f) are shown as ball-and-stick representations below them.

enantiomers are also referred to as *antipodes* or *opposite handed*. The molecules of an enantiomeric pair are individually chiral. Even though the elemental compositions of an enantiomeric pair are the same, the chemical samples composed of separate enantiomeric molecules often have different chemical and optical properties. In practice, a synthesized chemical sample can be composed of either one or both of the enantiomeric molecules. If a chemical sample contains equal numbers of enantiomeric molecules, then that sample is referred to as the *racemic mixture*. In contrast, if a chemical sample contains only one enantiomer, then that sample is referred to as an *enantiomerically pure*, or sometimes as an *optically pure*, sample. If a chemical sample contains both enantiomers, but the number of molecules of one enantiomer is greater than that of the other, then that sample is referred to as an *enantiomerically enriched* sample.

2.2 Configurations of Chiral Molecules

The previous section introduced the concept of enantiomers and enantiomerically pure samples. It is necessary to develop a transparent notation to label the enantiomers and enantiomerically pure samples so that they can be referred to unambiguously. The three-dimensional spatial orientation of atoms in a chiral molecule is specified as the *configuration*. The labels used for enantiomers specify their configurations and are referred to as the *configurational labels*. These configurational labels depend on the nature, or source, of chirality. Different types of chirality and the configurational labels recommended for a given type of chirality (Eliel et al. 2001; Nasipuri 2012; Moss 1996) are discussed in the following sections.

2.2.1 Central Chirality

In organic molecules, four different atoms connected to a central C atom are referred to as the *surrounding atoms* or *terminal atoms*. The source of chirality in such molecules is the central C atom surrounded by four different atoms. This type, or source, of chirality is referred to as the *central chirality*. The central atom is referred to as the chiral center or stereocenter and is sometimes highlighted with a star next to the central atom. The enantiomers of chiral molecules with central chirality are designated by the labels (*R*) and (*S*). These labels are derived using two rules: (a) sequence rule and (b) viewing rule. The assignment of priorities to the atoms that surround the central atom constitutes the sequence rule. A higher atomic number corresponds to a higher priority. After assigning the priorities, one has to view the molecule by keeping the lowest priority terminal atom away from the observer and trace the remaining terminal atoms from higher priority to lower priority, which constitutes the viewing rule. If this trace indicates a clockwise rotation, then the enantiomeric molecule is labeled as (*R*), where *R* represents *rectus*, from Latin for "right." If this trace indicates a counterclockwise rotation, then the enantiomeric molecule is labeled as (*S*), where *S* represents *sinister*, from Latin for "left."

In the case of CHFClBr, the priorities assigned to the surrounding atoms are Br > Cl > F > H. Keeping the lowest priority H atom in CHFClBr away from the observer, trace the remaining surrounding atoms Br, Cl, and F in the decreasing order of their priorities. The CHFClBr molecule with a clockwise trace (see Figure 2.6) is labeled as (*R*)-CHFClBr. The mirror image CHFClBr molecule, also shown in Figure 2.6, with a counterclockwise trace is labeled as (*S*)-CHFClBr. The enantiomerically pure sample of CHFClBr that contains only (*R*)-CHFClBr molecules is referred to as the (*R*)-CHFClBr sample. Similarly, the enantiomerically pure sample of CHFClBr that contains only (*S*)-CHFClBr molecules is referred to as the (*S*)-CHFClBr sample. A sample of CHFClBr that contains equal numbers of (*R*)-CHFClBr and (*S*)-CHFClBr molecules is referred to as the racemic CHFClBr sample.

In the case of molecules like CHFClBr, the assignment of priorities to surrounding atoms is straightforward. However, this is not the case for most organic molecules because some atoms connected to the central carbon atom can be the same, but each of them in turn could have extended connections to additional (secondary) atoms. The four atoms directly connected to the chiral central atom are referred to as the *primary substituents*, and those connected to

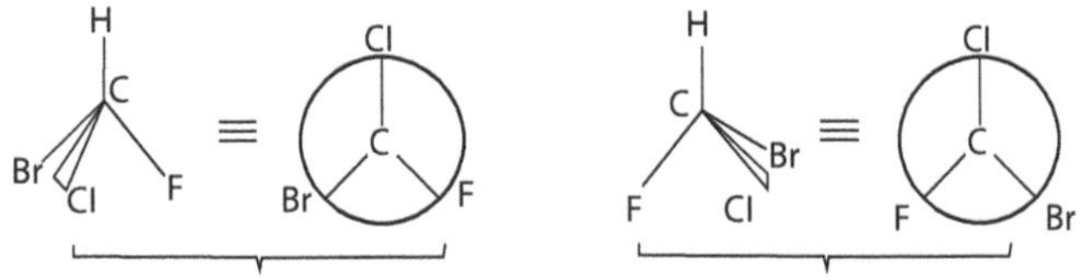

(*R*)-CHFClBr (*S*)-CHFClBr

FIGURE 2.6
Configurational labels for bromochlorofluoromethane. Keeping the lowest priority atom, H, away from the observer, so that H atom is behind the plane of paper, and tracing the remaining terminal atoms from high to low priority (bromine to fluorine) gives a clockwise trace or (*R*) configuration for the structure on left; the same procedure gives a counterclockwise trace or (*S*) configuration for the structure on right.

the primary substituents via extended connections are referred to as the *secondary substituents*. In such cases, ambiguities can arise in defining the priorities. To avoid such ambiguities, a set of rules, known as Cahn, Ingold, and Prelog (CIP) sequence rules (Cahn et al. 1966; Prelog and Helmchen 1982) for assigning the priorities for primary substituents are followed (Moss 1996). These can be summarized as follows: (a) The priorities of primary substituent atoms are determined by their atomic numbers (the higher the atomic number, the higher the priority). (b) For isotopes (such as H and D), higher mass is given the higher priority. (c) If two primary substituent atoms are the same, then their priorities are determined by the nature of secondary substituent atoms in their extended connections. Suppose that the central C atom is connected, as shown in Figure 2.7, to four different groups/atoms: (1) H, (2) CH_3, (3) CH_2OH, and (4) CH_2CH_2OH. The primary substituent atoms are H in group (1), C in group (2) with extended connections to three H (secondary substituent) atoms, C in group (3) with extended connections to three secondary substituent atoms (two H atoms and one O atom), and C in group (4) with extended connections to three secondary substituent atoms (two H atoms and one C atom). In groups (2)–(4), the primary substituents are all C atoms, so their priorities can only be established upon looking at their extended connections. Among (3) and (4), CH_2OH gets higher priority over CH_2CH_2OH because the primary substituent C, in the former, has extended connection to a higher priority secondary substituent atom O, while in the latter, the primary substituent C has an extended connection to a lower priority secondary substituent atom C. The secondary substituent atom O [of OH group in (3)] gets higher priority over the secondary substituent atom C [of $-CH_2CH_2OH$ group in (4)]. Both (3) and (4) have higher priorities over (2), because in (2) the primary substituent, atom C, has extended connections to three secondary substituent H atoms, which have lower priority over secondary substituent atoms C or O. Thus the priorities are 3 (CH_2OH) > 4 (CH_2CH_2OH) > 2 (CH_3) > 1 (H). If the priorities cannot be resolved by looking at secondary substituent atoms, then the analysis is extended outward to include tertiary substituent atoms, or beyond if necessary, until a resolution can

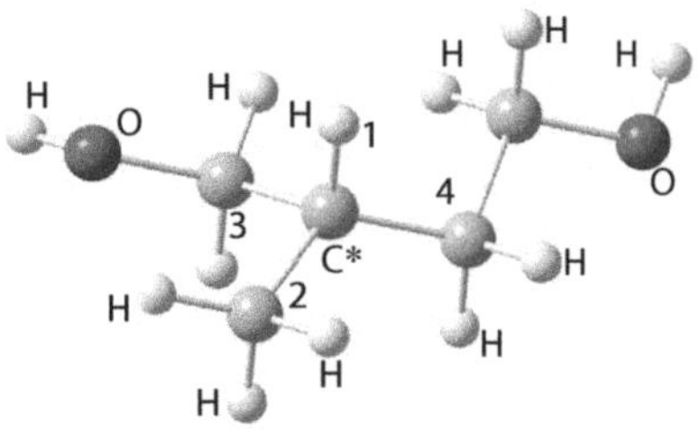

FIGURE 2.7
A chiral molecule with some primary substituent atoms as the same. Central carbon atom is labeled as C*, and the primary substituent atoms are labeled as 1–4. Note that 2, 3, and 4 are all C atoms. The priorities are: 3 > 4 > 2 > 1.

be found. If double and triple bonds are associated with identical secondary substituent atoms, then a triple bond gets higher priority over a double bond, which in turn gets higher priority over a single bond. The reader is encouraged to refer to the original article on CIP sequence rules (Cahn et al. 1966; Prelog and Helmchen 1982) for additional cases that are not discussed here.

A large portion of the older literature (especially in carbohydrate chemistry) used D- and L- designations for configurational labels (Shallenberger 1982), before (*R*)- and (*S*)- designations discussed above were introduced. The D- and L- designations are referred to as Fischer notation and are widely used in labeling carbohydrates and amino acids. The Fischer notation is based on glyceraldehyde, where a central carbon atom (or stereocenter) is connected to H, OH, CHO, and CH_2OH groups. As shown in Figure 2.8, the three carbon atoms chain, containing the stereocenter, in glyceraldehyde is depicted along a vertical line (with the easily oxidizable group, CHO, at the top and the CH_2OH group at the bottom), and the other two group connections to stereocenter are depicted by a horizontal line. The stereocenter is located at the intersection of these two lines (with the appearance of a cross). The glyceraldehyde molecule with the OH group depicted to the right, and H to the left, of the vertical line, is labeled as D-glyceraldehyde and that with the opposite depiction is referred to as L-glyceraldehyde. The bonds to stereocenter that are depicted horizontally represent bonds pointed above the plane of the paper, while those depicted vertically represent bonds pointed behind the plane of paper. Then, viewing the molecule such that the H atom is pointed away from the observer, the OH, CHO, and CH_2OH groups, in this order, represent a clockwise orientation in D-glyceraldehyde and a counterclockwise orientation in L-glyceraldehyde. The D- and L- labels for glyceraldehyde correspond, respectively, to (*R*)- and (*S*)- labels that we mentioned earlier in CIP notation. Compounds synthesized from D-glyceraldehyde, without change in the configuration at the stereocenter, are labeled as D-compounds.

For amino acids, view the molecule such that the H atom is pointed away from the observer. If the COOH, *R*, and NH_2 groups, in this order, represent

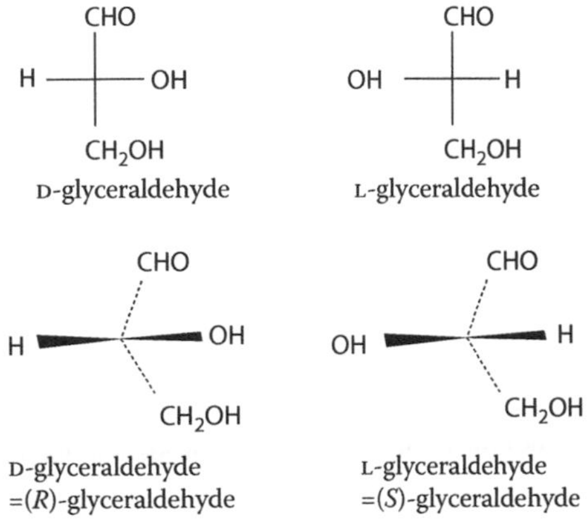

CHO
H — OH
CH_2OH
D-glyceraldehyde

CHO
OH — H
CH_2OH
L-glyceraldehyde

CHO
H — OH
CH_2OH
D-glyceraldehyde
=(*R*)-glyceraldehyde

CHO
OH — H
CH_2OH
L-glyceraldehyde
=(*S*)-glyceraldehyde

FIGURE 2.8
D- and L-glyceraldehyde and their relation with (*R*)- and (*S*)-notation.

a clockwise orientation, then amino acid is labeled as D-amino acid and that with a counterclockwise orientation is labeled as L-amino acid. If the *R* group, in CIP notation, has a lower priority than the COOH group, then the D- and L- labels for amino acids will also correspond, respectively, to the (*R*)- and (*S*)- labels in CIP notation. The D/L designations will not be discussed further, as this designation is no longer used due to ambiguities associated with these designations (Slocum et al. 1971).

Deviating from the current discussion, it is useful to note that the lowercase letters d- (for dextro) and l- (for levo) are also encountered in the literature, which do not have any connection to the D- and L- symbols discussed above but represent the rotation of plane-polarized light by chiral molecules (Eliel et al. 2001; Nasipuri 2012). We will address this topic in Chapter 3.

Central chirality can also arise when the central atom is other than carbon. Silicon also forms tetravalent compounds, just as carbon, so the discussion for carbon-containing chiral molecules can be extended to silicon-containing compounds. Other central atoms providing central chirality include nitrogen, sulfur, and phosphorous. Examples of these cases are shown in Figure 2.9. In the case of N and S as the central atoms, one of the surrounding groups around the central atom can be a lone pair of electrons. In applying the sequence rules for configurational labels, the lone pair of electrons (sometimes referred to as a phantom atom and associated with zero mass) is assigned the lowest priority. Note that chirality can also arise with boron as the central atom in molecules classified as propellers (Kaufmann and Boese 1990).

When multiple centers of chirality are present in a molecule, each center is given a configurational label as described above. For example, let us consider dimethyltartrate as shown in Figure 2.10. The four carbon atoms in the backbone of this molecule are labeled as C_1, C_2, C_3, and C_4, starting from the carbon

FIGURE 2.9
Central chirality associated with N, S, or P as the central atom: (a) Nitrogen atom connected to F, Cl, and Br atoms. With a lone pair of electrons on nitrogen as the fourth surrounding group, the configuration at the central N atom is (*S*); (b) sulfur atom connected to O, N, and C atoms. With a lone pair of electrons on sulfur as the fourth surrounding group (pointing behind the plane of paper, the configuration at the central S atom is (*S*); and (c) the phosphorous atom is connected to O, N, C (of phenyl group) and C (of *t*-butyl group). C (of *t*-butyl group), pointing behind the plane of paper, has the lowest priority of all four primary substituents, so the configuration at P atom is (*S*).

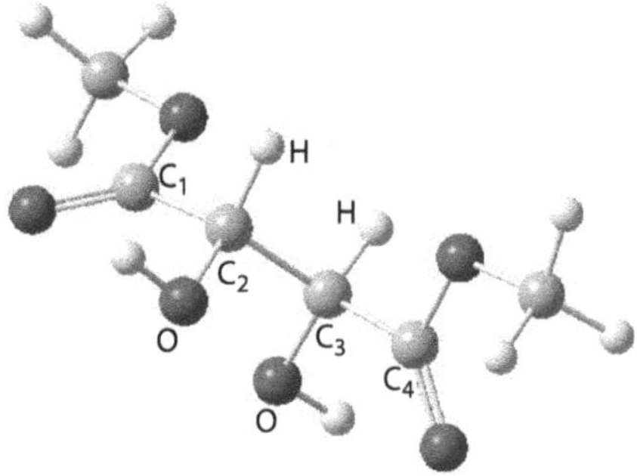

FIGURE 2.10
(2*S*,3*S*)-dimethyltartrate. The chiral centers are C_2 and C_3 each with (*S*) configuration.

atom of one of the two carboxylate ($COOCH_3$) groups. The two chiral centers in this molecule are C_2 and C_3. The first chiral center C_2 is surrounded by H, C_1, O, and C_3. The priority of C_1 is greater than C_3 because the former is connected to two secondary substituent O atoms (one through a double bond), yielding an (S) configuration at C_2. Similarly the second chiral center C_3 is surrounded by H, C_2, O, and C_4. The priority of C_4 is greater than C_2 because the former is connected to two secondary substituent O atoms (one through a double bond), also yielding an (*S*) configuration at C_3. To indicate that the molecule shown in Figure 2.10 has two chiral centers at positions 2 and 3, dimethyltartrate shown in Figure 2.10 is referred to as (2*S*,3*S*)-dimethyltartrate.

A molecule with *n* chiral centers has 2^n possible configurations. Of these, one-half are the mirror images of the other half. This leads to $2^n/2 = 2^{n-1}$ distinct configurations, called *diastereomers*. In the case of dimethyltartrate, there are four possible combinations for two chiral centers, namely (2*S*,3*S*)-, (2*R*,3*R*)-, (2*R*,3*S*)-, and (2*S*,3*R*)-. Of these, (2*S*,3*S*)- and (2*R*,3*R*)- are one enantiomeric pair and (2*S*,3*R*)- and (2*R*,3*S*)- are another enantiomeric pair. However, it should be noted that dimethyltartrate is a special case due to the equivalence of two carbon atoms C_2 and C_3 (substitutent atoms around C_2 are equivalent to those around C_3). As a consequence, in (2*S*,3*R*)- and (2*R*,3*S*)-configurations of dimethyltartrate, the chiral properties arising from one chiral center are nullified by those from the other. As a result, (2*S*,3*R*)- and (2*R*,3*S*)-dimethyltartrate behave, individually, as achiral molecules. For this reason, (2*S*,3*R*)- and (2*R*,3*S*)- dimethyltartrate are referred to as *meso* forms of dimethyltartrate. Note the distinction between meso and racemic: meso represents a single type of molecule with mutually canceling chiral centers within each molecule, while racemic represents equal numbers of enantiomeric molecules.

2.2.2 Axial Chirality

In the previous section, we have considered molecules with a chiral center, where a central atom is connected to four different atoms/groups. A molecule can also be chiral even if it does not have a chiral center. For example, consider the H_2O_2 molecule. If the H–O–O–H dihedral angle is zero or 180°, then the molecule would have plane of symmetry, so it cannot be chiral. However, if the H–O–O–H dihedral angle is neither zero nor 180°, then there is no plane of symmetry, rotation-reflection symmetry, or point of inversion, and therefore the H_2O_2 molecule with a dihedral angle other than 0° and 180°, is chiral. When a molecule is chiral due to restricted rotation around an axis (or bond), the source of chirality is referred to as *axial chirality*. Such restricted rotation around an axis/bond is also called *atropisomerism*. The configurational labels for chiral H_2O_2 molecules (see Figure 2.11) are determined by looking along

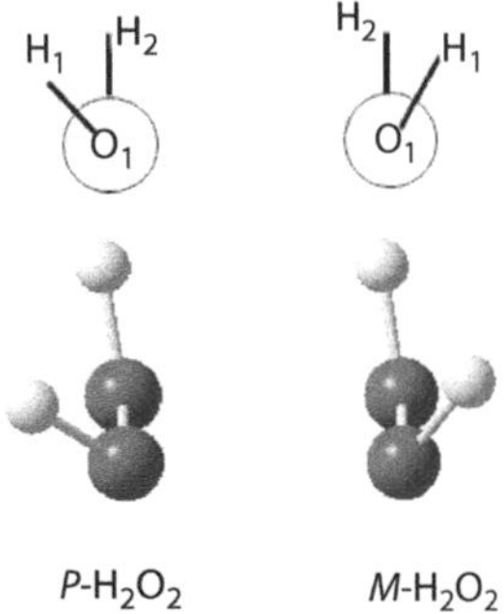

FIGURE 2.11
Enantiomeric molecules of H_2O_2. The central O–O bond is the axis of chirality. In the axial chirality nomenclature, H_2O_2 with a positive dihedral angle is labeled as (*aR*) and that with a negative dihedral angle is labeled as (*aS*). In helical chirality nomenclature, H_2O_2 with a positive dihedral angle (left) is labeled as *P*, and that with a negative dihedral angle (right) is labeled as *M*. The top two drawings are Newman projections when viewed along the O_1–O_2 bond, and the bottom two are ball-and-stick molecular representations.

the O–O bond and determining the shortest rotation pathway, clockwise or counterclockwise, needed to make the nearest H–O–O plane overlap with the farthest O–O–H plane. In the *axial chirality nomenclature,* the enantiomer with the shortest clockwise rotation pathway, or shortest clockwise rotation angle, is labeled as (*aR*) and that with the shortest counterclockwise rotation pathway, or the shortest counterclockwise rotation angle, is labeled as (*aS*). The letter "*a*" in front of *R* and *S* labels stands for "axial," representing the source of chirality. Alternate representations of these labels are (R_a) and (S_a), where the letter "*a*" is used as the subscript to *R*/*S*. In the *helical chirality nomenclature* (vide infra), clockwise rotation is labeled with *P*, and counterclockwise rotation is labeled with *M*, where *P* stands for plus helix and *M* stands for minus helix. For the H_2O_2 molecule, (*aR*) corresponds to *P* and (*aS*) corresponds to *M*, but this correspondence is not general (vide infra).

For chiral H_2O_2 molecules shown in Figure 2.11, the chirality arises only if the two O–H bonds are restricted not to be in the same plane. In reality, however, the individual O–H bonds in H_2O_2 are free to rotate around the O–O bond, so one cannot synthesize or isolate chiral H_2O_2 samples. Nevertheless, H_2O_2 is useful for illustrating the concept of axial and helical chirality.

H_2O_2 is also a special case in that, when viewing along the O–O bond, only one atom in the front and one atom in the rear are involved in determining the configurational labels. In several organic molecules, however, the source of axial chirality is the restricted rotation around a C–C bond. For example, consider a substituted biphenyl molecule shown in Figure 2.12 where the planes of two phenyl groups are nearly 90° to each other. Here the rotation around the central C–C′ bond (the bond connecting two phenyl rings) is restricted because of the presence of O–H groups in the ortho positions (positions adjacent to the C atoms of central C–C′ bond). To determine the

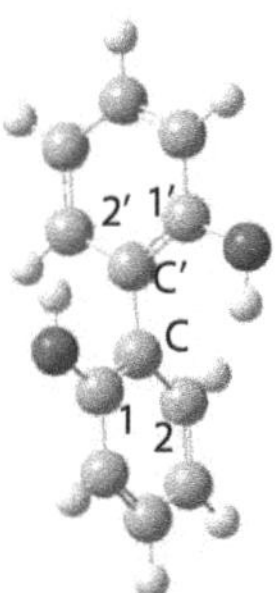

FIGURE 2.12
Axial chirality in a substituted biphenyl molecule. The central C–C′ bond connecting the two phenyl rings is the axis of chirality. The two atoms connected to each of the central C and C′ atoms are labeled as 1, 2 and 1′, 2′, respectively, using the priority rules (1 being higher priority). Viewing along the C–C′ bond and keeping 2′ away from the observer, the trace of 1, 2, and 1′ is counterclockwise, so the configurational label in axial nomenclature is (*aS*). In the helical chirality nomenclature, the shortest rotation pathway for bringing the 1–C–C′ plane to coincide with that of C–C′–1′ is clockwise, so the configurational label in helix nomenclature is *P*.

configurational label for this molecule, label the atoms in ortho positions to C (of central C–C′ bond) as 1 and 2 (with 1 for the higher priority atom) and, similarly, those in ortho positions to C′ as 1′ and 2′. In this biphenyl molecule, the ortho atoms 1 and 1′ get higher priority over the ortho atoms 2 and 2′, because the former are attached to the O and C atoms while the latter are attached to the H and C atoms. Then looking along the central C–C′ bond, and keeping 2′ (lowest priority atom attached to the farthest C′ atom) away from the observer, trace the atoms 1, 2, and 1′. Use the configurational labels (*aR*) or (R_a) for a clockwise trace and (*aS*) or (S_a) for a counterclockwise trace. The substituted biphenyl molecule, shown in Figure 2.12, has a counterclockwise trace, and therefore its configurational label is (*aS*) or (S_a). When *helical chirality nomenclature* is used for this molecule, we need to consider only the highest priority atoms connected to C and C′ atoms. These are 1 and 1′. Then find the shortest rotation pathway (clockwise or counterclockwise) to bring the plane of 1–C–C′ to coincide with that of C–C′–1′. This would result in a clockwise rotation for the molecule shown in Figure 2.12, and therefore its configurational label in a helical chirality nomenclature is *P*. For organic molecules with C–C bond as the axis of chirality, (*aS*) in axial nomenclature corresponds to *P* in helical chirality nomenclature and (*aR*) corresponds to *M*. Besides substituted biphenyls, other types of organic molecules that possess axial chirality include substituted biaryls, allenes, and spiranes.

2.2.3 Helical Chirality

Sometimes overcrowding in a molecule can result in the generation of chirality. For example, phenanthrene (the structure on the left in Figure 2.13), with three fused benzene rings, is a planar molecule and is achiral because it has a plane of symmetry. However, when two methyl groups are substituted for hydrogen atoms at carbon atoms C_1 and C_4, the respective benzene rings containing C_1 and C_4 twist in opposite directions to avoid crowding CH_3 groups (dimethylphenanthrene structure shown on the right in Figure 2.13). As a result, the terminal benzene rings are no longer in one plane, and the plane of symmetry is lost. Unlike the case of molecules with axial chirality, molecules with helical chiraity have a curved skeleton and the terminal rings of the curve twist in opposite direction resulting in the development of a helical twist. The configurational labels used for helical chirality are *P* and *M*, as discussed earlier. These labels are determined by the twist between the terminal rings. For dimethylphenanthrene, shown in Figure 2.13, viewing along C_2–C_3 bond, the C_1–C_2–C_3 plane has to be rotated counterclockwise to bring it to coincide with the plane of C_2–C_3–C_4. Therefore, the configurational label for dimethylphenanthrene shown in Figure 2.13 is *M*. Other molecules possessing helical chirality are substituted benzo[*c*]phenanthrenes (four benzene rings fused together to yield a curved skeleton); helicenes (five or more benzene rings fused together to yield a curved skeleton) are other familiar examples possessing helical chirality. While phenanthrene

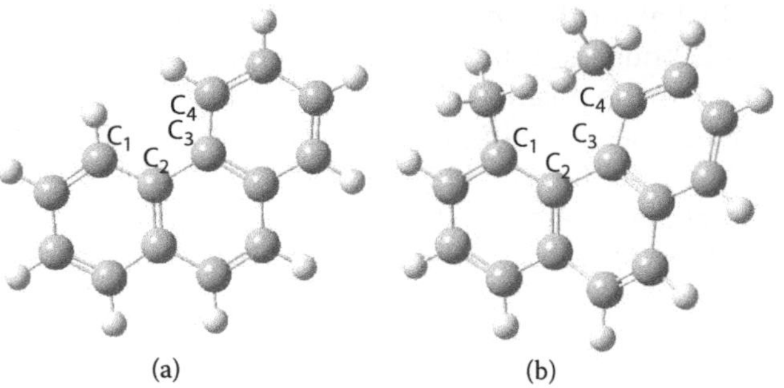

FIGURE 2.13
(a) Planar phenanthrene molecule. (b) Helical chirality in a dimethylphenanthrene molecule. The presence of methyl groups at C_1 and C_4 makes the respective phenyl rings deviate from planarity by ~30° (the dihedral angle C_1–C_2–C_3–C_4 shown here is −30°). Viewing along the C_2–C_3 bond, the C_1–C_2–C_3 plane has to be rotated counterclockwise to bring it to coincide with the C_2–C_3–C_4 plane, so the configurational label for dimethylphenanthrene shown here is *M*.

and benzo[*c*]phentharene need substituents to induce helical chirality, substituents are not needed to induce chirality in helicenes, because the curved skeleton formed by the fused benzene rings is long enough for the terminal rings to "bump" into each other, unless they twist in opposite directions. Triangulanes (de Meijere et al. 2004) and tetramantanes (Schreiner et al. 2009) fall under the class of helicenes, for which *P* and *M* conventions are used.

In addition to these crowded organic molecules, peptides and proteins adopt helical secondary structures. The structure often referred to as α-helix is in fact a *P*-helix (right-handed helix) and that as polyproline II has local left-handed, or *M*-helical, segments.

2.2.4 Planar Chirality

Molecules that contain a planar aromatic ring attached to a hoop-like handle are called *ansa* compounds (*ansa* in Latin means "handle"). If the planar ring cannot swivel through the "hoop," then such molecules are said to possess planar chirality. When the planar ring is a substituted benzene ring and the handle is a polymethylene chain attached to the para carbon atoms of benzene ring, this type of molecules is referred to as [*n*]paracyclophanes. Here "*n*" stands for the number of methylene groups that constitute the handle. To determine the configurational labels for molecules with planar chirality, the following procedure is used (see Figure 2.14). First, we need to choose the preferred side of benzene plane. The preferred side is the one that has a substituent of higher priority. If there is no preference, then either side may be used. Because of the carboxylic group at C_3 (see Figure 2.14), C_3 is the preferred side of benzene plane. Then we pick a pilot atom of the polymethyelne chain from that side. The pilot atom is the first carbon atom of the polymethylene chain that is connected to the

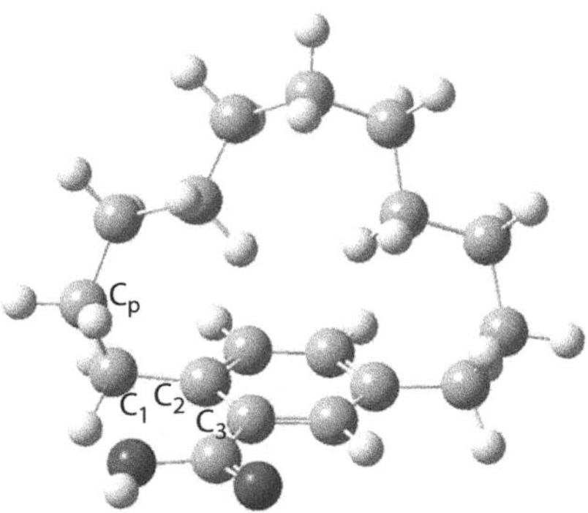

FIGURE 2.14
Planar chirality in paracyclophanes. The presence of a carboxylic group at C_3 prevents the swiveling of the benzene ring through the methylene group handle that connects the para positions of the benzene ring. The trace from C_1 to C_2 to C_3 is clockwise, and therefore the configurational label for paracyclophane displayed here is *P*.

carbon atom in the plane of the benzene ring but is not in the plane itself. The pilot atom in Figure 2.14 is labeled as C_p. Starting from the first carbon atom in the plane, we go along the bonds that lead us to the atoms of higher priority. This direction is C_1–C_2–C_3 as shown in Figure 2.14. Since the trace of C_1–C_2–C_3 has a clockwise rotation, the configurational label for paracyclophane, shown in Figure 2.14, is *P*. Sometimes *R* or *S* notation is also used as R_p or S_p, with subscript "p" identifying the planar chirality.

More complex structures can be found in this category, including those with two phenyl rings stacked above each other (and connected through methylene chains) and cyclophynes that have the benzene rings connected through acetylenic groups with triple bonds.

2.2.5 Propeller Chirality

Earlier we mentioned that boron compounds can be chiral because ligands attached to boron can possess a propeller twist that can lead to either a right-handed helical twist or a left-handed helical twist (Kaufmann and Boese 1990). However, in inorganic complexes, λ (lambda, for left-handed propeller twist) and Δ (delta, for right-handed propeller twist) designations are used for the enantiomers of bis and tridentate metal complexes (Connelly et al. 2005).

2.2.6 Other Sources of Chirality

The readers are encouraged to refer to the original articles in learning about molecules of unusual structures, namely cryptophans (Collet 1996), fullerenes (Thilgen et al. 1997; Powell et al. 2002), and transition metal complexes (Connelly et al. 2005).

2.3 Conformations

The specification of chiral molecular structure involves not only the absolute configuration but also the predominant conformation(s). Conformations can be divided into two broad groups: (a) bond conformations and (b) ring conformations.

(a) *Bond conformations:* A bond conformation defines the relative orientation of atoms, or groups, connected to a single covalent bond or to a double bond. The atoms, or groups, connected to these bonds are collectively referred to as ligands. Ligands connected to both sides of a single bond can be rotated relative to each other, and the resulting different relative orientations represent different conformations. Such rotations of ligands are not possible around double bonds, but the ligands could be physically placed at different relative dispositions.

For example, the chiral molecule (*S*)-chlorofluoromethanol (see Figure 2.15) can have different orientations, or conformations, for the hydroxyl hydrogen atom. These orientations differ in the torsional angle of H–O–C–X (where X is one of the other atoms attached to C) segment. These different conformations can have different energies, so some can be more energetically favorable than others. Note that these different orientations do not influence the configuration at the chiral center but can influence the chiroptical properties. Different labels/designations are used for specifying the conformers. For relative orientations of ligands attached to single bonds, the labels recommended by the International Union of Pure and Applied Chemistry are (McNaught and Wilkinson 1997; Cross and Klyne 1976) synperiplanar (sp), synclinal (sc), anticlinal (ac), and antiperiplanar (ap). The specific abbreviations and corresponding ranges for torsional angles are +sp (0° to +30°), −sp (0° to −30°), +sc (+30° to +90°), −sc (−30° to −90°), +ac (+90° to +150°), −ac (−90° to −150°), +ap (+150° to +180°), and −ap (−150° to −180°). Table 2.1 summarizes this information in a compact form. Three of the possible conformations of (*S*)-chlorofluoromethanol are shown in

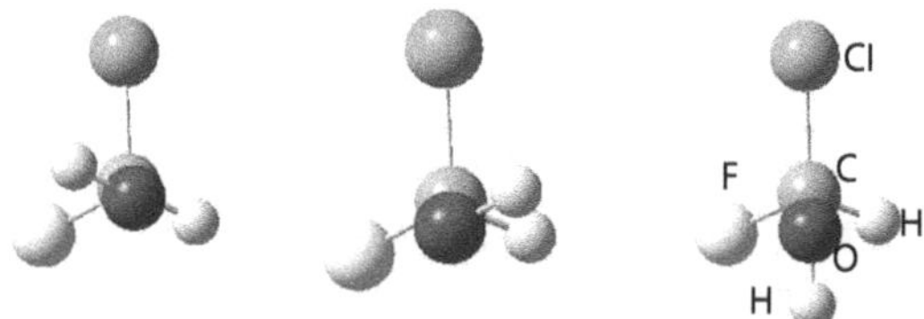

FIGURE 2.15
Three different orientations of a hydroxyl hydrogen atom in (*S*)-chlorofluoromethanol with conformational labels based on the H–O–C–Cl dihedral angle: synclinal with +60° torsional angle (+sc, left); −60° torsional angle (-sc, middle); and antiperiplanar, 180°, torsional angle (ap, right). Alternate designations used for these three conformations are, respectively, positive gauche (g+), negative gauche (g$^-$), and trans (t).

TABLE 2.1

Nomenclature for Conformations Arising from Rotation around a Single Bond

Torsional angle	Label[a]	Abbreviation	Torsional Angle	Label[a]	Abbreviation
0° to +30°	Synperiplanar	(+)-sp	0° to −30°	Synperiplanar	(−)-sp
30° to +90°	Synclinal[b]	(+)-sc	(−30°) to −90°	Synclinal[b]	(−)-sc
90° to +150°	Anticlinal	(+)-ac	(−90°) to −150°	Anticlinal	(−)-ac
150° to +180°	Antiperiplanar	(+)-ap	(−150°) to −180°	Antiperiplanar	(−)-ap

[a] IUPAC recommendations; see http://goldbook.iupac.org.
[b] Synclinal is synonymous to "gauche."

Figure 2.15. Another commonly used label, gauche, is synonymous to synclinal. Two other labels, *cis* (from Latin, meaning "on the same side") and *trans* (from Latin, meaning "across"), are also used to represent the relative orientation of ligands, around single bonds, with torsional angles of 0° and 180°, respectively (Panico et al. 1993).

For relative placement of groups attached to double bonds, the labels (see Table 2.2) used are cis and trans or Z (from the German *zusammen*, meaning "together") and E (from the German *entgegen*, meaning "opposite"). The Z/E labels were introduced to avoid some ambiguities that arise in using the cis/trans labels (Panico et al. 1993). In Z/E notation, the four ligands attached to C=C are given priorities (following CIP rules), and the relative disposition of the highest priority ligand on one end of the double bond is specified with respect to the highest priority ligand on the other end of the double bond. The Z and E conformations of (*S*)-1-(bromofluoromethoxy)-1-chloro-2-fluoroethene are shown in Figure 2.16.

(b) *Ring conformations:* Cyclic molecules have ring connections, and these rings can exist in different conformations: chair, boat, envelope, half-chair, twist boat, and so on (Nelson and Brammer 2011). These conformations differ in the way torsional segments in the rings are puckered. For example, six-membered ring molecules can exist in chair and boat conformations, each with substituents in axial or equatorial positions. Axial and equatorial orientations refer, respectively, to the perpendicular and parallel orientations of a chemical bond of interest to the mean plane of the ring. (*S*)-3-Chlorocyclohexanone in equatorial-chair, axial-chair, and boat conformation is shown in Figure 2.17. Five-membered ring molecules can exist with two different ring puckering conformations (see Figure 2.18). For larger rings, there will be more flexibility for ring conformations, and the reader is encouraged to review literature for such conformations.

TABLE 2.2

Nomenclature for Conformations Describing the Relationship of Atoms/Groups Attached to a Double Bond

Label[a]	Description	Label[b]	Description
Cis	Two atoms/groups lie on the same side of a plane	Z	Highest priority atom/group on one end of double bond is on the same side as the highest priority atom/group on the other end
Trans	Two atoms/groups lie on the opposite sides of a plane	E	Highest priority atom/group on one end of double bond is on the opposite side as the highest priority atom/group on the other end

IUPAC recommendations; see http://goldbook.iupac.org.

[a] *cis* and *trans* labels are also used for relationships of groups around single bonds.

[b] Z for German word *zusammen*, meaning "together"; E for German word *entgegen*, meaning "opposite."

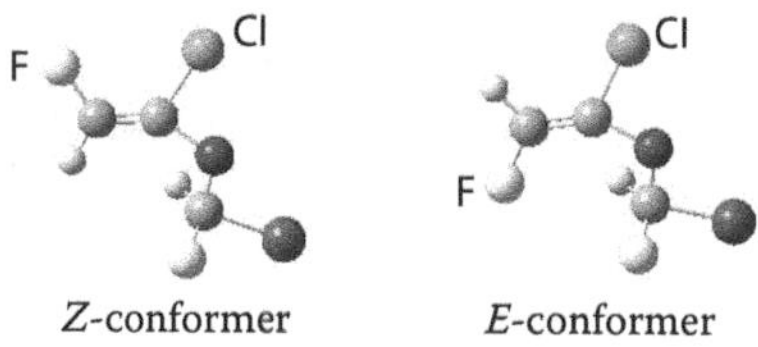

FIGURE 2.16
Z (left) and E (right) conformers of (S)-1-(bromofluoromethoxy)-1-chloro-2-fluoroethene.

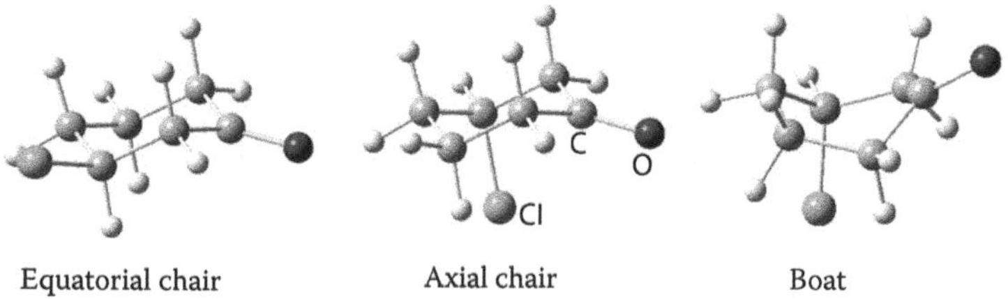

FIGURE 2.17
(S)-3-chlorocyclohexanone in equatorial chair conformation (left), axial chair conformation (middle), and in boat conformation (right).

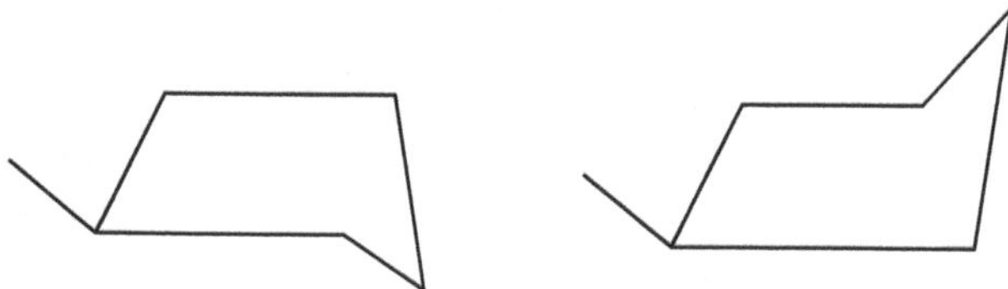

FIGURE 2.18
A five-membered ring with two different ring puckering conformations.

2.4 Biomolecular Structures

Naturally occurring proteins and peptides are made up of L-amino acids. The hydrogen bonding interactions among these amino acids result in different types of three-dimensional structures, referred to as the secondary structures. The most commonly found secondary structures are right-handed α-helix, β-sheet, and left-handed poly-proline-II-type helix. Thus these molecules have both central chirality (from constituent L-amino acids) as well as structural chirality (due to the formation of secondary structures). Similarly, nucleic acids form different types of structures (B-DNA, Z-DNA, etc). In these molecules, chirality arises from the helical stacking of nucleic acid bases, as well as from the central chirality in the sugar moieties contained in nucleic acids. A separate chapter is needed to provide the details associated with biomolecular structures. Therefore the reader is referred to standard books on biomolecular structures (Branden and Tooze 1999).

References

Branden, C., and J. Tooze. 1999. *Introduction to Protein Structure*, 2nd ed. New York, NY: Garland Publishing.

Cahn, R. S., C. Ingold, and V. Prelog. 1966. Specification of molecular chirality. *Angew. Chem. Int. Ed.* 5(4):385–415.

Collet, A. 1996. Cryptophans. In: J. L. Artwood, J. E. D. Davies, D. D. Macnicol, and F. Vogtle (eds.). *Comprehensive Supramolecular Chemistry*. New York, NY: Elsevier.

Connelly, N. G., T. Damhus, R. M. Hartshorn, and A. T. Hutton. 2005. *The Red Book-Nomenclature for Inorganic Chemistry: IUPAC Recommendations*. London, UK: The Royal Society of Chemistry.

Cotton, F. A. 1971. *Chemical Applications of Group Theory*, 2nd ed. New York, NY: Wiley-Interscience.

Cross, L. C., and W. Klyne. 1976. Rules for the nomenclature of organic chemistry. Section E: Stereochemistry. *Pure Appl. Chem.* 45:11–30.

de Meijere, A., H. Schill, S. I. Kozhushkov, R. Walsh, E. M. Muller, and H. Grubmuller. 2004. Cyclopropylidenes, bicyclopropylidenes, and vinylcarbenes—Some modes of formation and preparative applications. *Russ. Chem. Bull.* 53(5):947–959.

Eliel, E. L., S. H. Wilen, and M. P. Doyle. 2001. *Basic Organic Stereochemistry*. New York, NY: Wiley.

Kaufmann, D., and R. Boese. 1990. A borate propeller compound as chiral catalyst for an asymmetrically induced Diels-Alder reaction. *Angew. Chem. Int. Ed.* 29(5):545–546.

McNaught, A. D., and A. Wilkinson. 1997. *IUPAC. Compendium of Chemical Terminology*, 2nd ed. Oxford, UK: Blackwell Scientific Publications.

Moss, G. P. 1996. Basic terminology of stereochemistry. *Pure Appl. Chem.* 68(12):2193–2222.

Nasipuri, D. 2012. *Stereochemistry of Organic Compounds: Principles and Applications.* London, UK: New Academic Science.
Nelson, D. J., and C. N. Brammer. 2011. Toward consistent terminology for cyclohexane conformers in introductory organic chemistry. *J. Chem. Educ.* 88(3):292–294.
Panico, R., W. H. Powell, and J. C Richer. 1993. *"Recommendation 7.1.2." A Guide to IUPAC Nomenclature of Organic Compounds.* Oxford, UK: Blackwell Science.
Powell, W. H., F. Cozzi, G. P. Moss, C. Thilgen, R. J. R. Hwu, and A. Yerin. 2002. Nomenclature for the C-60-I-h and C-70-D-5h(6) fullerenes—(IUPAC recommendations 2002). *Pure Appl. Chem.* 74(4):629–695.
Prelog, V., and G. Helmchen. 1982. Basic principles of the CIP-system and proposals for a revision. *Angew. Chem. Int. Ed.* 21(8):567–583.
Schreiner, P. R., A. A. Fokin, H. P. Reisenauer, B. A. Tkachenko, E. Vass, M. M. Olmstead, D. Blaeser, R. Boese, J. E. P. Dahl, and R. M. K. Carlson. 2009. [123]Tetramantane: Parent of a new family of σ-helicenes. *J. Am. Chem. Soc.* 131(32):11292–11293.
Shallenberger, R. S. 1982. *Advanced Sugar Chemistry: Principles of Sugar Stereochemistry.* Westport, CT: Avi Publishing Co., Inc.
Slocum, D. W., D. Sugarman, and S. P. Tucker. 1971. 2 faces of D and L nomenclature. *J. Chem. Educ.* 48(9):597.
Thilgen, C., A. Herrmann, and F. Diederich. 1997. Configurational description of chiral fullerenes and fullerene derivatives with a chiral functionalization pattern. *Helv. Chim. Acta* 80(1):183–199.

3

Interaction of Circularly Polarized Light with Chiral Molecules

3.1 Introduction

In this chapter, we will consider optical rotation (OR), circular dichroism (CD), and circular difference Raman scattering, which are three different phenomena that result from the interaction of circularly polarized visible or infrared light with chiral molecules. An in-depth treatment of the underlining phenomena requires significant quantum mechanical and mathematical background. The goal of this chapter is to provide easy-to-follow arguments for beginning researchers and the expressions needed to progress to the next chapter. A good description of these topics can also be found in the literature (Barron 2004; Charney 1985; Fitts 1974; Polavarapu 1998; Tinoco and Cantor 1970). The readers with a preference to avoid the equations may skip to the end of this chapter where a brief summary of the important results is provided. Readers interested in quantum chemical formulations can find detailed information in Chapter 4.

3.2 Optical Rotation

When plane-polarized light of a certain wavelength λ traverses through an isotropic medium containing chiral molecules, the orientation of the polarization axis of light emerging from the chiral medium can be different from that entering the chiral medium. This phenomenon is called OR. To understand this phenomenon, let us recall from Chapter 1 that right circular polarization (RCP) can be generated from two orthogonal linearly polarized components with equal amplitudes and a phase difference of $\pi/2$. Similarly, left circular polarization (LCP) can be generated from those components with a difference of $3\pi/2$ or $-\pi/2$. The superposition of RCP and LCP gives linear polarization, which can be seen by combining Equations 1.6 and 1.8 of Chapter 1:

$$\begin{aligned}\mathbf{F}_{RCP} + \mathbf{F}_{LCP} &= F_0(\mathbf{u}\cos\Theta_t - \mathbf{v}\sin\Theta_t) + F_0(\mathbf{u}\cos\Theta_t + \mathbf{v}\sin\Theta_t) \\ &= 2F_0(\mathbf{u}\cos\Theta_t) = 2\mathrm{F}_x\end{aligned} \tag{3.1}$$

where $\Theta_t = 2\pi\nu t$. That means, linearly polarized light with its electric vector along x axis can be seen as the average of RCP and LCP:

$$\mathrm{F}_x = \frac{\mathbf{F}_{RCP} + \mathbf{F}_{LCP}}{2} \tag{3.2}$$

A visual depiction of this relation is shown in Figure 3.1. To understand the influence of chiral molecules on linearly polarized light traversing through an isotropic medium of chiral molecules, we have to incorporate (1) the dependence of light propagation on the length of travel in the medium, l (often referred to as the path length); and (2) different speeds for RCP and LCP through the isotropic chiral medium. To account for these two points, path length and refractive indices have to be introduced into the expressions for RCP and LCP light waves. Then,

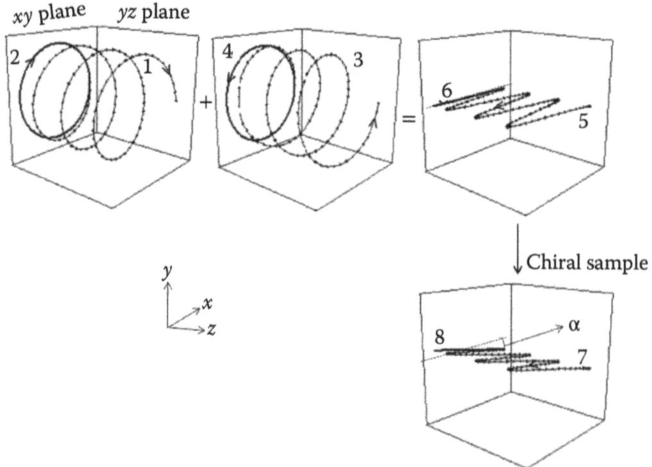

FIGURE 3.1
The leftmost figure depicts the right circularly polarized electric vector, labeled as 1 (and its projection in the xy plane labeled as 2), and the second figure depicts the left circularly polarized electric vector labeled as 3 (and its projection in the xy plane labeled as 4). The third figure depicts the resultant average of the first two, yielding the linearly polarized electric vector along the x axis, labeled as 5 (and its projection in the xy plane labeled as 6). The fourth figure depicts the linearly polarized electric vector rotated by the chiral sample with $\alpha = 10°$, labeled as 7 (and its projection in the xy plane, labeled as 8).

$$\mathbf{F}_{RCP} = F_0(\mathbf{u}\cos\Theta_R - \mathbf{v}\sin\Theta_R) \quad (3.3)$$

$$\mathbf{F}_{LCP} = F_0(\mathbf{u}\cos\Theta_L + \mathbf{v}\sin\Theta_L) \quad (3.4)$$

where

$$\Theta_R = 2\pi\left(\nu t - \frac{l n_R}{\lambda}\right) \quad (3.5)$$

$$\Theta_L = 2\pi\left(\nu t - \frac{l n_L}{\lambda}\right) \quad (3.6)$$

In the aforementioned equations, l is path length, that is, the length traveled by light in the sample medium; n_R and n_L are, respectively, the refractive indices for RCP and LCP. These refractive indices can be written as follows:

$$n_R = n - \Delta \quad (3.7)$$

$$n_L = n + \Delta \quad (3.8)$$

where

$$n = \frac{n_R + n_L}{2} \quad (3.9)$$

$$\Delta = \frac{n_L - n_R}{2} \quad (3.10)$$

Substituting Equations 3.7 through 3.10 in Equations 3.5 and 3.6, the following equations are obtained:

$$\Theta_R = 2\pi\left(\nu t - \frac{ln}{\lambda} + \frac{l\Delta}{\lambda}\right) \quad (3.11)$$

$$\Theta_L = 2\pi\left(\nu t - \frac{ln}{\lambda} - \frac{l\Delta}{\lambda}\right) \quad (3.12)$$

Since $\left(\frac{\Theta_R - \Theta_L}{2}\right) = \frac{2\pi l\Delta}{\lambda}$, Equations 3.11 and 3.12 can be rewritten in terms of Θ, where

$$\Theta = 2\pi\left(\nu t - \frac{nl}{\lambda}\right) \quad (3.13)$$

Then,

$$\Theta_R = \Theta + \frac{(\Theta_R - \Theta_L)}{2} = \Theta + \alpha \tag{3.14}$$

$$\Theta_L = \Theta - \frac{(\Theta_R - \Theta_L)}{2} = \Theta - \alpha \tag{3.15}$$

where α is the OR (in radians) given by the relation:

$$\alpha = \frac{(\Theta_R - \Theta_L)}{2} = \frac{2\pi l \Delta}{\lambda} = \frac{\pi l}{\lambda}(n_L - n_R) \tag{3.16}$$

In this equation, $(n_L - n_R)$ is the difference between refractive indices of left and right circularly polarized light waves and is referred to as the circular birefringence (CB). Since α is proportional to $(n_L - n_R)$, OR is often referred to as CB.

Substituting Equations 3.14 and 3.15 into Equations 3.3 and 3.4, for RCP and LCP, and using the expansions cos(A ± B) = cosAcosB ∓ sinAsinB and sin(A ± B) = sinAcosB ± cosAsinB, the following relations are obtained:

$$\mathbf{F}_{RCP} = F_0[\mathbf{u}\,(\cos\Theta\,\cos\alpha - \sin\Theta\,\sin\alpha) - \mathbf{v}\,(\sin\Theta\,\cos\alpha + \cos\Theta\,\sin\alpha)] \tag{3.17}$$

$$\mathbf{F}_{LCP} = F_0[\mathbf{u}\,(\cos\Theta\,\cos\alpha + \sin\Theta\,\sin\alpha) + \mathbf{v}\,(\sin\Theta\,\cos\alpha - \cos\Theta\,\sin\alpha)] \tag{3.18}$$

The average of these two equations becomes

$$(\mathbf{F}_{RCP} + \mathbf{F}_{LCP})/2 = F_0\,[\mathbf{u}\,\cos\Theta\,\cos\alpha - \mathbf{v}\,\cos\Theta\,\sin\alpha] = \mathrm{F}_x\,\cos\alpha - \mathrm{F}_y\,\sin\alpha \tag{3.19}$$

where $F_x = F_0\,\mathbf{u}\,\cos\Theta$ and $F_y = F_0\,\mathbf{v}\,\cos\Theta$.

For an isotropic medium of achiral molecules $\Theta_R = \Theta_L$, so α = 0 and Δ = 0. Then, Equation 3.19 becomes $(\boldsymbol{F}_{RCP} + \boldsymbol{F}_{LCP})/2 = F_0\mathbf{u}\,\cos\Theta = \mathrm{F}_x$. This result suggests, as mentioned earlier, that the average of RCP and LCP (equivalent to linearly polarized light) transmitting through isotropic achiral medium exits the sample without any change in the orientation of its polarization axis.

However, for an isotropic medium of chiral molecules, $\Theta_R \neq \Theta_L$, so α ≠ 0 and Δ ≠ 0. Then, the average of RCP and LCP transmitting through an isotropic chiral medium for a given distance *l* is equivalent to the resultant of linearly polarized waves along *x* and *y* axes, each with a different amplitude (modified by the terms cosα and sinα), but without phase difference (see Equation 3.19). As demonstrated in Chapter 1, the resultant linear polarization would be at a certain angle from the *x* axis. That angle here is α radians. For positive α, the resulting vector rotates from +*x* axis toward −*y* axis. A visual depiction of this OR phenomenon is provided in Figure 3.1.

The OR in radians is given by Equation 3.16 and that in degrees is obtained by multiplying the right-hand side of Equation 3.16 with 360/2π:

$$\alpha = \frac{\pi l}{\lambda}(n_L - n_R) \times \frac{360}{2\pi} = \frac{180 \times l}{\lambda}(n_L - n_R) \tag{3.20}$$

OR depends on the length that light traverses through the medium of chiral molecules and on the number of chiral molecules encountered by the traversing light (which in turn depends on the concentration of chiral molecules).

A more commonly used quantity is specific rotation, which in the modern literature is referred to as specific optical rotation (SOR). SOR is defined for neat liquid samples as

$$[\alpha] = \frac{\alpha}{\rho l} \tag{3.21}$$

where ρ is the density of liquid expressed as g·cc^{-1}; path length l is expressed in decimeters, dm (1 dm = 0.1 m = 10 cm) and [α] will be in units of deg·cc·g^{-1} dm^{-1}. For liquid solutions, this equation is written as follows:

$$[\alpha] = \frac{\alpha}{cl} \tag{3.22}$$

where the density of liquid in Equation 3.21 is replaced with concentration, which is expressed as mass, m (in g) of chiral solute per unit volume of solution (V in cc), that is, $c = \left(\frac{m}{V}\right)$ g/cc. Note the use of lower case "*c*" in Equation 3.22 to express concentration as g/cc of solution. Here also, [α] will be in units of deg cc g^{-1} dm^{-1}.

In synthetic chemistry community, concentration is expressed as grams of chiral solute in 100 cc of solution, and in that case, SOR is given as follows:

$$[\alpha] = 100 \times \frac{\alpha}{c' l} \tag{3.23}$$

where c' is concentration expressed as grams of chiral solute in 100 cc of solution.

Another quantity encountered in the literature is molar rotation, [Φ], in deg·cm^2·dmol^{-1} units, which is obtained by multiplying the SOR, [α], with M/100, where M is the molar mass of solute substance:

$$[\phi] = [\alpha] \times \frac{M}{100} \tag{3.24}$$

Measurement of SOR as a function of wavelength, λ, yields a dispersion curve that is referred to as optical rotatory dispersion.

In Chapter 2, we mentioned that d and l symbols are used to specify the chiral substances. These symbols specify the SOR of the chiral sample measured at sodium *D* line, whose wavelength is 589 nm. The SOR at sodium *D* line is designated as $[\alpha]_D$ or $[\alpha]_{589}$. If $[\alpha]_D$ is positive, then the sample is said to be dextrorotatory and is labeled as d-sample. If the $[\alpha]_D$ is negative, then the sample is said to be levorotatory and is labeled as l-sample. The d- and l- symbols are not to be confused with D and L symbols that are used in Chapter 2 for specifying the absolute configuration.

3.3 Circular Dichroism

When circularly polarized light of certain wavelength λ traverses through an isotropic medium containing chiral molecules, the intensity of light emerging from the chiral medium can be different from that entering the chiral medium due to the absorption of light. The light absorbed by chiral molecules differs for RCP and LCP. This differential absorption of circularly polarized light is called CD. To understand this phenomenon, we have to modify the amplitudes of RCP and LCP as they pass through the chiral medium, as follows:

$$\mathbf{F}_{\mathrm{RCP}} = F_0 e^{\frac{-2\pi k_R l}{\lambda}} (\mathbf{u}\cos\Theta_\mathrm{R} - \mathbf{v}\sin\Theta_\mathrm{R}) \tag{3.25}$$

$$\mathbf{F}_{\mathrm{LCP}} = F_0 e^{\frac{-2\pi k_L l}{\lambda}} (\mathbf{u}\cos\Theta_\mathrm{L} + \mathbf{v}\sin\Theta_\mathrm{L}) \tag{3.26}$$

In these equations, k_L and k_R are referred to as the absorption indices for LCP and RCP, respectively. Designating,

$$F_\mathrm{R} = F_0 \mathrm{e}^{\frac{-2\pi k_R l}{\lambda}} \tag{3.27}$$

$$F_\mathrm{L} = F_0 \mathrm{e}^{\frac{-2\pi k_L l}{\lambda}} \tag{3.28}$$

and using Equations 3.14 and 3.15, the sum $\boldsymbol{F}_{\mathrm{RCP}} + \boldsymbol{F}_{\mathrm{LCP}}$ becomes

$$\begin{aligned}\mathbf{F}_{\mathrm{RCP}} + \mathbf{F}_{\mathrm{LCP}} &= [(F_\mathrm{R} + F_\mathrm{L})\,\mathbf{u}\cos\Theta - (F_\mathrm{R} - F_\mathrm{L})\,\mathbf{v}\sin\Theta]\cos\alpha \\ &\quad - [(F_\mathrm{R} - F_\mathrm{L})\,\mathbf{u}\sin\Theta + (F_\mathrm{R} + F_\mathrm{L})\,\mathbf{v}\cos\Theta]\sin\alpha\end{aligned} \tag{3.29}$$

The two terms on the right side of this equation are plotted separately in Figure 3.2.

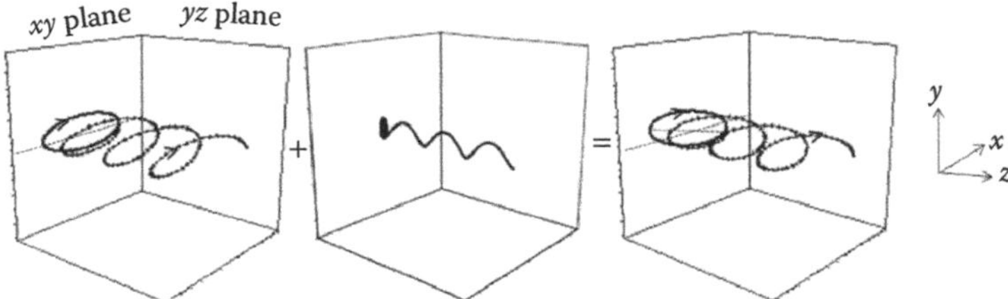

FIGURE 3.2
The leftmost figure corresponds to $[(F_R + F_L)\mathbf{u}\cos\theta - (F_R - F_L)\mathbf{v}\sin\theta]\cos\alpha$, the middle figure corresponds to $[-(F_R - F_L)\mathbf{u}\sin\theta - (F_R + F_L)\mathbf{v}\cos\theta]\sin\alpha$ (see Equation 3.29), and both represent elliptically polarized light with a clockwise electric vector. The major axes of their ellipses are along the x and y axes, respectively. The figure on the right corresponds to the sum of these two terms, showing that the major axis of the resulting ellipse is rotated from the x axis. These figures were simulated with $(F_R + F_L) = 0.75$, $(F_R - F_L) = 0.25$, and $\alpha = 10°$.

The first term on the right-hand side of Equation 3.29 represents elliptically polarized light (see Chapter 1 and Table 1.2), because the phase difference between x and y components is $\pi/2$ and the amplitudes of x and y components are not equal. Since the amplitude of x component, $(F_R + F_L)\cos\alpha$, is greater than that of y component, $(F_R - F_L)\cos\alpha$, this ellipse will have x axis as the major axis and y axis as the minor axis (see left most panel in Figure 3.2).

The second term on the right-hand side of Equation 3.29 represents another ellipse with $(F_R + F_L)\sin\alpha$ as the amplitude of y component and $(F_R - F_L)\sin\alpha$ as the amplitude of x component. Since the amplitude of y component is greater than that of x component, this ellipse will have y axis as the major axis and x axis as the minor axis (see the middle panel in Figure 3.2).

When the contributions of these two terms are added, the resultant is an ellipse rotated from x axis (see the right most panel in Figure 3.2).

The path length used in CD experiments is usually small, so α at those path lengths is small (see Equation 3.20). Even for $\alpha = 10°$, it can be seen from Figure 3.2 that the contribution from second term in Equation 29 is small. For smaller α, $\sin\alpha \sim 0$ and $\cos\alpha \sim 1$. Then, in the limit of small α, Equation 3.29 becomes

$$\mathbf{F}_{RCP} + \mathbf{F}_{LCP} = (F_R + F_L)\mathbf{u}\cos\Theta - (F_R - F_L)\mathbf{v}\sin\Theta \quad (3.30)$$

As discussed in Chapter 1, Equation 3.30 represents the right-handed elliptically polarized electric vector, with $(F_R + F_L)$ as the amplitude along unit vector $\mathbf{u}$ and $(F_R - F_L)$ as the amplitude along unit vector $\mathbf{v}$. Thus, the major axis of the ellipse is oriented along the x axis.

Most books do not address what happens when both OR and CD are present at the same time for a given sample. However, this can be visualized from Figure 3.2. In the absence of OR, the presence of CD converts the linearly polarized light (represented as $\mathbf{F}_{RCP} + \mathbf{F}_{LCP}$) into elliptically polarized light with its major axis aligned along the x axis (see Equation 3.30). In the presence of OR, this polarization ellipse rotates by α, the OR.

In the OR measurement, if the sample also exhibits CD, then one measures the rotation of the major axis of polarization ellipse. In the CD measurement, one may safely ignore the occurrence of OR for two reasons: (1) Observed rotations are usually of the order of less than a degree (note that even for quartz calibration devices of 10 cm long, the observed rotations are only a few degrees) and (2) the path lengths used in CD measurements are usually small (in the range of 0.1–1 mm), compared with those used in OR measurements (in the range of 50–200 mm). Under such conditions, setting cosα term ~1 and sinα term ~0 in Equation 3.29 is a good approximation. For these reasons, Equation 3.30 is widely used for the consideration of CD.

The ellipticity, η, associated with an ellipse (see Appendix 1) is defined as follows:

$$\tan\eta = \frac{F_R - F_L}{F_R + F_L} = \frac{e^{-2\pi k_R l/\lambda} - e^{-2\pi k_L l/\lambda}}{e^{-2\pi k_R l/\lambda} + e^{-2\pi k_L l/\lambda}} \tag{3.31}$$

Multiplying the numerator and denominator of the right-hand side with $e^{\pi l(k_L + k_R)/\lambda}$,

$$\tan\eta = \frac{e^{\pi l(k_L - k_R)/\lambda} - e^{-\pi l(k_L - k_R)/\lambda}}{e^{\pi l(k_L - k_R)/\lambda} + e^{-\pi l(k_L - k_R)/\lambda}} = \tanh\left(\frac{\pi l}{\lambda}(k_L - k_R)\right) \tag{3.32}$$

Note that in the right-hand side of the aforementioned equation, the identity, $\frac{e^{\beta} - e^{-\beta}}{e^{\beta} + e^{-\beta}} = \tanh\beta$, has been used. The expansion of $\tanh x = x - x^3/3 + \ldots$ for $x < 1$, suggests that for small values of the argument, $\tanh x \sim x$. Similarly, the expansion of $\tan x = x + x^3/3 + \ldots$ for $x < 1$, indicates that ellipticity can be seen to be proportional to the difference in absorption indices for LCP and RCP. That is,

$$\eta \approx \frac{\pi l}{\lambda}(k_L - k_R) \tag{3.33}$$

The complementarity of OR and ellipticity can be seen from Equations 3.16 and 3.33. While OR is a measure of the CB, ellipticity is a measure of the CD.

Equation 3.33 can be recast in terms of molar extinction coefficients, which are the quantities often derived from spectroscopic measurements. When monochromatic light of intensity I_o passes through an isotropic medium, the intensity of light exiting the medium, I, is given by Beer's law as follows:

$$I = I_0 e^{-2.303\varepsilon Cl} = I_0 10^{-\varepsilon Cl} \tag{3.34}$$

where ε is called decadic molar extinction coefficient and C is the sample concentration (in mol·L^{-1}). Note that the uppercase "C" used here for concentrations (mol·L^{-1}) is different from the lowercase "c" used for concentrations (in g·cc^{-1}) to calculate SOR (Equation 3.22). The absorbance, A, often measured in spectroscopy experiments is as follows:

$$A = \log_{10} \frac{I_0}{I} = \varepsilon C l \tag{3.35}$$

For RCP and LCP, ε is written, respectively, as ε_R and ε_L. These extinction coefficients can be related to absorption indices mentioned earlier. Note that the intensity of light is equal to the square of the amplitude of its electric vector. Then the intensity can be written from the amplitude, $F = F_0 e^{\frac{-2\pi kl}{\lambda}}$ as

$$I = I_0 e^{\frac{-4\pi kl}{\lambda}} \tag{3.36}$$

Combining this equation with Beer's law Equation 3.34, one obtains

$$2.303 \varepsilon C l = \frac{4\pi kl}{\lambda} \tag{3.37}$$

Then, the absorption index k can be seen to be

$$k = \frac{2.303 \varepsilon C \lambda}{4\pi} \tag{3.38}$$

For left and right circularly polarized light components, Equation 3.38 is modified by adding subscripts L and R, respectively (for both absorption index, k, and molar extinction coefficient, ε). Substituting Equation 3.38 in Equation 3.33, ellipticity in radians becomes

$$\eta \approx \frac{2.303Cl}{4} (\varepsilon_L - \varepsilon_R) = \frac{2.303}{4} \times \Delta A \tag{3.39}$$

where $\Delta A = (\varepsilon_L - \varepsilon_R)Cl$. Ellipticity in degrees is obtained by multiplying the right-hand side of this equation with 360/2π:

$$\eta \approx \frac{2.303}{4} \Delta A \times \frac{360}{2\pi} \tag{3.40}$$

With ellipticity in degrees, differential absorption for LCP and RCP, a unitless quantity, becomes

$$\Delta A = \eta \times \frac{4 \times 2\pi}{2.303 \times 360} = \frac{\eta}{32.988} \tag{3.41}$$

When ellipticity is expressed in millidegrees, ΔA becomes

$$\Delta A = \frac{\eta}{32988} \tag{3.42}$$

and differential molar extinction coefficient becomes

$$\Delta\varepsilon = \frac{\eta}{32988 \times C \times l} \tag{3.43}$$

Most commercial CD instruments (especially those operating in the visible spectral region) report the measured CD spectra in millidegrees, so the aforementioned equation is often used for converting to $\Delta\varepsilon$. When concentration is expressed in mol L^{-1} and path length in cm, the units for $\Delta\varepsilon$ are L mol^{-1} cm^{-1}.

Molar ellipticity $[\Theta]$, defined in deg cm^2 mol^{-1} units, can be written from the aforementioned equation as follows:

$$[\theta] = \frac{\eta}{Cl} = 32988 \times \Delta\varepsilon \tag{3.44}$$

when η is expressed in millidegrees, concentration in mol L^{-1}, and path length in cm. However, it is common practice to express molar ellipticity in deg cm^2 $dmol^{-1}$, in which case the aforementioned equation modifies to

$$[\theta] = \frac{\eta}{Cl} = 3298.8 \times \Delta\varepsilon \tag{3.45}$$

We have not made any distinction between CD associated with electronic and vibrational transitions up to this point, as the level of treatment provided in this chapter does not allow that distinction. However, the quantum chemical treatments (Chapters 4 and 5) do provide clear distinction between CD associated with electronic and vibrational transitions.

3.4 Vibrational Raman Optical Activity (VROA)

Unlike the case of OR (or SOR), absorption and CD measurements, which are based on light transmission through the sample, Rayleigh and Raman spectra are derived from the light scattered by the sample. The origin of scattered light is the electric dipole moment induced in the molecules by the intense laser light of frequency ν, incident on the sample. The induced electric dipole moment oscillates at the frequency of incident light. The oscillating-induced electric dipole moment emits radiation, which serves as the source of scattered light. The frequency of scattered light can be different from that of incident light, because of the beat frequencies between oscillating electric field (emitted by the induced dipoles) and molecular vibrations. The induced electric dipole moment component μ_α, in the presence of electric field component F_β incident on the sample is given by the relation

$$\mu_\alpha = \alpha_{\alpha\beta} F_\beta = \alpha_{\alpha\beta} F_\beta^0 \cos 2\pi\nu t \tag{3.46}$$

where $\alpha_{\alpha\beta}$ is a component of molecular electric dipole–electric dipole polarizability tensor (see Chapter 4, Equation 4.67). In Equation 3.46, the time varying electric field F_β oscillating at frequency ν is written as $F_\beta^0 \cos 2\pi\nu t$. The components of molecular electric dipole–electric dipole polarizability tensor can be expanded in normal vibrational coordinates of the molecule as

$$\alpha_{\alpha\beta} = \alpha_{\alpha\beta}^0 + \sum_k \left(\frac{\partial \alpha_{\alpha\beta}}{\partial Q_k} \right) Q_k^\circ \cos 2\pi\nu_k t \tag{3.47}$$

where Q_k is the normal coordinate of kth vibration, Q_k° is maximum vibrational amplitude, and ν_k is kth vibrational frequency. Substituting Equation 3.47 into Equation 3.46, the following equation is obtained:

$$\mu_\alpha = \alpha_{\alpha\beta}^0 F_\beta^0 \cos 2\pi\nu t + F_\beta^0 \sum_k \frac{\partial \alpha_{\alpha\beta}}{\partial Q_k} Q_k^\circ \cos 2\pi\nu_k t \cos 2\pi\nu t \tag{3.48}$$

Writing the cosine product cosAcosB as $\frac{1}{2}\left[\cos(A+B) + \cos(A-B)\right]$, the oscillating-induced electric dipole moment becomes

$$\mu_\alpha = \alpha_{\alpha\beta}^0 F_\beta^0 \cos 2\pi\nu t + \sum_k \frac{F_\beta^0}{2} \left(\frac{\partial \alpha_{\alpha\beta}}{\partial Qk} \right) Q_k^\circ \left[\cos 2\pi t(\nu + \nu_k) + \cos 2\pi t(\nu - \nu_k)\right] \tag{3.49}$$

From Equation 3.49, one can see that the induced electric dipole moment oscillates at frequencies ν as well as $\nu \pm \nu_k$. The former is responsible for Rayleigh light scattering and the latter for vibrational Raman light scattering. The frequency term $\nu - \nu_k$ corresponds to Stokes vibrational Raman transition and $\nu + \nu_k$ to anti-Stokes vibrational Raman transition. In other words, in the light scattered by the sample, scattered photons have energies of not only $h\nu$ but also $h(\nu - \nu_k)$ and $h(\nu + \nu_k)$. The detection of these scattered photons of energies $h\nu$, $h(\nu - \nu_k)$, and $h(\nu + \nu_k)$ provides, respectively, the Rayleigh, Stokes Raman, and anti-Stokes Raman spectra.

The intensity of light emitted by an electric dipole oriented along z axis oscillating at frequency ν is given as $\frac{2\pi^3 c\mu_z^2 \sin^2\theta}{\lambda^4 r^2}$ (Long 1977), where Θ is the angle between the direction of the dipole and the direction of observation, r is the distance from the point of observation to the location of the dipole and $\lambda = \frac{c}{\nu}$; θ will be 90° when the dipole is oriented along the z axis and observation is along the y axis. The induced electric dipole moment μ_z comes from $\left(\frac{\partial \alpha_{z\beta}}{\partial Q_k} \right) F_\beta^0$ and $\left(F_\beta^0\right)^2$ can be converted to incident light intensity I^0.

Then, it can be noticed that the scattered Raman intensity depends on the incident laser light intensity I^0 and the inverse fourth power of its wavelength $\left(\frac{1}{\lambda^4}\right)$. Scattered Raman intensities are often reported as differential cross sections (the intensity of scattered light per unit solid angle, Ω), $\frac{d\sigma}{d\Omega}$, where σ is the scattering cross section (scattered intensity divided by incident intensity). After taking spatial averaging and Boltzmann populations among different vibrational levels, Raman intensity for kth vibration in a given scattering geometry is represented as follows:

$$\frac{\mathrm{d}\sigma_k}{\mathrm{d}\Omega} = \left(I_\alpha^\gamma\right)_k = K\frac{\left(\bar{\nu} \mp \bar{\nu}_k\right)^4}{\bar{\nu}_k\left(\pm 1 \mp e^{\frac{\mp hc\bar{\nu}_k}{bT}}\right)} S_k \tag{3.50}$$

where K is a constant. In Equation 3.50, the subscript α and superscript γ of $\left(I_\alpha^\gamma\right)_k$ represent, respectively, the polarization of scattered light and polarization of incident laser light; $\bar{\nu}$ is the incident laser wavenumber, and $\bar{\nu}_k$ is the wavenumber of kth vibration, both expressed in cm^{-1}; "b" is Boltzmann constant (normally symbolized as k, but not here due to conflicts with kth vibration); The upper sign in "$\mp$" and "$\pm$" of Equation 3.50 corresponds to Stokes vibrational Raman transitions and the lower sign to anti-Stokes vibrational Raman transitions. S_k is scattering geometry-dependent Raman activity given as follows:

$$S_k = \left(a_1\bar{\alpha}_k^2 + a_2\beta_k^2\right) \tag{3.51}$$

where a_1 and a_2 are constants that depend on the geometry and polarizations used for the experimental scattering measurements, $\bar{\alpha}_k$ is the mean of $\left(\frac{\partial\alpha_{\alpha\beta}}{\partial Q_k}\right)$ tensor and β_k^2 is the anisotropy of that tensor. These two quantities are given as follows:

$$\bar{\alpha}_k^2 = \left(\frac{1}{9}\right)\left(\frac{\partial\alpha_{xx}}{\partial Q_k} + \frac{\partial\alpha_{yy}}{\partial Q_k} + \frac{\partial\alpha_{zz}}{\partial Q_k}\right)^2 \tag{3.52}$$

$$\begin{aligned}\beta_k^2 = \left(\frac{1}{2}\right)\Bigg\{&\left(\frac{\partial\alpha_{xx}}{\partial Q_k} - \frac{\partial\alpha_{yy}}{\partial Q_k}\right)^2 + \left(\frac{\partial\alpha_{xx}}{\partial Q_k} - \frac{\partial\alpha_{zz}}{\partial Q_k}\right)^2 + \left(\frac{\partial\alpha_{yy}}{\partial Q_k} - \frac{\partial\alpha_{zz}}{\partial Q_k}\right)^2 \\ &+ 6\left[\left(\frac{\partial\alpha_{xy}}{\partial Q_k}\right)^2 + \left(\frac{\partial\alpha_{xz}}{\partial Q_k}\right)^2 + \left(\frac{\partial\alpha_{yz}}{\partial Q_k}\right)^2\right]\Bigg\}\end{aligned} \tag{3.53}$$

Note that $\alpha_{\alpha\beta}$ tensor elements are in units of volume (Å^3, where 1 Å = 10^{-8} cm) or in Bohr^3 (Bohr = 0.52918 Å); Q_k are in Å amu$^{-1/2}$ (where amu stands for atomic mass unit) or in Bohr amu$^{-1/2}$, so the Raman activities are in units of Å^4 amu^{-1} (or in Bohr^4 amu^{-1}; see Appendix 8 for unit conversions).

The description mentioned earlier provides a succinct background for vibrational Raman scattering by typical molecules. Light scattering arising from the interaction of chiral molecules with circularly polarized light involves the induced magnetic dipole moment, m_α, and electric quadrupole moment, $\Theta_{\alpha\beta}$, along with the electric dipole moment mentioned earlier. Induced moments can arise from an electric field, magnetic field, and their time derivatives, associated with the incident light (Barron and Buckingham 1971).

$$\mu_\alpha = \mu_\alpha^0 + G_{\alpha\beta} F_\beta + \omega^{-1} G'_{\alpha\beta} \dot{B}_\beta + \ldots \tag{3.54}$$

$$m_\alpha = m_\alpha^0 + G_{\beta\alpha} F_\beta - \omega^{-1} G'_{\beta\alpha} \dot{F}_\beta + \ldots \tag{3.55}$$

$$\Theta_{\alpha\beta} = \Theta_{\alpha\beta}^0 + A_{\alpha\beta\gamma} F_\gamma \tag{3.56}$$

where $G_{\alpha\beta}$ and $G'_{\alpha\beta}$ are, respectively, the real and imaginary parts of the electric dipole–magnetic dipole polarizability tensor (see Chapter 4, Equation 4.77), and $A_{\alpha\beta\gamma}$ is the electric dipole–electric quadrupole polarizability tensor (see Chapter 5, Equation 5.38). Just as the oscillating-induced electric dipole moment, the oscillating magnetic dipole and quadrupole moments can also contribute to scattered radiation. For describing the scattered radiation in this situation, it is necessary to derive the electric field generated by these oscillating-induced moments and use that electric field in the Stokes parameters (see Appendix 1). These Stokes parameters for scattered light will not be presented here, as the equations involved for analysis with Stokes parameters are necessarily complex. Therefore, interested readers should consult the original literature (Barron 1976; Barron and Buckingham 1975). A summary, without going through the complicated equations, is provided below.

In the case of Raman scattering, molecular vibrations modulate the molecular electric dipole polarizability tensors (see Equation 3.47). Therefore, the normal coordinate derivatives of electric dipole polarizability tensors, $\frac{\partial \alpha_{\alpha\beta}}{\partial Q_k}, \frac{\partial G'_{\alpha\beta}}{\partial Q_k}, \frac{\partial A_{\alpha\beta\gamma}}{\partial Q_k}$, are the ones that are responsible for Raman scattering from chiral molecules. The interference terms (i.e., the sum of appropriate product terms) between $\left(\frac{\partial \alpha_{\alpha\beta}}{\partial Q_k}\right)$ and $\left(\frac{\partial G'_{\alpha\beta}}{\partial Q_k}\right)$ and between $\left(\frac{\partial \alpha_{\alpha\beta}}{\partial Q_k}\right)$ and $\left(\frac{\partial A_{\alpha\beta\gamma}}{\partial Q_k}\right)$ tensor elements are needed to generate a difference by the chiral molecules. This differential response can be seen in two different ways: (1) in response to alternating right and left circularly polarized incident laser light, the vibrational Raman intensities with a given linear polarization can be different and (2) in response to unpolarized incident laser light the vibrational Raman intensities with right and left circularly polarizations can be different. These differences are referred to as VROA.

The Raman circular intensity difference scattering, $I_\alpha^\gamma - I_\beta^\delta$, represents VROA, where superscripts on I represent the polarization of incident light and the subscripts represent that of scattered light. As mentioned earlier, VROA can be observed with either alternating right and left circularly polarized incident laser light or with unpolarized incident laser light.

For observing VROA, three different scattering geometries can be used (see Figure 3.3). These geometries include (1) 90° scattering geometry, where scattered light at 90° to the incident light propagation is collected; (2) 180° backscattering, where backscattered light is collected and analyzed; and (3) 0° forward scattering, where light scattered in the forward direction is collected and analyzed. In each of these geometries, different combinations of incident and scattered light polarizations can be utilized.

When the incident laser light is circularly polarized and scattered light is analyzed for a linear polarization, this arrangement is referred to as incident circular polarization (ICP)-VROA. For 90° scattering geometry, ICP-VROA measures the circular intensity difference, $I_\alpha^R - I_\alpha^L$, where superscripts R and L represent, respectively, right and left circularly polarized incident light; subscript α represents the direction of scattered linear polarization. The measurement is referred to as ICP-depolarized VROA (Barron, et al. 1973; Hug et al. 1975) when scattered light with polarization parallel to the scattering plane is collected (see Figure 3.4) and as ICP-polarized VROA (Barron et al. 1989; Hug and Surbeck 1979) when scattered light with polarization perpendicular to the scattering plane is collected. When the axis of linear

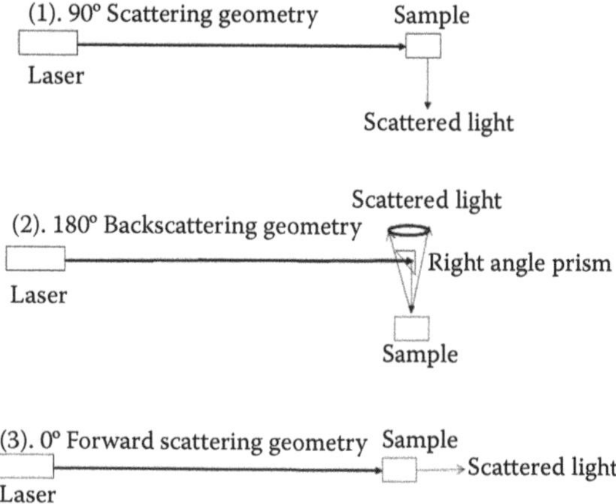

FIGURE 3.3
Different scattering geometries possible for vibrational Raman optical activity (VROA) measurements: (1) 90° scattering geometry, (2) 180° backscattering geometry, and (3) 0° forward scattering geometry.

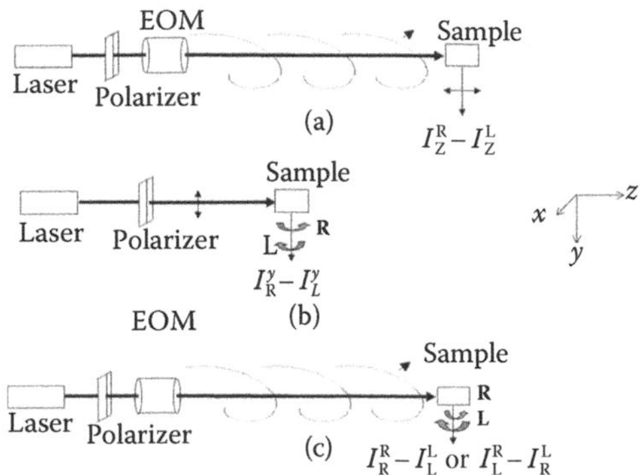

FIGURE 3.4
Some of the different scattering geometries for VROA measurements in 90° scattering geometry. (a) Incident circular polarization-depolarized VROA: The incident laser light is modulated between right and left circular polarizations using an electro-optic modulator (EOM) and scattered light is analyzed for linear polarization in the *yz* scattering plane. (b) Scattered circular polarization VROA: The incident laser light is linearly polarized with polarization in the *yz* scattering plane and scattered light is analyzed for right and left circular polarizations. (c) Dual circular polarization VROA: The incident laser light is modulated between right and left circular polarizations using EOM and scattered light is analyzed for right and left circular polarizations.

polarization analyzer in the scattered light is placed at 35.3° from *xy* plane (or 54.7° from *yz* plane), the resulting measurement is referred to as magic angle VROA (Hecht and Barron 1989), and this measurement will not have contribution from the normal coordinate derivatives of electric dipole–electric quadrupole polarizability tensor elements.

When the scattered light is analyzed for circular polarizations, while the incident light is linearly polarized (or unpolarized), this arrangement is referred to as scattered circular polarization (SCP)-VROA (see Figure 3.4), represented as $I_R^\alpha - I_L^\alpha$ (Spencer et al. 1988). For 90° SCP-VROA measurement, incident light polarization in the scattering plane is less susceptible to artifacts. For 180° SCP-VROA measurement, incident light is unpolarized.

When the incident laser light is circularly polarized and scattered light is also analyzed for circular polarizations, this arrangement (see Figure 3.4) is referred to as dual circular polarization (DCP)-VROA (Nafie and Freedman 1989). In DCP-VROA, one can analyze the scattered light for the same circular polarization as that of ICP; that is, $I_R^R - I_L^L$ or for opposite circular polarization to that of ICP; that is, $I_L^R - I_R^L$. These two arrangements are referred to, respectively, as DCP-I and DCP-II.

The Raman circular intensity difference scattering for *k*th vibration is given by the following relation:

$$\left(I_\alpha^\gamma - I_\beta^\delta\right)_k = K\frac{\left(\bar{\nu} \mp \bar{\nu}_k\right)^4}{\bar{\nu}_k\left(\pm 1 \mp e^{\frac{\mp hc\bar{\nu}_k}{bT}}\right)}P_k \tag{3.57}$$

In Equation 3.57, P_k is scattering geometry-dependent Raman circular intensity difference activity, which is given as follows:

$$P_k = \frac{1}{c}\left(a_3\bar{\alpha}_k\bar{G}_k' + a_4\gamma_k^2 + a_5\delta_k^2\right) \tag{3.58}$$

where constants a_3, a_4, and a_5 depend on the geometry and polarizations used for the experimental measurements, and $\bar{\alpha}_k\bar{G}_k'$, γ_k^2, and δ_k^2 represent the mean and anisotropy of interference between $\left(\frac{\partial\alpha_{\alpha\beta}}{\partial Q_k}\right)$ and optical activity tensors $\left(\frac{\partial G_{\alpha\beta}'}{\partial Q_k}\right)$ and $\left(\frac{\partial A_{\alpha\beta\gamma}}{\partial Q_k}\right)$ as defined below:

$$\bar{\alpha}_k\bar{G}_k' = \left(\frac{1}{9}\right)\left(\frac{\partial\alpha_{xx}}{\partial Q_k} + \frac{\partial\alpha_{yy}}{\partial Q_k} + \frac{\partial\alpha_{zz}}{\partial Q_k}\right)\left(\frac{\partial G_{xx}'}{\partial Q_k} + \frac{\partial G_{yy}'}{\partial Q_k} + \frac{\partial G_{zz}'}{\partial Q_k}\right) \tag{3.59}$$

$$\gamma_k^2 = \left(\frac{1}{2}\right)\left\{\begin{aligned} &\left(\frac{\partial\alpha_{xx}}{\partial Q_k} - \frac{\partial\alpha_{yy}}{\partial Q_k}\right)\left(\frac{\partial G_{xx}'}{\partial Q_k} - \frac{\partial G_{yy}'}{\partial Q_k}\right) + \left(\frac{\partial\alpha_{xx}}{\partial Q_k} - \frac{\partial\alpha_{zz}}{\partial Q_k}\right)\left(\frac{\partial G_{xx}'}{\partial Q_k} - \frac{\partial G_{zz}'}{\partial Q_k}\right) + \left(\frac{\partial\alpha_{yy}}{\partial Q_k} - \frac{\partial\alpha_{zz}}{\partial Q_k}\right)\left(\frac{\partial G_{yy}'}{\partial Q_k} - \frac{\partial G_{zz}'}{\partial Q_k}\right) + \\ &3\left[\begin{aligned} &\left(\frac{\partial\alpha_{xy}}{\partial Q_k}\right)\left(\frac{\partial G_{xy}'}{\partial Q_k} + \frac{\partial G_{yx}'}{\partial Q_k}\right) + \left(\frac{\partial\alpha_{xz}}{\partial Q_k}\right)\left(\frac{\partial G_{xz}'}{\partial Q_k} + \frac{\partial G_{zx}'}{\partial Q_k}\right) + \\ &\left(\frac{\partial\alpha_{yz}}{\partial Q_k}\right)\left(\frac{\partial G_{yz}'}{\partial Q_k} + \frac{\partial G_{zy}'}{\partial Q_k}\right)\end{aligned}\right]\end{aligned}\right\} \tag{3.60}$$

$$\delta_k^2 = \left(\frac{\omega}{2}\right)\left\{\begin{aligned} &\left(\frac{\partial\alpha_{yy}}{\partial Q_k} - \frac{\partial\alpha_{xx}}{\partial Q_k}\right)\left(\frac{\partial A_{zxy}}{\partial Q_k}\right) + \left(\frac{\partial\alpha_{xx}}{\partial Q_k} - \frac{\partial\alpha_{zz}}{\partial Q_k}\right)\left(\frac{\partial A_{yzx}}{\partial Q_k}\right) + \left(\frac{\partial\alpha_{zz}}{\partial Q_k} - \frac{\partial\alpha_{yy}}{\partial Q_k}\right)\left(\frac{\partial A_{xyz}}{\partial Q_k}\right) + \\ &\left(\frac{\partial\alpha_{xy}}{\partial Q_k}\right)\left(\frac{\partial A_{yyz}}{\partial Q_k} - \frac{\partial A_{zyy}}{\partial Q_k} + \frac{\partial A_{zxx}}{\partial Q_k} - \frac{\partial A_{xxz}}{\partial Q_k}\right) + \left(\frac{\partial\alpha_{xz}}{\partial Q_k}\right)\left(\frac{\partial A_{yzz}}{\partial Q_k} - \frac{\partial A_{zzy}}{\partial Q_k} + \frac{\partial A_{xxy}}{\partial Q_k} - \frac{\partial A_{yxx}}{\partial Q_k}\right) \\ &+ \left(\frac{\partial\alpha_{yz}}{\partial Q_k}\right)\left(\frac{\partial A_{zzx}}{\partial Q_k} - \frac{\partial A_{xzz}}{\partial Q_k} + \frac{\partial A_{xyy}}{\partial Q_k} - \frac{\partial A_{yxx}}{\partial Q_k}\right)\end{aligned}\right\} \tag{3.61}$$

It should be noted (see Appendix 8, Section A8.13) that $\alpha_{\alpha\beta}$ are in units of $Å^3$ (or Bohr^3), and $\omega^{-1}G_{\alpha\beta}'$ (where $\omega = 2\pi\nu$ is the circular frequency of incident laser light) and $A_{\alpha\beta\gamma}$ elements are in units of $Å^4$ (or Bohr^4). The normal coordinate derivatives, $\left(\frac{\partial\alpha_{\alpha\beta}}{\partial Q_k}\right)$, are in units of $Å^2$ amu$^{-1/2}$; $\omega^{-1}\left(\frac{\partial G_{\alpha\beta}'}{\partial Q_k}\right)$ and $\left(\frac{\partial A_{\alpha\beta\gamma}}{\partial Q_k}\right)$ are in units of $Å^3$ amu$^{-1/2}$. Therefore, $\omega^{-1}\,\bar{\alpha}_k\bar{G}_k'$, $\omega^{-1}\,\gamma_k^2$, and $\omega^{-1}\,\delta_k^2$ are in units of $Å^5$ amu^{-1} (or Bohr^5 amu^{-1}).

The Raman circular intensity difference activity becomes

$$\begin{aligned} P_k &= \frac{\omega}{c}\left(a_3\omega^{-1}\bar{\alpha}_k\bar{G}_k' + a_4\omega^{-1}\gamma_k^2 + a_5\omega^{-1}\delta_k^2\right) \\ &= \frac{2\pi}{\lambda}\left(a_3\omega^{-1}\bar{\alpha}_k\bar{G}_k' + a_4\omega^{-1}\gamma_k^2 + a_5\omega^{-1}\delta_k^2\right)\end{aligned} \tag{3.62}$$

where λ is the wavelength of incident laser light, and P_k is obtained in units of $Å^4$ amu^{-1} (see Appendix 8, Section A8.13, for unit conversions).

The Raman circular intensity difference activity, P_k, is approximately four orders of magnitude smaller than that of normal Raman activity, S_k (Equation 3.51), because of the presence of λ in the denominator of Equation 3.62.

To eliminate the dependence on incident laser intensity, associated constants and instrumental parameters, it is customary to report dimensionless circular intensity differential (CID), Δ, as the ratio of circular intensity difference, $I_\alpha^\gamma - I_\beta^\delta$, to the circular intensity sum, $I_\alpha^\gamma + I_\beta^\delta$. For kth vibration,

$$\Delta_k = \left(\frac{I_\alpha^\gamma - I_\beta^\delta}{I_\alpha^\gamma + I_\beta^\delta} \right)_k \tag{3.63}$$

Appropriate expressions for CIDs (Barron 1976) in selected scattering arrangements, adopted from Polavarapu (1998), are summarized in Table 3.1. In the expressions provided for Δ_k, the numerator is proportional to Raman circular intensity difference activity (Equation 3.62), while the denominator is proportional to corresponding Raman activity (Equation 3.51). Analogous expressions for additional scattering arrangements have been summarized elsewhere (Polavarapu 1998).

TABLE 3.1

Expressions for Raman Circular Intensity Differentials Associated with Selected Scattering Geometries for Vibrational Raman Optical Activity Measurements

Scattering Geometry	Abbreviation[a]		Δ, CID[a]
90°	Depolarized ICP (90°, d) SCP (90°, d)	$\left(\frac{I_z^R - I_z^L}{I_z^R + I_z^L}\right)_k \left(\frac{I_R^y - I_L^y}{I_R^y + I_L^y}\right)_k$	$\dfrac{\frac{16}{3}\left(\frac{2\pi}{\lambda}\right)\left(3\omega^{-1}\gamma_k^2 - \omega^{-1}\delta_k^2\right)}{8\beta_k^2}$
	Polarized ICP (90°, p) SCP (90°, p)	$\left(\frac{I_x^R - I_x^L}{I_x^R + I_x^L}\right)_k, \left(\frac{I_R^x - I_L^x}{I_R^x + I_L^x}\right)_k$	$\dfrac{\frac{8}{3}\left(\frac{2\pi}{\lambda}\right)\left(45\bar{\alpha}_k\omega^{-1}\bar{G}'_k + 7\omega^{-1}\gamma_k^2 + \omega^{-1}\delta_k^2\right)}{\frac{4}{3}\left(45\bar{\alpha}_k^2 + 7\beta_k^2\right)}$
	Magic angle ICP (90°, m)	$\left(\frac{I_*^R - I_*^L}{I_*^R + I_*^L}\right)_k$	$\dfrac{\frac{40}{3}\left(\frac{2\pi}{\lambda}\right)\left(9\bar{\alpha}_k\omega^{-1}\bar{G}'_k + 2\omega^{-1}\gamma_k^2\right)}{\frac{20}{3}\left(9\bar{\alpha}_k^2 + 2\beta_k^2\right)}$
	Dual circular polarization-I DCP-I (90°)	$\left(\frac{I_R^R - I_L^L}{I_R^R + I_L^L}\right)_k$	$\dfrac{\frac{8}{3}\left(\frac{2\pi}{\lambda}\right)\left(45\bar{\alpha}_k\omega^{-1}\bar{G}'_k + 13\omega^{-1}\gamma_k^2 - \omega^{-1}\delta_k^2\right)}{\frac{2}{3}\left(45\bar{\alpha}_k^2 + 13\beta_k^2\right)}$

(Continued)

TABLE 3.1 (*Continued*)

Expressions for Raman Circular Intensity Differentials Associated with Selected Scattering Geometries for Vibrational Raman Optical Activity Measurements

Scattering Geometry	Abbreviation[a]		Δ, CID[a]
180°	ICP (180, u) SCP (180, u)	$\left(\frac{I_u^R - I_u^L}{I_u^R + I_u^L}\right)_k \left(\frac{I_R^u - I_L^u}{I_R^u + I_L^u}\right)_k$	$\frac{\frac{32}{3}\left(\frac{2\pi}{\lambda}\right)\left(3\omega^{-1}\gamma_k^2 + \omega^{-1}\delta_k^2\right)}{\frac{4}{3}\left(45\bar{\alpha}_k^2 + 7\beta_k^2\right)}$
	Dual circular polarization-I DCP-I (180°)	$\left(\frac{I_R^R - I_L^L}{I_R^R + I_L^L}\right)_k$	$\frac{\frac{64}{3}\left(\frac{2\pi}{\lambda}\right)\left(3\omega^{-1}\gamma_k^2 + \omega^{-1}\delta_k^2\right)}{\frac{4}{3}\left(12\beta_k^2\right)}$

Source: Polavarapu, P.L., *Vibrational Spectra: Principles and Applications with Emphasis on Optical Activity*, Elsevier, New York, 1998.

[a] Superscripts on I represent the polarization of incident laser light, and subscripts represent that of scattered light (see text). The z axis represents incident light propagation direction, the y axis represents scattered light propagation direction, and the yz plane represents the scattering plane. ICP, SCP, and DCP represent, respectively, incident circular polarization, scattered circular polarization, and dual circular polarization experimental arrangements (see text). The definitions for terms in the third column are given in the text. The letters "d," "p," "m," and "u" represent, respectively, depolarized, polarized, magic angle, and unpolarized. $\bar{\alpha}_k\omega^{-1}\bar{G}'_k$, $\omega^{-1}\gamma_k^2$, and $\omega^{-1}\delta_k^2$ are in units of A^5 amu^{-1} (Equation 3.59 through 3.61); $\bar{\alpha}_k^2$ and β_k^2 are in units of A^4 amu^{-1} (Equations 3.52 and 3.53).

3.5 Summary

The experimentally observed optical rotation property arises from circular birefringence, that is, the difference in refractive indices for left and right circularly polarized light as described by Equations 3.16 and 3.20. The physical consequence of circular birefringence is that the plane of linearly polarized light, when passed through an isotropic chiral sample, will be rotated. The terms "optical rotation" and "circular birefringence" are used interchangeably for the same phenomenon. The experimentally observed circular dichroism property arises from the difference in absorption indices for left and right circularly polarized lights, as described by Equation 3.43. The physical consequence of this circular differential absorbance is that the circularly polarized light, when passed through a chiral sample, will emerge as elliptically polarized light. The measured circular dichroism property reflects the ellipticity of emerging light. The Raman circular intensity difference scattering is a phenomenon that arises from the interference of electric dipole–electric dipole polarizability with electric dipole–magnetic dipole polarizability and electric dipole–electric quadrupole polarizability tensor elements (see Equation 3.58). This chapter provided the description of optical rotation, circular dichroism, and Raman optical activity phenomena, without involving quantum mechanical details. However, to fully understand these phenomena, quantum mechanical details will have to be considered, which are provided in Chapter 4.

References

Barron, L. D. 1976. Rayleigh and Raman scattering of polarized light. *Mol. Spectrosc. (Chem. Soc., London)* 4:96–124.

Barron, L. D. 2004. *Molecular Light Scattering and Optical Activity*, 2nd ed. Cambridge, UK: Cambridge University Press.

Barron, L. D., M. P. Bogaard, and A. D. Buckingham. 1973. Differential Raman scattering of right and left circularly polarized light by asymmetric molecules. *Nature* 241(5385):113–114.

Barron, L. D., and A. D. Buckingham. 1971. Rayleigh and Raman scattering from optically active molecules. *Mol. Phys.* 20(6):1111–1119.

Barron, L. D, and A. D. Buckingham. 1975. Rayleigh and Raman optical activity. *Ann. Rev. Phys. Chem.* 26:381–396.

Barron, L. D., L. Hecht, and S. M. Blyth. 1989. Polarized Raman optical activity of menthol and related molecules. *Spectrochim. Acta, Part A* 45(3):375–379.

Charney, E. 1985. *The Molecular Basis of Optical Activity*. Malabar, FL: Krieger Publishing Co.

Fitts, D. D. 1974. *Vector Analysis in Chemistry*. New York, NY: McGraw Hill.

Hecht, L., and L. D. Barron. 1989. Magic angle Raman optical activity—beta pinene and nopinone. *Spectrochim. Acta, Part A* 45(6):671–674.

Hug, W., and H. Surbeck. 1979. Vibrational Raman optical activity spectra recorded in perpendicular polarization. *Chem. Phys. Lett.* 60(2):186–192.

Hug, W., S. Kint, G. F. Bailey, and J. R. Scherer. 1975. Raman circular intensity differential spectroscopy. Spectra of (−)-alpha-pinene and (+)-slpha-phenylethylamine. *J. Am. Chem. Soc.* 97(19):5589–5590.

Long, D.A. 1977. *Raman Spectroscopy*. New York, NY: McGraw-Hill.

Nafie, L. A., and T. B. Freedman. 1989. Dual circular polarization Raman optical activity. *Chem. Phys. Lett.* 154(3):260–266.

Polavarapu, P.L. 1998. *Vibrational Spectra: Principles and Applications with Emphasis on Optical Activity*. New York, NY: Elsevier.

Spencer, K. M., T. B. Freedman, and L. A. Nafie. 1988. Scattered circular polarization Raman optical activity. *Chem. Phys. Lett.* 149(4):367–374.

Tinoco, I., Jr., and C. R. Cantor. 1970. Application of optical rotatory dispersion and circular dichroism in biochemical analysis. In: D. Glick (ed). *Methods of Biochemistry Analysis*. Vol. 18. New York, NY: John Wiley & Sons, pp. 81–203.

4

Quantum Chemical Formulation of Chiroptical Spectroscopy

In developing the quantum chemical formalisms for chiroptical spectroscopy, one cannot avoid the involved mathematical expressions. This chapter, as it deals with one of the most mathematically intense topics, may not be palatable for those with limited mathematical background. Nevertheless, derivations are provided in greater detail, so that beginning researchers willing to explore can follow them with ease. Much of the quantum chemical formulation presented in this chapter is based on the lecture notes that the author has prepared (Polavarapu 1998) for a quantum chemistry course, taught at Vanderbilt University, drawn from the popular textbooks (Atkins 1983; Levine 1975). Those who do not want to wade through a multitude of expressions may skip to the end of the chapter where a summary is provided.

The outline of this chapter is as follows: In Section 4.1, the interaction between light and matter is considered from a classical point of view, and the expression for energy in the presence of external fields is obtained. Time-independent perturbation theory is then introduced in Section 4.2 to obtain first-order correction to energy and wavefunction in the presence of external perturbations. Using a static electric field as an external perturbation, these results are applied to derive the expression for static electric dipole–electric dipole polarizability, and it is shown that second-order correction to energy is required for this polarizability to contribute to the energy in the presence of static electric fields. Time-dependent perturbation theory is introduced in Section 4.3 to obtain the expressions for time-dependent coefficients and wavefunctions. Using time-varying electric and magnetic fields as perturbations, these results are used in Section 4.4 to obtain the expressions for dynamic electric dipole–electric dipole and electric dipole–magnetic dipole polarizabilities. All of this information serves as the background needed in Section 4.5 for formulating the quantum chemical interaction of molecules with circularly polarized light (CPL) and to obtain the expressions for optical rotation, absorption, and circular dichroism. The relation of these expressions to experimentally determined quantities is presented. Finally, Section 4.6 introduces the transition polarizabilities as the source for scattered radiation.

4.1 Interaction of Molecules with Electric Field

A particle with electric charge q placed in an electrostatic potential, V^e at a distance vector $\boldsymbol{r}$, experiences the interaction energy of $qV^e(\boldsymbol{r})$; V^e has the units of energy per unit charge. This potential can be expanded in Taylor series (Struve 1989) around an arbitrary origin ($\boldsymbol{r} = 0$) as

$$V^e(r) = V_o^e + \left(\frac{\partial V^e}{\partial x}\right)_o x + \left(\frac{\partial V^e}{\partial y}\right)_o y + \left(\frac{\partial V^e}{\partial z}\right)_o z +$$
$$\frac{1}{2}\left[\left(\left(\frac{\partial^2 V^e}{\partial x^2}\right)_o x^2 + \left(\frac{\partial^2 V^e}{\partial y^2}\right)_o y^2 + \left(\frac{\partial^2 V^e}{\partial z^2}\right)_o z^2\right) + \right. \tag{4.1}$$
$$\left. 2\left(\frac{\partial^2 V^e}{\partial x \partial y}\right)_o xy + 2\left(\frac{\partial^2 V^e}{\partial x \partial z}\right)_o xz + 2\left(\frac{\partial^2 V^e}{\partial y \partial z}\right)_o yz\right] + \ldots.$$

This long expression can be shortened using the abbreviations R_α ($\alpha = x, y,$ or z) for $x, y,$ and z and using Einstein summation convention (where summation over $\alpha/\beta = x, y,$ and z is implied):

$$V^e(r) = V_o^e + \left(\frac{\partial V^e}{\partial R_\alpha}\right)_o R_\alpha + \frac{1}{2}\left(\frac{\partial^2 V^e}{\partial R_\alpha \partial R_\beta}\right)_o R_\alpha R_\beta + \ldots. \tag{4.2}$$

Then, the interaction energy E_{int} for one particle system in an electrostatic potential becomes

$$E_{int} = q\left[V_o^e + \left(\frac{\partial V^e}{\partial R_\alpha}\right)_o R_\alpha + \frac{1}{2}\left(\frac{\partial^2 V^e}{\partial R_\alpha \partial R_\beta}\right)_o R_\alpha R_\beta + \ldots\right] \tag{4.3}$$

Noting that the electric field component $F_\alpha = -\left(\frac{\partial V^e}{\partial R_\alpha}\right)_o$, interaction energy becomes

$$E_{int} = q\left[V_o^e - F_\alpha R_\alpha - \frac{1}{2}\left(\frac{\partial F_\alpha}{\partial R_\beta}\right)_o R_\alpha R_\beta + \ldots\right] \tag{4.4}$$

For a molecule with a collection of particles, each with charge q_A at distance vector $\boldsymbol{R}_A$, the preceding equation is summed over all particles:

$$E_{int} = \sum_A q_A V_o^e - F_\alpha \sum_A q_A R_{A\alpha} - \frac{1}{2}\sum_A q_A \left(\frac{\partial F_\alpha}{\partial R_{A\beta}}\right)_o R_{A\alpha} R_{A\beta} + \ldots \tag{4.5}$$

Assuming that the electric field gradient is uniform throughout the molecule and introducing the expressions for electric dipole moment μ_α and electric quadrupole moment $\Theta_{\alpha\beta}$

$$\mu_\alpha = \sum_A q_A R_{A\alpha} \tag{4.6}$$

$$\Theta_{\alpha\beta} = \frac{1}{2}\sum_A q_A R_{A\alpha} R_{A\beta} \tag{4.7}$$

the interaction energy becomes

$$E_{int} = \sum_A q_A V_o^e - \mu_\alpha F_\alpha - \Theta_{\alpha\beta}\left(\frac{\partial F_\alpha}{\partial R_\beta}\right)_o + ... \tag{4.8}$$

A traceless definition for electric quadrupole moment (Barron and Buckingham 1975)

$$\Theta_{\alpha\beta} = \frac{1}{2}\sum_A q_A \left(3R_{A\alpha}R_{A\beta} - R_A^2\delta_{\alpha\beta}\right) \tag{4.9}$$

is commonly used (note that Kronecker delta function, $\delta_{\alpha\beta} = 1$ for $\alpha = \beta$ and zero otherwise), in which case the interaction energy is written as

$$E_{int} = \sum_A q_A V_o^e - \mu_\alpha F_\alpha - \frac{1}{3}\Theta_{\alpha\beta}\left(\frac{\partial F_\alpha}{\partial R_\beta}\right)_o + ... \tag{4.10}$$

Designating the derivative $\left(\frac{\partial F_\alpha}{\partial R_\beta}\right)_o$ as $F_{\alpha\beta}$, and noting that for neutral molecules, the sum of charges is zero, total energy for a neutral molecule is written as

$$E = E_o + E_{int} = E_o - \mu_\alpha F_\alpha - \frac{1}{3}\Theta_{\alpha\beta}F_{\alpha\beta} + ... \tag{4.11}$$

where E_o is the energy in the absence of interaction with electric field. This equation indicates that the energy of a neutral molecule in the presence of electric field changes not only because of the interaction between electric field and molecular electric dipole moment but also from the interaction between the electric field gradient and molecular quadrupole moment. In addition to these two contributions, a contribution from the interaction with molecular polarizability is also present, which can be seen by expanding the energy as a function of electric field:

$$E = E_o + \left(\frac{\partial E}{\partial F_\alpha}\right)_o F_\alpha + \frac{1}{2}\left(\frac{\partial^2 E}{\partial F_\alpha^2}\right)_o F_\alpha^2 + ... \tag{4.12}$$

Comparing the coefficient of F_α in Equations 4.11 and 4.12, one can see that $\left(\frac{\partial E}{\partial F_\alpha}\right)_o = -\mu_\alpha$. The second derivative $\left(\frac{\partial^2 E}{\partial F_\alpha^2}\right)_o$ is equal to the negative of molecular electric dipole polarizability $\alpha_{\alpha\alpha}$. Then, the total energy is written as

$$E = E_o - \mu_\alpha F_\alpha - \frac{1}{2}\alpha_{\alpha\alpha} F_\alpha^2 - \frac{1}{3}\Theta_{\alpha\beta} F_{\alpha\beta} + \dots \tag{4.13}$$

If the interactions are considered to be weak, then the individual interaction energy terms of the preceding equation will be used as the basis for introducing perturbations for evaluating the quantum chemical energy of the system (vide infra) in the presence of external perturbations.

4.2 Time-Independent Perturbation Theory

A molecule, in the absence of external fields, in a stationary state n is represented by the wavefunction ψ_n^o and energy E_n^o, which are, respectively, the eigenfunctions and eigenvalues of the unperturbed Hamiltonian H^o. The superscript "o" identifies the absence of external perturbations. The Schrodinger equation for this system is given as

$$H^o \psi_n^o = E_n^o \psi_n^o \tag{4.14}$$

and the energy is given as

$$E_n^o = \frac{\int (\psi_n^o)^* H \psi_n^o \, d\tau}{\int (\psi_n^o)^* \psi_n^o \, d\tau} = \frac{\langle \psi_n^o | H^o | \psi_n^o \rangle}{\langle \psi_n^o | \psi_n^o \rangle} \tag{4.15}$$

In the bracket notation, $\langle \psi_n^o |$ represents $(\psi_n^o)^*$, and $| \psi_n^o \rangle$ represents ψ_n^o, where $(\psi_n^o)^*$ represents the complex conjugate of ψ_n^o. The complex conjugate of a function is obtained by replacing "i" in the original function with "$-i$" (note that $i^2 = -1$). For real wavefunctions, original functions are same as their complex conjugates. For a normalized wavefunction, $\langle \psi_n^o | \psi_n^o \rangle = 1$.

To see how this energy changes in the presence of a weak time-independent perturbation, the Hamiltonian is modified as (Atkins 1983)

$$H = H^o + H' + H'' + \dots \tag{4.16}$$

where H' and H'' are referred to as the first-order and second-order corrections to the unperturbed Hamiltonian. The resulting wavefunctions and energies are written similarly as

$$\psi_n = \psi_n^o + \psi_n' + \psi_n'' + \dots \tag{4.17}$$

$$E_n = E_n^o + E_n' + E_n'' + \dots \tag{4.18}$$

where ψ_n' and ψ_n'' are, respectively, the first-order and second-order corrections to the unperturbed wavefunction ψ_n^o; E_n' and E_n'' represent, respectively, the first-order and second-order corrections to the unperturbed energy E_n^o.

Using the Schrodinger equation for the perturbed system,

$$H\psi_n = E_n\psi_n \tag{4.19}$$

and substituting Equations 4.16 through 4.18 into Equation 4.19 yields

$$\begin{aligned}\left(H^o + H' + H'' + ..\right)\left(\psi_n^o + \psi_n' + \psi_n'' + ...\right) = \left(E_n^o + E_n' + E_n'' + ...\right)\times \\ \left(\psi_n^o + \psi_n' + \psi_n'' + ...\right)\end{aligned} \tag{4.20}$$

The terms of zero order, first order, and second order can be separated to yield

$$H^o\psi_n^o = E_n^o\psi_n^o \tag{4.21}$$

$$H^o\psi_n' + H'\psi_n^o = E_n^o\psi_n' + E_n'\psi_n^o \tag{4.22}$$

$$H^o\psi_n'' + H'\psi_n' + H''\psi_n^o = E_n^o\psi_n'' + E_n'\psi_n' + E_n''\psi_n^o \tag{4.23}$$

where Equations 4.21 through 4.23 represent, respectively, the unperturbed, first-order and second-order equations.

4.2.1 First-Order Corrections

The first-order equation, Equation 4.22, can be rearranged as

$$\left(H^o - E_n^o\right)\psi_n' = \left(E_n' - H'\right)\psi_n^o \tag{4.24}$$

The first-order correction to wavefunction can be written as a linear combination of the unperturbed orthonormal wavefunctions as

$$\psi_n' = \sum_{m \# n} c_m\psi_m^o \tag{4.25}$$

where the summation extends over all eigenfunctions of the Hamiltonian, except n; c_m is the coefficient to be determined shortly. Substitution of Equation 4.25 in Equation 4.24 leads to

$$\left(H^o - E_n^o\right)\left(\sum_{m \# n} c_m\psi_m^o\right) = \left(E_n' - H'\right)\psi_n^o \tag{4.26}$$

Upon left-multiplying Equation 4.26 with ψ_n^o, on both sides, and integrating, one obtains

$$\sum_{m\#n} c_m\left\{\left\langle\psi_n^o\middle|H^o\middle|\psi_m^o\right\rangle - E_n^o\left\langle\psi_n^o\middle|\psi_m^o\right\rangle\right\} = E_n'\left\langle\psi_n^o\middle|\psi_n^o\right\rangle - \left\langle\psi_n^o\middle|H'\middle|\psi_n^o\right\rangle \quad (4.27)$$

In this equation, $\left\langle\psi_n^o\middle|H^o\middle|\psi_m^o\right\rangle = E_m^o\left\langle\psi_n^o\middle|\psi_m^o\right\rangle = 0$; $E_n^o\left\langle\psi_n^o\middle|\psi_m^o\right\rangle = 0$; and $\left\langle\psi_n^o\middle|\psi_n^o\right\rangle = 1$, because wavefunctions are orthonormal. Then, the expression for first-order correction to the energy becomes

$$E_n' = \left\langle\psi_n^o\middle|H'\middle|\psi_n^o\right\rangle = H_{nn}' \quad (4.28)$$

That means, the integral of perturbing Hamiltonian with unperturbed wavefunction in state n yields the first-order correction to energy of that state. The first-order correction to wavefunction can be obtained similarly by left-multiplying Equation 4.26 with $\psi_k^o(k \neq n)$ and integrating the result:

$$\sum_{m\#n} c_m\left\{\left\langle\psi_k^o\middle|H^o\middle|\psi_m^o\right\rangle - E_n^o\left\langle\psi_k^o\middle|\psi_m^o\right\rangle\right\} = E_n'\left\langle\psi_k^o\middle|\psi_n^o\right\rangle - \left\langle\psi_k^o\middle|H'\middle|\psi_n^o\right\rangle \quad (4.29)$$

The left-hand side of Equation 4.29 equals $c_m\left(E_m^o - E_n^o\right)$ for $m = k$ and vanishes for $m \# k$. On the right-hand side of Equation 4.29, $E_n'\left\langle\psi_k^o\middle|\psi_n^o\right\rangle = 0$, because wavefunctions are orthonormal. Then, the coefficient c_m becomes

$$c_m = \frac{-\left\langle\psi_m^o\middle|H'\middle|\psi_n^o\right\rangle}{\left(E_m^o - E_n^o\right)} = \frac{\left\langle\psi_m^o\middle|H'\middle|\psi_n^o\right\rangle}{\left(E_n^o - E_m^o\right)} = \frac{H_{mn}'}{\left(E_n^o - E_m^o\right)} \quad (4.30)$$

where the abbreviation

$$H_{mn}' = \left\langle\psi_m^o\middle|H'\middle|\psi_n^o\right\rangle \quad (4.31)$$

was introduced. The energy and wavefunction corrected to first order can then be written as follows:

$$E_n = E_n^o + E_n' = E_n^o + H_{nn}' \quad (4.32)$$

$$\psi_n = \psi_n^o + \psi_n' = \psi_n^o + \sum_{m\#n} c_m\psi_m^o = \psi_n^o + \sum_{m\#n} \frac{H_{mn}'}{\left(E_n^0 - E_m^0\right)}\psi_m^o \quad (4.33)$$

where H_{nn}' and H_{mn}' are defined in Equations 4.28 and 4.31.

4.2.2 Molecules in Static Electric Field

The first-order perturbation theory described earlier can be used to see what happens when a molecule is placed in an external static electric field. In the presence of a static electric field, the interaction energy can be inferred, from Equation 4.13, to be $-\mu_\alpha F_\alpha$. Then, in the presence of weak static electric field, the first-order correction to the Hamilton is given as $H' = -\mu_\alpha F_\alpha$. Substituting this relation for first-order correction to energy (see Equation 4.32),

$$E_n = E_n^{\mathrm{o}} - \langle \psi_n | \hat{\mu}_\alpha | \psi_n \rangle F_\alpha = E_n^{\mathrm{o}} - \mu_{\alpha,nn} F_\alpha \tag{4.34}$$

where μ_α inside the integral is written as an operator by placing a caret on top of μ_α. Equation 4.34 is the counterpart of classical expression $E = E_{\mathrm{o}} - \mu_\alpha F_\alpha$ (which represents the first two terms on the right-hand side of Equations 4.11 and 4.13 when the interaction with quadrupole moment is not included). Unlike in Equation 4.13, the contribution from molecular electric dipole polarizability to energy is not present in Equation 4.34 because, that contribution arises from second-order correction to energy (vide infra). In arriving at Equation 4.34, the energy corrections are limited to first order.

However, the molecular electric dipole polarizability contribution to electric dipole moment can be determined using the first-order correction to wavefunction. The electric dipole moment is determined by evaluating the integral $\langle \psi_n | \hat{\mu}_{\alpha'} | \psi_n \rangle$. Upon substituting the wavefunction corrected to first order, the α'-component of electric dipole moment integral becomes

$$\begin{aligned}
\langle \psi_n | \hat{\mu}_{\alpha'} | \psi_n \rangle = \mu_{\alpha'} &= \left\langle \psi_n^{\mathrm{o}} + \sum_{m \# n} \frac{H'_{mn}}{(E_n^{\mathrm{o}} - E_m^{\mathrm{o}})} \psi_m^{\mathrm{o}} \middle| \hat{\mu}_{\alpha'} \middle| \psi_n^{\mathrm{o}} + \sum_{m \# n} \frac{H'_{mn}}{(E_n^{\mathrm{o}} - E_m^{\mathrm{o}})} \psi_m^{\mathrm{o}} \right\rangle \\
&= \langle \psi_n^{\mathrm{o}} | \hat{\mu}_{\alpha'} | \psi_n^{\mathrm{o}} \rangle + \left\langle \sum_{m \# n} \frac{H'_{mn}}{(E_n^{\mathrm{o}} - E_m^{\mathrm{o}})} \psi_m^{\mathrm{o}} \middle| \hat{\mu}_{\alpha'} \middle| \psi_n^{\mathrm{o}} \right\rangle + \left\langle \psi_n^{\mathrm{o}} \middle| \hat{\mu}_{\alpha'} \middle| \sum_{m \# n} \frac{H'_{mn}}{(E_n^{\mathrm{o}} - E_m^{\mathrm{o}})} \psi_m^{\mathrm{o}} \right\rangle \\
&+ \left\langle \sum_{m \# n} \frac{H'_{mn}}{(E_n^{\mathrm{o}} - E_m^{\mathrm{o}})} \psi_m^{\mathrm{o}} \middle| \hat{\mu}_{\alpha'} \middle| \sum_{m \# n} \frac{H'_{mn}}{(E_n^{\mathrm{o}} - E_m^{\mathrm{o}})} \psi_m^{\mathrm{o}} \right\rangle
\end{aligned} \tag{4.35}$$

In this equation, $H'_{mn} = \langle \psi_m^{\mathrm{o}} | -\hat{\mu}_\alpha F_\alpha | \psi_n^{\mathrm{o}} \rangle = \langle \psi_m^{\mathrm{o}} | -\hat{\mu}_\alpha | \psi_n^{\mathrm{o}} \rangle F_\alpha = -\mu_{\alpha,mn} F_\alpha$ (see Equation 4.31).

Ignoring the second-order term (the last term on the right-hand side of Equation 4.35), the following equation is obtained:

$$\begin{aligned}
\mu_{\alpha'} &= \langle \psi_n^{\mathrm{o}} | \hat{\mu}_{\alpha'} | \psi_n^{\mathrm{o}} \rangle + \sum_{m \# n} \frac{H'_{mn}}{(E_n^{\mathrm{o}} - E_m^{\mathrm{o}})} \langle \psi_m^{\mathrm{o}} | \hat{\mu}_{\alpha'} | \psi_n^{\mathrm{o}} \rangle \\
&+ \sum_{m \# n} \frac{H'_{mn}}{(E_n^{\mathrm{o}} - E_m^{\mathrm{o}})} \langle \psi_n^{\mathrm{o}} | \hat{\mu}_{\alpha'} | \psi_m^{\mathrm{o}} \rangle \\
&= \mu_{\alpha'}^{\mathrm{o}} + \sum_{m \# n} \frac{(\mu_{\alpha',mn} + \mu_{\alpha',nm}) H'_{mn}}{(E_n^{\mathrm{o}} - E_m^{\mathrm{o}})} = \mu_{\alpha'}^{\mathrm{o}} \\
&+ \sum_{m \# n} \frac{-(\mu_{\alpha',mn} + \mu_{\alpha',nm}) \mu_{\alpha,mn}}{(E_n^{\mathrm{o}} - E_m^{\mathrm{o}})} F_\alpha = \mu_{\alpha'}^{\mathrm{o}} + \alpha_{\alpha'\alpha} F_\alpha
\end{aligned} \tag{4.36}$$

In this equation, the static molecular electric dipole polarizability, $\alpha_{\alpha'\alpha}$ is defined as

$$\alpha_{\alpha'\alpha} = \sum_{m\#n} \frac{-(\mu_{\alpha',mn} + \mu_{\alpha',nm})\mu_{\alpha,mn}}{(E_n^o - E_m^o)} = -\sum_{m\#n} \frac{2\mu_{\alpha',nm}\mu_{\alpha,mn}}{(E_n^o - E_m^o)} = \sum_{m\#n} \frac{2\mu_{\alpha',nm}\mu_{\alpha,mn}}{(E_m^o - E_n^o)} \tag{4.37}$$

$\mu_{\alpha'}^0$ is the electric dipole moment of the molecule in state n in the absence of electric field and $\mu_{\alpha',mn} = \langle \psi_m^o | \hat{\mu}_{\alpha'} | \psi_n^o \rangle$. The relation $\langle \psi_m^o | \hat{\mu}_{\alpha'} | \psi_n^o \rangle = \langle \psi_n^o | \hat{\mu}_{\alpha'} | \psi_m^o \rangle$ was also used in the preceding equation. Note that the subscript α' is the component of the electric dipole moment that is being evaluated, and the subscript α is the component of the electric field that is perturbing the system.

The preceding expression (Equation 4.37) for molecular polarizability $\alpha_{\alpha'\alpha}$ contains two electric dipole moment integrals $\mu_{\alpha',nm}$ and $\mu_{\alpha,mn}$. For this reason, $\alpha_{\alpha'\alpha}$ is often explicitly referred to as electric dipole–electric dipole polarizability.

The difference, $\langle \psi_n | \hat{\mu}_{\alpha'} | \psi_n \rangle - \langle \psi_n^o | \hat{\mu}_{\alpha'} | \langle \psi_n^o \rangle = \mu_{\alpha'} - \mu_{\alpha'}^o = \Delta\mu_{\alpha'}$, is the induced electric dipole moment and depends on the molecular electric dipole–electric dipole polarizability and applied electric field. In the expanded form, the induced electric dipole moment is written, explicitly, as $\Delta\mu_{\alpha'} = \alpha_{\alpha'\alpha}F_\alpha = \alpha_{\alpha'x}F_x + \alpha_{\alpha'y}F_y + \alpha_{\alpha'z}F_z$ *for* $\alpha' = x, y$, or z.

The generalized form of induced electric dipole moment is

$$\Delta\mu_\alpha = \alpha_{\alpha\beta}F_\beta \tag{4.38}$$

and that of static electric dipole–electric dipole molecular polarizability is

$$\alpha_{\alpha\beta} = \sum_{m\#n} \frac{2\mu_{\alpha,nm}\mu_{\beta,mn}}{(E_m^o - E_n^o)} \tag{4.39}$$

4.2.3 Second-Order Corrections

The second-order equation, Equation 4.23, can be rearranged as

$$(H^o - E_n^o)\psi_n'' + (H' - E_n')\psi_n' + (H'' - E_n'')\psi_n^o = 0 \tag{4.40}$$

The second-order correction to wavefunction can also be written as a linear combination of the unperturbed wavefunctions. For clarity, however, we will use different symbols for coefficients here than those used in first-order correction (Equation 4.25):

$$\psi_n'' = \sum_{m\#n} d_m \psi_m^o \tag{4.41}$$

Upon substituting Equations 4.25 and 4.41 into Equation 4.40, left multiplying with ψ_n^o and integrating

$$\left\langle \psi_n^o \middle| H^o \middle| \sum_{m \# n} d_m \psi_m^o \right\rangle - E_n^o \left\langle \psi_n^o \middle| \sum_{m \# n} d_m \psi_m^o \right\rangle + \left\langle \psi_n^o \middle| H' \middle| \sum_{m \# n} c_m \psi_m^o \right\rangle$$
$$-E_n' \left\langle \psi_n^o \middle| \sum_{m \# n} c_m \psi_m^o \right\rangle + \left\langle \psi_n^o \middle| H'' \middle| \psi_n^o \right\rangle - E_n'' \left\langle \psi_n^o \middle| \psi_n^o \right\rangle = 0 \quad (4.42)$$

The integrals, $\left\langle \psi_n^o \middle| H^o \middle| \sum_{m \# n} d_m \psi_m^o \right\rangle$, $\left\langle \psi_n^o \middle| \sum_{m \# n} d_m \psi_m^o \right\rangle$, and $\left\langle \psi_n^o \middle| \sum_{m \# n} c_m \psi_m^o \right\rangle$, are zero by virtue of orthogonality of wavefunctions. The surviving part of Equation 4.42 is

$$\left\langle \psi_n^o \middle| H' \middle| \sum_{m \# n} c_m \psi_m^o \right\rangle + \left\langle \psi_n^o \middle| H'' \middle| \psi_n^o \right\rangle - E_n'' = 0 \quad (4.43)$$

This equation can be rearranged as

$$E_n'' = \sum_{m \# n} c_m \left\langle \psi_n^o \middle| H' \middle| \psi_m^o \right\rangle + \left\langle \psi_n^o \middle| H'' \middle| \psi_n^o \right\rangle = \sum_{m \# n} c_m H'_{nm} + H''_{nn} \quad (4.44)$$

Substituting for c_m from Equation 4.30, the second-order correction to energy becomes

$$E_n'' = \sum_{m \# n} \frac{H'_{mn} H'_{nm}}{\left(E_n^o - E_m^o\right)} + H''_{nn} \quad (4.45)$$

The total energy to second order is obtained by substituting Equations 4.28 and 4.45 in Equation 4.18:

$$E = E_n^o + H'_{nn} + \sum_{m \# n} \frac{H'_{mn} H'_{nm}}{\left(E_n^o - E_m^o\right)} + H''_{nn} \quad (4.46)$$

The wavefunction corrected to second order is not necessary for the current purpose and therefore not discussed but can be obtained similarly.

Going back to the static electric field perturbation, $H' = -\mu_\alpha F_\alpha$, and not considering the second-order term H''_{nn}, the energy of the molecule to second order becomes

$$E_n = E_n^o - \mu_{\alpha,nn} F_\alpha + \sum_{m \# n} \frac{\mu_{\alpha,mn} \mu_{\alpha,nm}}{\left(E_n^o - E_m^o\right)} F_\alpha^2 = E_n^o - \mu_{\alpha,nn} F_\alpha - \sum_{m \# n} \frac{\mu_{\alpha,mn} \mu_{\alpha,nm}}{\left(E_m^o - E_n^o\right)} F_\alpha^2 \quad (4.47)$$

which is the counterpart of classical expression $E = E_o - \mu_\alpha F_\alpha - \frac{1}{2} \alpha_{\alpha\alpha} F_\alpha^2$ given in Equation 4.13.

4.3 Time-Dependent Perturbation Theory

The time-dependent wavefunction, $\psi_n^o(t) = \psi_n^o e^{\frac{-iE_n^o t}{\hbar}}$, with $\hbar = \frac{h}{2\pi}$, satisfies the Schrõdinger equation:

$$H^o \psi_n^o(t) = E_n^o \psi_n^o(t) = i\hbar \frac{\partial \psi_n^o(t)}{\partial t} \tag{4.48}$$

In the presence of external time-dependent perturbation, the first-order correction to Hamiltonian is written as

$$H(t) = H^o + H'(t)$$

and the perturbed wavefunction for kth state can be written as a sum of all unperturbed state wavefunctions each multiplied with a time-dependent coefficient:

$$\psi_k(t) = \sum_n c_n(t)\psi_n^o(t) = \sum_n c_n(t)\psi_n^o e^{\frac{-iE_n^o t}{\hbar}} \tag{4.49}$$

In this equation, $c_n(t)$ represents a time-dependent coefficient to be determined later. The Schrodinger equation now can be written as

$$H\psi_k(t) = i\hbar \frac{\partial \psi_k(t)}{\partial t} \tag{4.50}$$

with

$$\begin{aligned} H\psi_k(t) &= \left[H^o + H'(t)\right]\left[\sum_n c_n(t)\psi_n^o(t)\right] \\ &= \sum_n c_n(t)H^o\psi_n^o(t) + \sum_n c_n(t)H'(t)\psi_n^o(t) \end{aligned} \tag{4.51}$$

$$i\hbar \frac{\partial \psi_k(t)}{\partial t} = i\hbar\left[\sum_n c_n(t)\frac{\partial \psi_n^o(t)}{\partial t}\right] + i\hbar\left[\sum_n \frac{\partial c_n(t)}{\partial t}\psi_n^o(t)\right] \tag{4.52}$$

The first term on the right-hand side of Equation 4.51 is equal to the first term on the right-hand side of Equation 4.52, by virtue of Equation 4.48. Then, the second term of Equation 4.51 is equal to the second term of Equation 4.52. That is,

$$\sum_n c_n(t)H'(t)\psi_n^o(t) = i\hbar\left[\sum_n \frac{\partial c_n(t)}{\partial t}\psi_n^o(t)\right] \tag{4.53}$$

Left-multiplying Equation 4.53 with ψ_m^o and integrate. Then, the right-hand side of Equation 4.53 is nonzero only for $n = m$, because of the orthogonality of wavefunctions. Then,

$$\sum_n c_n(t)H'_{mn}(t)e^{\frac{-iE_n^o t}{\hbar}} = i\hbar\frac{\partial c_m(t)}{\partial t}e^{\frac{-iE_m^o t}{\hbar}} \tag{4.54}$$

where $\psi_n^o(t) = \psi_n^o e^{\frac{-iE_n^o t}{\hbar}}$ and the abbreviation $\langle\psi_m^o|H'(t)|\psi_n^o\rangle = H'_{mn}(t)$ were used. Rearrange Equation 4.54 to obtain an expression for the time variation of the coefficient for state m:

$$\frac{\partial c_m(t)}{\partial t} = \frac{1}{i\hbar}\sum_n c_n(t)H'_{mn}(t)e^{\frac{-iE_n^o t}{\hbar}}e^{\frac{iE_m^o t}{\hbar}} = \frac{1}{i\hbar}\sum_n c_n(t)H'_{mn}(t)e^{i\omega_{mn}t} \tag{4.55}$$

where $\omega_{mn} = \left(\frac{E_m^o - E_n^o}{\hbar}\right)$. Integrate Equation 4.55 from $t = 0$ to a certain time and use the following assumptions (Atkins 1983): (a) at $t = 0$, the system is in the initial state so the coefficient for the initial state is 1 and is 0 for other states. Then, the integration of left-hand side of Equation 4.55 gives the time evolving coefficient for state m:

$$\int_0^t \frac{\partial c_m(t)}{\partial t}dt = c_m(t) - c_m(0) = c_m(t) \tag{4.56}$$

(b) for a weak perturbation, and at very short time after perturbation was applied, coefficients remain close to their starting values. Then, on the right-hand side of Equation 4.55, assume that $c_n(t) \sim c_n(0) = 1$ for the starting state n and ~0 for other states. Then, the time evolution of the coefficient for state m is given as

$$c_m(t) = \frac{1}{i\hbar}\int_0^t c_n(t)H'_{mn}(t)e^{i\omega_{mn}t} = \frac{1}{i\hbar}\int_0^t H'_{mn}(t)e^{i\omega_{mn}t} \tag{4.57}$$

where the subscript "n" represents the initial state and "m" represents a higher energy state. This is an important equation that will be useful to determine a variety of molecular properties.

4.4 Molecules in Dynamic Fields

To formulate the expressions pertinent to chiroptical spectroscopy, we need to consider the influence of electric and magnetic fields that are time dependent. These two perturbations are considered separately.

4.4.1 Electric Field Perturbation

The molecular electric dipole–electric dipole polarizability (Equation 4.37), determined earlier using static field perturbation, is referred to as the static electric dipole–electric dipole polarizability. The corresponding quantity to be determined using dynamic electric field perturbation is referred to as the dynamic electric dipole–electric dipole polarizability.

4.4.1.1 Dynamic Electric Dipole–Electric Dipole Polarizability

For dynamic electric field perturbation, consider the time-varying perturbation given by the relation

$$\begin{aligned} H'(t) &= -\mu_\alpha F_\alpha(t)\left(1-e^{-kt}\right) = -\mu_\alpha F_\alpha^0 \cos\omega t\left(1-e^{-kt}\right) \\ &= -\mu_\alpha F_\alpha^0\left(\frac{e^{i\omega t}+e^{-i\omega t}}{2}\right)\left(1-e^{-kt}\right) \end{aligned} \tag{4.58}$$

In this equation, $k = 1/\tau$ and τ is the time constant. The $\left(1-e^{-kt}\right)$ term is a damping factor used to ensure that the transient oscillations have subsided long after the perturbation is turned on. Using this perturbation, the time-dependent coefficient, Equation 4.57, becomes

$$\begin{aligned} c_m(t) &= \frac{1}{i\hbar}\int_0^t H'_{mn}(t)e^{i\omega_{mn}t}dt = \frac{-\mu_{\alpha,mn}F_\alpha^0}{2i\hbar}\int_0^t\left(e^{i\omega t}+e^{-i\omega t}\right)\left(1-e^{-kt}\right)e^{i\omega_{mn}t}dt \\ &= \frac{-\mu_{\alpha,mn}F_\alpha^0}{2i\hbar}\int_0^t\left(e^{i(\omega_{mn}+\omega)t}+e^{i(\omega_{mn}-\omega)t}\right)\left(1-e^{-kt}\right)dt \\ &= \frac{-\mu_{\alpha,mn}F_\alpha^0}{2i\hbar}\left[\int_0^t\left(e^{i(\omega_{mn}+\omega)t}+e^{i(\omega_{mn}-\omega)t}\right)dt\right. \\ &\qquad \left. -\int_0^t\left(e^{(i(\omega_{mn}+\omega)-k)t}+e^{(i(\omega_{mn}-\omega)-k)t}\right)dt\right] \\ &= \frac{-\mu_{\alpha,mn}F_\alpha^0}{2i\hbar}\left[\frac{e^{i(\omega_{mn}+\omega)t}-1}{i(\omega_{mn}+\omega)}+\frac{e^{i(\omega_{mn}-\omega)t}-1}{i(\omega_{mn}-\omega)}\right] \\ &\quad +\frac{\mu_{\alpha,mn}F_\alpha^0}{2i\hbar}\left[\frac{e^{(i(\omega_{mn}+\omega)-k)t}-1}{i(\omega_{mn}+\omega)-k}+\frac{e^{(i(\omega_{mn}-\omega)-k)t}-1}{i(\omega_{mn}-\omega)-k}\right] \end{aligned} \tag{4.59}$$

For $(\omega_{mn} \pm \omega) >> k$, and for $t >> 1/k$, use the approximation that $i(\omega_{mn} \pm \omega) - k \sim i(\omega_{mn} \pm \omega)$ in the denominator and $e^{-kt} \sim 0$ in the numerator, of Equation 4.59. Then,

$$c_m(t) = \frac{-\mu_{\alpha,mn} F_\alpha^0}{2i\hbar} \times \left[\frac{e^{i(\omega_{mn}+\omega)t} - 1}{i(\omega_{mn}+\omega)} + \frac{e^{i(\omega_{mn}-\omega)t} - 1}{i(\omega_{mn}-\omega)}\right] + \frac{\mu_{\alpha,mn} F_\alpha^0}{2i\hbar}\left[\frac{-1}{i(\omega_{mn}+\omega)} + \frac{-1}{i(\omega_{mn}-\omega)}\right]$$
$$= \frac{\mu_{\alpha,mn} F_\alpha^0}{2\hbar} \times \left[\frac{(\omega_{mn}-\omega)\left(e^{i(\omega_{mn}+\omega)t} - 1\right) + (\omega_{mn}+\omega)\left(e^{i(\omega_{mn}-\omega)t} - 1\right) + (\omega_{mn}-\omega) + (\omega_{mn}+\omega)}{\left(\omega_{mn}^2 - \omega^2\right)}\right] \tag{4.60}$$

Taking $e^{i\omega_{mn}t}$ as the common term and writing $e^{\pm i\omega t}$ as $\cos\omega t \pm i\sin\omega t$, $c_m(t)$ becomes

$$c_m(t) = \frac{\mu_{\alpha,mn} F_\alpha^0 e^{i\omega_{mn}t}}{2\hbar} \times \left[\frac{(\omega_{mn}-\omega)(\cos\omega t + i\sin\omega t) + (\omega_{mn}+\omega)(\cos\omega t - i\sin\omega t)}{\left(\omega_{mn}^2 - \omega^2\right)}\right]$$
$$+ \frac{\mu_{\alpha,mn} F_\alpha^0}{2\hbar} \times \left[\frac{-(\omega_{mn}-\omega) - (\omega_{mn}+\omega) + (\omega_{mn}-\omega) + (\omega_{mn}+\omega)}{\left(\omega_{mn}^2 - \omega^2\right)}\right] \tag{4.61}$$
$$= \frac{\mu_{\alpha,mn} F_\alpha^0 e^{i\omega_{mn}t}}{\hbar\left(\omega_{mn}^2 - \omega^2\right)}\left[\omega_{mn}\cos\omega t - i\omega\sin\omega t\right]$$

Since this coefficient is a complex quantity, we will also need its complex conjugate to determine the time-dependent dipole moments below. The complex conjugate of $c_m(t)$ is

$$c_m^*(t) = \frac{\mu_{\alpha,mn}^* F_\alpha^0 e^{-i\omega_{mn}t}}{\hbar\left(\omega_{mn}^2 - \omega^2\right)}\left[\omega_{mn}\cos\omega t + i\omega\sin\omega t\right] \tag{4.62}$$

The time-dependent wavefunction for initial state n can be written as

$$\psi_n(t) = \sum_m c_m(t)\psi_m^o(t) = \psi_n^o e^{\frac{-iE_n^o t}{\hbar}} + \sum_{m \# n} c_m(t)\psi_m^o e^{\frac{-iE_m^o t}{\hbar}} \tag{4.63}$$

Then, time-dependent electric dipole moment becomes

$$\mu_{\alpha'}(t) = \langle \psi_n(t)|\hat{\mu}_{\alpha'}|\psi_n(t)\rangle = \int \left(\psi_n^{o} e^{\frac{iE_n^{o}t}{\hbar}} + \sum_{m\#n} c_m^{*}(t)\psi_m^{o} e^{\frac{iE_m^{o}t}{\hbar}} \right) \hat{\mu}_{\alpha'} \times$$

$$\left(\psi_n^{o} e^{\frac{-iE_n^{o}t}{\hbar}} + \sum_{m\#n} c_m(t)\psi_m^{o} e^{\frac{-iE_m^{o}t}{\hbar}} \right) d\tau$$

$$= \langle \psi_n^{o}|\hat{\mu}_{\alpha'}|\psi_n^{o}\rangle + \sum_{m\#n} c_m^{*}(t) e^{i\omega_{mn}t} \langle \psi_m^{o}|\hat{\mu}_{\alpha'}|\psi_n^{o}\rangle \tag{4.64}$$

$$+ \sum_{m\#n} c_m(t) e^{-i\omega_{mn}t} \langle \psi_n^{o}|\hat{\mu}_{\alpha'}|\psi_m^{o}\rangle + \text{second-order term.}$$

Ignore the second-order term and substitute for $c_m(t)$ and it complex conjugate from Equations 4.61 and 4.62. Then,

$$\mu_{\alpha'}(t) = \mu_{\alpha'}^{0} +$$

$$\frac{F_\alpha^0}{\hbar} \sum_{m\#n} \left[\frac{\mu_{\alpha',mn}\mu_{\alpha,mn}^{*}[\omega_{mn}\cos\omega t + i\omega\sin\omega t] + \mu_{\alpha',nm}\mu_{\alpha,mn}[\omega_{mn}\cos\omega t - i\omega\sin\omega t]}{(\omega_{mn}^2 - \omega^2)} \right] \tag{4.65}$$

Case I: $\mu_{\alpha',mn}\mu_{\alpha,mn}^{*} = \mu_{\alpha',nm}\mu_{\alpha,mn}$

Hermitian electric dipole moment operators satisfy $\langle \psi_m^{o}|\hat{\mu}_{\alpha'}|\psi_n^{o}\rangle = \langle \psi_n^{o}|\hat{\mu}_{\alpha'}|\psi_m^{o}\rangle$ and $\langle \psi_m^{o}|\hat{\mu}_{\alpha}|\psi_n^{o}\rangle^{*} = \langle \psi_m^{o}|\hat{\mu}_{\alpha}|\psi_n^{o}\rangle$. Then, the $i\sin\omega t$ terms cancel, yielding

$$\mu_{\alpha'}(t) = \mu_{\alpha'}^{0} + \frac{F_\alpha^0}{\hbar} \sum_{m\#n} \frac{2\omega_{mn}\mu_{\alpha',nm}\mu_{\alpha,mn}\cos\omega t}{(\omega_{mn}^2 - \omega^2)}$$

$$= \mu_{\alpha'}^{0} + \sum_{m\#n} \frac{2\omega_{mn}\mu_{\alpha',nm}\mu_{\alpha,mn}}{\hbar(\omega_{mn}^2 - \omega^2)} F_\alpha(t) = \mu_{\alpha'}^{0} + \alpha_{\alpha'\alpha}(\omega) F_\alpha(t) \tag{4.66}$$

where $\alpha_{\alpha'\alpha}(\omega)$ is the dynamic electric dipole–electric dipole polarizability tensor element:

$$\alpha_{\alpha'\alpha}(\omega) = \sum_{m\#n} \frac{2\omega_{mn}\mu_{\alpha',nm}\mu_{\alpha,mn}}{\hbar(\omega_{mn}^2 - \omega^2)} \tag{4.67}$$

When $\omega_{mn} >> \omega$,

$$\alpha_{\alpha'\alpha}(\omega) \approx \sum_{m\#n} \frac{2\mu_{\alpha',nm}\mu_{\alpha,mn}}{\hbar\omega_{mn}} = \sum_{m\#n} \frac{2\mu_{\alpha',nm}\mu_{\alpha,mn}}{(E_m^{o} - E_n^{o})} = \alpha_{\alpha'\alpha} \tag{4.68}$$

That means, for perturbing frequency ω far less than the transition frequency ω_{mn}, dynamic electric dipole–electric dipole polarizability, $\alpha_{\alpha'\alpha}(\omega)$, reduces to static electric dipole–electric dipole polarizability, $\alpha_{\alpha'\alpha}$ (see Equation 4.39).

Case II: $\mu_{\alpha',mn}\mu_{\alpha,mn}^{*} = -\mu_{\alpha',nm}\mu_{\alpha,mn}$

This situation is relevant for resonance conditions, where $\alpha_{\alpha\beta} = -\alpha_{\beta\alpha}$. Then, the $\cos\omega t$ terms in Equation 4.65 cancel, yielding

$$\mu_{\alpha'}(t) = \mu^0_{\alpha'} - \frac{F^0_\alpha}{\hbar}\sum_{m \# n}\frac{2\mu_{\alpha',nm}\mu_{\alpha,mn} i\omega \sin\omega t}{\left(\omega^2_{mn} - \omega^2\right)} \tag{4.69}$$

The applied electric filed is $F_\alpha(t) = F^0_\alpha \cos\omega t$, and its time derivative is $\frac{dF_\alpha(t)}{dt} = \dot{F}_\alpha(t) = -\omega F^0_\alpha \sin\omega t$. Using this time derivative, Equation 4.69 can be rewritten as

$$\mu_{\alpha'}(t) = \mu^0_{\alpha'} - \sum_{m \# n}\frac{2i\mu_{\alpha',nm}\mu_{\alpha,mn}}{\hbar\left(\omega^2_{mn} - \omega^2\right)}F^0_\alpha \omega \sin\omega t = \mu^0_{\alpha'} + \sum_{m \# n}\frac{2i\omega\mu_{\alpha',nm}\mu_{\alpha,mn}}{\hbar\left(\omega^2_{mn} - \omega^2\right)}\frac{\dot{F}_\alpha(t)}{\omega} \tag{4.70}$$

$$= \mu^0_{\alpha'} + \omega^{-1}\alpha'_{\alpha'\alpha}(\omega)\dot{F}_\alpha(t).$$

In the preceding equation, $\alpha'_{\alpha'\alpha}(\omega)$ is defined as

$$\alpha'_{\alpha'\alpha}(\omega) = \sum_{m \# n}\frac{2i\omega\mu_{\alpha',nm}\mu_{\alpha,mn}}{\hbar\left(\omega^2_{mn} - \omega^2\right)} = -\sum_{m \# n}\frac{2\omega}{\hbar\left(\omega^2_{mn} - \omega^2\right)}\operatorname{Im}\left[\mu_{\alpha',nm}\mu_{\alpha,mn}\right] \tag{4.71}$$

where Im[..] stands for the imaginary part of [..]. Combining the cases I and II, the molecular electric dipole moment in the presence of time-dependent electric filed perturbation is obtained as

$$\mu_{\alpha'}(t) = \mu^0_{\alpha'} + \alpha_{\alpha'\alpha}(\omega)F_\alpha(t) + \omega^{-1}\alpha'_{\alpha'\alpha}(\omega)\dot{F}_\alpha(t) \tag{4.72}$$

4.4.2 Magnetic Field Perturbation

4.4.2.1 Dynamic Electric Dipole–Magnetic Dipole Polarizability

For magnetic field given as, $B_\alpha(t) = B^0_\alpha \cos\omega t$, the perturbing Hamiltonian is written as $H'(t) = -m_\alpha B_\alpha(t)\left(1 - e^{-kt}\right)$, where m_α is the magnetic dipole moment and B_α is the magnetic field. Replacing μ_α with m_α and F^0_α with B^0_α, in equations of Section 4.4.1, the time-dependent electric dipole moment given by Equation 4.65 now becomes

$$\mu_{\alpha'}(t) = \mu^0_{\alpha'} +$$

$$\frac{B^0_\alpha}{\hbar}\sum_{m \# n}\left[\frac{\mu_{\alpha',mn}m^*_{\alpha,mn}\left[\omega_{mn}\cos\omega t + i\omega\sin\omega t\right] + \mu_{\alpha',nm}m_{\alpha,mn}\left[\omega_{mn}\cos\omega t - i\omega\sin\omega t\right]}{\left(\omega^2_{mn} - \omega^2\right)}\right] \tag{4.73}$$

Case I: $\mu_{\alpha',mn}m^*_{\alpha,mn} = \mu_{\alpha',nm}m_{\alpha,mn}$

Under this condition, the $i\sin\omega t$ terms cancel, yielding

$$\mu_{\alpha'}(t) = \mu^0_{\alpha'} + \frac{B^0_\alpha}{\hbar}\sum_{m \# n}\frac{2\omega_{mn}\mu_{\alpha',nm}m_{\alpha,mn}\cos\omega t}{\left(\omega^2_{mn} - \omega^2\right)}$$

$$= \mu^0_{\alpha'} + \sum_{m \# n}\frac{2\omega_{mn}\mu_{\alpha',nm}m_{\alpha,mn}}{\hbar\left(\omega^2_{mn} - \omega^2\right)}B_\alpha(t) = \mu^0_{\alpha'} + G_{\alpha'\alpha}(\omega)B_\alpha(t) \tag{4.74}$$

where

$$G_{\alpha'\alpha}(\omega)=\sum_{m\#n}\frac{2\omega_{mn}\mu_{\alpha',nm}m_{\alpha,mn}}{\hbar\left(\omega_{mn}^2-\omega^2\right)} \tag{4.75}$$

Case II: $\mu_{\alpha',mn}m^*_{\alpha,mn}=-\mu_{\alpha',nm}m_{\alpha,mn}$

Electric dipole moment operator is Hermitian, so $\left\langle\psi_m^o\middle|\hat{\mu}_{\alpha'}\middle|\psi_n^o\right\rangle=\left\langle\psi_n^o\middle|\hat{\mu}_{\alpha'}\middle|\psi_m^o\right\rangle$. Magnetic dipole moment operator is imaginary, so $\left\langle\psi_m^o\middle|\hat{m}_\alpha\middle|\psi_n^o\right\rangle^*=\left\langle\psi_n^o\middle|\hat{m}_\alpha\middle|\psi_m^o\right\rangle=-\left\langle\psi_m^o\middle|\hat{m}_\alpha\middle|\psi_n^o\right\rangle$. Then, the $\cos\omega t$ terms cancel, yielding

$$\begin{aligned}\mu_{\alpha'}(t)&=\mu_{\alpha'}^0-\sum_{m\#n}\frac{2i\mu_{\alpha',nm}m_{\alpha,mn}}{\hbar\left(\omega_{mn}^2-\omega^2\right)}B_\alpha^0\omega\sin\omega t=\mu_{\alpha'}^0+\sum_{m\#n}\frac{2i\omega\mu_{\alpha',nm}m_{\alpha,mn}}{\hbar\left(\omega_{mn}^2-\omega^2\right)}\frac{\dot{B}_\alpha(t)}{\omega}\\&=\mu_{\alpha'}^0+\omega^{-1}G'_{\alpha'\alpha}(\omega)\dot{B}_\alpha(t),\end{aligned} \tag{4.76}$$

where dynamic electric dipole–magnetic dipole polarizability, $G'_{\alpha'\alpha}(\omega)$, is given as

$$G'_{\alpha'\alpha}(\omega)=\sum_{m\#n}\frac{2i\omega\mu_{\alpha',nm}m_{\alpha,mn}}{\hbar\left(\omega_{mn}^2-\omega^2\right)}=-\sum_{m\#n}\frac{2\omega\,\mathrm{Im}\left\{\mu_{\alpha',nm}m_{\alpha,mn}\right\}}{\hbar\left(\omega_{mn}^2-\omega^2\right)} \tag{4.77}$$

Combining the cases I and II, the electric dipole moment in time-dependent magnetic field is written (Barron 2004) as

$$\mu_{\alpha'}(t)=\mu_{\alpha'}^0+G_{\alpha'\alpha}(\omega)B_\alpha(t)+\omega^{-1}G'_{\alpha'\alpha}(\omega)\dot{B}_\alpha(t) \tag{4.78}$$

It will be useful to note (for later chapters) that, when $\mu_{\alpha',mn}$ is expressed in esu cm, $m_{\alpha,mn}$ is expressed in esu cm^2 s^{-1}, ω in cyc s^{-1}, and Plank's constant in J s, Equation 4.77 can be brought to the units of cm^4 (see Appendix 8).

In a similar manner, one can find the magnetic dipole moment in the presence of a time-dependent electric field, by replacing $\mu_{\alpha'}$ with $m_{\alpha'}$ in Equation 4.65 and carrying out the manipulations as before. This workout will lead to (Barron 2004)

$$m_{\alpha'}(t)=m_{\alpha'}^0+G_{\alpha\alpha'}(\omega)F_\alpha(t)-\omega^{-1}G'_{\alpha\alpha'}(\omega)\dot{F}_\alpha(t) \tag{4.79}$$

4.5 Interaction of Molecules with CPL

Even though the title of Chapter 3 is akin to that of this section, there is a significant difference in the approaches used in these two places. In Chapter 3, the speeds of left circular polarization (LCP) and right circular polarization (RCP) light components in a chiral medium are tacitly assumed to be different, so refractive indices could be assigned differently (see Equations 3.7 and 3.8), and it was sufficient to consider the electric vectors associated with LCP and RCP. In contrast, both electric and magnetic fields associated with LCP and RCP are considered here. In conjunction with perturbation theory developed in Sections 4.3 and 4.4, insight is gained into why the refractive indices of chiral molecules for LCP and RCP are different. Moreover, the fundamental quantum chemical relations needed for predicting the chiroptical properties are obtained in this section.

4.5.1 Electric and Magnetic Fields of CPL

4.5.1.1 Electric Field

In Chapter 1, we have seen that circularly polarized electric vector propagating along the z-axis is represented (see Equations 1.6 and 1.8) as

$$\mathbf{F}^{\pm} = F_0\left(\mathbf{u}\cos\Theta \pm \mathbf{v}\sin\Theta\right) \tag{4.80}$$

with + sign for left CPL and − sign for right CPL and $\Theta = 2\pi\left(\nu t - \frac{z}{\lambda}\right)$ Using $\omega = 2\pi\nu$, and $\nu\lambda = c$, Θ can be rewritten as $\Theta = \omega\left(t - \frac{z}{c}\right)$. The x- and y-components of CPL are

$$F_x^{\pm}(t) = F_0\cos\Theta \tag{4.81}$$

$$F_y^{\pm}(t) = \pm F_0\sin\Theta \tag{4.82}$$

where explicit time dependence is shown by adding t in parenthesis on the left-hand side.

4.5.1.2 Magnetic Field

To obtain the expression associated with the magnetic field of CPL, we make use of the Maxwell relation for electromagnetic radiation, $-\nabla\times\mathbf{F} = \dot{\mathbf{B}}$, where a dot on top of B indicates its time derivative. From this relation, noting that there is no z-component of electric field for light propagating along the z-axis, individual components of $\dot{\mathbf{B}}$ can be written as

$$\dot{B}_x(t) = -\left(\frac{\partial}{\partial y}F_z(t) - \frac{\partial}{\partial z}F_y(t)\right) = \frac{\partial}{\partial z}F_y(t) \tag{4.83}$$

$$\dot{B}_y(t) = -\left(\frac{\partial}{\partial z}F_x(t) - \frac{\partial}{\partial x}F_z(t)\right) = -\frac{\partial}{\partial z}F_x(t) \tag{4.84}$$

For CPL, the x- and y-components of $\dot{\mathbf{B}}$ become

$$\dot{B}_x^{\pm}(t) = \frac{\partial}{\partial z} F_y^{\pm}(t) = \frac{\partial\left(\pm F_0 \sin\omega\left(t - \frac{z}{c}\right)\right)}{\partial z} = (\pm F_0 \cos\Theta) \times \left(\frac{-\omega}{c}\right) \tag{4.85}$$

$$= \mp \frac{\omega F_0}{c} \cos\Theta = \mp \omega B_0 \cos\Theta$$

$$\dot{B}_y^{\pm}(t) = \frac{-\partial}{\partial z} F_x^{\pm}(t) = \frac{-\partial\left(F_0 \cos\omega\left(t - \frac{z}{c}\right)\right)}{\partial z} = -(-F_0 \sin\Theta) \times \left(\frac{-\omega}{c}\right) \tag{4.86}$$

$$= \frac{-\omega F_0}{c} \sin\Theta = -\omega B_0 \sin\Theta.$$

A useful relation between time derivatives of magnetic field and electric field of CPL can be obtained from Equations 4.85 and 4.86 as follows:

$$\dot{\mathbf{B}}^{\pm}(t) = \omega B_0(\mp \mathbf{u}\cos\Theta - \mathbf{v}\sin\Theta) = -\omega B_0(\pm \mathbf{u}\cos\Theta + \mathbf{v}\sin\Theta)$$
$$= \mp \omega B_0(\mathbf{u}\cos\Theta \pm \mathbf{v}\sin\Theta) = \mp \frac{\omega}{c} \mathbf{F}^{\pm}(t) \tag{4.87}$$

The magnetic field components can be derived by integrating their time derivatives, so

$$B_x^{\pm}(t) = \int \dot{B}_x^{\pm}(t)dt = \mp \int \omega B_o \cos\omega\left(t - \frac{z}{c}\right)dt = \mp B_0 \sin\Theta \tag{4.88}$$

$$B_y^{\pm}(t) = \int \dot{B}_y^{\pm}(t)dt = -\int \omega B_o \sin\omega\left(t - \frac{z}{c}\right)dt = B_0 \cos\Theta \tag{4.89}$$

Therefore, the magnetic field of CPL can be written from Equations 4.88 and 4.89 as

$$\mathbf{B}^{\pm}(t) = B_0\left(\mp \mathbf{u}\sin\Theta + \mathbf{v}\cos\Theta\right) = -B_0\left(\pm \mathbf{u}\sin\Theta - \mathbf{v}\cos\Theta\right) \tag{4.90}$$

The same relation can also be obtained from the relation, $\mathbf{k} \times \mathbf{F}^{\pm} = c\mathbf{B}^{\pm}$, where $\boldsymbol{k}$ is the unit vector along the z-axis. This is because the x-component of $\mathbf{k} \times \mathbf{F}^{\pm}$ is $-F_y^{\pm}(t) = \mp F_0 \sin\Theta = cB_x^{\pm}(t)$ and the y-component of $\mathbf{k} \times \mathbf{F}^{\pm}$ is $F_x^{\pm}(t) = F_0 \cos\Theta = cB_y^{\pm}(t)$.

4.5.2 Electric Dipole Moment Induced by Electric and Magnetic Fields of CPL

Using the electric and magnetic fields associated with CPL, we can write the perturbing Hamiltonian for molecules interacting with CPL. For this purpose, separate the x- and y-field components of CPL and let them interact with corresponding components of dipole moments. In this process, we only need to consider the field dependence on time and can drop the z/c term. For the sake of convenience in tracking, the maximum amplitudes of electric and magnetic field vector components will be labeled with its direction. That is, F_0 and B_0 used in the previous section will now be labeled specifically with directional components as F_x^0, F_y^0, B_x^0, and B_y^0. Then, for the electric field, the x-component is written as $F_x^{\pm}(t) = F_x^0 \cos\omega t = \frac{F_x^0}{2}\left(e^{i\omega t} + e^{-i\omega t}\right)$ and y-component as $F_y^{\pm}(t) = \pm F_y^0 \sin\omega t = \pm\frac{F_y^0}{2i}\left(e^{i\omega t} - e^{-i\omega t}\right)$. Similarly for the magnetic field, the x-component is $B_x^{\pm}(t) = \mp B_x^0 \sin\omega t = \mp\frac{B_x^0}{2i}\left(e^{i\omega t} - e^{-i\omega t}\right)$ and the y-component is $B_y^{\pm}(t) = B_y^0 \cos\omega t = \frac{B_y^0}{2}\left(e^{i\omega t} + e^{-i\omega t}\right)$. Using these components, the perturbing Hamiltonian is written as

$$\begin{aligned} H'^{\pm}(t) &= \left[-\mu_x F_x^{\pm}(t) - \mu_y F_y^{\pm}(t) - m_x B_x^{\pm}(t) - m_y B_y^{\pm}(t)\right]\left(1-e^{-kt}\right) \\ &= \frac{1}{2}\left[-\mu_x F_x^{o}\left(e^{i\omega t} + e^{-i\omega t}\right) - \mu_y F_y^{o}\left(\frac{\pm 1}{i}\right)\left(e^{i\omega t} - e^{-i\omega t}\right)\right. \\ &\quad \left. - m_x B_x^{o}\left(\mp\frac{1}{i}\right)\left(e^{i\omega t} - e^{-i\omega t}\right) - m_y B_y^{o}\left(e^{i\omega t} + e^{-i\omega t}\right)\right]\left(1-e^{-kt}\right) \\ &= \frac{1}{2}\left[-\mu_x F_x^{o}\left(e^{i\omega t} + e^{-i\omega t}\right) \pm i\mu_y F_y^{o}\left(e^{i\omega t} - e^{-i\omega t}\right)\right. \\ &\quad \left. \mp i m_x B_x^{o}\left(e^{i\omega t} - e^{-i\omega t}\right) - m_y B_y^{o}\left(e^{i\omega t} + e^{-i\omega t}\right)\right]\left(1-e^{-kt}\right) \end{aligned} \tag{4.91}$$

where $\left(1-e^{-kt}\right)$ term is used to suppress the oscillations (vide supra) and $1/i$ is written as $i/i^2 = -i$. Using this perturbation, the time evolution of the coefficient for state m is given, following Equation 4.57, as

$$c_m^{\pm}(t) = \frac{1}{i\hbar}\int_0^t H'^{\pm}_{mn}(t) e^{i\omega_{mn} t}\, dt \tag{4.92}$$

The perturbation terms involving $-\mu_\alpha F_\alpha^0\left(e^{i\omega t} + e^{-i\omega t}\right)\left(1-e^{-kt}\right)$ and $-m_\alpha B_\alpha^0\left(e^{i\omega t} + e^{-i\omega t}\right)\left(1-e^{-kt}\right)$ have been evaluated earlier individually in Sections 4.4.1 and 4.4.2, so we can transport those results to here. The perturbation terms involving $i\left(e^{i\omega t} - e^{-i\omega t}\right)\left(1-e^{-kt}\right)$ remain to be worked out. For this purpose, we will consider the perturbation term given by $\frac{i\mu_y F_y^0}{2}\left(e^{i\omega t} - e^{-i\omega t}\right)\left(1-e^{-kt}\right)$, extend the results to that given by $\frac{i m_x B_x^0}{2}\left(e^{i\omega t} - e^{-i\omega t}\right)\left(1-e^{-kt}\right)$, and then combine the results of all four individual perturbation terms.

For the perturbation represented by $\frac{i\mu_\alpha F_\alpha^0}{2}\left(e^{i\omega t} - e^{-i\omega t}\right)\left(1-e^{-kt}\right)$,

$$c_m(t)=\frac{F_\alpha^0\mu_{\alpha,mn}}{2\hbar}\int_0^t\left(e^{i\omega t}-e^{-i\omega t}\right)\left(1-e^{-kt}\right)e^{i\omega_{mn}t}\,dt$$

$$=\frac{F_\alpha^0\mu_{\alpha,mn}}{2\hbar}\int_0^t\left(e^{i(\omega_{mn}+\omega)t}-e^{i(\omega_{mn}-\omega)t}\right)\left(1-e^{-kt}\right)dt$$

$$=\frac{F_\alpha^0\mu_{\alpha,mn}}{2\hbar}\left[\frac{e^{i(\omega_{mn}+\omega)t}-1}{i(\omega_{mn}+\omega)}-\frac{e^{i(\omega_{mn}-\omega)t}-1}{i(\omega_{mn}-\omega)}\right. \tag{4.93}$$

$$\left.-\frac{e^{(i(\omega_{mn}+\omega)-k)t}-1}{i(\omega_{mn}+\omega)-k}+\frac{e^{(i(\omega_{mn}-\omega)-k)t}-1}{i(\omega_{mn}-\omega)-k}\right]$$

Invoke the approximation, $(\omega_{mn}\pm\omega)>>k$, so$\left(i(\omega_{mn}\pm\omega)-k\right)\approx i(\omega_{mn}\pm\omega)$ and (b) $t>>1/k$, so $e^{-kt}\sim 0$, as was done in obtaining Equations 4.60 and 4.61. Then,

$$c_m(t)=\frac{F_\alpha^0\mu_{\alpha,mn}}{2\hbar i\left(\omega_{mn}^2-\omega^2\right)}\left[e^{i\omega_{mn}t}\left[(\omega_{mn}-\omega)e^{i\omega t}-(\omega_{mn}+\omega)e^{-i\omega t}\right]-\right.$$

$$\left.(\omega_{mn}-\omega)+(\omega_{mn}+\omega)+(\omega_{mn}-\omega)-(\omega_{mn}+\omega)\right]$$

$$=\frac{F_\alpha^0\mu_{\alpha,mn}}{2\hbar i\left(\omega_{mn}^2-\omega^2\right)}e^{i\omega_{mn}t}\left[(\omega_{mn}-\omega)(\cos\omega t+i\sin\omega t)-\right. \tag{4.94}$$

$$\left.(\omega_{mn}+\omega)(\cos\omega t-i\sin\omega t)\right]$$

$$=\frac{F_\alpha^0\mu_{\alpha,mn}}{2\hbar i\left(\omega_{mn}^2-\omega^2\right)}e^{i\omega_{mn}t}\left(2i\omega_{mn}\sin\omega t-2\omega\cos\omega t\right)$$

$$=\frac{-iF_\alpha^0\mu_{\alpha,mn}}{\hbar\left(\omega_{mn}^2-\omega^2\right)}e^{i\omega_{mn}t}\left(i\omega_{mn}\sin\omega t-\omega\cos\omega t\right)$$

The complex conjugate of this coefficient becomes

$$c_m^*(t)=\frac{iF_\alpha^0\mu_{\alpha,mn}^*}{\hbar\left(\omega_{mn}^2-\omega^2\right)}e^{-i\omega mnt}\left(-i\omega_{mn}\sin\omega t-\omega\cos\omega t\right) \tag{4.95}$$

Following the procedure used in obtaining Equations 4.64 and 4.65, the time-dependent electric dipole moment under the current perturbation becomes

$$\mu_{\alpha'}(t)=\mu_{\alpha'}^0+$$

$$\frac{F_\alpha^0}{\hbar}\sum_{m\#n}\left[\frac{\mu_{\alpha',mn}\mu_{\alpha,mn}^*\left[\omega_{mn}\sin\omega t-i\omega\cos\omega t\right]+\mu_{\alpha',nm}\mu_{\alpha,mn}\left[\omega_{mn}\sin\omega t+i\omega\cos\omega t\right]}{\left(\omega_{mn}^2-\omega^2\right)}\right] \tag{4.96}$$

To obtain the corresponding result for perturbation represented by the term $\frac{im_\alpha B_\alpha^0}{2}\left(e^{i\omega t}-e^{-i\omega t}\right)\left(1-e^{-kt}\right)$, replace F_α^0 with B_α^0, $\mu_{\alpha,mn}$ with $m_{\alpha,mn}$, and $\mu^*_{\alpha,mn}$ with $m^*_{\alpha,mn}$. That is,

$$\mu_{\alpha'}(t)=\mu_{\alpha'}^0+\frac{B_\alpha^0}{\hbar}\sum_{m\#n}\left[\frac{\mu_{\alpha',mn}m^*_{\alpha,mn}\left[\omega_{mn}\sin\omega t-i\omega\cos\omega t\right]+\mu_{\alpha',nm}m_{\alpha,mn}\left[\omega_{mn}\sin\omega t+i\omega\cos\omega t\right]}{\left(\omega_{mn}^2-\omega^2\right)}\right] \quad (4.97)$$

Now we have all four contributions needed to obtain the result under perturbating Hamiltonian, $H'^{\pm}(t)$. Combining Equations 4.65, 4.73, 4.96, and 4.97, and remembering to use the component x or y for α, the following equation is obtained for time-dependent electric dipole moment:

$$\begin{aligned}\mu_{\alpha'}^{\pm}(t)=\mu_{\alpha'}^0&+\frac{F_x^0}{\hbar}\sum_{m\#n}\left[\frac{\mu_{\alpha',mn}\mu^*_{x,mn}\left[\omega_{mn}\cos\omega t+i\omega\sin\omega t\right]+\mu_{\alpha',nm}\mu_{x,mn}\left[\omega_{mn}\cos\omega t-i\omega\sin\omega t\right]}{\left(\omega_{mn}^2-\omega^2\right)}\right]\\&\pm\frac{F_y^0}{\hbar}\sum_{m\#n}\left[\frac{\mu_{\alpha',mn}\mu^*_{y,mn}\left[\omega_{mn}\sin\omega t-i\omega\cos\omega t\right]+\mu_{\alpha',nm}\mu_{y,mn}\left[\omega_{mn}\sin\omega t+i\omega\cos\omega t\right]}{\left(\omega_{mn}^2-\omega^2\right)}\right]\\&\mp\frac{B_x^0}{\hbar}\sum_{m\#n}\left[\frac{\mu_{\alpha',mn}m^*_{x,mn}\left[\omega_{mn}\sin\omega t-i\omega\cos\omega t\right]+\mu_{\alpha',nm}m_{x,mn}\left[\omega_{mn}\sin\omega t+i\omega\cos\omega t\right]}{\left(\omega_{mn}^2-\omega^2\right)}\right]\\&+\frac{B_y^0}{\hbar}\sum_{m\#n}\left[\frac{\mu_{\alpha',mn}m^*_{y,mn}\left[\omega_{mn}\cos\omega t+i\omega\sin\omega t\right]+\mu_{\alpha',nm}m_{y,mn}\left[\omega_{mn}\cos\omega t-i\omega\sin\omega t\right]}{\left(\omega_{mn}^2-\omega^2\right)}\right].\end{aligned} \quad (4.98)$$

Case I: When $\mu_{\alpha',mn}\mu^*_{\alpha,mn}=\mu_{\alpha',nm}\mu_{\alpha,mn}$ and $\mu_{\alpha',mn}m^*_{\alpha,mn}=\mu_{\alpha',nm}m_{\alpha,mn}$,

$$\begin{aligned}\mu_{\alpha'}^{\pm}(t)=\mu_{\alpha'}^0&+\sum_{m\#n}\frac{2\mu_{\alpha',nm}\mu_{x,mn}\omega_{mn}}{\hbar\left(\omega_{mn}^2-\omega^2\right)}F_x^0\cos\omega t\pm\sum_{m\#n}\frac{2\mu_{\alpha',nm}\mu_{y,mn}\omega_{mn}}{\hbar\left(\omega_{mn}^2-\omega^2\right)}F_y^0\sin\omega t\\&\mp\sum_{m\#n}\frac{2\mu_{\alpha',nm}m_{x,mn}\omega_{mn}}{\hbar\left(\omega_{mn}^2-\omega^2\right)}B_x^0\sin\omega t+\sum_{m\#n}\frac{2\mu_{\alpha',nm}m_{y,mn}\omega_{mn}}{\hbar\left(\omega_{mn}^2-\omega^2\right)}B_y^0\cos\omega t\end{aligned} \quad (4.99)$$

Substituting the expressions for $\alpha_{\alpha'\alpha}(\omega)$, $G_{\alpha'\alpha}(\omega)$, and $F_\alpha^{\pm}(t)$ (see Equations 4.67, 4.68, 4.74, 4.75, 4.81, and 4.82), we obtain

$$\mu_{\alpha'}^{\pm}(t)=\mu_{\alpha'}^0+\alpha_{\alpha'x}(\omega)F_x^{\pm}(t)+\alpha_{\alpha'y}(\omega)F_y^{\pm}(t)+G_{\alpha'x}(\omega)B_x^{\pm}(t)+G_{\alpha'y}(\omega)B_y^{\pm}(t). \quad (4.100)$$

Case II: When $\mu_{\alpha',mn}\mu^*_{\alpha,mn}=-\mu_{\alpha',nm}\mu_{\alpha,mn}$ and $\mu_{\alpha',mn}m^*_{\alpha,mn}=-\mu_{\alpha',nm}m_{\alpha,mn}$

$$\mu_{\alpha'}^{\pm}(t)=\mu_{\alpha'}^{0}+\sum_{m\#n}\frac{2i\omega\mu_{\alpha',nm}\mu_{x,mn}}{\hbar\left(\omega_{mn}^{2}-\omega^{2}\right)}\left(-F_{x}^{0}\sin\omega t\right)\pm\sum_{m\#n}\frac{2i\omega\mu_{\alpha',nm}\mu_{y,mn}}{\hbar\left(\omega_{mn}^{2}-\omega^{2}\right)}F_{y}^{0}\cos\omega t$$
$$\mp\sum_{m\#n}\frac{2i\omega\mu_{\alpha',nm}m_{x,mn}}{\hbar\left(\omega_{mn}^{2}-\omega^{2}\right)}B_{x}^{0}\cos\omega t+\sum_{m\#n}\frac{2i\omega\mu_{\alpha',nm}m_{y,mn}}{\hbar\left(\omega_{mn}^{2}-\omega^{2}\right)}\left(-B_{y}^{0}\sin\omega t\right) \tag{4.101}$$

Substituting the expressions for $\alpha'_{\alpha'\alpha}(\omega)$, $G'_{\alpha'\alpha}(\omega)$, and $\dot{F}_{\alpha}^{\pm}(t)$ (see Equations 4.71, 4.77, 4.81, and 4.82), we obtain

$$\mu_{\alpha'}^{\pm}(t)=\mu_{\alpha'}^{0}+\alpha'_{\alpha'x}(\omega)\frac{\dot{F}_{x}^{\pm}(t)}{\omega}+\alpha'_{\alpha'y}(\omega)\frac{\dot{F}_{y}^{\pm}(t)}{\omega}+G'_{\alpha'x}(\omega)\frac{\dot{B}_{x}^{\pm}(t)}{\omega}$$
$$+G_{\alpha'y}(\omega)\frac{\dot{B}_{y}^{\pm}(t)}{\omega} \tag{4.102}$$

Combining cases I and II, the general expression is written as

$$\mu_{\alpha'}^{\pm}(t)=\mu_{\alpha'}^{0}+\alpha_{\alpha'\alpha}(\omega)F_{\alpha}^{\pm}(t)+\alpha'_{\alpha'\alpha}(\omega)\frac{\dot{F}_{\alpha}^{\pm}(t)}{\omega}+G_{\alpha'\alpha}(\omega)B_{\alpha}^{\pm}(t)+G'_{\alpha'\alpha}(\omega)\frac{\dot{B}_{\alpha}^{\pm}(t)}{\omega} \tag{4.103}$$

Earlier we noted that $\dot{B}^{\pm}=\mp\frac{\omega}{c}\mathbf{F}^{\pm}$ (see Equation 4.87), so the preceding expression becomes

$$\mu_{\alpha'}^{\pm}(t)=\mu_{\alpha'}^{0}+\alpha_{\alpha'\alpha}(\omega)F_{\alpha}^{\pm}(t)+\alpha'_{\alpha'\alpha}(\omega)\frac{\dot{F}_{\alpha}^{\pm}(t)}{\omega}+G_{\alpha'\alpha}(\omega)B_{\alpha}^{\pm}(t)\mp G'_{\alpha'\alpha}(\omega)\frac{F_{\alpha}^{\pm}(t)}{c}. \tag{4.104}$$

The $\alpha'_{\alpha'\alpha}(\omega)$ and $G_{\alpha'\alpha}(\omega)$ tensor elements are important in special cases and need not be of concern here. For the present consideration, we only need to consider the terms $\alpha_{\alpha'\alpha}(\omega)$ and $G'_{\alpha'\alpha}(\omega)$. Then,

$$\mu_{\alpha'}^{\pm}(t)=\mu_{\alpha'}^{0}+\left(\alpha_{\alpha'\alpha}(\omega)\mp\frac{G'_{\alpha'\alpha}(\omega)}{c}\right)F_{\alpha}^{\pm}(t) \tag{4.105}$$

Equation 4.105 is of fundamental importance because it shows that the electric dipole moments induced in molecules by left and right CPL can be different when $G'_{\alpha'\alpha}$ is nonzero. As a consequence, the refractive indices for left and right CPL can be different, which was tacitly assumed to explain the optical rotation of chiral molecules in Chapter 3.

4.5.3 Specific Optical Rotation

The difference in refractive indices for left and right CPL, $(n_L - n_R)$, is related to the optical rotation parameter β, as follows (Eyring et al. 1961):

$$(n_L - n_R)=\frac{16\pi^2 N_1}{\lambda}\beta \tag{4.106}$$

where N_1 is the number density. If N_1 is expressed as molecules/cc and wavelength λ in cm, β will have units of cm^4. The optical rotation parameter β is the negative trace of $\omega^{-1}G'_{\alpha'\alpha}(\omega)$ tensor (Equation 4.77) with a factor of "c" (speed of light), to account for the difference in the way $m_{\alpha,mn}$ is expressed in the definition for β and $\omega^{-1}G'_{\alpha'\alpha}(\omega)$:

$$\beta = \frac{c}{3}\sum_{m\# n}\frac{2\,\mathrm{Im}\left\{\mu_{x,nm}m_{x,mn} + \mu_{y,nm}m_{y,mn} + \mu_{z,nm}m_{z,mn}\right\}}{\hbar\left(\omega_{mn}^2 - \omega^2\right)} \tag{4.107}$$

In the expressions for $\omega^{-1}G'_{\alpha'\alpha}(\omega)$, $m_{\alpha,mn}$ is expressed in esu cm^2 s^{-1}, but in Equation 4.107, $m_{\alpha,mn}$ is expressed in esu cm. This modification is convenient to express the product, $\mu_{\alpha,\mathrm{nm}}\, m_{\alpha,mn}$ in commonly reported esu^{2}cm^2 units (see Chapter 5). Nevertheless, β is in units of cm^4, as is $\omega^{-1}G'_{\alpha'\alpha}(\omega)$ (see Appendix 8, Section A8.13).

Using Equations 3.20 and 4.106, optical rotation per unit length in units of deg becomes

$$\frac{\alpha}{l} = \frac{360}{2\pi}\times\frac{\pi}{\lambda}\times\frac{16\pi^2 N_1}{\lambda}\beta = \frac{360}{2\pi}\times\frac{16\pi^3 N_1}{\lambda^2}\beta \tag{4.108}$$

The specific optical rotation (Equation 3.21) for neat liquid samples in deg cc g^{-1} dm^{-1} units now becomes

$$[\alpha] = \frac{\alpha}{\rho l} = \frac{10\text{ cm}}{\text{dm}}\times\frac{360}{2\pi}\times\frac{16\pi^3 N_1}{\rho\lambda^2}\beta = \frac{1800\times 16\times\pi^2 N_A}{\lambda^2 M}\beta \tag{4.109}$$

where N_1/ρ can be replaced with N_A/M (with M representing the molar mass and N_A representing the Avogadro number). For liquid solutions also one obtains the same relation

$$[\alpha] = \frac{\alpha}{cl} = \frac{10\text{ cm}}{\text{dm}}\times\frac{360}{2\pi}\times\frac{16\pi^3 N_1}{c\lambda^2}\beta = \frac{1800\times 16\times\pi^2 N_A}{\lambda^2 M}\beta \tag{4.110}$$

where c (not to be confused with speed of light) is the concentration expressed as g/cc and N_1/c for liquid solutions can also be replaced with N_A/M.

It is useful to note here that in Equation 4.107, the Im $\{\mu_{x,nm}m_{x,mn} + \mu_{y,nm}m_{y,mn} + \mu_{z,nm}m_{z,mn}\}$ term, in the numerator of the right-hand side, represents rotational strength (vide infra), which is a measure of the integrated circular dichroism band intensity. Therefore, specific optical rotation depends on the circular dichroism band intensities of all electronic transitions. This point will be emphasized in the calculations of specific optical rotation (see Chapter 5).

4.5.4 Transition Rates for Molecules Interacting with CPL

The probability of finding the system in an excited state, m, upon exposure to electromagnetic radiation is given by the product $c_m(t)c_m^*(t)$, where these coefficients are defined earlier. For convenience in formulating this product for CPL, the perturbing Hamiltonian in the presence of CPL can be rewritten, without the damping term, as

$$H'^{\pm}(t)=\frac{\left(e^{i\omega t}+e^{-i\omega t}\right)}{2}\left[-\mu_x F_x^0 - m_y B_y^0\right]+i\frac{\left(e^{i\omega t}-e^{-i\omega t}\right)}{2}\left[\pm\mu_y F_y^0 \mp m_x B_x^0\right] \quad (4.111)$$

Using this perturbing Hamiltonian, the time-dependent coefficient for state m becomes

$$\begin{aligned}
c_m^{\pm}(t) &= \frac{1}{i\hbar}\int_0^t H'^{\pm}_{mn}(t)e^{i\omega_{mn}t}dt \\
&= \frac{1}{2i\hbar}\int_0^t\Big[\left(e^{i\omega t}+e^{-i\omega t}\right)\left[-\mu_{x,mn}F_x^0 - m_{y,mn}B_y^0\right] \\
&\quad + i\left(e^{i\omega t}-e^{-i\omega t}\right)\left[\pm\mu_{y,mn}F_y^0 \mp m_{x,mn}B_x^0\right]\Big]e^{i\omega_{mn}t}dt \\
&= \frac{\left[-\mu_{x,mn}F_x^0 - m_{y,mn}B_y^0\right]}{2i\hbar}\int_0^t\left(e^{i(\omega_{mn}+\omega)t}+e^{i(\omega_{mn}-\omega)t}\right)dt \\
&\quad + \frac{\left[\pm\mu_{y,mn}F_y^0 \mp m_{x,mn}B_x^0\right]}{2\hbar}\int_0^t\left(e^{i(\omega_{mn}+\omega)t}-e^{i(\omega_{mn}-\omega)t}\right)dt
\end{aligned} \quad (4.112)$$

The integrals in Equation 4.112 can be written as

$$\begin{aligned}
&\int_0^t\left(e^{i(\omega_{mn}+\omega)t}\pm e^{i(\omega_{mn}-\omega)t}\right)dt \\
&= \frac{e^{i(\omega_{mn}+\omega)t}-1}{i(\omega_{mn}+\omega)} \pm \frac{e^{i(\omega_{mn}-\omega)t}-1}{i(\omega_{mn}-\omega)}
\end{aligned} \quad (4.113)$$

The second term is dominant for the absorption process (where ω_{mn}~ω), so we can ignore the first term. Then,

$$\begin{aligned}
c_m^{\pm}(t) &= \frac{\left[-\mu_{x,mn}F_x^0 - m_{y,mn}B_y^0\right]}{2i\hbar}\left[\frac{e^{i(\omega_{mn}-\omega)t}-1}{i(\omega_{mn}-\omega)}\right] - \\
&\quad \frac{\left[\pm\mu_{y,mn}F_y^0 \mp m_{x,mn}B_x^0\right]}{2\hbar}\left[\frac{e^{i(\omega_{mn}-\omega)t}-1}{i(\omega_{mn}-\omega)}\right] \\
&= \frac{\left[\mu_{x,mn}F_x^0 + m_{y,mn}B_y^0\right]}{2\hbar}\left[\frac{e^{i(\omega_{mn}-\omega)t}-1}{(\omega_{mn}-\omega)}\right] + \\
&\quad i\frac{\left[\pm\mu_{y,mn}F_y^0 \mp m_{x,mn}B_x^0\right]}{2\hbar}\left[\frac{e^{i(\omega_{mn}-\omega)t}-1}{(\omega_{mn}-\omega)}\right]
\end{aligned}$$

$$= \frac{\left(e^{i(\omega_{mn}-\omega)t}-1\right)}{2\hbar(\omega_{mn}-\omega)}\left\{\left(\mu_{x,mn}F_x^0 + m_{y,mn}B_y^0\right) + i\left(\pm\mu_{y,mn}F_y^0 \mp m_{x,mn}B_x^0\right)\right\}$$
$$= \frac{\left(e^{i(\omega_{mn}-\omega)t}-1\right)}{2\hbar(\omega_{mn}-\omega)}\left\{\left(\mu_{x,mn}F_x^0 \pm i\mu_{y,mn}F_y^0\right) + \left(m_{y,mn}B_y^0 \mp i m_{x,mn}B_x^0\right)\right\} \tag{4.114}$$

For CPL, the amplitudes $F_x^0 = F_y^0 = F^0$ and $B_x^0 = B_y^0 = B^0$. Therefore,

$$c_m^{\pm}(t) = \frac{\left(e^{i(\omega_{mn}-\omega)t}-1\right)}{2\hbar(\omega_{mn}-\omega)}\left\{\left(\mu_{x,mn} \pm i\mu_{y,mn}\right)F^0 + \left(m_{y,mn} \mp i m_{x,mn}\right)B^0\right\} \tag{4.115}$$

The complex conjugate of this coefficient becomes

$$\left(c_m^{\pm}(t)\right)^* = \frac{\left(e^{-i(\omega_{mn}-\omega)t}-1\right)}{2\hbar(\omega_{mn}-\omega)}\left\{\left(\mu_{x,mn}^* \mp i\mu_{y,mn}^*\right)F^0 + \left(m_{y,mn}^* \pm i m_{x,mn}^*\right)B^0\right\} \tag{4.116}$$

To obtain the probability of transition to state m, $P_m^{\pm}(t) = c_m^{\pm}(t)\left(c_m^{\pm}(t)\right)^*$, first evaluate the following two products:

$$\left(e^{i(\omega_{mn}-\omega)t}-1\right)\left(e^{-i(\omega_{mn}-\omega)t}-1\right) = 1 - e^{i(\omega_{mn}-\omega)t} - e^{-i(\omega_{mn}-\omega)t} + 1$$
$$= 2\left[1-\cos(\omega_{mn}-\omega)t\right] = 4\sin^2\frac{(\omega_{mn}-\omega)t}{2} \tag{4.117}$$

$$\left\{\left(\mu_{x,mn} \pm i\mu_{y,mn}\right)F^0 + \left(m_{y,mn} \mp i m_{x,mn}\right)B^0\right\}\left\{\left(\mu_{x,mn}^* \mp i\mu_{y,mn}^*\right)F^0 + \left(m_{y,mn}^* \pm i m_{x,mn}^*\right)B^0\right\} =$$
$$\left[\mu_{x,mn}\mu_{x,mn}^* \mp i\mu_{x,mn}\mu_{y,mn}^* \pm i\mu_{y,mn}\mu_{x,mn}^* + \mu_{y,mn}\mu_{y,mn}^*\right]\left(F^0\right)^2 +$$
$$\left[m_{y,mn}m_{y,mn}^* \pm i m_{y,mn}m_{x,mn}^* \mp i m_{x,mn}m_{y,mn}^* + m_{x,mn}m_{x,mn}^*\right]\left(B^0\right)^2 +$$
$$\left[\mu_{x,mn}m_{y,mn}^* \pm i\mu_{x,mn}m_{x,mn}^* \pm i\mu_{y,mn}m_{y,mn}^* - \mu_{y,mn}m_{x,mn}^*\right]\left(F^0B^0\right) +$$
$$\left[m_{y,mn}\mu_{x,mn}^* \mp i m_{y,mn}\mu_{y,mn}^* \mp i m_{x,mn}\mu_{x,mn}^* - m_{x,mn}\mu_{y,mn}^*\right]\left(B^0F^0\right) \tag{4.118}$$

Here, we use the relations, $\mu_{\alpha,mn}^* = \mu_{\alpha,mn} = \mu_{\alpha,nm}$ and $m_{\alpha,mn}^* = -m_{\alpha,mn} = m_{\alpha,nm}$. Then, $\mp i\mu_{x,mn}\mu_{y,mn}^* \pm i\mu_{y,mn}\mu_{x,mn}^* = 0$, $\pm i m_{y,mn}m_{x,mn}^* \mp i m_{x,mn}m_{y,mn}^* = 0$, $\mu_{x,mn}m_{y,mn}^* + m_{y,mn}\mu_{x,mn}^* = 0$; and $-\mu_{y,mn}m_{x,mn}^* - m_{x,mn}\mu_{y,mn}^* = 0$. Substituting these relations into the product, $P_m^{\pm}(t) = c_m^{\pm}(t)\left(c_m^{\pm}(t)\right)^*$, the expression for $P_m^{\pm}(t)$ becomes

$$P_m^{\pm}(t) = \frac{1}{(\omega_{mn}-\omega)^2} \times 4\sin^2\frac{(\omega_{mn}-\omega)t}{2} \times V_{mn}^{\pm 2} \tag{4.119}$$

where

$$V_{mn}^{\pm 2} = \frac{1}{4\hbar^2}\Big\{\big[\mu_{x,nm}\mu_{x,mn} + \mu_{y,nm}\mu_{y,mn}\big]\big(F^0\big)^2 + \big[m_{x,nm}m_{x,mn} + m_{y,nm}m_{y,mn}\big]\big(B^0\big)^2 \mp 2i\big[\mu_{x,nm}m_{x,mn} + \mu_{y,nm}m_{y,mn}\big]\big(F^0B^0\big)\Big\} \quad (4.120)$$

The term "$\sin^2 \frac{(\omega_{mn}-\omega)t}{2}$" can be eliminated by considering a band of perturbing frequencies (Atkins 1983; Levine 1975). With ρ^ν representing the frequency density of states

$$P^{\pm}(t) = V_{mn}^{\pm 2} t^2 \int_{-\infty}^{\infty} \frac{\sin^2 \frac{(\omega_{mn}-\omega)t}{2}}{\left(\frac{(\omega_{mn}-\omega)t}{2}\right)^2} \rho^\nu \, d\nu \quad (4.121)$$

Using $\omega = 2\pi\nu$, and designating $x = \frac{(\omega-\omega_{mn})t}{2} = \frac{2\pi(\nu-\nu_{mn})t}{2}$, $dx = \pi t d\nu$, the preceding equation becomes

$$P^{\pm}(t) = \frac{V_{mn}^{\pm 2} t^2 \rho^\nu}{\pi t} \int_{-\infty}^{\infty} \left(\frac{\sin x}{x}\right)^2 dx = V_{mn}^{\pm 2} t \rho^\nu \quad (4.122)$$

In the expression for $V_{mn}^{\pm 2}$, the terms $\frac{(F^0)^2}{4}, \frac{(B^0)^2}{4}, \frac{(F^0B^0)}{4}$ can be converted to energy contained in electromagnetic wave, yielding $\varepsilon\pi/V$, as follows: The energy contained in electromagnetic wave, $\mathbf{F}^{\pm}(t) = F^0(\mathbf{u}\cos\omega t \pm \mathbf{v}\sin\omega t)$, is given as

$$\varepsilon = \frac{1}{4\pi}\left\langle F^{\pm}(t)\right\rangle^2 = \frac{1}{4\pi}\left[\frac{\int_0^{2\pi}\left(F^0\cos\omega t\right)^2 dt}{\int_0^{2\pi} dt} + \frac{\int_0^{2\pi}\left(F^0\sin\omega t\right)^2 dt}{\int_0^{2\pi} dt}\right] = \frac{1}{4\pi}\left[\frac{(F^0)^2\frac{\pi}{\omega}}{\frac{2\pi}{\omega}} + \frac{(F^0)^2\frac{\pi}{\omega}}{\frac{2\pi}{\omega}}\right] = \frac{(F^0)^2}{4\pi} \quad (4.123)$$

Energy in a certain volume V is $\frac{(F^0)^2}{4\pi}V$. Then, $\frac{(F^0)^2}{4} = \frac{\varepsilon\pi}{V}$, where ε/V is the energy density (energy ε in volume V). Then, the transition probability becomes

$$P^{\pm}(t) = \frac{\pi t \rho^E}{\hbar^2}\left\{\left[\mu_{x,nm}\mu_{x,mn} + \mu_{y,nm}\mu_{y,mn}\right] + \left[m_{x,nm}m_{x,mn} + m_{y,nm}m_{y,mn}\right] \mp 2i\left[\mu_{x,nm}m_{x,mn} + \mu_{y,nm}m_{y,mn}\right]\right\}$$

$$= \frac{\pi t \rho^E}{\hbar^2}\left\{\left[\mu_{x,nm}\mu_{x,mn} + \mu_{y,nm}\mu_{y,mn}\right] + \left[m_{x,nm}m_{x,mn} + m_{y,nm}m_{y,mn}\right] \pm 2\,\mathrm{Im}\left[\mu_{x,nm}m_{x,mn} + \mu_{y,nm}m_{y,mn}\right]\right\} \tag{4.124}$$

$$= \frac{\pi t \rho^E}{\hbar^2}\left\{\boldsymbol{\mu}_{nm}\bullet\boldsymbol{\mu}_{mn} + \mathbf{m}_{nm}\bullet\mathbf{m}_{mn} \pm 2\,\mathrm{Im}\left[\boldsymbol{\mu}_{nm}\bullet\mathbf{m}_{mn}\right]\right\}$$

where the energy density of states $\rho^E = \rho^\nu\varepsilon/V$. The transition rate is $dP^{\pm}(t)/dt = W^{\pm}$. Averaging over all orientations (applicable to isotropic samples), one obtains

$$W^{\pm} = \frac{2\pi\rho^E}{3\hbar^2}\left\{\boldsymbol{\mu}_{nm}\bullet\boldsymbol{\mu}_{mn} + \mathbf{m}_{nm}\bullet\mathbf{m}_{mn} \pm 2\,\mathrm{Im}\left[\boldsymbol{\mu}_{nm}\bullet\mathbf{m}_{mn}\right]\right\} \tag{4.125}$$

This equation is of fundamental importance, as it shows that transition rates for left and right CPL can be different when the third term on the right-hand side of Equation 4.125 is nonzero, which occurs for chiral molecules. The difference in transition rates for left and right CPL is written as

$$W^{+} - W^{-} = \frac{2\pi\rho^E}{3\hbar^2}\times 4\,\mathrm{Im}\left[\boldsymbol{\mu}_{nm}\bullet\mathbf{m}_{mn}\right] = \frac{2\pi\rho^E}{3\hbar^2}\times 4R_{mn} \tag{4.126}$$

where

$$R_{mn} = \mathrm{Im}\left[\boldsymbol{\mu}_{nm}\bullet\mathbf{m}_{mn}\right] \tag{4.127}$$

In this equation, R_{mn} is called rotational strength (also some times as rotatory strength) and determines the integrated circular dichroism band intensity (vide infra).

For the absorption process from electric dipole transitions, the relevant transition rate is

$$W = \frac{2\pi\rho^E}{3\hbar^2}\left\{\boldsymbol{\mu}_{nm}\bullet\boldsymbol{\mu}_{mn}\right\} = \frac{2\pi\rho^E}{3\hbar^2}D_{mn} \tag{4.128}$$

where the term

$$D_{mn} = \left\{\boldsymbol{\mu}_{nm}\bullet\boldsymbol{\mu}_{mn}\right\} \tag{4.129}$$

is called dipole transition strength (or simply dipole strength) and determines the integrated absorption band intensity (vide infra).

In the presentation so far, we have considered a general molecular transition from state n to excited state m. Therefore, the expression for rotational strength R_{mn} is general enough that it can be used for circular dichroism associated with both electronic and vibrational transitions. Similarly, the expression for dipole strength D_{mn} is general enough that can be used for absorption associated with both electronic and vibrational transitions. However, in the practical implementation of the rotational strength expression for vibrational circular dichroism, one faces difficulties that have been overcome recently. These aspects are discussed in Chapter 5.

In the case of electronic transitions, another quantity that is often used in representing absorption band intensities is a unit-less quantity called oscillator strength, f_{mn}, that is related to dipole strength:

$$f_{mn} = \frac{8\pi^2 m_e \nu_{mn}}{3e^2 h} D_{mn} \tag{4.130}$$

where m_e is mass of electron, e is electron charge, and ν_{mn} is transition frequency (see Appendix 8 for unit conversions).

4.5.5 Relation to Experimental Absorption and Circular Dichroism

The transition rates given earlier yield the rate at which a molecule absorbs a photon of energy $h\nu$. Let $n(\nu)$ be the number density (number of molecules per unit volume) absorbing in the range ν to ν +dν, such that $\int n(\nu)d\nu = N$, the total number density. Restricting the transition rate to absorption process associated with electric dipole transitions (see Equation 4.128), the rate of change of energy density in the range, ν to ν +dν, is

$$\frac{d}{dt}\left(\rho^E d\nu\right) = -(h\nu)\times\frac{2\pi\rho^E}{3\hbar^2} D_{mn}\, n(\nu)d\nu \tag{4.131}$$

That is,

$$\frac{d}{dt}\rho^E = -(h\nu)\times\frac{2\pi\rho^E}{3\hbar^2} D_{mn}\, n(\nu) \tag{4.132}$$

When the liquid solution of a sample is exposed to light, absorption of a photon of energy $h\nu$ from the incident light causes a decrease in the intensity of light exiting the sample. This decrease depends both on the concentration of the sample and width of the sample (through which light propagates and is often referred to as pathlength). For a pathlength dl, the aforementioned rate of change is equal to the change in intensity of light with pathlength. That is, $\frac{dI(\nu)}{dl}$ [see Appendix 17 of Atkins (1983)]. Then,

$$\frac{dI(\nu)}{dl} = -(h\nu)\times\frac{2\pi\rho^{E}}{3\hbar^{2}}D_{mn}\,n(\nu) \tag{4.133}$$

Denoting the intensity of light at frequency ν as $I(\nu) = \rho^{E}c$, the preceding equation becomes

$$\frac{dI(\nu)}{dl} = -(h\nu)\times\frac{2\pi I(\nu)}{3c\hbar^{2}}D_{mn}\,n(\nu) \tag{4.134}$$

or

$$\frac{dI(\nu)}{I(\nu)} = -(h\nu)\times\frac{2\pi}{3c\hbar^{2}}D_{mn}\,n(\nu)dl \tag{4.135}$$

When light passes through a sample solution of concentration C and path length l, the decrease in light intensity is proportional to the intensity of light, sample concentration, and path length. Therefore,

$$dI(\nu) = -\alpha(\nu)I(\nu)Cdl \tag{4.136}$$

The proportionality constant $\alpha(\nu)$ is called the molar absorption coefficient. Integration of this equation yields

$$I(\nu) = I^{0}(\nu)e^{-A(\nu)} = I^{0}(\nu)e^{-\alpha(\nu)Cl} \tag{4.137}$$

where I^{0} is the intensity of light before entering the sample cell, I is the intensity of light exiting the sample, and absorbance $A(\nu)$ is

$$A(\nu) = \ln\frac{I^{0}(\nu)}{I(\nu)} = \alpha(\nu)Cl \tag{4.138}$$

Comparing Equation 4.136 with Equation 4.135 and integrating over the band width

$$\begin{aligned}\int\frac{\alpha(\nu)}{\nu}d\nu &= \int\frac{2\pi h}{3cC\hbar^{2}}D_{mn}n(\nu)d\nu \\ &= \frac{2\pi h}{3cC\hbar^{2}}D_{mn}\int n(\nu)d\nu = \frac{2\pi h}{3c\hbar^{2}}D_{mn}\frac{N}{C} = \frac{2\pi hN_{A}}{3c\hbar^{2}}D_{mn}\end{aligned} \tag{4.139}$$

where N_A is the Avogadro number. On the left-hand side, ν in the denominator can be approximated as the band center, ν^{o}, so $\int\frac{\alpha(\nu)}{\nu^{0}}d\nu = \frac{1}{\nu^{0}}\int\alpha(\nu)d\nu$. Then, one can obtain the relation for integrated molar absorption coefficient, A_{mn}, for transition from n to m, as

$$A_{mn} = \int \alpha(\nu)d\nu = \frac{2\pi h N_A \nu^0}{3c\hbar^2} D_{mn} = \frac{8\pi^3 N_A \nu^0}{3ch} D_{mn} \tag{4.140}$$

Thus, dipole strength D_{mn} represents the integrated molar absorption coefficient.

Repeating the preceding process for CPL absorption (see Equation 4.126) and taking the difference, $A^+ - A^-$, the difference in integrated molar absorption coefficients for LCP and RCP becomes

$$\Delta A_{mn} = \int \{\alpha^+(\nu) - \alpha^-(\nu)\} d\nu = 4 \times \left(\frac{8\pi^3 N_A \nu^0}{3ch} \right) R_{mn} \tag{4.141}$$

Thus, rotational strength R_{mn} represents the difference in integrated molar absorption coefficients for LCP and RCP.

The ratio, $\Delta A/A$, called dissymmetry factor, g, becomes

$$g_{mn} = \frac{\Delta A_{mn}}{A_{mn}} = \frac{4R_{mn}}{D_{mn}} \tag{4.142}$$

When the logarithm in Equation 4.138 is converted from base e to base 10, $I(\nu) = I^0(\nu)10^{-\varepsilon(\nu)Cl}$ (where $\varepsilon(\nu)$ is called decadic molar absorption coefficient), and $\ln a = 2.303 \log_{10} a$, the absorbance can be rewritten as

$$A(\nu) = 2.303 \log_{10} \left(\frac{I^0(\nu)}{I(\nu)} \right) = 2.303\varepsilon(\nu)Cl \tag{4.143}$$

From the preceding two equations, the relation between the molar absorption coefficient and the decadic molar absorption coefficient becomes

$$\alpha(\nu) = 2.303\varepsilon(\nu) \tag{4.144}$$

This relation can be substituted in the expressions for A_{mn} and ΔA_{mn}, and the integrated decadic molar absorption coefficient can be determined as

$$\int \varepsilon(\nu)d\nu = \frac{1}{2.303} \int \alpha(\nu)d\nu \tag{4.145}$$

and

$$\int \{\varepsilon^+(\nu) - \varepsilon^-(\nu)\} d\nu = \frac{1}{2.303} \int \{\alpha^+(\nu) - \alpha^-(\nu)\} d\nu \tag{4.146}$$

4.6 Transition Polarizabilities

The expression for time-dependent electric dipole moment (Equation 4.72) indicates that the molecular electric dipole moment oscillates at the frequency of the electric field associated with incident light. An oscillating electric dipole moment emits radiation, so $\alpha_{\alpha'\alpha}(\omega)F_\alpha(t)$ term serves as the source for elastic scattering or Rayleigh scattered light. To consider the origin of inelastic scattering, where the frequency of scattered light differs from that of incident light, one needs to consider the time-dependent electric dipole transition moment, $\langle\psi_m(t)|\hat{\mu}_{\alpha'}|\psi_n(t)\rangle$. Note that this integral is different from the one considered for the electric dipole moment in Equation 4.64, where the initial and final states were the same. The evaluation of this integral follows the procedure used for the time-dependent electric dipole moment in Section 4.4, except that coefficients used in the expansion of $\psi_n(t)$ and $\psi_m(t)$ have to be tracked carefully. Then, one obtains the following relation for electric dipole transition moment:

$$\langle\psi_m(t)|\hat{\mu}_{\alpha'}|\psi_n(t)\rangle = \left\{\mu^0_{\alpha,mn} + [\alpha_{\alpha'\alpha}]_{mn} F_\alpha(t) + \omega^{-1}\left[\alpha'_{\alpha'\alpha}\right]_{mn} \dot{F}_\alpha(t)\right\} e^{i\omega_{mn}t} \tag{4.147}$$

where transition polarizabilities, $[\alpha_{\alpha'\alpha}]_{mn}$ and $\left[\alpha'_{\alpha'\alpha}\right]_{mn}$, are given by the following expressions:

$$[\alpha_{\alpha'\alpha}]_{mn} = \sum_{j\neq n} \frac{2\pi\omega_{jn}}{h\left(\omega_{jn}^2 - \omega^2\right)} \mu_{\alpha',mj}\mu_{\alpha,jn} + \sum_{j\neq m} \frac{2\pi\omega_{jm}}{h\left(\omega_{jm}^2 - \omega^2\right)} \mu_{\alpha',jn}\mu_{\alpha,mj} \tag{4.148}$$

$$\left[\alpha'_{\alpha'\alpha}\right]_{mn} = \sum_{j\neq n} \frac{2\pi i\omega}{h\left(\omega_{jn}^2 - \omega^2\right)} \mu_{\alpha',mj}\mu_{\alpha,jn} - \sum_{j\neq m} \frac{2\pi i\omega}{h\left(\omega_{jm}^2 - \omega^2\right)} \mu_{\alpha',jn}\mu_{\alpha,mj} \tag{4.149}$$

Note that when the initial and final states are the same, the transition polarizabilities $[\alpha_{\alpha'\alpha}]_{mn}$ and $\left[\alpha'_{\alpha'\alpha}\right]_{mn}$ reduce to the dynamic polarizabilities, $\alpha_{\alpha'\alpha}(\omega)$ and $\alpha'_{\alpha'\alpha}(\omega)$ (Equations 4.67 and 4.71) discussed earlier. When $F_\alpha(t)$ is expressed as $\frac{F^0_\alpha}{2}\left(e^{i\omega t} + e^{-i\omega t}\right)$, it can be seen that oscillations of the electric transition dipole moment occur at $\omega \pm \omega_{mn}$, which will be the source for radiation scattered at frequencies different from that of incident light (see Equation 3.48 for the corresponding classical explanation).

4.7 Summary

Despite long and tedious expressions involved, the essence of this chapter is that left and right CPL components interact with chiral molecules differently. This difference in interaction is responsible for the observed properties of optical rotation and circular dichroism. The electric dipole moments induced in chiral molecules by left and right CPL are seen to be different through Equation 4.105, which is of fundamental importance. The differences in induced electric dipole moments explain the difference in propagation speeds for left and right CPL in chiral media, resulting in the existence of optical rotation phenomenon. The specific optical rotation depends on the parameter β (see Equation 4.107 through 4.110), which depends on the rotational strengths associated with all possible electronic transitions. Rotational strength is the dot product of electric and magnetic dipole transition moment vectors associated with a transition. Circular dichroism depends on the rotational strength (see Equation 4.127) of a particular transition, either electronic or vibrational, and this phenomenon can be observed when the electric dipole transition moment vector is not orthogonal to the magnetic dipole transition moment vector. The specific optical rotation is seen to be dependent on the circular dichroism intensities of all possible electronic transitions. Absorption intensity, associated with either electronic or vibrational transitions, depends on the dipole strength (which is the square of the electric dipole transition moment) of a particular transition (see Equation 4.129). The transition polarizabilities explain the quantum chemical origin of Raman scattering and Raman optical activity parameters discussed in Chapter 3. The calculations of specific optical rotation, circular dichroism, and vibrational Raman optical activity properties are described in Chapter 5.

References

Atkins, P. W. 1983. *Molecular Quantum Mechanics*. Oxford, UK: Oxford University Press.

Barron, L. D. 2004. *Molecular Light Scattering and Optical Activity*, 2nd ed. Cambridge, UK: Cambridge University Press.

Barron, L. D., and A. D. Buckingham. 1975. Rayleigh and Raman optical activity. *Ann. Rev. Phys. Chem.* 26:381–396.

Eyring, H., J. Walter, and G. E. Kimball. 1961. *Quantum Chemistry*. New York, NY: Wiley.

Levine, I. N. 1975. *Molecular Spectroscopy*. Chichester, UK: Wiley.

Polavarapu, P. L. 1998. *Vibrational Spectra: Principles and Applications with Emphasis on Optical Activity*. New York, NY: Elsevier.

Struve, W. S. 1989. *Fundamentals of Molecular Spectroscopy*. New York, NY: John Wiley and Sons.

5

Prediction of Chiroptical Spectra

5.1 Introduction

To make use of chiroptical spectra for deducing the molecular structural information, a connection between the experimentally measured spectra and structure is needed. This connection is provided by the theoretical models describing the chiroptical phenomena. These theoretical models may arise from fundamentally rigorous principles or approximate concepts. Therefore, spectra can be predicted for a given molecular structure at different levels of approximations and reliability. In this process, one may choose to adopt one or more of three different approaches: (a) spectral predictions using modern quantum chemical theories, (b) spectral predictions using approximate models, and (c) extending the known spectra-structure correlations to unknown structures. The latter approach is more suitable for spectral interpretations rather than for predictions, and therefore, spectra-structure correlations are discussed in Chapter 10. The predictions of chiroptical spectra using quantum chemical methods and approximate models are discussed in this chapter.

5.2 Spectral Predictions Using Quantum Chemical Methods

The development of quantum chemical theories for predicting the chiroptical spectroscopic properties, the availability of commercial and freeware computer codes implementing these methods (Gaussian09 2013; ADF 2014; DALTON 2015; CADPAC 2001; PSI4 2014), and the availability of computers with faster processors for carrying out these calculations led to a routine adoption of the quantum chemical methods for spectral predictions. The close match obtained between experimental observations and corresponding quantum chemical predictions for a majority of the studied molecules is providing impetus for the use of quantum chemical predictions in interpreting the experimental spectra. These observations are attracting many research laboratories to tackle new research problems using chiroptical spectroscopy.

5.2.1 Optical Rotatory Dispersion

The specific optical rotation (SOR) was given, in Chapter 4, in terms of optical rotation parameter β (see Equations 4.109 and 4.110). This equation is rewritten here for SOR as

$$[\alpha(\lambda)] = \frac{1800 \times 16 \times \pi^2 N_A}{\lambda^2 M}\beta \tag{5.1}$$

where (see Equation 4.107)

$$\beta = \frac{c}{3}\sum_{m \# n} \frac{2\,\mathrm{Im}\{\mu_{x,nm}m_{x,mn} + \mu_{y,nm}m_{y,mn} + \mu_{z,nm}m_{z,mn}\}}{\hbar\left(\omega_{mn}^2 - \omega^2\right)} = \frac{-\omega^{-1}G'_{\alpha\alpha}}{3} \tag{5.2}$$

N_A represents the Avogadro number, and M the molar mass of solute molecules (see the statements following Equation 4.107 in Chapter 4 for units of β). Although not explicitly shown, β is also wavelength dependent.

The main problem in using the above equation is the involvement of the sum over excited states for calculating β. This limitation was overcome (Amos 1982) by obtaining the static limit (i.e., $\omega \to 0$ or $\lambda \to \infty$) value of $\omega^{-1}G'_{\alpha\beta}$ as the overlap of the electric field and magnetic field gradients of the ground state wave function:

$$\omega^{-1}G'_{\alpha\beta} = -\frac{h}{4\pi}\mathrm{Im}\left\{\left\langle \frac{\partial\psi_n}{\partial F_\alpha}\middle|\frac{\partial\psi_n}{\partial B_\beta}\right\rangle\right\} \tag{5.3}$$

Initial predictions (Polavarapu 1997) of quantum chemical predictions of SOR were based on this static method. The linear response or frequency-dependent (dynamic) method (Helgaker et al. 1994) removed the static limit restriction, and $\omega^{-1}G'_{\alpha\beta}$ could be obtained as a function of wavelength. This development enabled the calculation of optical rotatory dispersion (ORD) in the nonresonant wavelength region (Polavarapu and Zhao 1998).

It should be noted that ORD and electronic circular dichroism (ECD) are not independent of each other. To see the connection between ORD and ECD, it is more convenient to deal with molar rotation, $[\phi(\lambda)]$ (in units of deg cm^2/dmol), which is defined in Chapter 3 (see Equation 3.24) as

$$[\phi(\lambda)] = [\alpha(\lambda)] \times \frac{M}{100} \tag{5.4}$$

Upon substituting the Equations 5.1 and 5.2 into Equation 5.4, and substituting the constants, the expression (Polavarapu 2007) for molar rotation is obtained as

$$
\begin{aligned}
\left[\phi(\lambda)\right] &= \frac{M}{100} \times \frac{1800 \times 16\pi^2 N_A}{\lambda^2 M} \times \frac{1}{3}\frac{4\pi c}{h} \sum_{m \# n} \frac{1}{4\pi^2 c^2 \left(\dfrac{1}{\lambda_{mn}^2} - \dfrac{1}{\lambda^2}\right)} \\
&\quad \times \mathrm{Im}\left\{\langle \psi_n^{\circ} | \mu_\alpha | \psi_m^{\circ} \rangle \langle \psi_m^{\circ} | \mathrm{m}_\alpha | \psi_n^{\circ} \rangle\right\} \\
&= 96 \times \frac{22}{7} \times \frac{6.022 \times 10^{23}}{(6.62607 \times 10^{-27})(2.9979 \times 10^{10})} \sum_{m \# n} \frac{\lambda_{mn}^2}{\left(\lambda^2 - \lambda_{mn}^2\right)} \\
&\quad \times \mathrm{Im}\left\{\langle \psi_n^{\circ} | \mu_\alpha | \psi_m^{\circ} \rangle \langle \psi_m^{\circ} | \mathrm{m}_\alpha | \psi_n^{\circ} \rangle\right\} \\
&= 91.5 \times 10^{40} \sum_{m \# n} \frac{\lambda_{mn}^2}{\left(\lambda^2 - \lambda_{mn}^2\right)} \mathrm{Im}\left\{\langle \psi_n^{\circ} | \mu_\alpha | \psi_m^{\circ} \rangle \langle \psi_m^{\circ} | \mathrm{m}_\alpha | \psi_n^{\circ} \rangle\right\}
\end{aligned}
\tag{5.5}
$$

Using the definition for rotational strength (see Equation 4.127), in units of 10^{-40} esu^{2}cm^2, the expression for molar rotation becomes

$$
\left[\phi(\lambda)\right] = 91.5 \sum_{m \# n} \frac{\lambda_{mn}^2}{\lambda^2 - \lambda_{mn}^2} R_{mn} \tag{5.6}
$$

Equation 5.6 is referred to as the sum-over-state (SOS) expression for molar rotation. This expression indicates that molar rotation can be determined from the knowledge of the rotational strengths of all electronic transitions. An orbital-based analysis method to determine the major contributors in this summation has been proposed (Caricato 2015).

However, the SOS expression given by Equation 5.6 can only be used for the nonresonant region ($\lambda \neq \lambda_{mn}$). To avoid singularity at $\lambda = \lambda_{mn}$, the denominator in the above equation has to be modified to include a damping term that is related to the bandwidth of electronic transitions (Norman et al. 2004; Barron 2004). Incorporation of this damping factor allows one to see (Polavarapu 2007) the connection between ORD and ECD spectra.

Noting that $\mathrm{Im}\left\{\langle \psi_n^{\circ} | \mu_\alpha | \psi_m^{\circ} \rangle \langle \psi_m^{\circ} | \mathrm{m}_\alpha | \psi_n^{\circ} \rangle\right\} = -\mathrm{Im}\left\{\langle \psi_m^{\circ} | \mu_\alpha | \psi_n^{\circ} \rangle \langle \psi_n^{\circ} | \mathrm{m}_\alpha | \psi_m^{\circ} \rangle\right\}$, Equation 5.6 can be written as

$$
\begin{aligned}
[\phi(\lambda)] &= 91.5 \sum_{m \# n} \frac{\lambda_{mn}^2}{\lambda^2 - \lambda_{mn}^2} \frac{1}{2}\left[\mathrm{Im}\left\{\langle \psi_n^{\circ} | \mu_\alpha | \psi_m^{\circ} \rangle \langle \psi_m^{\circ} | \mathrm{m}_\alpha | \psi_n^{\circ} \rangle\right\} - \mathrm{Im}\left\{\langle \psi_m^{\circ} | \mu_\alpha | \psi_n^{\circ} \rangle \langle \psi_n^{\circ} | \mathrm{m}_\alpha | \psi_m^{\circ} \rangle\right\}\right] \\
&= \frac{91.5}{2} \sum_{m \# n} \lambda_{mn} \left[\frac{\mathrm{Im}\left\{\langle \psi_n^{\circ} | \mu_\alpha | \psi_m^{\circ} \rangle \langle \psi_m^{\circ} | \mathrm{m}_\alpha | \psi_n^{\circ} \rangle\right\}}{\lambda - \lambda_{mn}} + \frac{\mathrm{Im}\left\{\langle \psi_m^{\circ} | \mu_\alpha | \psi_n^{\circ} \rangle \langle \psi_n^{\circ} | \mathrm{m}_\alpha | \psi_m^{\circ} \rangle\right\}}{\lambda + \lambda_{mn}}\right]
\end{aligned}
\tag{5.7}
$$

To eliminate the singularity in the above equation, the damping terms are to be included in the denominator, yielding a complex quantity:

$$[\tilde{\phi}(\lambda)] = \frac{91.5}{2}\sum_{m\#n}\lambda_{mn}\left[\frac{\mathrm{Im}\{\langle\psi_n^o|\mu_\alpha|\psi_m^o\rangle\langle\psi_m^o|\mathrm{m}_\alpha|\psi_n^o\rangle\}}{\lambda-\lambda_{mn}-i\sigma_m}+\frac{\mathrm{Im}\{\langle\psi_m^o|\mu_\alpha|\psi_n^o\rangle\langle\psi_n^o|\mathrm{m}_\alpha|\psi_m^o\rangle\}}{\lambda+\lambda_{mn}+i\sigma_m}\right] \tag{5.8}$$

where σ_m is the *half-width*, at half the maximum height of the band representing the electronic transition from state n to m. On the right-hand side of the above equation, multiplying the first term with $\frac{\lambda-\lambda_{mn}+i\sigma_m}{\lambda-\lambda_{mn}+i\sigma_m}$ and the second term with $\frac{\lambda+\lambda_{mn}-i\sigma_m}{\lambda+\lambda_{mn}-i\sigma_m}$, one obtains

$$\begin{aligned}\left[\tilde{\phi}(\lambda)\right] &= \frac{91.5}{2}\sum_{m\#n}\lambda_{mn}\left[\frac{(\lambda-\lambda_{mn}+i\sigma_m)R_{mn}}{(\lambda-\lambda_{mn})^2+\sigma_m^2}-\frac{(\lambda+\lambda_{mn}-i\sigma_m)R_{mn}}{(\lambda+\lambda_{mn})^2+\sigma_m^2}\right]\\ &= \left[\phi(\lambda)\right]+i\left[\theta(\lambda)\right]\end{aligned} \tag{5.9}$$

In Equation 5.9, the real part $\phi(\lambda)$ gives the molar rotation while the imaginary part $\theta(\lambda)$ yields molar ellipticity (see Equations 3.44 and 3.45) representing ECD, in units of deg cm^2/dmol.

The SOS expression, useful for calculating molar rotation in the resonant region, can also be obtained in an alternate form as follows. The $\frac{1}{\lambda^2-\lambda_{mn}^2}$ term in Equation 5.6 can be written as $f + ig$ (Barron 2004), where

$$f = \frac{\lambda^2-\lambda_{mn}^2}{(\lambda^2-\lambda_{mn}^2)^2+(2\sigma_m\lambda)^2} \tag{5.10}$$

and

$$g = \frac{2\sigma_m\lambda}{(\lambda^2-\lambda_{mn}^2)^2+(2\sigma_m\lambda)^2} \tag{5.11}$$

where σ_m is the *half-width* at half the maximum height of the band, representing the transition from state n to m. The real part, f, contributes to molar rotation while the imaginary part contributes to ECD.

Another way of seeing the connection between ORD and ECD is through the Kramers–Kronig (KK) transform (see Appendix 2), which offers the means to convert ECD spectrum into ORD and vice versa (Moscowitz 1960) (Polavarapu 2005) (Rudolph and Autschbach 2008):

$$[\phi(\lambda)] = \frac{2}{\pi}\int_0^\infty [\theta(\mu)]\frac{\mu}{(\lambda^2-\mu^2)}\,d\mu \tag{5.12}$$

This relation is strictly valid only when ECD spectrum covering the entire region of electronic transitions is available. However, the calculation, or measurement, of rotational strengths of all electronic transitions is nearly impossible, especially when large molecules are dealt with. Therefore, the integral in Equation 5.12 is truncated to a finite wavelength region.

Recently, a complex propagator approach to obtain quantum mechanical prediction of ECD spectral intensity distribution and ORD simultaneously (Krykunov et al. 2006; Jiemchooroj and Norman 2007) in a single calculation at discrete wavelength intervals has been developed. This approach is particularly useful for large biomolecules, where the calculation of excitation energies and rotational strengths for hundreds of electronic transitions becomes unwieldy.

In summary, the quantum chemical calculations of ORD or SOR can be carried out using different theoretical approaches.

1. In the static method (Amos 1982), SOR can be calculated at only one wavelength that is much longer than that of first electronic transition.
2. In the linear response or dynamic method (Helgaker et al. 1994), the wavelength-dependent SOR can be calculated in the nonresonant regions.
3. SOS expression allows approximate prediction of ORD in the nonresonant regions, from the knowledge of rotational strengths of electronic transitions.
4. ORD can also be derived from ECD using the KK transform (Moscowitz 1960) (Polavarapu 2005) (Rudolph and Autschbach 2008).
5. Using the complex propagator approach (Jiemchooroj and Norman 2007) (Krykunov et al. 2006), both ORD and ECD can be calculated simultaneously at desired wavelength resolution, without the need for ECD spectral simulations.

The static, SOS, and KK methods are approximate in nature, so these methods are to be avoided. The dynamic and complex propagator approaches are the preferred ones. The dynamic method is available in most programs (Gaussian 09 2013; DALTON 2015; PSI4 2014; ADF 2014), while the complex propagator approach is available in selected programs (DALTON 2015).

5.2.1.1 Practical Considerations

Role of theoretical level. For predicting SORs, different theoretical levels have been used. A theoretical level is defined by the combination of chosen quantum chemical method and basis set.

The quantum chemical methods include (a) Hartree–Fock (HF), (b) complete active space self-consistent filed (CASSCF) (Polavarapu et al. 2000), (c) a variation–perturbation method including electronic correlation (Pericou-Cayere et al. 1998), (d) density functional theory (DFT) (Yabana and Bertsch 1999; Cheeseman et al. 2000; Grimme 2001; Srebro et al. 2011), and (e) coupled cluster (CC) (Ruud and Helgaker 2002; Crawford and Stephens 2008; Pedersen et al. 2004) methods. Predictions of specific rotations using HF or DFT can be obtained routinely using the standard programs, while other methods (such as CASSCF and CC) are handled by those trained in quantum chemical methods. The HF theory does not include the effects of electron correlation, while DFT attempts to recover these effects using density functionals. The DFT method is most widely used for routine calculations.

The choice of basis set plays an important role in the reliability of SOR predictions. While smaller basis sets such as 6–31G (d) have been used in the early stages of the development of SOR calculations, these basis sets are not considered to be reliable enough in the current times. Reliable SOR predictions generally require much larger basis sets such as 6–311++G(2d,2p), aug-cc-pVDZ, or aug-cc-pVTZ. Additional details on this topic can be found in recent reviews (Crawford 2012; Autschbach 2012; Vaccaro 2012; Autschbach et al. 2011).

Role of molecular vibrations. The accuracy of theoretical predictions for individual molecules could be tested using the SORs for vapor phase samples (Müller et al. 2000; Lahiri et al. 2014). These studies indicated that the inclusion of the influence of molecular vibrations on specific rotation is important. The influence of anharmonic potential energy surface (PES) and of nonlinear property surface could be studied by expressing SOR as a function of vibrational normal coordinates (Ruud et al. 2001; Wiberg et al. 2003; Ruud and Zanasi 2005; Mort and Autschbach 2005, 2006). However, low-frequency torsional vibrations showed extreme sensitivity toward SORs. Instead of omitting the contributions from such vibrations, these vibrations are treated separately by solving the Schrödinger equation for those low frequency vibrations separately (Crawford and Allen 2009; Mort and Autschbach 2008; Lahiri et al. 2012, 2015). The negative sign of SOR for (*R*)-(-)-carvone, measured in different solvents, could not be reproduced without the inclusion of vibrational corrections (Lambert et al. 2012).

5.2.2 Electronic Absorption and Circular Dichroism Spectra

Quantum chemical predictions of electronic absorption (EA) and ECD spectra can be obtained in two different ways:

1. In the first method, dipole strengths (see Chapter 4, Equation 4.129) or dimensionless oscillator strengths (see Chapter 4, Equation 4.130) and rotational strengths (see Chapter 4, Equation 4.127) associated with individual electronic transitions are calculated (Pedersen et al. 1999; Autschbach et al. 2002; Diedrich and Grimme 2003; Pecul et al. 2004; Stephens et al. 2004). For simulating the predicted EA spectra, dipole strengths or dimensionless oscillator strengths are converted to integrated molar absorption coefficients (see Chapter 4, Equation 4.140), and spectral distribution is simulated using a chosen band shape (see Appendix 3 for spectral simulation). For obtaining the ECD spectra, rotational strengths are converted to differential integrated molar absorption coefficients (see Chapter 4, Equation 4.141), and spectral distribution is simulated using a chosen band shape (see Appendix 3 for spectral simulation).

2. In the second method, instead of calculating the dipole and rotational strengths for individual transitions, it is possible to calculate the ECD spectral intensity distribution at desired wavelength resolution (Norman et al. 2004; Jiemchooroj and Norman 2007; Krykunov et al. 2006), as mentioned in the previous section on ORD, using a complex propagator approach. This approach requires assuming the lifetimes of excited states but avoids the need for spectral simulations. However, spectral simulation using dipole/rotational strengths and discrete wavelength interval spectra using complex propagator methods yield equivalent results only when Lorentzian spectral intensity distribution is used in the spectral simulations (because Equations 5.10 and 5.11 represent Lorentzian functions).

5.2.2.1 Practical Considerations

Role of theoretical level. The advances made in quantum chemical methods are now permitting routine predictions of EA and ECD spectra. However, due to the difficulties in representing the excited electronic states accurately, and the sensitivity of ECD predictions to the approximations involved in the method employed, the general trend is to use the highest level of affordable theory (in terms of computational resources available). Most studies are conducted with hydrid functionals such as B3LYP and Coulomb-attenuated functionals such as CAM-B3LYP. Larger basis sets are usually needed for reliable ECD predictions. The reader can find a detailed coverage on this topic in recent reviews (Crawford 2012; Autschbach 2012; Goerigk et al. 2012; Autschbach et al. 2011; Pescitelli and Bruhn 2016).

Role of molecular vibrations. The vibronic features become evident when experimental measurements are obtained for vapor phase samples. Much progress has been made in investigating the vibrational fine structure associated with EA and ECD bands within the Franck–Condon and Herzberg–Teller approaches (Neugebauer et al. 2005; Nooijen 2006; Pescitelli et al. 2013; Barone et al. 2014).

Role of temperature. When conformational equilibrium exists between different conformers, the temperature-dependent ECD spectra can be gainfully used to determine the enthalpy and entropy differences between the conformers. One such application was reported by measuring temperature-dependent ECD spectra for (3*R*)-methylcyclopentanone in 34 different solvents (Al-Basheer et al. 2007).

5.2.3 Vibrational Absorption and Circular Dichroism Spectra

The expressions for dipole strength and rotational strength, needed to predict vibrational absorption (VA) and vibrational circular dichroism (VCD) spectra, are similar to those for EA and ECD spectra except that the wavefunctions used here are for transitions that take place from a ground vibrational state to an excited vibrational state, both belonging to the same ground electronic state. These differences require additional considerations as described below.

The total wavefunction is written as a product of the electronic, vibrational, and rotational wavefunctions, $\psi_n^{el}\psi_v^n\psi_r^{n,v}$, where ψ_n^{el} is the wavefunction for the nth electronic state, ψ_v^n is the wavefunction for the vth vibrational state belonging to the nth electronic state, and $\psi_r^{n,v}$ is the rotational wavefunction belonging to the nth electronic and vth vibrational states. Although the rotational-vibrational transitions can be seen in the vapor phase VCD spectra (Polavarapu 1989a), this is not so for liquid samples. Therefore, for liquid samples, the rotational states accompanying a vibrational transition are not resolved, and the rotational wavefunction will only serve the purpose of averaging over all orientations (Wilson et al. 1980) of an isotropic sample. Then, the total wavefunction can be replaced with the product $\psi_n^{el}\psi_v^n$. The vibrational wavefunction is written as the product of (3N-6) harmonic oscillator wavefunctions ϕ_i. That is,

$$\psi_v^n = \varphi_1\varphi_2\varphi_3..\varphi_{(3N-6)} = \prod_{i=1}^{(3N-6)} \varphi_i \tag{5.13}$$

The electric dipole moment operator can be expanded in dimensionless vibrational normal coordinates, q_i as

$$\hat{\mu}_\alpha = \hat{\mu}_\alpha^0 + \sum_i \frac{\partial \mu_\alpha}{\partial q_i}\hat{q}_i + ... \tag{5.14}$$

Using this expansion and the above-mentioned product wavefunction, $\psi_n^{el}\psi_v^n$, the electric dipole transition moment integral becomes

$$\begin{aligned}
\mu_{\alpha,vv'} &= \left\langle \psi_n^{el}\psi_{v'}^n \middle| \hat{\mu}_\alpha \middle| \psi_n^{el}\psi_v^n \right\rangle = \left\langle \psi_n^{el}\psi_{v'}^n \middle| \hat{\mu}_\alpha^0 + \sum_i \frac{\partial \mu_\alpha}{\partial q_i}\hat{q}_i + .. \middle| \psi_n^{el}\psi_v^n \right\rangle \\
&= \left\langle \psi_n^{el}\psi_{v'}^n \middle| \hat{\mu}_\alpha^0 \middle| \psi_n^{el}\psi_v^n \right\rangle + \left\langle \psi_n^{el}\psi_{v'}^n \middle| \sum_i \frac{\partial \mu_\alpha}{\partial q_i}\hat{q}_i \middle| \psi_n^{el}\psi_v^n \right\rangle + .. \\
&= \left\langle \psi_{v'}^n \middle| \psi_v^n \right\rangle \left\langle \psi_n^{el} \middle| \hat{\mu}_\alpha^0 \middle| \psi_n^{el} \right\rangle + \sum_i \frac{\partial \mu_\alpha}{\partial q_i} \left\langle \psi_n^{el}\psi_{v'}^n \middle| \hat{q}_i \middle| \psi_n^{el}\psi_v^n \right\rangle + ...
\end{aligned} \tag{5.15}$$

The first term on the right-hand side of the above equation is zero due to the orthogonality of vibrational wavefunctions. The second term can be written as

$$\begin{aligned}
\sum_i \frac{\partial \mu_\alpha}{\partial q_i} \left\langle \psi_n^{el}\psi_{v'}^n \middle| \hat{q}_i \middle| \psi_n^{el}\psi_v^n \right\rangle &= \sum_i \left(\frac{\partial \mu_\alpha}{\partial q_i} \right) \left\langle \psi_n^{el} \middle| \psi_n^{el} \right\rangle \left\langle \psi_{v'}^n \middle| \hat{q}_i \middle| \psi_v^n \right\rangle \\
&= \sum_i \left(\frac{\partial \mu_\alpha}{\partial q_i} \right) \left\langle \varphi_1'\varphi_2'...\varphi_{(3N-6)}' \middle| \hat{q}_i \middle| \varphi_1\varphi_2...\varphi_{(3N-6)} \right\rangle
\end{aligned} \tag{5.16}$$

When one vibrational normal mode is excited and others are in a ground state, the right-hand side of the above equation can be simplified. For kth mode in vibrational excited state,

$$\sum_i \left(\frac{\partial \mu_\alpha}{\partial q_i}\right)\left\langle \varphi_1'\varphi_2'\ldots\varphi_{(3N-6)}' \middle| \hat{q}_i \middle| \varphi_1\varphi_2\ldots\varphi_{(3N-6)} \right\rangle = \left(\frac{\partial \mu_\alpha}{\partial q_k}\right)\left\langle \varphi_k' \middle| \hat{q}_k \middle| \varphi_k \right\rangle \tag{5.17}$$

The summation in the above equation disappears, because $\langle \varphi_i | q_i | \varphi_i \rangle = 0$. The right-hand side of the above equation is nonzero when the vibrational quantum numbers of the initial and excited states of kth vibration differ by 1. Designating φ_k along with its vibrational quantum number, υ_k, as $\varphi_k^{\upsilon_k}$ and φ_k' as $\varphi_k^{\upsilon_k+1}$, the electric dipole transition moment integral for υ_k to υ_k+1 vibrational transition becomes

$$\mu_{\alpha,k} = \left(\frac{\partial \mu_\alpha}{\partial q_k}\right)\left\langle \varphi_k^{\upsilon_k+1} \middle| \hat{q}_k \middle| \varphi_k^{\upsilon_k} \right\rangle = \left(\frac{\partial \mu_\alpha}{\partial q_k}\right)\sqrt{\left(\frac{\upsilon_k+1}{2}\right)} \tag{5.18}$$

and the dipole strength (see Chapter 4, Equation 4.129) for kth vibrational transition becomes

$$D_{\upsilon_k,\upsilon_k+1} = \mu_k \bullet \mu_k = \left(\frac{\partial \mu}{\partial q_k}\right)^2 \left(\frac{\upsilon_k+1}{2}\right) \tag{5.19}$$

Using this expression for dipole strength, the change in intensity of light with pathlength (see Chapter 4, Equation 4.135) becomes

$$\begin{aligned} \frac{dI(\nu)}{I(\nu)} &= -(h\nu)\times\frac{2\pi}{3c\hbar^2} D_{\upsilon_k,\upsilon_k+1} n(\nu)\,dl \\ &= -(h\nu)\times\frac{\pi}{3c\hbar^2}\left(\frac{\partial \mu}{\partial q_k}\right)^2 (\upsilon_k+1) n(\nu)\,dl \end{aligned} \tag{5.20}$$

where $n(\nu)$ is the number density of molecules (number of molecules per unit volume) absorbing at ν. Unlike in electronic states (where all molecules can be considered to be in the ground electronic state), the difference in energies of vibrational levels need not be large enough to have all molecules in the ground vibrational state. For this reason, $n(\nu)$ is to be replaced with the difference, $n_{\upsilon_k}(\nu) - n_{\upsilon_k+1}(\nu)$, where $n_{\upsilon_k}(\nu)$ and $n_{\upsilon_k+1}(\nu)$, respectively, represent the number density of molecules in a vibrational state with quantum number υ_k and υ_k+1:

$$\frac{dI(\nu)}{I(\nu)} = -(h\nu)\times\frac{\pi}{3c\hbar^2}\left(\frac{\partial \mu}{\partial q_k}\right)^2 (\upsilon_k+1)\left(n_{\upsilon_k}(\nu) - n_{\upsilon_k+1}(\nu)\right) dl \tag{5.21}$$

In the harmonic oscillator approximation, transitions between any two successive vibrational levels will appear at the same frequency as that between the two lowest vibrational levels. Thus, a fundamental VA band, with $\upsilon_k = 0$, in the experimental spectrum should also be considered to contain hot transitions with $\upsilon_k = 1, 2, ..,$ and so on besides the fundamental transition. Then, $(\upsilon_k + 1)(n_{\upsilon_k}(\nu) - n_{\upsilon_k+1}(\nu))$ needs to be summed over all υ_k:

$$\sum_{\upsilon_k=0}^{\infty} (\upsilon_k + 1)(n_{\upsilon_k}(\nu) - n_{\upsilon_k+1}(\nu)) = n(\nu) \sum_{\upsilon_k=0}^{\infty} (\upsilon_k + 1)(x_{\upsilon_k} - x_{\upsilon_k+1}) \quad (5.22)$$

In this equation, $n_{\upsilon_k}(\nu) = x_{\upsilon_k} n(\nu)$ and $n_{\upsilon_k+1}(\nu) = x_{\upsilon_k+1} n(\nu)$ with x_{υ_k} and x_{υ_k+1} representing the fractional number densities of molecules in vibrational states with quantum number υ_k and $\upsilon_k + 1$ has been used. It can be easily seen that $\sum_{\upsilon_k=0}^{\infty} (\upsilon_k + 1)(x_{\upsilon_k} - x_{\upsilon_k+1}) = 1$, because

$$\begin{aligned} \sum_{\upsilon_k=0}^{\infty} (\upsilon_k + 1)(x_{\upsilon_k} - x_{\upsilon_k+1}) &= 1 \times (x_0 - x_1) + 2 \times (x_1 - x_2) + 3 \times (x_2 - x_3) + .. \\ &= x_0 + x_1 + x_2 + ... = 1. \end{aligned} \quad (5.23)$$

Thus, for υ_k to $\upsilon_k + 1$ vibrational transition, Equation 5.21 becomes

$$\frac{dI(\nu)}{I(\nu)} = -(h\nu) \times \frac{\pi}{3c\hbar^2} \left(\frac{\partial \mu}{\partial q_k} \right)^2 n(\nu)\, dl \quad (5.24)$$

Comparing this equation with, $dI(\nu) = -\alpha(\nu) I(\nu) C dl$ and integrating (see Chapter 4, Equations 4.136 through 4.140), the integrated molar absorption coefficient becomes

$$A_k = \int \alpha(\nu) d\nu = \frac{8\pi^3 N_A \nu_k^0}{3ch} D_k \quad (5.25)$$

This equation gives the integrated molar absorption coefficient in units of $cm^2 mol^{-1} s^{-1}$. However, the experimental infrared spectra are presented as a function of wavenumber, $\bar{\nu}$, and integrated molar absorption coefficients are expressed in units of km mol^{-1}. In that case, the above equation is modified as

$$A_k = \int \alpha(\bar{\nu}) d\bar{\nu} = \frac{8\pi^3 N_A \nu_k^0}{3c^2 h} D_k = \frac{8\pi^3 N_A \bar{\nu}_k^0}{3ch} D_k \quad (5.26)$$

where $\bar{\nu}_k^0$ is the center position of kth vibrational band in cm^{-1} (see Appendix 8 for Unit conversions). For some applications, it is necessary to extract the values of electric dipole moment derivatives. For such cases, the substitution of Equation 5.19 for fundamental transition in Equation 5.26 gives

$$A_k = \frac{4\pi^3 N_A \bar{\nu}_k^0}{3ch} \left(\frac{\partial \mu}{\partial q_k} \right)^2 \quad (5.27)$$

Using the conversion from dimensionless normal coordinate q_k to normal coordinate Q_k,

$$q_k^2 = \frac{4\pi^2 c\bar{\nu}_k^0}{h} Q_k^2 = \alpha_k Q_k^2 \tag{5.28}$$

$$A_k = \frac{4\pi^3 N_A \bar{\nu}_k^0}{3ch} \times \frac{h}{4\pi^2 c\bar{\nu}_k^0} \times \left(\frac{\partial \mu}{\partial Q_k}\right)^2 = \frac{N_A \pi}{3c^2}\left(\frac{\partial \mu}{\partial Q_k}\right)^2 \tag{5.29}$$

Using the appropriate constants, one obtains the simple relation for A_k in the units of km mol^{-1} (see Appendix 8) as

$$A_k = 42.3 \times \left(\frac{\partial \mu}{\partial Q_k}\right)^2 \tag{5.30}$$

Proceeding in the same manner (see Chapter 4, Equation 4.141), the integrated VCD intensity associated with the $\upsilon_k = 0-1$ transition of the kth vibration is obtained as

$$\Delta A_k = \int \{\alpha^+(\bar{\nu}) - \alpha^-(\bar{\nu})\} d\bar{\nu} = 4 \times \left(\frac{8\pi^3 \bar{\nu}_k^0 N_A}{3hc}\right) R_k \tag{5.31}$$

To obtain the relation with electric and magnetic dipole moment derivatives, the following relation (Polavarapu 1998) can be used for $\upsilon_k = 0-1$ transition of kth vibration:

$$R_k = \mathrm{Im}[\mu_k \bullet \mathbf{m}_k] = \frac{h}{4\pi}\left(\frac{\partial \mu_\alpha}{\partial Q_k}\right)\left(\frac{\partial m_\alpha}{\partial \dot{Q}_k}\right) \tag{5.32}$$

where the magnetic dipole moment is expanded in terms of normal coordinate velocities, $\dot{Q}_k$. Then,

$$\begin{aligned}\Delta A_k &= \int \{\alpha^+(\bar{\nu}) - \alpha^-(\bar{\nu})\} d\bar{\nu} = 4 \times \left(\frac{8\pi^3 \bar{\nu}_k^0 N_A}{3hc}\right) \times \frac{h}{4\pi}\left(\frac{\partial \mu_\alpha}{\partial Q_k}\right)\left(\frac{\partial m_\alpha}{\partial \dot{Q}_k}\right) \\ &= \frac{8\pi^2 \bar{\nu}_k^0 N_A}{3c}\left(\frac{\partial \mu_\alpha}{\partial Q_k}\right)\left(\frac{\partial m_\alpha}{\partial \dot{Q}_k}\right)\end{aligned} \tag{5.33}$$

The sum of VCD intensities associated with all fundamental vibrations is related to the permanent electric dipole moment (Polavarapu 1986a), while the frequency-weighted sum is related to inverse force constant matrix elements (Polavarapu 1987a).

In practice, VA and VCD calculations are carried out in two steps: (1) evaluate the Cartesian derivatives, $\left(\frac{\partial \mu_\alpha}{\partial X_{A\beta}}\right)$ and $\left(\frac{\partial m_\alpha}{\partial \dot{X}_{A\beta}}\right)$, where $X_{A\beta}$ is the βth component of the displacement vector of atom A and $\dot{X}_{A\beta}$ is the corresponding velocity component. $\left(\frac{\partial \mu_\alpha}{\partial X_{A\beta}}\right)$ are called the atomic polar tensor (APT) elements (Person and Newton 1974); and $\left(\frac{\partial m_\alpha}{\partial \dot{X}_{A\beta}}\right)$ are called the atomic axial tensor (AAT) elements (Cheeseman et al. 1996) and (2) transform these Cartesian derivatives to normal coordinates using **S** vectors (Wilson et al. 1980):

$$\frac{\partial \mu_\alpha}{\partial Q_k} = \sum_{A,\beta} \left(\frac{\partial \mu_\alpha}{\partial X_{A\beta}}\right)\left(\frac{\partial X_{A\beta}}{\partial Q_k}\right) = \sum_{A,\beta} \left(\frac{\partial \mu_\alpha}{\partial X_{A\beta}}\right) S^k_{A\beta} \tag{5.34}$$

$$\begin{aligned}\frac{\partial m_\alpha}{\partial \dot{Q}_k} &= \sum_{A,\beta}\left(\frac{\partial m_\alpha}{\partial \dot{X}_{A\beta}}\right)\left(\frac{\partial \dot{X}_{A\beta}}{\partial \dot{Q}_k}\right) = \sum_{A,\beta}\left(\frac{\partial m_\alpha}{\partial \dot{X}_{A\beta}}\right)\left(\frac{\partial X_{A\beta}}{\partial Q_k}\right)\\ &= \sum_{A,\beta}\left(\frac{\partial m_\alpha}{\partial \dot{X}_{A\beta}}\right) S^k_{A\beta}\end{aligned} \tag{5.35}$$

The **S** vectors are obtained by diagonalizing the mass-weighted Cartesian force constant matrix. It is necessary that these **S** vectors be obtained at the geometry that corresponds to the minimum on the PES. In principle, it is possible to do the calculations for APTs and AATs, at a different theoretical level than that used for **S** vectors, although it is not advised or commonly practiced.

One of the vexing problems in the early stages of VCD development was that in the Born–Oppenheimer approximation, the electronic contribution to the magnetic dipole transition moment, for transitions among vibrational states that belong to the ground electronic state, vanishes. Since m is an imaginary operator, the hermiticity property requires $\langle \psi^{el}_n | \hat{m}^{el}_\alpha | \psi^{el}_n \rangle = \left[\langle \psi^{el}_n | \hat{m}^{el}_\alpha | \psi^{el}_n \rangle\right]^* = 0$, for molecules in nondegenerate electronic states. This problem has now been resolved using the magnetic field perturbation (MFP) method, where the wavefunction perturbed by a magnetic field is needed. These advances (Galwas 1983; Stephens 1985; Buckingham et al. 1987) paved the way for successful predictions of VCD spectra in the mid-infrared region. Both commercial and freeware quantum mechanical programs (Gaussian09 2013; ADF 2014; DALTON 2015; PSI4 2014) are available for predicting dipole strengths and rotational strengths for fundamental vibrational transitions using the MFP approach. These programs are designed to be simple enough for nonspecialists to use. Some savings in computational time can be realized using an alternate formula for the Cartesian nuclear displacement derivatives of magnetic dipole moment as the frequency derivative at zero frequency of a linear response function involving nuclear displacements and magnetic field (Coriani et al. 2011).

An alternate approach to regaining an electronic contribution to the magnetic dipole transition moment is to incorporate corrections to Born–Oppenheimer ground-state orbitals using vibronic coupling (VC) theory (Nafie and Freedman 1983; Nafie 2004). This approach, recently implemented (Scherrer et al. 2013) using DFT and referred to as the nuclear velocity perturbation method, was stated to give results equivalent to those obtained with the MFP method.

The predicted dipole strengths and rotational strengths are converted, respectively, to integrated absorption and circular dichroism intensities, as described earlier. It is inconvenient to experimentally determine the integrated intensities for all fundamental vibrational bands, and therefore, the predicted integrated intensities cannot be easily compared with corresponding experimental quantities. Instead, the predicted integrated intensities are converted to predicted spectra using spectral simulations (Appendix 3) with the assumption of band shapes (usually Lorentzian), and these predicted spectra are compared (see Chapter 8) with the experimentally observed spectra.

5.2.3.1 Practical Considerations

Role of theoretical level. The advances made in quantum chemical methods are now permitting routine predictions of VA and VCD spectra. Since excited electronic states are not involved, except for situations such as inorganic complexes with low-lying electronic states, VCD calculations in general can be carried out at lower levels of theory than those used for other chiroptical spectroscopic methods. Even though the calculations for vibrational normal modes and atomic polar and axial tensors can be carried out at different levels of theory, this is not the general practice for VCD predictions. Extensive investigations have suggested that a combination of B3LYP functional with a 6–31G* basis set is the minimal level required for VCD predictions. Sufficiently higher accuracy predictions require the use of B3LYP or B3PW91 functional in combination with TZVP or cc-pVTZ basis sets (Stephens et al. 2008). Additional details can be found in recent reviews (Ruud 2012).

Most VCD calculations have been undertaken in the harmonic oscillator approximation. Due to the absence of anharmonic contributions, the calculated and experimental vibrational frequencies often differ and a frequency scaling factor is often used. Recent developments to incorporate anharmonic contributions to both vibrational frequencies as well as VCD intensities are facilitating the VCD predictions much more precisely (Barone et al. 2014; Bloino and Barone 2012).

Role of electronic transitions. The low-lying electronic transitions can influence the VCD intensities of certain bands, and this influence is considered to originate from VC. Such studies are beginning to emerge in recent years (Domingos et al. 2014; Merten et al. 2012; He et al. 2001).

5.2.4 Vibrational Raman and Raman Optical Activity spectra

For vibrational Raman scattering, as for VCD, initial and final vibrational states belong to the same ground electronic state. However, the nature of the scattering process involves the influence of higher electronic states that appear indirectly through appropriate molecular polarizability derivatives.

Assume that the total wave function can be written as a product of electronic, vibrational, and rotational wave functions (see Section 5.2.3), and that the rotational part of the integrals can be replaced with classical averaging for all molecular orientations. Then, the transition electric dipole polarizability can be written as $\langle \psi^n_{\upsilon'} | \hat{\alpha}_{\alpha\beta} | \psi^n_{\upsilon} \rangle$, where superscript n represents the ground electronic state and subscripts ν and ν' represent the vibrational states; $\alpha_{\alpha\beta}$ is the electric dipole–electric dipole polarizability tensor (see Chapter 4, Equations 4.39, 4.67, and 4.68); and the caret over $\alpha_{\alpha\beta}$ indicates the operator property. Expanding $\hat{\alpha}_{\alpha\beta}$ in a Taylor series as a function of dimensionless normal coordinates, q_i,

$$\hat{\alpha}_{\alpha\beta} = \hat{\alpha}^0_{\alpha\beta} + \sum_i \left(\frac{\partial \alpha_{\alpha\beta}}{\partial q_i} \right) \hat{q}_i + \ldots \tag{5.36}$$

and writing the vibrational wavefunction as the product of harmonic oscillator wave functions (see Section 5.2.3), the transition electric dipole polarizability integral is written as

$$\langle \psi^n_{\upsilon'} | \hat{\alpha}_{\alpha\beta} | \psi^n_{\upsilon} \rangle = \left(\frac{\upsilon_i + 1}{2} \right)^{\frac{1}{2}} \left(\frac{\partial \alpha_{\alpha\beta}}{\partial q_i} \right) \tag{5.37}$$

The dimensionless normal coordinates, q_i, can be converted to normal coordinate Q_i using Equation 5.28. Vibrational Raman spectral intensities are determined by the mean and anisotropy of the polarizability derivative tensor $\left(\frac{\partial \alpha_{\alpha\beta}}{\partial Q_i} \right)$ (see Chapter 3, Equations 3.52 and 3.53). The prediction of spectral intensities in the vibrational Raman optical activity (VROA) spectra requires (see Chapter 3, Equations 3.57 through 3.61) three different polarizability derivative tensors, namely $\left(\frac{\partial \alpha_{\alpha\beta}}{\partial Q_i} \right)$, $\left(\frac{\omega^{-1} \partial G'_{\alpha\beta}}{\partial Q_i} \right)$, and $\left(\frac{\partial A_{\alpha\beta\gamma}}{\partial Q_i} \right)$, where $\omega^{-1} G'_{\alpha\beta}$ and $A_{\alpha\beta\gamma}$ are, respectively, the electric dipole–magnetic dipole and electric dipole–electric quadrupole polarizability tensors. The quantum mechanical expressions for

$\alpha_{\alpha\beta}$ and $\omega^{-1}G'_{\alpha\beta}$ are given, respectively, by Equations 4.67 and 4.77, while that for $A_{\alpha\beta\gamma}$ is as follows:

$$A_{\alpha\beta\gamma} = \frac{4\pi}{h}\sum_{n\neq m}\frac{\omega_{mn}}{\omega_{mn}^2-\omega^2}\operatorname{Re}\left\{\left\langle\psi_n^o\middle|\hat{\mu}_\alpha\middle|\psi_m^o\right\rangle\left\langle\psi_m^o\middle|\hat{\Theta}_{\beta\gamma}\middle|\psi_n^o\right\rangle\right\} \tag{5.38}$$

In Equation 5.38, $\hat{\Theta}$ is the electric quadrupole moment operator. In practice, one determines the Cartesian polarizability derivatives $\left(\frac{\partial\alpha_{\alpha\beta}}{\partial X_A}\right)$, $\left(\frac{\omega^{-1}\partial G'_{\alpha\beta}}{\partial X_A}\right)$, and $\left(\frac{\partial A_{\alpha\beta\gamma}}{\partial X_A}\right)$, where X_A is a Cartesian displacement of atom A. These Cartesian polarizability derivatives, referred to as Cartesian Raman and optical activity tensors, satisfy interesting translational and rotational sum rules (Polavarapu 1982; Polavarapu 1990b). The Cartesian derivatives are then converted to normal coordinate derivatives, $\left(\frac{\partial\alpha_{\alpha\beta}}{\partial Q_i}\right)$, $\left(\frac{\omega^{-1}\partial G'_{\alpha\beta}}{\partial Q_i}\right)$, and $\left(\frac{\partial A_{\alpha\beta\gamma}}{\partial Q_i}\right)$, using **S** vectors (Wilson et al. 1980). These normal coordinate derivative tensors are then used with Equations 3.57 through 3.62 to calculate vibrational Raman and VROA intensities and spectra simulated using chosen band shapes (see Appendix 3). The sum of frequency-weighted vibrational Raman/VROA intensities over all fundamental vibrations is related to the inverse force constant matrix elements (Polavarapu 1987b).

Analytic quantum mechanical procedures to evaluate the above-mentioned derivatives were not available in the early stages. For this reason, all three derivative tensors were evaluated numerically by calculating the tensors $\alpha_{\alpha\beta}$, $\omega^{-1}G'_{\alpha\beta}$, and $A_{\alpha\beta\gamma}$ at equilibrium and displaced nuclear geometries. The VROA calculation with numerical derivatives can be carried out in two parts: one part involving the calculation of **S** vectors and another part involving the calculation of Cartesian polarizability derivative tensors, $\left(\frac{\partial\alpha_{\alpha\beta}}{\partial X_A}\right)$, $\left(\frac{\omega^{-1}\partial G'_{\alpha\beta}}{\partial X_A}\right)$, and $\left(\frac{\partial A_{\alpha\beta\gamma}}{\partial X_A}\right)$. These two parts can be carried out at two different theoretical levels, if desired. The first part involving the calculation of **S** vectors is done at a theoretical level where molecular geometry is also optimized. For the second part, a different theoretical level may be used, but the same molecular geometry as that used in the first part must be used.

The first ab initio VROA calculations (Bose et al. 1989) (Polavarapu 1990b) used this approach by evaluating $\omega^{-1}G'_{\alpha\beta}$ in the static limit (Amos 1982) and non-London orbitals.

The gauge origin independence and wavelength dependence for derivative tensors $\left(\frac{\omega^{-1}\partial G'_{\alpha\beta}}{\partial X_A}\right)$ were later achieved (Helgaker et al. 1994) using the London orbitals and the dynamic method. Since then several VROA calculations have been performed using this method at the HF level, although sometimes normal coordinates obtained at the density functional theoretical level were used. However, recent implementation (Ruud et al. 2002) of DFT methods for calculating the needed tensors made it possible to predict VROA spectra with improved accuracy. The absolute configuration of bromochlorofluoromethane was determined for the first time in this manner (Costante et al. 1997; Polavarapu 2002). The availability of analytic approaches (Frisch et al. 1986; Amos 1986; Quinet and Champagne 2001; Quinet et al. 2005; Liégeois et al. 2007) to calculate the derivatives $\left(\frac{\partial\alpha_{\alpha\beta}}{\partial X_A}\right)$, $\left(\frac{\omega^{-1}\partial G'_{\alpha\beta}}{\partial X_A}\right)$, and $\left(\frac{\partial A_{\alpha\beta\gamma}}{\partial X_A}\right)$ has provided a huge savings of computer time and facilitated wider applications for VROA. Improved accuracy in the predictions has also been offered by the implementation of the CC method (Crawford and Ruud 2011). Even though the analytic methods expedite the VROA calculations, the numerical method of evaluating the needed derivative tensors is still useful for dealing with large molecules, because calculations at displaced geometries can be parceled out to different nodes (Simmen et al. 2012) and carried out in parallel (if access to hundreds of nodes is not an issue). The recent reviews (Ruud and Thorvaldsen 2009; Parchansky et al. 2014; Mutter, Zielinski, Popelier, et al. 2015) summarize the developments in VROA predictions.

5.2.4.1 Practical Considerations

Role of theoretical level. The advances made in quantum mechanics are now permitting routine predictions of vibrational Raman and VROA spectra. Unlike in the case of VCD, higher electronic states are involved in the Raman scattering process, and therefore demands for reliable prediction of VROA are much higher than those for VCD.

While initial calculations were carried out using the HF method (Bose, Barron, and Polavarapu 1989) and a few calculations using multiconfiguration self-consistent field (Pecul and Rizzo 2003) and CC (Crawford and Ruud 2011) methods have also been reported, most current VROA calculations are carried out using DFT. Investigation of various functionals suggested a preference for the B3LYP functional (Reiher et al. 2005; Daněček et al. 2007).

It was found that the use of a special form of the 3–21++G basis set with diffuse *p* functions on hydrogens, with an exponent of 0.2, referred to as rarefied diffuse polarization function and shell augmented (rDPS), is essential and contribution of the core electrons is not important to accurately predict the VROA spectra (Zuber and Hug 2004). A benchmark study (Cheeseman and Frisch 2011) indicated that the basis set requirements for VROA tensors are different from those for vibrational normal modes, and the combination of aug(sp)-cc-pVDZ (which is the aug-cc-pVDZ basis set after removing diffuse *d* functions on all atoms, while keeping the diffuse *s* and *p* sets) and cc-pVTZ, respectively, for VROA tensors and vibrational normal modes, is appropriate for intermediate sized molecules. For large size molecules, corresponding recommended combination was rDPS and 6–31G*. Additional details can be found in recent reviews (Ruud 2012).

5.2.4.2 Resonance ROA Spectra

The theory of resonance VROA involving a single electronic excited state has been developed (Nafie 1996). Ab initio computational methods to predict resonance VROA have also been developed (Jensen et al. 2007). The influence of a second resonant electronic state was also investigated (Luber et al. 2010). A near-resonance theory, which includes vibronic levels, has been developed more recently (Nafie 2008).

5.2.4.3 Surface-Enhanced VROA Spectra

As the experimental methods to measure surface-enhanced VROA are being pursued, theoretical methods to predict surface-enhanced VROA are also being developed (Novák et al. 2012; Chulhai and Jensen 2014).

5.2.5 Role of Solvation on Chiroptical Spectra

When experimental spectra measured for liquid solutions are analyzed, it is necessary for the theoretical calculations to account for the solvent influence. Solvents can be divided into three categories: (a) nonpolar solvents that do not participate in hydrogen bonding (e.g., CCl_4), (b) polar solvents that do not participate in hydrogen bonding (e.g., $CHCl_3$), and (c) polar solvents that do participate in hydrogen bonding (e.g., CH_3OH, H_2O and $(CH_3)_2SO$). For the first two categories, the solvent influence can be incorporated implicitly. For the third category, however, the explicit solvation model is necessary.

Implicit solvation models: The implicit solvation can be incorporated using a polarizable continuum model (PCM), which represents the solvent as a dielectric cavity, (Mennucci et al. 2002) or using the conductor-like screening model (COSMO), which represents the solvent as a conductor (Klamt and Schuurmann 1993). For polar solvents that do not participate in hydrogen bonding, the implicit solvation model is often sufficient. These implicit solvation models, however, fail to mimic the hydrogen bonding influence that is prevalent when solvents capable of hydrogen bonding with solute molecules are used for experimental measurements.

Explicit solvation models: In the explicit solvation model, a certain limited number of solvent molecules, sufficient to represent the direct interaction with the solute molecule, can be placed around the solute molecule, and the whole cluster of these molecules is used for quantum chemical calculation of a chosen chiroptical property. However, since one can orient the solvent molecules in numerous ways around the solute molecule, a large number of solvent orientations are possible. Optimization of structures for all these orientations is necessary, which can be very tedious. To incorporate long-range solvation effects, this cluster can then be treated with PCM/COSMO. It is hoped that this combination of explicit and implicit solvation models can help realize the full solvation effect.

However, a few solvent molecules to represent the direct interaction with solute may not always be sufficient, especially when spectra are measured in aqueous solutions. In such cases, inclusion of enough solvent molecules to represent a complete solvation sphere may be necessary. Since the quantum chemical calculations on such large solute–solvent clusters is a time-intensive process, hybrid calculations that treat the solute molecule at the quantum chemical level and solvent molecules at the molecular mechanics level have been developed. These methods are referred to as QM/MM approaches.

The ambiguities in generating ad hoc orientations of the solvent molecules around solute can be avoided by undertaking the molecular dynamics (MD) simulations of solute molecule in a cage of solvent molecules, use the time-evolving snapshots of molecular structures to calculate the desired chiroptical property at the QM/MM level and take an appropriate average of these values. While this approach has been adopted for incorporating solvent effects on SOR and ECD properties (Mukhopadhyay et al. 2007; Kundrat and Autschbach 2008, 2009), the same procedure for VCD and VROA is problematic. This is because vibrational frequencies and normal modes have the proper meaning only when the geometry used is at the minimum of PES. The geometries of snapshots obtained in MD simulations may not correspond to the minimum of PES, and therefore some of the vibrational frequencies calculated at those geometries can be imaginary. One way to get around this problem is to partially optimize the snapshot structures of the solute by allowing bond lengths and angles (but not dihedral angles) to vary, do the vibrational frequency calculations at these partially optimized geometries, and exclude the structures that have imaginary vibrational frequencies. When a large number of solvent molecules are involved, it will be unrealistic to perform vibrational calculation for the entire cluster, so the QM/MM approach (Cheeseman et al. 2011; Mutter, Zielinski, Cheeseman, et al. 2015; Urago et al. 2014; Zielinski et al. 2015) is often used.

Cho and coworkers (Yang and Cho 2009) used MD simulations to predict VCD spectra in a different way by Fourier transforming the cross-correlation function of the time-dependent electric and magnetic dipole moments. This approach was also used to predict VCD for a α-helical peptide in the THz frequency (30–170 cm^{-1}) range (Choi and Cho 2014).

The methods used for representing the solvation effects in all four branches of chiroptical spectroscopy were reviewed (Pecul 2012).

The influence of various solvents on the SOR of (–)-α-methylbenzylamine (Fischer et al. 2006) and (–)-carvone (Lambert et al. 2012) has been investigated in terms of the solvent parameters. The influence of solvent on the ECD spectra of (+)-carvone (Lambert et al. 2012) was investigated by comparing the spectra measured for both gas and solution phases.

Methyloxirane is one of the few molecules that received broader attention as its experimental SOR in water solvent could not be satisfactorily reproduced by theoretical predictions. The solute–solvent clusters are believed to be responsible for the experimental observations and treating the system with an ad hoc number of surrounding water molecules was not considered to be sufficient. More recently, Barone and coworkers reported a novel approach in treating the methyloxirane–water system (Lipparini et al. 2013). The solute molecule was treated at the DFT CAM-B3LYP level with the aug-cc-pVDZ basis set, and MD snapshots of solute–water clusters were analyzed. After extracting the solute–solvent cluster structures from thousands of MD snapshots, within 12–16 Angstroms radius of the sphere from the center of the solute molecule, SORs were calculated for the extracted cluster structures using the QM/FQ/PCM approach (where FQ represents floating charges). Vibrational contributions of solute molecule to a specific rotation were then added to the averaged specific rotations for cluster structures. In this way, they were able to obtain a good agreement between the experimental and calculated ORD of methyloxirane in water. This study demonstrated the care needed in modeling the solvent environment for explaining the observed SORs.

The effect of solvation on the ECD and SOR of austdiol, a fungal metabolite, in methanol was investigated recently (Tedesco et al. 2014). This molecule has two OH groups that can have short-range interactions via hydrogen bonding with the solvent. These short-range interactions could not be described by implicit solvation models. While standard calculations without solvation dynamics were found not to reproduce the experimental ECD spectrum, the calculated ECD spectrum obtained as an average of 100 clusters taken from ab initio MD trajectory was found to closely reproduce the experimental ECD spectrum. This study provides a good example for the importance of properly describing the solvation effects on ECD spectra.

Implicit solvation models were used in VCD studies on organic molecules in nonhydrogen bonding solvents. Explicit solvation models were used for VCD studies on organic molecules in solvents capable of forming hydrogen bonds. For amino acids and peptides, where H_2O or D_2O was the solvent, explicit solvation models, in combination with MD simulations, were used. The recent reviews can be consulted for individual molecules investigated (Pecul 2012; Perera et al 2016). In recent studies that included the explicit solvation model for dibromonaphthol in $(CD_3)_2SO$ solvent and an adduct of dimethylfumarate with anthracene in $CDCl_3$, the importance of the dynamic nature of the solvent in the solute–solvent clusters has been emphasized (Heshmat et al. 2014; Passarello et al. 2014). Specifically, solvent molecules in the direct vicinity of solute are not fixed in their positions but fluctuate. This is reflected by the shallow PES for the parameters representing the distance between solvent and solute molecules and relative orientation of the solvent molecule with respect to the solute molecule. These solvent fluctuations appear as large amplitude, or low-frequency, vibrations.

The analysis of vibrational Raman and VROA spectra of proline zwitterionic forms in H_2O and D_2O solutions indicated that the solvent environment was found to modulate the properties of the hydrophobic part of the molecule indirectly by interacting with the ionic group (Kapitán et al. 2006). The use of explicit water molecules and continuum solvent treatment was recommended for the simulation of VROA, as well as VCD, spectra of amino acids, peptides, and proteins in aqueous solution (Jalkanen et al. 2008). The solvent effects were found to be important particularly for the prediction of VROA features for low-frequency modes. In a simultaneous investigation of implicit and explicit solvation models, the hydrogen-bonded water molecules of the inner hydration shell were found to be important in reproducing the VROA spectral intensities (Hopmann et al. 2011). For correctly reproducing the VROA spectrum of methyl-β-ᴅ-glucoside in water, a full MD simulation with QM/MM calculation was found to be necessary and the implicit solvation model to be insufficient (Cheeseman et al. 2011). Similarly reproducing the VROA spectrum of ᴅ-glucuronic acid and *N*-acetyl-ᴅ-glucosamine in water (Mutter, Zielinski, Cheeseman, et al. 2015) and of β-ᴅ-xylose in water (Zielinski et al. 2015), the combination of MD simulations with QM/MM calculations was undertaken. To reproduce the VROA spectrum of a cyclic peptide, cyclo(ʟ-Ala-Gly), in water, a large number of MD snapshots and the averaging of VROA spectra of these snapshot structures were found to be necessary (Urago et al. 2014). The agreement between calculated and experimental VROA spectra of tris-(ethylendiamine)-rhodium(III) chloride in water was found to improve when interactions with explicit solvent molecules are incorporated (Humbert-Droz et al. 2014), although the lowest energy conformers were not the preferred ones.

5.2.6 Approximate Methods for Large Molecules: Cartesian Tensor Transfer and Fragmentation

While the quantum chemical calculations for molecules containing up to 50–100 atoms has become routine, performing reliable quantum chemical calculations for even larger molecules such as peptides and proteins is not practical. In such cases, two different approaches have been used.

In the Cartesian tensor transfer method (CTTM), the force constants and Cartesian derivative tensors obtained for smaller peptide units are transferred to a large peptide of interest (Bouř et al. 1997). This CTTM has been widely used for VCD predictions, but some limitations were suggested more recently (Yamamoto et al. 2012; Bieler et al. 2011). One deficiency of the CTTM method has always been that the sum rules that Cartesian tensors are supposed to obey (Person and Newton 1974; Polavarapu 1990a, b; Cheeseman et al. 1996) were never verified.

In the fragmentation method (Choi et al. 2005), the dipole moment derivatives of individual fragments are calculated in the local coordinate system and, after rotation to the global coordinate system, combined using the eigenvectors of the Hessian matrix. Using the geometric derivatives of the necessary polarizability tensors formulated in the atomic orbital basis, the fragmentation method was also advanced for VROA predictions (Thorvaldsen et al. 2012).

There are two drawbacks in these fragmentation methods: (a) the orientation and the positions of the fragments used for modeling calculation will not coincide with those of the corresponding parts in the larger parent molecule, which requires appropriate rotation and translational of the molecular tensors; and (b) the above-mentioned sum rules of Cartesian tensors are not verified in these fragmentation methods also. These drawbacks are eliminated in the "molecules in molecules fragmentation" method discussed in the next section.

5.2.7 Accurate Calculations on Large Molecules: Molecules-in-Molecules Fragmentation Method

The molecules-in-molecules (MIM) fragmentation method (Mayhall and Raghavachari 2011) is a more fundamental approach than the CTTM and fragment methods discussed in the previous section. Full-fledged quantum chemical calculations scale up as ~N^5 to N^8, where N is the number of atoms in the molecule, which makes these calculations for large molecules nearly impossible. If a large molecule can be sliced up into n smaller subunits, then a full-fledged quantum chemical calculation will scale up as ~$n \times N_s^5$ to ~$n \times N_s^8$, where N_s is the number of atoms in the subunit, bringing the calculations into a manageable range. However, these calculations on subunits are to be weaved back together properly to obtain the properties of larger full molecule. The MIM method just does that by slicing up a large molecule into smaller subunits by breaking C–C bonds between connecting units and attaching H atoms to the dangling bonds. Then, the properties of subunits are properly treated by projecting the contributions from attached H atoms onto host and supporting atoms, before combining them to obtain the properties of the full large molecule. This method has been successfully applied to predict VA spectra of 25 small-to-large peptides (Jose and Raghavachari 2015) and Raman spectra of 21 large linear and cage molecules (Jovan Jose and Raghavachari 2015). The predictions obtained with the MIM fragmentation method for VCD spectra of 10 carbohydrates, perhydrotriphenylene, and cryptophane-A are highly encouraging (Jovan Jose et al. 2015). Similarly, the benchmark MIM-VROA studies on 10 carbohydrates and the good comparison obtained for experimental and MIM-predicted VROA spectra for D-Maltose, α-cyclodextrin, and cryptophan-A (Jose and Raghavachari 2016) indicate an encouraging prognosis for predicting the chiroptical properties of very large molecules using quantum mechanical theories.

The sum rules for the Cartesian tensors mentioned earlier are satisfied in this MIM method. Also the rotations and translations of molecular tensors are not needed in this method because the fragmentation scheme here retains their original orientations and positions in the larger parent molecule.

5.3 Spectral Predictions Using Approximate Models

5.3.1 Optical Rotatory Dispersion

Early predictions of SOR depended on van't Hoff's principle of optical superposition (van't Hoff 1875), according to which the SOR of a chiral compound can be written as a sum of the contributions from individual chiral centers in the molecule. Subsequently this concept was extended (Whiffen 1956) for dihedral segments in a molecule and for both chiral centers and dihedral segments (Brewster 1959). In contrast, the semi-classical models based on the concept of bond polarizabilities (Kirkwood 1937) and atom polarizabilites (Applequist 1973) make use of electromagnetic theory. These models were popular in the earlier decades, but they are not widely used now because more reliable quantum chemical methods are now available for routine predictions.

5.3.2 Electronic Circular Dichroism

Empirical rules (the so-called sector rules) have been widely used in the past (Eliel et al. 2001; Harada and Nakanishi 1983; Lightner and Gurst 2000). These sector rules are seldom used in the current times. The exciton chirality or exciton coupling (EC) model (Harada and Nakanishi 1983; Berova et al. 2007; Berova et al. 2010; Superchi et al. 2004) has been the most widely used model for relating the ECD spectra to the molecular structure. The same model is also referred to as the coupled oscillator (CO) model when applied to VCD. An extensive description of this model is presented in Appendix 4.

Exciton coupling, exciton chirality, or the CO model (see Appendix 4): A chromophore is a chemical group where electronic transition is localized and is responsible for EA in the UV-visible region. The EC/CO model considers the dipolar coupling interaction between electric dipole transition moments from two chromophores. If these two chromophores are identical, the energies of excited electronic states of the chromophores are considered to be identical in the absence of any interaction. If the electric dipole transition moments of these two chromophores interact through dipolar coupling, then the excited states, written as linear combinations of unperturbed states, will be split in energy, and this model is referred to as the degenerate EC or degenerate CO model. In this model, the transitions from the ground electronic state to the two split excited states result in energy-separated transitions with an oppositely signed ECD of equal magnitude. This energy-separated, oppositely signed ECD pattern is referred to as a *bisignate couplet*. If positive ECD appears at longer wavelengths and negative ECD appears at shorter wavelengths, then such bisignate couplets are referred to as *positive couplets*. If negative ECD appears at longer wavelengths and positive ECD appears at shorter wavelengths, then such bisignate couplets are referred to as *negative couplets*. The theory behind the above-mentioned dipolar coupling indicates that the *P* chiral arrangement of the transition moments of chromophores leads to a positive ECD couplet, and the *M* chiral arrangement of the transition moments of chromophores leads to a negative ECD couplet. This is the essence of the EC/CO model, and its visual presentation is shown in Appendix 4.

If the two considered chromophores are not identical, then the above-mentioned model is referred to as the nondegenerate EC or nondegenerate CO model. Here, the ECD intensities of two components need not be equal. It is not necessary to restrict the number of chromophoric transitions to two, and several coupled transitions can also be considered (see Appendix 4).

The EC/CO model may work well if dipolar coupling is the dominant mechanism of interaction between two identical chromophores, and there are no other transitions that might interact with the two transitions under consideration. Otherwise, the model predictions can fail.

Note that the basic operating principle behind the EC/CO model applies to both ECD and VCD interpretations. An attractive feature of the EC/CO model is that the predictions of this model can be easily visualized.

Since the EC/CO model has a theoretical basis (Appendix 4), the EC/CO model has been characterized in the literature as a nonempirical method. However, such categorization should be qualified by indicating that the underlining theory of the EC/CO model is one of approximate and conceptual nature.

Devoe's polarizability model (DeVoe 1964, 1965; Superchi et al. 2004): This is another approach that has been widely applied over the years for interpreting the experimentally observed ECD spectra. There are limitations in applying this model as well.

5.3.3 Vibrational Circular Dichroism

The EC/CO model originally developed for ECD spectral interpretations was also formulated for VCD spectral interpretations (Holzwarth and Chabay 1972) and has been widely used (Birke et al. 1992; Gulotta et al. 1989; Su and Keiderling 1980; Bour and Keiderling 1992). The details of the EC/CO model for VCD are similar to that for ECD (see Appendix 4).

In a different approach, a model specifically based on vibrational properties has been worked out for A-B-B-A-type segments. In this approach, the displacements of atoms during symmetric and antisymmetric A–B stretching vibrations could de deduced (Polavarapu 1986b) based solely on Eckart–Sayvetz conditions (Wilson et al. 1980) and imposing C_2 symmetry. Using these displacements and bond charges, the expressions for dipole and rotational strengths for symmetric and anti-symmetric stretching vibrations have been derived. The relation of predicted VCD signs with the handedness of the A-B-B-A segment is same as that in the EC/CO model (Polavarapu et al. 1987). This approach has been extended to bending vibrations of A-B-B-A segments as well (Polavarapu 1987c). However, it should be remembered that predictions obtained with such simplified models do not often reproduce the experimental observations very well, and even if they did, it would not be for the right reason. This is because the vibrations of real molecules encompass many more chemical bonds, and the influence from different chemical bonds cannot be faithfully represented by the simplified models.

Semi-classical models based on atomic charge concepts and bond moment concepts have been developed for interpreting VCD spectra (Abbate et al. 1981; Escribano and Barron 1988; Polavarapu 1989b; Moskovits and Gohin 1982). However, these methods do not have sufficient accuracy to rely on their predictions. For this reason, these methods are not currently used and therefore not described here.

Based on quantum chemical principles, the localized molecular orbital (LMO) (Nafie and Walnut 1977) and VC (Nafie and Freedman 1983; Dutler and Rauk 1989) models have also been developed. LMO and VCD models enjoyed some success in reproducing the experimental VCD data (Yang and Rauk 1992; Pickard et al. 1992; Polavarapu and Bose 1991) but were later abandoned due to the development of fundamentally rigorous formulations of VCD predictions (vide supra).

5.3.4 Vibrational Raman Optical Activity

A two-group model, much like the EC/CO model, developed for VROA, has been very useful for gaining insight into the generation of the VROA mechanism (Barron and Buckingham 1974). Semi-classical models for VROA based on bond polarizability (Barron 2004) and atom-dipole interaction (Prasad and Nafie 1979) concepts have been developed, but these methods do not have sufficient accuracy for obtaining reliable predictions. For this reason, these methods are not in current use.

5.4 Summary

Quantum chemical methods have been developed to reliably predict ORD, ECD, VCD, and VROA. The availability of user-friendly quantum chemical software (Gaussian09 2013; ADF 2014; DALTON 2015; CADPAC 2001; PSI4 2014) and computers with faster processors and large data storage space are making quantum chemical predictions of chiroptical properties for large molecules rather routine. The widespread use of chiroptical spectroscopic methods for chiral molecular structure determination in recent years is mostly because of these quantum chemical developments that have taken place in the last two decades.

References

Abbate, S., L. Laux, J. Overend, and A. Moscowitz. 1981. A charge flow model for vibrational rotational strengths. *J. Chem. Phys.* 75(7):3161–3164.

ADF. 2014. Amsterdam Density Functional Molecular Modeling Suite. http://www.scm.com, Amsterdam, Netherlands.

Al-Basheer, W., R. M. Pagni, and R. N. Compton. 2007. Spectroscopic and theoretical investigation of (R)-3-Methylcyclopentanone. The effect of solvent and temperature on the distribution of conformers. *J. Phys. Chem. A* 111 (12):2293–2298.

Amos, R. D. 1986. Calculation of polarizability derivatives using analytic gradient methods. *Chem. Phys. Lett.* 124(4):376–381.

Amos, R. D. 1982. Electric and magnetic properties of CO, HF, HCl, and CH_3F. *Chem. Phys. Lett.* 87(1):23–26.

Applequist, J. 1973. On the polarizability theory of optical rotation. *J. Chem. Phys.* 58 4251–4259.

Autschbach, J. 2012. Ab initio electronic circular dichroism and optical rotatory dispersion: From organic molecules to transition metal complexes. In: N. Berova, P. L. Polavarapu, K. Nakanishi and R. W. Woody (eds). *Comprehensive Chiroptical Spectroscopy*. Vol. 1, New York, NY: Wiley.

Autschbach, J., L. Nitsch-Velasquez, and M. Rudolph. 2011. Time-dependent density functional response theory for electronic chiroptical properties of chiral molecules. In: R. Naaman, D. N. Beratan, and D. H. Waldeck (eds). *Electronic and Magnetic Properties of Chiral Molecules and Supramolecular Architectures*. Berlin, Germany: Springer-Verlag.

Autschbach, J., T. Ziegler, S. J. A. van Gisbergen, and E. J. Baerends. 2002. Chiroptical properties from time-dependent density functional theory. I. Circular dichroism spectra of organic molecules. *J. Chem. Phys.* 116 (16):6930–6940.

Barone, V., A. Baiardi, and J. Bloino. 2014. New developments of a multifrequency virtual spectrometer: Stereo-electronic, dynamical, and environmental effects on chiroptical spectra. *Chirality* 26(9):588–600.

Barron, L. D. 2004. *Molecular Light Scattering and Optical Activity*, 2nd ed. Cambridge, UK: Cambridge University Press.

Barron, L. D., and A. D. Buckingham. 1974. Simple two-group model for Rayleigh and Raman optical activity. *J. Am. Chem. Soc.* 96 (15):4769–4773.

Berova, N., L. Di Bari, and G. Pescitelli. 2007. Application of electronic circular dichroism in configurational and conformational analysis of organic compounds. *Chem. Soc. Rev.* 36:914–931.

Berova, N., G. A. Ellestad, and N. Harada. 2010. Characterization by circular dichroism spectroscopy. In: L. Mander and H.-W. B. Liu (eds). *Comprehensive Natural Products II: Chemistry and Biology*. Oxford, UK: Elsevier.

Bieler, N. S., M. P. Haag, C. R. Jacob, and M. Reiher. 2011. Analysis of the Cartesian tensor transfer method for calculating vibrational spectra of polypeptides. *J. Chem. Theory Comput.* 7(6):1867–1881.

Birke, S. S., I. Agbaje, and M. Diem. 1992. Experimental and computational infrared CD studies of prototypical peptide conformations. *Biochemistry* 31:450–455.

Bloino, J., and V. Barone. 2012. A second-order perturbation theory route to vibrational averages and transition properties of molecules: General formulation and application to infrared and vibrational circular dichroism spectroscopies. *J. Chem. Phys.* 136 (12):124108.

Bose, P. K., L. D. Barron, and P. L. Polavarapu. 1989. Ab initio and experimental vibrational Raman optical activity in (+)-(R)-methylthiirane. *Chem. Phys. Lett.* 155 (4–5):423–429.

Bour, P., and T. A. Keiderling. 1992. Computational evaluation of the coupled oscillator model in the vibrational circular dichroism of selected small molecules. *J. Am. Chem. Soc.* 114:9100–9105.

Bouř, P., J. Sopková, L. Bednárová, P. Maloň, and T. A. Keiderling. 1997. Transfer of molecular property tensors in Cartesian coordinates: A new algorithm for simulation of vibrational spectra. *J. Comput. Chem.* 18(5):646–659.

Brewster, J. H. 1959. A useful model of optical activity. I. Open chain compounds. *J. Am. Chem. Soc.* 81:5475–5483.

Buckingham, A. D., P. W. Fowler, and P. A. Galwas. 1987. Velocity-dependent property surfaces and the theory of vibrational circular dichroism. *Chem. Phys.* 112:1–14.

CADPAC. 2001. Cambridge Analytical Derivatives Package. http://www-theor.ch.cam.ac.uk/software/cadpac.html, Cambridge, UK.

Caricato, M. 2015. Orbital analysis of molecular optical activity based on configuration rotatory strength. *J. Chem. Theory Comput.* 11(4):1349–1353.

Cheeseman, J. R., M. J. Frisch, F. J. Devlin, and P. J. Stephens. 1996. Ab initio calculation of atomic axial tensors and vibrational rotational strengths using density functional theory. *Chem. Phys. Lett.* 252:211–220.

Cheeseman, J. R., and M. J. Frisch. 2011. Basis set dependence of vibrational Raman and Raman optical activity intensities. *J. Chem. Theory Comput.* 7 (10):3323–3334.

Cheeseman, J. R., M. J. Frisch, F. J. Devlin, and P. J. Stephens. 2000. Hartree–Fock and density functional theory ab initio calculation of optical rotation using GIAOs: Basis set dependence. *J. Phys. Chem. A* 104(5):1039–1046.

Cheeseman, J. R., M. S. Shaik, P. L. A. Popelier, and E. W. Blanch. 2011. Calculation of Raman optical activity spectra of methyl-β-D-glucose incorporating a full molecular dynamics simulation of hydration effects. *J. Am. Chem. Soc.* 133 (13):4991–4997.

Choi, J. H., and M. Cho. 2014. Terahertz chiroptical spectroscopy of an α-helical polypeptide: A molecular dynamics simulation study. *J. Phys. Chem. B* 118 (45):12837–12843.

Choi, J. H., J. S. Kim, and M. Cho. 2005. Amide I vibrational circular dichroism of polypeptides: Generalized fragmentation approximation method. *J. Chem. Phys.* 122 (17):174903.

Chulhai, D. V., and L. Jensen. 2014. Simulating surface-enhanced Raman optical activity using atomistic electrodynamics-quantum mechanical models. *J. Phys. Chem. A* 118 (39):9069–9079.

Coriani, S., A. J. Thorvaldsen, K. Kristensen, and P. Jorgensen. 2011. Variational response-function formulation of vibrational circular dichroism. *Phys. Chem. Chem. Phys.* 13 (10):4224–4229.

Costante, J., L. Hecht, P. L. Polavarapu, A. Collet, and L. D. Barron. 1997. Absolute configuration of bromochlorofluoromethane from experimental and ab initio theoretical vibrational Raman optical activity. *Angew. Chem. Int. Ed.* 36(8):885–887.

Crawford, T. D. 2012. High-accuracy quantum chemistry and chiroptical properties. In: N. Berova, P. L. Polavarapu, K. Nakanishi and R. W. Woody (eds). *Comprehensive Chiroptical Spectroscopy*. Vol. 1. New York, NY: Wiley.

Crawford, T. D., and W. D. Allen. 2009. Optical activity in conformationally flexible molecules: A theoretical study of large-amplitude vibrational averaging in (R)-3-chloro-1-butene. *Mol. Phys.* 107 (8–12):1041–1057.

Crawford, T. D., and K. Ruud. 2011. Coupled-cluster calculations of vibrational Raman optical activity spectra. *ChemPhysChem* 12 (17):3442–3448.

Crawford, T. D., and P. J. Stephens. 2008. Comparison of time-dependent density-functional theory and coupled cluster theory for the calculation of the optical rotations of chiral molecules. *J. Phys. Chem. A* 112(6):1339–1345.

DALTON. 2015. A Molecular Electronic Structure Program. http://daltonprogram.org.

Daněček, P., J. Kapitán, V. Baumruk, L. Bednárová, V. Kopecký, and P. Bouř. 2007. Anharmonic effects in IR, Raman, and Raman optical activity spectra of alanine and proline zwitterions. *J. Chem. Phys.* 126 (22):224513.

DeVoe, H. 1964. Optical properties of molecular aggregates. I. Classical model of electronic absorption and refraction. *J. Chem. Phys.* 41(2):393–400.

DeVoe, H. 1965. Optical properties of molecular aggregates. II. Classical theory of the refraction, absorption, and optical activity of solutions and crystals. *J. Chem. Phys.* 43(9):3199–3208.

Diedrich, C., and S. Grimme. 2003. Systematic investigation of modern quantum chemical methods to predict electronic circular dichroism spectra. *J. Phys. Chem. A* 107 (14):2524–2539.

Domingos, S. R., A. Huerta-Viga, L. Baij, S. Amirjalayer, D. A. Dunnebier, A. J. Walters, M. Finger, L. A. Nafie, B. de Bruin, W. J. Buma, and S. Woutersen. 2014. Amplified vibrational circular dichroism as a probe of local biomolecular structure. *J. Am. Chem. Soc.* 136(9):3530–3535.

Dutler, R., and A. Rauk. 1989. Calculated infrared absorption and vibrational circular dichroism intensities of oxirane and its deuterated analogs. *J. Am. Chem. Soc.* 111 (18):6957–6966.

Eliel, E. L., S. H. Wilen, and M. P. Doyle. 2001. *Basic Organic Stereochemistry.* New York, NY: Wiley.

Escribano, J. R., and L. D. Barron. 1988. Valence optical theory of vibrational circular dichroism and Raman optical activity. *Mol. Phys.* 65(2):327–344.

Fischer, A. T., R. N. Compton, and R. M. Pagni. 2006. Solvent effects on the optical rotation of (S)-(−)-α-Methylbenzylamine. *J. Phys. Chem. A* 110 (22):7067–7071.

Frisch, M. J., Y. Yamaguchi, J. F. Gaw, H. F. Schaefer, and J. S. Binkley. 1986. Analytic Raman intensities from molecular electronic wave functions. *J. Chem. Phys.* 84(1):531–532.

Galwas, P. A. 1983. On the distribution of optical polarization in molecules. Ph.D. thesis. Cambridge, UK: Cambridge University.

Gaussian 09. 2013. Gaussian Inc., Willingford, CT, USA.

Goerigk, L., H. Kruse, and S. Grimme. 2012. Theoretical electronic circular dichroism spectroscopy of large organic and supramolecular systems. In: N. Berova, P. L. Polavarapu, K. Nakanishi, and R. W. Woody (eds). *Comprehensive Chiroptical Spectroscopy*. Vol. 1. New York, NY: Wiley.

Grimme, S. 2001. Calculation of frequency dependent optical rotation using density functional response theory. *Chem. Phys. Lett.* 339 (5–6):380–388.

Gulotta, M., D. J. Goss, and M. Diem. 1989. IR vibrational CD in model deoxyoligonucleotides: Observation of the B → Z phase transition and extended coupled oscillator intensity calcuations. *Biopolymers* 28 (12):2047–2058.

Harada, N., and K. Nakanishi. 1983. *Circular Dichroic Spectroscopy. Exciton Coupling in Organic Stereochemistry*. Mill Valley, CA: University Science Books.

He, Y., X. Cao, L. A. Nafie, and T. B. Freedman. 2001. Ab initio VCD calculation of a transition-metal containing molecule and a new intensity enhancement mechanism for VCD. *J. Am. Chem. Soc.* 123 (45):11320–11321.

Helgaker, T., K. Ruud, K. L. Bak, P. Jorgensen, and J. Olsen. 1994. Vibrational Raman optical activity calculations using London atomic orbitals. *Faraday Discuss.* 99(0):165–180.

Heshmat, M., E. J. Baerends, P. L. Polavarapu, and V. P. Nicu. 2014. The importance of large-amplitude motions for the interpretation of mid-infrared vibrational absorption and circular dichroism spectra: 6,6′-Dibromo-[1,1′-binaphthalene]-2,2′-diol in Dimethyl Sulfoxide. *J. Phys. Chem. A* 118:4766–4777.

Holzwarth, G., and I. Chabay. 1972. Optical activity of vibrational transitions: A coupled oscillator model. *J. Chem. Phys.* 57:1632–1635.

Hopmann, K. H., K. Ruud, M. Pecul, A. Kudelski, M. Dračínský, and P. Bouř. 2011. Explicit versus implicit solvent modeling of Raman optical activity spectra. *J. Phys. Chem. B* 115 (14):4128–4137.

Humbert-Droz, M., P. Oulevey, L. M. Lawson Daku, S. Luber, H. Hagemann, and T. Burgi. 2014. Where does the Raman optical activity of [Rh(en)3]3+ come from? Insight from a combined experimental and theoretical approach. *Phys. Chem. Chem. Phys.* 16 (42):23260–23273.

Jalkanen, K. J., I. M. Degtyarenko, R. M. Nieminen, X. Cao, L. A. Nafie, F. Zhu, and L. D. Barron. 2008. Role of hydration in determining the structure and vibrational spectra of L-alanine and N-acetyl L-alanine N′-methylamide in aqueous solution: a combined theoretical and experimental approach. *Theor. Chem. Acc.* 119 (1–3):191–210.

Jensen, L., J. Autschbach, M. Krykunov, and G. C. Schatz. 2007. Resonance vibrational Raman optical activity: A time-dependent density functional theory approach. *J. Chem. Phys.* 127 (13):134101-1:134101-11.

Jiemchooroj, A., and P. Norman. 2007. Electronic circular dichroism spectra from the complex polarization propagator. *J. Chem. Phys.* 126 (13):134101-1:134101-7.

Jose, K. V., and K. Raghavachari. 2015. Evaluation of energy gradients and infrared vibrational spectra through molecules-in-molecules fragment-based approach. *J. Chem. Theory Comput.* 11(3):950–961.

Jose, K. V., and K. Raghavachari. 2016. Raman optical activity spectra for large molecules through molecules-in-molecules fragment-based approach. *J. Chem. Theory Comput.* 12 (2):585–594.

Jovan Jose, K. V., D. Beckett, and K. Raghavachari. 2015. Vibrational circular dichroism spectra for large molecules through molecules-in-molecules fragment-based approach. *J. Chem. Theory Comput.* 11:4238–4247.

Jovan Jose, K. V., and K. Raghavachari. 2015. Molecules-in-molecules fragment-based method for the evaluation of Raman spectra of large molecules. *Mol. Phys.* 113 (19–20):3057–3066.

Kapitán, J., V. Baumruk, V. Kopecký, R. Pohl, and P. Bouř. 2006. Proline zwitterion dynamics in solution, glass, and crystalline state. *J. Am. Chem. Soc.* 128 (41):13451–13462.

Kirkwood, J. G. 1937. On the theory of optical rotatory power. *J. Chem. Phys.* 5:479–491.

Klamt, A., and G. Schuurmann. 1993. COSMO: A new approach to dielectric screening in solvents with explicit expressions for the screening energy and its gradient. *J. Chem. Soc., Perkin Trans.* 2(5):799–805.

Krykunov, M., M. D. Kundrat, and J. Autschbach. 2006. Calculation of circular dichroism spectra from optical rotatory dispersion, and vice versa, as complementary tools for theoretical studies of optical activity using time-dependent density functional theory. *J. Chem. Phys.* 125(19):194110–194113.

Kundrat, M. D., and J. Autschbach. 2008. Ab initio and density functional theory modeling of the chiroptical response of glycine and alanine in solution using explicit solvation and molecular dynamics. *J. Chem. Theory Comput.* 4(11):1902–1914.

Kundrat, M. D., and J. Autschbach. 2009. Modeling of the chiroptical response of chiral amino acids in solution using explicit solvation and molecular dynamics. *J. Chem. Theory Comput.* 5(4):1051–1060.

Lahiri, P., K. B. Wiberg, and P. H. Vaccaro. 2012. A tale of two carenes: Intrinsic optical activity and large-amplitude nuclear displacement. *J. Phys. Chem. A* 116(38):9516–9533.

Lahiri, P., K. B. Wiberg, and P. H. Vaccaro. 2015. Intrinsic optical activity and large-amplitude displacement: conformational flexibility in (R)-Glycidyl Methyl Ether. *J. Phys. Chem. A* 119(30):8311–8327.

Lahiri, P., K. B. Wiberg, P. H. Vaccaro, M. Caricato, and T. D. Crawford. 2014. Large solvation effect in the optical rotatory dispersion of norbornenone. *Angew. Chem.* 126(5):1410–1413.

Lambert, J., R. N. Compton, and T. D. Crawford. 2012. The optical activity of carvone: A theoretical and experimental investigation. *J. Chem. Phys.* 136(11):114512.

Liégeois, V., K. Ruud, and B. Champagne. 2007. An analytical derivative procedure for the calculation of vibrational Raman optical activity spectra. *J. Chem. Phys.* 127 (20):204105-1:204105-6.

Lightner, D. A., and J. E. Gurst. 2000. *Organic Conformational Analysis and Stereochemistry from Circular Dichroism Spectroscopy*. New York, NY: Wiley-VCH.

Lipparini, F., F. Egidi, C. Cappelli, and V. Barone. 2013. The optical rotation of methyloxirane in aqueous solution: A never ending story? *J. Chem. Theory Comput.* 9(4):1880–1884.

Luber, S., J. Neugebauer, and M. Reiher. 2010. Enhancement and de-enhancement effects in vibrational resonance Raman optical activity. *J. Chem. Phys.* 132(4):044113.

Mayhall, N. J., and K. Raghavachari. 2011. Molecules-in-molecules: An extrapolated fragment-based approach for accurate calculations on large molecules and materials. *J. Chem. Theory Comput.* 7(5):1336–1343.

Mennucci, B., J. Tomasi, R. Cammi, J. R. Cheeseman, M. J. Frisch, F. J. Devlin, S. Gabriel, and P. J. Stephens. 2002. Polarizable continuum model (PCM) calculations of solvent effects on optical rotations of chiral molecules. *J. Phys. Chem. A* 106:6102–6113.

Merten, C., K. Hiller, and Y. Xu. 2012. Effects of electron configuration and coordination number on the vibrational circular dichroism spectra of metal complexes of trans-1,2-diaminocyclohexane. *Phys. Chem. Chem. Phys.* 14(37):12884–12891.

Mort, B. C., and J. Autschbach. 2005. Magnitude of zero-point vibrational corrections to optical rotation in rigid organic molecules: A time-dependent density functional study. *J. Phys. Chem. A* 109(38):8617–8623.

Mort, B. C., and J. Autschbach. 2006. Temperature dependence of the optical rotation of fenchone calculated by vibrational averaging. *J. Phys. Chem. A* 110(40):11381–11383.

Mort, B. C., and J. Autschbach. 2008. A pragmatic recipe for the treatment of hindered rotations in the vibrational averaging of molecular properties. *ChemPhysChem* 9(1):159–170.

Moscowitz, A. 1960. Theory and analysis of optical rotatory dispersion curves. In: C. Djerassi (ed). *Optical rotatory dispersion*. New York, NY: McGraw Hill.

Moskovits, M., and A. Gohin. 1982. Vibrational circular dichroism: Effect of charge fluxes and bond currents. *J. Phys. Chem.* 86(20):3947–3950.

Mukhopadhyay, P., G. Zuber, P. Wipf, and D. N. Beratan. 2007. Contribution of a solute's chiral solvent imprint to optical rotation. *Angew. Chem. Int. Ed.* 46(34):6450–6452.

Müller, T., K. B. Wiberg, and P. H. Vaccaro. 2000. Cavity ring-down polarimetry (CRDP): A new scheme for probing circular birefringence and circular dichroism in the gas phase. *J. Phys. Chem. A* 104(25):5959–5968.

Mutter, S. T., F. Zielinski, J. R. Cheeseman, C. Johannessen, P. L. A. Popelier, and E. W. Blanch. 2015. Conformational dynamics of carbohydrates: Raman optical activity of d-glucuronic acid and N-acetyl-d-glucosamine using a combined molecular dynamics and quantum chemical approach. *Phys. Chem. Chem. Phys.* 17(8):6016–6027.

Mutter, S. T., F. Zielinski, P. L. A. Popelier, and E. W. Blanch. 2015. Calculation of Raman optical activity spectra for vibrational analysis. *Analyst.* 140:2944–2956.

Nafie, L. A., and T. H. Walnut. 1977. Vibrational circular dichroism theory: A localized molecular orbital model. *Chem. Phys. Lett.* 49:441–446.

Nafie, L. A. 1996. Theory of resonance Raman optical activity: The single electronic state limit. *Chem. Phys.* 205(3):309–322.

Nafie, L. A. 2004. Theory of cibrational circular dichroism and infrared absorption: Extension to molecules with low-lying excited electronic states. *J. Phys. Chem. A* 108(35):7222–7231.

Nafie, L. A., and T. B. Freedman. 1983. Vibronic coupling theory of infrared vibrational transitions. *J. Chem. Phys.* 78(12):7108–7116.

Nafie, L. A. 2008. Theory of Raman scattering and Raman optical activity: Near resonance theory and levels of approximation. *Theor. Chem. Acc.* 119(1–3):39–55.

Neugebauer, J., E. J. Baerends, M. Nooijen, and J. Autschbach. 2005. Importance of vibronic effects on the circular dichroism spectrum of dimethyloxirane. *J. Chem. Phys.* 122(23):234305-1:234305-7.

Nooijen, M. 2006. Investigation of Herzberg–Teller Franck–Condon approaches and classical simulations to include effects due to vibronic coupling in circular dichroism spectra: The case of dimethyloxirane continued. *Int. J. Quantum Chem.* 106(12):2489–2510.

Norman, P., K. Ruud, and T. Helgaker. 2004. Density-functional theory calculations of optical rotatory dispersion in the nonresonant and resonant frequency regions. *J. Chem. Phys.* 120(11):5027–5035.

Novák, V., J. Šebestík, and P. Bouř. 2012. Theoretical modeling of the surface-enhanced Raman optical activity. *J. Chem. Theory Comput.* 8(5):1714–1720.

Parchansky, V., J. Kapitan, and P. Bour. 2014. Inspecting chiral molecules by Raman optical activity spectroscopy. *RSC Adv.* 4(100):57125–57136.

Passarello, M., S. Abbate, G. Longhi, S. Lepri, R. Ruzziconi, and V. P. Nicu. 2014. Importance of C*–H based modes and large amplitude motion effects in vibrational circular dichroism spectra: The case of the chiral adduct of dimethyl fumarate and anthracene. *J. Phys. Chem. A* 118(24):4339–4350.

Pecul, M. 2012. Modeling of solvation effects on chiroptical spectra. In: N. Berova, P. L. Polavarapu, K. Nakanishi, and R. W. Woody (eds). *Comprehensive Chiroptical Spectroscopy*. Vol. 1. New York, NY: Wiley.

Pecul, M., and A. Rizzo. 2003. Raman optical activity spectra: Basis set and electron correlation effects. *Mol. Phys.* 101(13):2073–2081.

Pecul, M., K. Ruud, and T. Helgaker. 2004. Density functional theory calculation of electronic circular dichroism using London orbitals. *Chem. Phys. Lett.* 388(1–3):110–119.

Pedersen, T. B., H. Koch, L. Boman, and A. M. J. Sánchez de Merás. 2004. Origin invariant calculation of optical rotation without recourse to London orbitals. *Chem. Phys. Lett.* 393(4–6):319–326.

Pedersen, T. B., H. Koch, and K. Ruud. 1999. Coupled cluster response calculation of natural chiroptical spectra. *J. Chem. Phys.* 110(6):2883–2892.

Perera, A. S., J. Thomas, M. R. Poopari, and Y. Xu. 2016. The clusters-in-a liquid approach for solvation: New insights from conformer specific gas phase spectroscopy and vibrational optical activity spectroscopy. *Frontiers in Chemistry* 4(9):1–17.

Pericou-Cayere, M., M. Rerat, and A. Dargelos. 1998. Theoretical treatment of the electronic circular dichroism spectrum and the optical rotatory power of H2S2. *Chem. Phys.* 226(3):297–306.

Person, W. B., and J. H. Newton. 1974. Dipole moment derivatives and infrared intensities. I. Polar tensors. *J. Chem. Phys.* 61(3):1040–1049.

Pescitelli, G., V. Barone, L. Di Bari, A. Rizzo, and F. Santoro. 2013. Vibronic coupling dominates the electronic circular dichroism of the benzene chromophore 1Lb band. *J. Org. Chem.* 78(15):7398–7405.

Pescitelli, G., and O. Bruhn. 2016. Good computational practice in the assignment of absolute configurations by TDDFT calculations of ECD spectra. *Chirality* 28(6):466–74.

Pickard, S. T., H. E. Smith, P. L. Polavarapu, T. M. Black, A. Rauk, and D. Yang. 1992. Synthesis, experimental, and ab initio theoretical vibrational circular dichroism, and absolute configurations of substituted oxiranes. *J. Am. Chem. Soc.* 114(17):6850–6857.

Polavarapu, P. L. 1982. Atomic Raman tensors and intensity sum rule. *J. Mol. Spectrosc.* 93(2):450–452.

Polavarapu, P. L. 1986a. A sum rule for vibrational circular dichroism intensities of fundamental transitions. *J. Chem. Phys.* 84(1):542–543.

Polavarapu, P. L. 1986b. Vibrational circular dichroism of A_2B_2 molecules. *J. Chem. Phys.* 85:6245–6246.

Polavarapu, P. L. 1987a. A frequency weighted sum rule for vibrational circular dichroism intensities. *J. Chem. Phys.* 87 (11):6775–6776.

Polavarapu, P. L. 1987b. A frequency-weighted sum rule for vibrational Raman optical activity. *Chem. Phys. Lett.* 139(6):558–562.

Polavarapu, P. L. 1987c. Absorption and circular dichroism due to bending vibrations of A_2B_2 molecules with C_2 symmetry. *J. Chem. Phys.* 87:4419–4422.

Polavarapu, P. L. 1989a. Rotational—vibrational circular dichroism. *Chem. Phys. Lett.* 161(6):485–490.

Polavarapu, P. L., ed. 1989b. Vibrational optical activity. In: H. D. Bist, J. R. During, and J. F. Sullivan (eds). *Vibrational Spectra and Structure*. Amsterdam: Elsevier.
Polavarapu, P. L. 1990a. Ab initio vibrational Raman and Raman optical activity spectra. *J. Phys. Chem.* 94 (21):8106–8112.
Polavarapu, P. L. 1990b. Sum rules for Cartesian polarizability derivative tensors. *Chem. Phys. Lett.* 174(5):511–516.
Polavarapu, P. L. 1997. Ab initio molecular optical rotations and absolute configurations. *Mol. Phys.* 91(3):551–554.
Polavarapu, P. L. 1998. *Vibrational Spectra: Principles and Applications with Emphasis on Optical Activity*. New York, NY: Elsevier.
Polavarapu, P. L. 2002. The absolute configuration of bromochlorofluoromethane. *Angew. Chem. Int. Ed.* 41(23):4544–4546.
Polavarapu, P. L. 2005. Kramers–Kronig transformation for optical rotatory dispersion studies. *J. Phys. Chem. A* 109(32):7013–7023.
Polavarapu, P. L. 2007. Renaissance in chiroptical spectroscopic methods for molecular structure determination. *Chem. Rec.* 7(2):125–136.
Polavarapu, P. L., and Chunxia C. Zhao. 1998. Ab initio predictions of anomalous optical rotatory dispersion. *J. Am. Chem. Soc.* 121(1):246–247.
Polavarapu, P. L., and P. K. Bose. 1991. Ab initio localized molecular orbital predictions of vibrational circular dichroism: Trans-1,2-dideuteriocyclopropane. *J. Phys. Chem.* 95(4):1606–1608.
Polavarapu, P. L., C. S. Ewig, and T. Chandramouly. 1987. Conformations of tartaric acid and its esters. *J. Am. Chem. Soc.* 109:7382–7386.
Polavarapu, P. L., D. K. Chakraborty, and K. Ruud. 2000. Molecular optical rotation: An evaluation of semiempirical models. *Chem. Phys. Lett.* 319(5–6):595–600.
Prasad, P. L., and L. A. Nafie. 1979. The atom dipole interaction model of Raman optical activity: Reformulation and comparison to the general two-group model. *J. Chem. Phys.* 70(12):5582–5588.
PSI4. 2014. A Open-Source Suite of Ab Initio Quantum Chemistry Programs, USA.
Quinet, O., and B. Champagne. 2001. Time-dependent Hartree–Fock schemes for analytical evaluation of the Raman intensities. *J. Chem. Phys.* 115(14):6293–6299.
Quinet, O., V. Liégeois, and B. Champagne. 2005. TDHF evaluation of the dipole–quadrupole polarizability and its geometrical derivatives. *J. Chem. Theory Comput.* 1(3):444–452.
Reiher, M., V. Liégeois, and K. Ruud. 2005. Basis set and density functional dependence of vibrational Raman optical activity calculations. *J. Phys. Chem. A* 109(33):7567–7574.
Rudolph, M., and J. Autschbach. 2008. Fast generation of nonresonant and resonant optical rotatory dispersion curves with the help of circular dichroism calculations and Kramers-Kronig transformations. *Chirality* 20:995–1008.
Ruud, K. 2012. Ab initio methods for vibrational circular dichroism and Raman optical activity. In: N. Berova, P. L. Polavarapu, K. Nakanishi, and R. W. Woody (eds). *Comprehensive Chiroptical Spectroscopy*. Vol. 1. New York, NY: Wiley.
Ruud, K., and T. Helgaker. 2002. Optical rotation studied by density-functional and coupled-cluster methods. *Chem. Phys. Lett.* 352(5–6):533–539.
Ruud, K., T. Helgaker, and P. Bouř. 2002. Gauge-origin independent density-functional theory calculations of vibrational Raman optical activity. *J. Phys. Chem. A* 106(32):7448–7455.

Ruud, K., P. R. Taylor, and P. O. Åstrand. 2001. Zero-point vibrational effects on optical rotation. *Chem. Phys. Lett.* 337(1–3):217–223.

Ruud, K., and A. J. Thorvaldsen. 2009. Theoretical approaches to the calculation of Raman optical activity spectra. *Chirality* 21(1E):E54–E67.

Ruud, K., and R. Zanasi. 2005. The importance of molecular vibrations: The sign change of the optical rotation of methyloxirane. *Angew. Chem. Int. Ed.* 44(23):3594–3596.

Scherrer, A., R. Vuilleumier, and D. Sebastiani. 2013. Nuclear velocity perturbation theory of vibrational circular dichroism. *J. Chem. Theory Comput.* 9(12):5305–5312.

Simmen, B., T. Weymuth, and M. Reiher. 2012. How many chiral centers can raman optical activity spectroscopy distinguish in a molecule? *J. Phys. Chem. A* 116(22):5410–5419.

Srebro, M., N. Govind, W. A. de Jong, and J. Autschbach. 2011. Optical rotation calculated with time-dependent density functional theory: The OR45 benchmark. *J. Phys. Chem. A* 115(40):10930–10949.

Stephens, P. J., D. M. McCann, F. J. Devlin, J. R. Cheeseman, and M. J. Frisch. 2004. Determination of the absolute configuration of [32](1,4) barrelenophanedicarbonitrile using concerted time-dependent density functional theory calculations of optical rotation and electronic circular dichroism. *J. Am. Chem. Soc.* 126(24):7514–7521.

Stephens, P. J. 1985. Theory of vibrational circular dichroism. *J. Phys. Chem.* 89:748–752.

Stephens, P. J., F. J. Devlin, and J. J. Pan. 2008. The determination of the absolute configurations of chiral molecules using vibrational circular dichroism (VCD) spectroscopy. *Chirality* 20(5):643–663.

Su, C. N., and T. A. Keiderling. 1980. Conformation of dimethyl tartrate in solution. Vibrational circular dichroism results. *J. Am. Chem. Soc.* 102:511–515.

Superchi, S., E. Giorgio, and C. Rosini. 2004. Structural determinations by circular dichroism spectra analysis using coupled oscillator methods: An update of the applications of the DeVoe polarizability model. *Chirality* 16(7):422–451.

Tedesco, D., R. Zanasi, B. Kirchner, and C. Bertucci. 2014. Short-range solvation effects on chiroptical properties: A time-dependent density functional theory and ab initio molecular dynamics computational case study on austdiol. *J. Phys. Chem. A* 118(50):11751–11757.

Thorvaldsen, A. J., B. Gao, K. Ruud, M. Fedorovsky, G. Zuber, and W. Hug. 2012. Efficient calculation of ROA tensors with analytical gradients and fragmentation. *Chirality* 24(12):1018–1030.

Urago, H., T. Suga, T. Hirata, H. Kodama, and M. Unno. 2014. Raman optical activity of a cyclic dipeptide analyzed by quantum chemical calculations combined with molecular dynamics simulations. *J. Phys. Chem. B* 118(24):6767–6774.

Vaccaro, P. H. 2012. Optical rotation and intrinsic optical activity. In: N. Berova, P. L. Polavarapu, K. Nakanishi, and R. W. Woody (eds). *Comprehensive Chiroptical Spectrosocpy.* Vol. 1. New York, NY: Wiley.

van't Hoff, G. H. 1875. Sur les formules de structure dans l'espace. *Bull. Soc. Chim. Fr.* 23:295–301.

Whiffen, D. H. 1956. Optical rotation and geometrical structure. *Chem. Ind.* 964–968.

Wiberg, K. B., P. H. Vaccaro, and J. R. Cheeseman. 2003. Conformational effects on optical rotation. 3-substituted 1-butenes. *J. Am. Chem. Soc.* 125(7):1888–1896.

Wilson, E. B., J. C. Decius, and P. C. Cross. 1980. *Molecular Vibrations: The Theory of Infrared and Raman Vibrational Spectra.* New York, NY: Dover Publications.

Yabana, K., and G. F. Bertsch. 1999. Application of the time-dependent local density approximation to optical activity. *Phys. Rev. A* 60(2):1271–1279.

Yamamoto, S., X. Li, K. Ruud, and P. Bouř. 2012. Transferability of various molecular property tensors in vibrational spectroscopy. *J. Chem. Theory Comput.* 8(3):977–985.

Yang, D., and A. Rauk. 1992. Vibrational circular dichroism intensities: Ab initio vibronic coupling theory using the distributed origin gauge. *J. Chem. Phys.* 97(9):6517–6534.

Yang, S., and M. Cho. 2009. Direct calculations of vibrational absorption and circular dichroism spectra of alanine dipeptide analog in water: Quantum mechanical/molecular mechanical molecular dynamics simulations. *J. Chem. Phys.* 131(13):135102.

Zielinski, F., S. T. Mutter, C. Johannessen, E. W. Blanch, and P. L. A. Popelier. 2015. The Raman optical activity of [small beta]-d-xylose: Where experiment and computation meet. *Phys. Chem. Chem. Phys.* 17:21799–21809.

Zuber, G., and W. Hug. 2004. Rarefied basis sets for the calculation of optical tensors. 1. The importance of gradients on hydrogen atoms for the Raman scattering tensor. *J. Phys. Chem. A* 108(11):2108–2118.

6

Conformational Analysis

Before embarking on using the predicted chiroptical properties to interpret corresponding experimental data, it is necessary to ensure that the complete conformational space of a given molecule has been explored, and that all possible conformations have been evaluated for obtaining the predicted properties. The identification of predominant conformations and determination of Boltzmann population weights (see Appendix 5) for these conformers is a prerequisite for utilizing the predicted properties. A number of conformational search programs are commercially available (Conflex 2010; HyperChem 2014; MacroModel 2008; Spartan 2014) to facilitate the determination of predominant conformations. The main theme of conformational search programs is to rotate the rotatable groups around single bonds, pucker the rings (when present), and utilize built-in molecular mechanics force fields to determine the lowest energy conformers. However, it is necessary to be aware of limitations in using these programs, as discussed in Section 6.1.

The conformational search programs with the theme mentioned above may not be practical when solvent environment is to be modeled by including the explicit solvent molecules surrounding the solute molecule because of several reasons: (a) the increased number of degrees of freedom for relative orientations of solvent molecules around a solute molecule makes the conformational search difficult, if not impossible; (b) molecular mechanics–based force fields are not appropriate to account for solute–solvent interactions. In such cases, molecular dynamics (MD) simulations as described in Section 6.2 provide a better approach for determining various conformations possible.

6.1 Conformational Analysis of Solute Molecule

The information discussed in this section applies to all four areas of chiroptical spectroscopy: electronic circular dichroism, optical rotatory dispersion, vibrational circular dichroism, and vibrational Raman optical activity. Manual conformational search, where different conformations are generated by manually rotating the atoms around single bonds, used to be the common approach in the older literature. In the current literature, the use of commercial programs that carry out automated conformational search has become routine. The automated conformational analysis should involve rotation of atoms around all rotatable (single) bonds, puckering of rings, and identification of the absolute configurations of structures generated. However, each program will likely have a different workflow and inherent limitations. As a result, it is not uncommon to find different number of conformations with different programs. Moreover, the conformations obtained for the individual enantiomers of chiral molecules may not have perfect mirror image relationships. Therefore, it is advisable to exercise caution in the use of commercial conformational search programs. This chapter discusses some important points that one needs to be aware of during automated conformational search.

6.1.1 Designation of Bonds

The first step in an automated conformational search is to provide a starting structure as the input for the conformational search program to manipulate on. This structure is provided in the form of an input file. From this input file, the chosen program should recognize all rotatable (single) bonds. Each program may recognize the nature of bonds (single, double, etc.) based on different criterion: (a) information provided in the input file, (b) the valence of each atom, or (c) default (built-in) definitions for bond lengths. Figure 6.1 shows an example where a single bond between C and O atoms of the C-OCH_3 group is depicted with partial double-bond character (pointed by arrows). The files created with "mol" extension contain the information on the type of bonds present in the molecule (Dalby et al. 1992). If a "mol" file was created by a program that identifies the bonds based on default bond lengths, then a bond that was meant to be a single bond, but its length falls outside the criterion for a single bond, can be saved as one with no bond at all or as a partial double bond or higher order bond. It is imperative to know how the chosen conformational search program handles such cases, to make sure that the rotation of atoms/groups around some, what should have been, single bonds are not excluded during the conformational search. The molecular mechanics–based conformational search programs have the bond force constants defined based on the type of bonds. A misidentification of the type of bonds may not carry out the energy minimization as intended. Therefore, the user needs to verify that the structure that was input into the conformational search program has been received by that program as intended by the user.

6.1.2 Ring-Puckering Angles

Besides the rotation of atoms around rotatable bonds, conformational freedom also arises from the puckering of rings in molecules that contain rings. Although the general definitions for ring puckering have been formulated (Cremer and Pople 1975), the conformational search drivers may not use these general definitions of ring puckering. An example of a five-membered ring

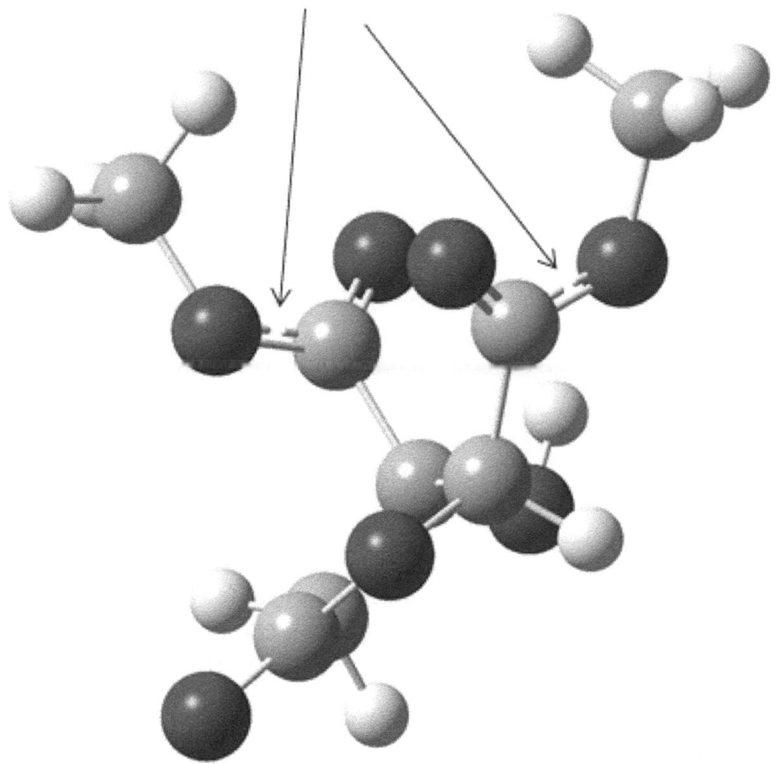

FIGURE 6.1
A single bond between C and O atoms of the C-OCH_3 group is depicted with partial double-bond characters (pointed to by arrows).

molecule, garcinia acid dimethylester (Polavarapu et al. 2010), with opposite ring-puckering angles is shown in Figure 6.2. If all output structures contain ring-puckering angles of the same sign as that contained in the input structure, then it is an indication that the conformational driver may not have been able to alter the sign of the ring-puckering angles. In such cases, it is necessary to build a different initial structure with an oppositely signed puckering angle and rerun the conformational search. It is a good practice to try different input structures with starting ring-puckering angles of opposite signs and explore whether the output structures contain both positive and negative puckering angles (Polavarapu 2012). In addition, it is a safe practice to use the potential energy surface (PES) scan options embedded in the quantum chemical programs, in addition to using the conformational search programs (Polavarapu et al. 2012). When a PES scan is performed, a chosen dihedral angle can be varied at the desired increment and a complete energy landscape as a function of that dihedral angle can be obtained. The troughs in PES identify the geometries with the lowest energies.

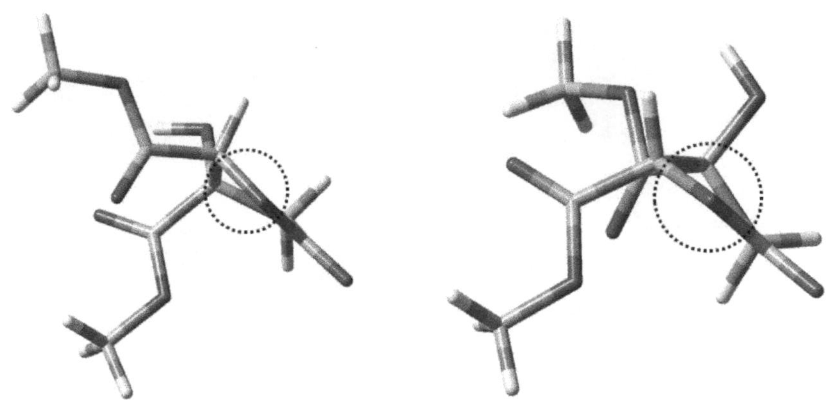

FIGURE 6.2
Opposite ring-puckering angles (highlighted by dotted circles) in five-membered ring molecules.

For example, an additional ring-puckering conformation that was not identified with a conformational search program has been identified through the potential energy scans for schizozygane alkaloids (Stephens et al. 2007) and for hibiscus acid disodium salt (Polavarapu et al. 2012).

6.1.3 Symmetries of Generated Conformations

By the very nature of the conformational search methods, where atoms have to be displaced from their initial positions, the symmetry present in the input structure need not be conserved during the conformational search. For example, an ethane molecule in staggered conformation will have D_{3d} symmetry, whereas the same molecule in the eclipsed conformation will have D_{3h} symmetry. A conformational search program is expected to find both of these conformations. For a starting structure with C_2 symmetry, the conformational search programs will invariably generate the conformations with no symmetry. Therefore, it is imperative that symmetries of the conformations generated by a conformational search program are assessed.

6.1.4 Structures with Equivalent Energies

A conformational search program may provide options to sort and save the conformations found using a criterion such as energy or a structural parameter. If the energy criterion is chosen, then some structures that happen to have the same energy, but different structural parameters, may not be saved as different possible conformers, and instead only one of such structures may be saved. If a different criterion based on a geometrical parameter is chosen, then the structures absent in the previous criterion can be recognized as possible conformers. A good example to verify this point is to find all possible conformations of H_2O_2 where the structures with opposite dihedral angles, which are actually the stereoisomers, will have the same energy (see Figure 6.3). The structures sorted by a dihedral angle may identify the structures with opposite dihedral angles, whereas the structures sorted by energy may not.

6.1.5 Atropisomers

The H_2O_2 example given above is a case where one starts with a stereoisomer and finds the opposite stereoisomer during a conformational search. The rotation around the O–O bond in H_2O_2 can go through the structures that have *P* and *M* configurations during one full cycle. Although the stereoisomers of H_2O_2 cannot be isolated in laboratory experiments, they serve as good examples to recognize the importance of identifying such instances. This is because, in chiroptical spectroscopy, it is common to deal with atropisomers (see Chapter 2, Section 2.2.2), where rotation around an axial bond leads to the opposite configurations. Since the main function of conformational search programs is to rotate the atoms around rotatable single bonds, the structures with opposite absolute configurations are to be expected in the process of identifying the conformers associated with atropisomers, as noted for Vanol (Polavarapu 2012). Therefore, one should not assume that all of the structures generated by a conformational search program will have the same absolute configuration as that of the input structure. For a sulfonate molecule (Smuts et al. 2013) shown in Figure 6.4, the rotation around

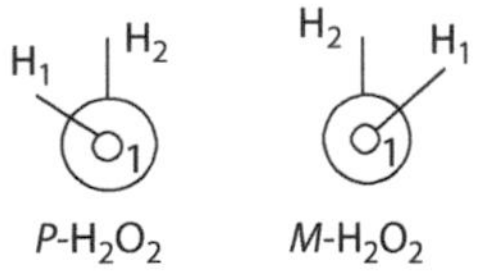

H_1 H_2 O_1 H_2 H_1 O_1

P-H_2O_2 *M*-H_2O_2

FIGURE 6.3
Two stereoisomers of H_2O_2 that may be identified as conformers with oppositely signed dihedral angles. Since they have the same energy, energy criterion will not identify them separately.

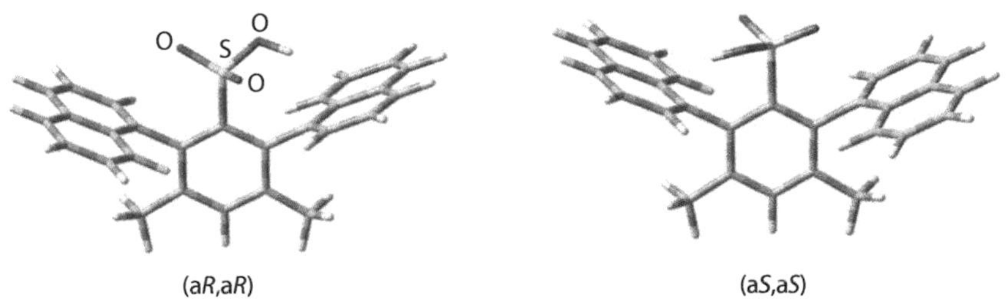

FIGURE 6.4
Two different stereoisomers, (a*R*,a*R*) and (a*S*,a*S*), that could be generated via rotation of binaphthyl groups around the C–C bond connecting the binaphthyl groups to the benzene group. Two other possible diastereomers, (a*R*,a*S*) and (a*S*,a*R*), are not shown. Note that the double bonds in the binaphthyl and benzene rings and the sulfonate group are not shown.

the C–C bonds connecting the benzene ring to the two binaphthyl groups leads to four possible diastereomers: (a*R*,a*R*), (a*R*,a*S*), (a*S*,a*R*), and (a*S*,a*S*) (see Chapter 2, Figure 2.12 for axial chirality notation). In addition, the S–O bonds can be exchanged in three possible locations and the H–O–S–O dihedral angle can be changed to three different orientations. Thus, each diastereomer can have nine possible conformations. The conformations associated with each diastereomer are to be grouped separately before quantum chemical calculations can be undertaken for those structures. The structure output by the conformational search programs is often imported in automated processes into the quantum mechanical programs for subsequent calculations without human intervention. If this is done without sorting the structures according to the nature of their absolute configurations, there is a risk of adding the chiroptical spectra that belong to different stereoisomers.

6.1.6 Molecular Ions and Radicals

Salts are often used as a means of crystallizing a sample and the differences in their solution, and solid-state structures are of interest. Also the salt form, providing a polar head group and sometimes behaving as a surfactant, is more appropriate for generating organized assemblies, such as micelles. Thus a conformational search for charged chiral surfactants is of general interest (Polavarapu and Vijay 2012). A molecule, ion, or radical is viewed in the quantum chemical description as a collection of nuclei and electrons, and there is no need to invoke the concept of chemical bonds. However, the description of a molecule, ion, or radical in molecular mechanics depends on invoking the presence of chemical bonds (single bond, double bond, etc.) connecting the atoms, because a force constant is assigned to each chemical bond based on the nature of that chemical bond. The presence of additional, or a fewer number of, electrons in a chemical entity is not automatically accounted for in molecular mechanics–based programs As a result, the user needs to verify how ionic bonds, or electrostatic interactions between ionic species, and radicals are accommodated in the chosen conformational search program. If ionic species are not accommodated properly, then the conformational search for a molecular ion or a molecule composed of oppositely charged ions (as in salts and zwitterions) may not be carried out correctly (Polavarapu 2012). Similarly, a conformational search for radicals may not be carried out correctly.

6.1.7 Dimeric Structures

The conformational analysis of molecules that exist in dimeric form is nontrivial, as most commercial conformational analysis programs do not recognize dimeric structure as one assembly. One commercial program (MacroModel 2008), however, is known (Goldsmith et al. 2003) to handle the conformational analysis for dimeric structures.

6.1.8 Enantiomer Structures

For molecules with multiple sources of chirality, separate conformational analyses for enantiomers need not yield an identical number of conformations with mirror image structures. Therefore, it will be prudent to carry out separate conformational analysis for enantiomers to identify possible conformers that may have been missed in the conformational analysis of a single enantiomer. It is useful to note that some internal coordinate interaction force constants will have the opposite signs (Polavarapu 1993) for enantiomers.

6.1.9 Boltzmann Populations

The information obtained from conformational search programs on possible conformations becomes the starting point for the subsequent quantum chemical calculations. The generated structures need to be verified first for difference in the nature of their absolute configurations. If structures of differing absolute configurations are found, then they need to be sorted into different groups by their absolute configurations, and Boltzmann populations (see Appendix 5) of structures in each group are to be separately determined. The energies provided by molecular mechanics–based conformational search programs are approximate in nature and should not be considered to be quantitatively accurate. As a result, Boltzmann populations determined using the energies provided by molecular mechanics–based conformational search programs can be grossly incorrect. The structures, sorted by their absolute configurations, generated from these conformational search programs are to be optimized using the quantum chemical methods. The starting point for these quantum chemical methods is the use of a B3LYP (or a suitable) density functional and the 6-31G* basis set. Some of the conformations obtained with molecular mechanics programs may turn out to be not stable when investigated with quantum chemical methods. It is necessary to ensure that the structures optimized with the quantum chemical methods are at the minimum of the PES by doing the vibrational frequency calculation at the optimized structures. The absence of imaginary frequency for a given optimized structure indicates that this structure is stable. The populations of such stable structures can be determined (see Appendix 5) using either electronic energies or Gibbs free energies. Although Gibbs free energies take entropy contributions into account and are considered to be the appropriate ones to use, there is some discussion on the reliability of low-frequency vibrations predicted using quantum chemical methods in the harmonic approximation (Dos Santos et al. 2002). As a result, a clear answer as to which of these two (electronic vs. Gibbs) energies should be used for determining the populations is not obvious. It would be useful to determine the populations using both of these energies to verify whether there are any significant differences between them (Polavarapu et al. 2010). If there is no significant difference between the two sets of populations, then the above discussion becomes moot; otherwise, additional investigations would become necessary to select one set of populations.

6.2 Conformational Analysis of Solute–Solvent Clusters

As discussed in Chapter 5 (Section 5.2.5), determination of all possible orientations of the solvent molecules around a solute molecule is difficult, with commercial conformational search programs, especially when a large number of solvent molecules are involved. This situation can be handled with MD simulations, where time evolution of solute molecule in a solvent bath is determined using Newtonian dynamics. A series of snapshots taken during the time evolution represent different conformations of the system. Several recent papers on this subject (see Chapter 5, Section 5.2.5 for references), as applied to chiroptical spectroscopy, can be consulted for intricacies involved in these MD simulations.

6.3 Summary

Conformational analysis for solute molecules is a prerequisite in analyzing the predicted chiroptical properties. The use of a single commercially available conformational search program may not be adequate for identifying all possible conformations of chiral molecules. Therefore, it is advisable to adopt a multipronged approach (Polavarapu et al. 2011) that includes a manual conformation search and PES scans in addition to using a commercially available conformational search program. However, in the case of solute–solvent clusters, MD simulations may be required for the involved conformational analysis.

References

Conflex. 2010. High performance conformational analysis. http://www.conflex.us/.

Cremer, D., and J. A. Pople. 1975. General definition of ring puckering coordinates. *J. Am. Chem. Soc.* 97(6):1354–1358.

Dalby, A., J. G. Nourse, W. D. Hounshell, A. K. I. Gushurst, D. L. Grier, B. A. Leland, and J. Laufer. 1992. Description of several chemical structure file formats used by computer programs developed at Molecular Design Limited. *J. Chem. Inf. Comput. Sci.* 32(3):244–255.

Dos Santos, H. F., W. R. Rocha, and W. B. De Almeida. 2002. On the evaluation of thermal corrections to gas phase ab initio relative energies: Implications to the conformational analysis study of cyclooctane. *Chem. Phys.* 280(1–2):31–42.

Goldsmith, M. -R., N. Jayasuriya, D. N. Beratan, and P. Wipf. 2003. Optical rotation of noncovalent aggregates. *J. Am. Chem. Soc.* 125:15696–15697.

HyperChem. 2014. Hypercube, Inc. http://www.hyper.com/.

MacroModel 9.6. 2008. Schrodinger LLC, NewYork, NY.

Polavarapu, P. L. 1993. Chiral force constants. Recommendations for the presentation of internal coordinate force constants. *J. Computational Chem.* 14, 751–752.

Polavarapu, P. L. 2012. Molecular structure determination using chiroptical spectroscopy: Where we may go wrong? *Chirality* 24(11):909–920.

Polavarapu, P. L., E. A. Donahue, K. C. Hammer, V. Raghavan, G. Shanmugam, I. Ibnusaud, D. S. Nair, C. Gopinath, and D. Habel. 2012. Chiroptical spectroscopy of natural products: Avoiding the aggregation effects of chiral carboxylic acids. *J. Nat. Prod.* 75(8):1441–1450.

Polavarapu, P. L., E. A. Donahue, G. Shanmugam, G. Scalmani, E. K. Hawkins, C. Rizzo, I. Ibnusaud, G. Thomas, D. Habel, and D. Sebastian. 2011. A single chiroptical spectroscopic method may not be able to establish the absolute configurations of diastereomers: Dimethylesters of hibiscus and garcinia acids. *J. Phys. Chem. A* 115(22):5665–5673.

Polavarapu, P. L., G. Scalmani, E. K. Hawkins, C. Rizzo, N. Jeirath, I. Ibnusaud, D. Habel, D. Sadasivan Nair, and S. Haleema. 2010. Importance of solvation in understanding the chiroptical spectra of natural products in solution phase: Garcinia acid dimethyl ester. *J. Nat. Prod.* 74(3):321–328.

Polavarapu, P. L., and R. Vijay. 2012. Chiroptical spectroscopy of surfactants. *J. Phys. Chem. A* 116(21):5112–5118.

Smuts, J. P., X. -Q. Hao, Z. Han, C. Parpia, M. J. Krische, and D. W. Armstrong. 2013. Enantiomeric separations of chiral sulfonic and phosphoric acids with barium-doped cyclofructan selectors via an ion interaction mechanism. *Anal. Chem.* 86(2):1282–1290.

Spartan. 2014. Wavefunction Inc. https://www.wavefun.com/products/spartan.html.

Stephens, P. J., J. -J. Pan, F. J. Devlin, M. Urbanová, and J. Hájíček. 2007. Determination of the absolute configurations of natural products via density functional theory calculations of vibrational circular dichroism, electronic circular dichroism and optical rotation: The schizozygane alkaloid schizozygine. *J. Org. Chem.* 72(7):2508–2524.

7

Experimental Methods for Measuring Chiroptical Spectra

7.1 Introduction

The experimental methods differ significantly for measuring optical rotatory dispersion (ORD), electronic circular dichroism (ECD), vibrational circular dichroism (VCD), and vibrational Raman optical activity (VROA) spectra. Commercial instruments are available for measuring each of these four spectra, which lead to both advantages and disadvantages. The main advantage results from the availability of readily usable instruments (although at a significant financial commitment), eliminating the need to build instruments in individual laboratories and facilitating scientists with different research backgrounds in pursuing research in these areas. The main disadvantage accrues when commercial instruments are used as black boxes and potential artifacts (i.e., spurious signals) are not recognized. The goal in this chapter is to give an overview of the experimental methods and to point out the potential pitfalls that one should be aware of.

7.2 Optical Rotatory Dispersion

7.2.1 Optical Rotation of Liquid Solutions

The experimentally measured optical rotation (OR) is converted and reported as a specific optical rotation (SOR), which in the older literature used to be referred to as a specific rotation (SR).

A plot of SOR as a function of wavelength is referred to as the ORD curve/spectrum. The instruments for measuring OR are referred to as polarimeters or spectropolarimeters, and the concepts involved in these measurements are very simple (see Figure 7.1). OR measurements are routinely performed in the visible spectral region (Djerassi 1960; Crabbé 1965). In constructing a polarimeter, the first step is to generate the light component of appropriate

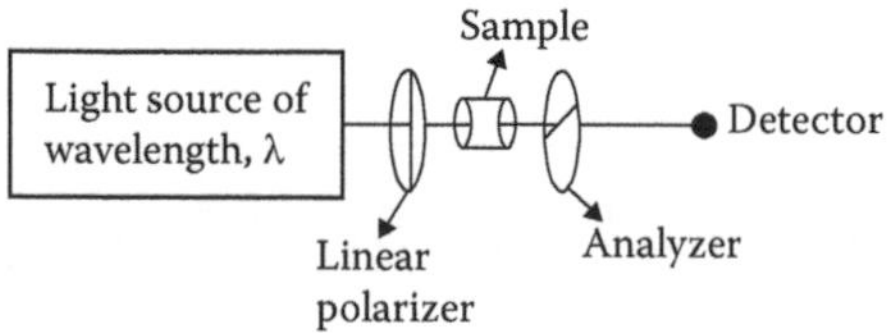

FIGURE 7.1
Schematic of a polarimeter for optical rotation measurements.

wavelength in the visible spectral region. Three different types of research-grade commercial polarimeters are in current use, and these instruments differ in the way the visible light wavelengths are selected. (a) Dispersive scanning polarimeters use a broadband light source, and a scanning monochromator equipped with diffraction grating. The underlying principles of a monochromator can be found in Appendix 6. These instruments can continuously change the probing light component by using a servo-motor controlled rotation of the grating, and OR can be measured as a function of wavelength at a desired wavelength resolution. The wavelength resolution is determined by the components (such as groove density of grating, focal length, and slit width) of the monochromator used. These instruments are commercially available (e.g., from JASCO Inc.) and are usually expensive. (b) Discrete wavelength polarimeters use a broadband light source with narrow band-pass wavelength filters, where a few band-pass filters, each transmitting around a different wavelength, are placed on a rotating wheel. The desired wavelength is then obtained by rotating the electronically controlled wheel so that the appropriate band-pass filter is brought into the light beam. These instruments are manufactured by a handful of companies (e.g., JASCO Inc., Rudolph Research Analytical and Rudolph Research) and can measure OR at discrete wavelengths (preselected by the band-pass filters) and are usually inexpensive. The most commonly used narrow band-pass filters are for use at wavelengths 633, 589, 546, 435, 405, and 365 nm. (c) Laser polarimeters use a monochromatic laser light source. These polarimeters are commercially available (e.g., from PDR-Chiral) and are limited to measurements at the wavelength of the laser used (unless multiple lasers of different wavelengths or a tunable laser can be accommodated in the instrument).

The selected light component (dispersed from a monochromator or broadband light filtered with a narrow band-pass filter or monochromatic laser light) is linearly polarized using an appropriate linear polarizer, referred to as the input polarizer. Glan–Thomson calcite polarizers offering better rejection of unwanted polarization are made up of two calcite prisms cemented together and are used as linear polarizers. High-quality polaroid polarizers are also used for achieving the linearly polarized light. The linearly polarized light component passes through the sample held in an appropriate polarimeter cell, then through a linear polarization analyzer (similar to the input polarizer), referred to as the output analyzer, and finally to an appropriate light detector.

When the output analyzer axis (A) is perpendicular to that of input polarizer (P), and the sample is optically inactive or without any sample in the beam, the light is prevented from reaching the detector (see Figure 7.2). This position is referred to as the null point. When a chiral sample is introduced into the sample cell (S), linear polarization of the light component can be rotated by the sample (see Chapter 3, Figure 3.1), so the analyzer position no longer represents the null point. The analyzer axis must then be rotated to a new position to get the null point. The shortest angle α by which the analyzer axis is rotated to find this null point is the measured OR. The angle is positive if the rotation is clockwise and negative if the rotation is counterclockwise when viewed into the incoming light (Eliel et al. 2001). The SOR is then deduced from the measured OR using Equation 3.21.

Calibration of polarimeters: Calibrated quartz crystal plates, traceable to the National Institute of Standards and Technology (NIST), are used to verify the accuracy of measured OR. Samples of pure sucrose and dextrose, with rotation value specified by the NIST, may also be used. One of these calibration methods should be used to verify the correct functioning of the instrument periodically. Certified quartz plates or sucrose samples are usually supplied by the commercial manufacturers of polarimeters.

Amount of sample needed for OR measurements: The measured OR of a sample depends not only on the length of light travel through the sample (i.e., sample cell path length) and concentration but also on the wavelength of light, molecular structure, concentration, solvent, and temperature. Therefore, the

(I). Without sample in the light beam

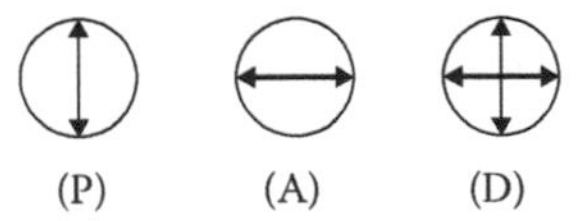

(II). With sample in the light beam

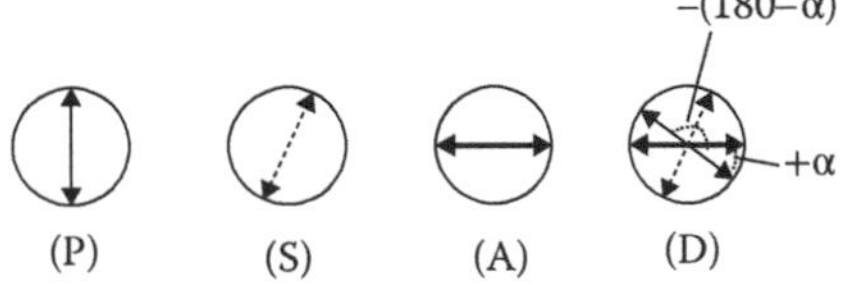

FIGURE 7.2
Orientations of light polarization before and after passing through the sample and of analyzer. (I) When there is no sample in the beam, the polarization axes of the input polarizer (P) and the output analyzer (A) are perpendicular to each other, corresponding to the null point at the detector (D). (II) When there is a chiral sample in the beam, the polarization axis of input light (P) is rotated by the sample (S), and the analyzer axis (A) needs to be rotated by α to reach the null point at the detector (D). The orientation of the analyzer axis at the null point will be at 90° to the polarization axis exiting the sample (S).

combination of concentration and cell path length, needed to register the OR above noise level, will have to be determined through trial measurements for individual cases. Low sample concentrations in the range of mg/mL are typically used for OR measurements. Volumes of the order of 1 mL and more are usually required to fill the polarimeter cells.

Polarimeter cells: Polarimeter cells equipped with inexpensive glass windows for measurements in the visible region and with fused silica windows for measurement at wavelengths shorter than visible wavelengths are normally used. The maximum possible amount of the sample can be held in the path of the light, when long cylindrical polarimeter cells (see Figure 7.3) are used. Polarimeter cells of different lengths, varying from 25 to 200 mm, are commercially available. However, a 0.5 dm (5 cm) length cell is the most commonly used. Better accuracy in the measured values can be realized when longer cells are used. However, longer cells can result in increased absorption of visible light (especially for colored samples) and then, increased detector noise due to a smaller amount of light reaching the detector. In such cases, one has to use shorter length cells. The measured rotation is independent of the bore diameter of the cell, but a smaller diameter may block the light thereby increasing the noise in measurement. A larger diameter bore, sufficient to pass the light unobstructed, minimizes the noise. However, larger diameter bore cells require larger amounts of solution to fill the cell. Figure 7.3 shows the schematics of different types of polarimeter cells used for OR measurements. The formation of bubbles is a practical problem in filling the cylindrical cells (especially cells with narrower bore) with a liquid solution. If the formed bubble stays in the light path, the OR reading can become unstable. Therefore, one must view through the cell along the cylindrical axis and ensure the absence of bubbles in the cell, if the cell is not made of a transparent glass body. Cells with a bubble trap (Figure 7.3b) or cells equipped with a central funnel (Figure 7.3c) are therefore preferred. In the case of cells with a central funnel, care must be taken to avoid the concentration variation caused by evaporation during the measurement by using a suitable cover for the funnel.

Wavelength dependence: It is a common practice for synthetic chemists to measure the OR values at the sodium D line (589 nm). However, the OR magnitude generally increases as the wavelength decreases, and therefore, it can be advantageous to measure OR values at shorter wavelengths. The variation

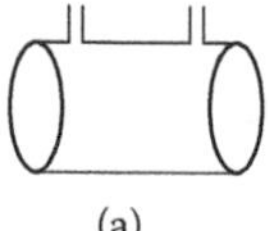
(a)

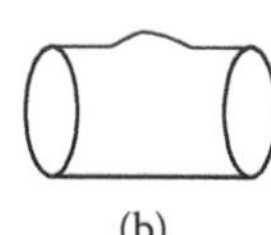
(b)

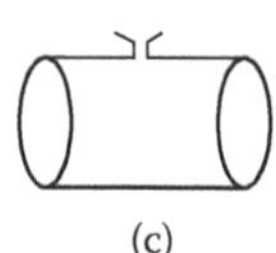
(c)

FIGURE 7.3
Different cylindrical sample cells, available for optical rotation measurements, with (a) inlet and outlet, (b) bubble trap, and (c) central funnel.

of OR α with wavelength, λ, is often represented by Drude's equation (Lowry 1964; Lakhtakia 1990):

$$\alpha(\lambda) = \sum_{m} \frac{k_m}{\lambda^2 - \lambda_m^2}, \tag{7.1}$$

where the summation is over electronic transitions (each of wavelength λ_m) and k_m is a constant. Note that $\alpha(\lambda)$ goes to infinity at $\lambda = \lambda_m$, so Drude's equation is applicable at wavelengths away from the locations of electronic transitions. The theoretical basis for Equation 7.1 can be found in the quantum chemical derivation of the analogous equation in Chapter 5 (see Equation 5.6).

With discrete wavelength polarimeters, OR can be measured at several wavelengths, converted to SOR and a coarse ORD spectrum derived. Discrete wavelength ORD spectrum of the aqueous solution of ammonium-D-camphor-10-sulfonate at 1.2 mg/mL concentration, obtained from OR measured at six different wavelengths, is shown in Figure 7.4. The dispersive scanning polarimeters, in contrast, can measure the continuous ORD spectrum at a desired wavelength resolution.

Solvent dependence: The magnitude of SOR can vary with the solvent for different reasons. One is the difference in refractive indices of different solvents. However, changes in magnitudes resulting from the differences in refractive index are usually small. A change in magnitude and, sometimes, also the sign of SOR occurs more often due to predominance of different conformations of the sample molecules in different solvents and/or strong interactions (such as hydrogen bonding) between solute and solvent. Such changes can be seen in the ORD spectra of dimethyl-L-tartrate in three different solvents, shown in Figure 7.5. It will be extremely useful to routinely measure the OR in different solvents.

Temperature dependence: The populations of individual conformations of conformationally labile compounds can change with temperature, and therefore SOR can change with temperature. Even for a sample with single conformer molecules, the SOR can still change with temperature, because SOR is inherently dependent on temperature due to the influence from molecular

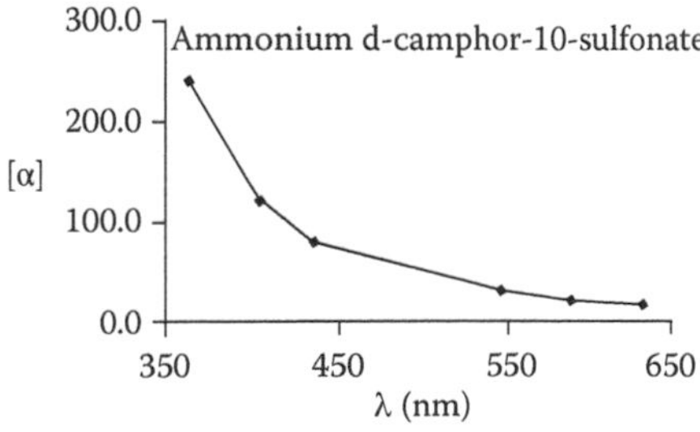

FIGURE 7.4
Discrete wavelength ORD spectrum of ammonium-d-camphor-10-sulfonate showing wavelength dependence of specific rotation. (From Polavarapu, P. L., *Chirality*, 18(9), 723–732, 2006.)

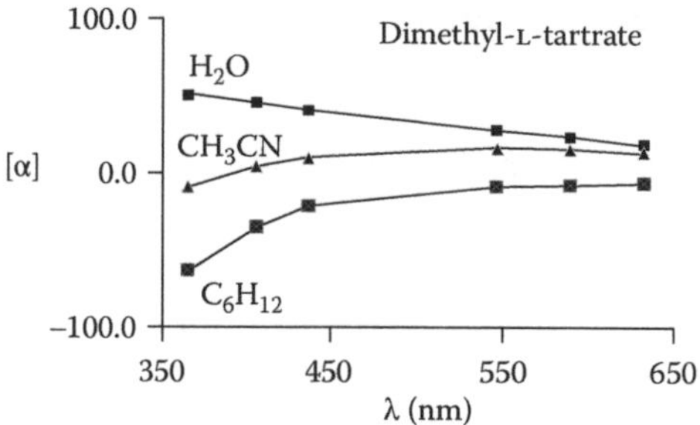

FIGURE 7.5
Discrete wavelength ORD spectra of dimethyl-L-tartrate in different solvents showing solvent dependence. (From Polavarapu, P. L., *Chirality*, 18(9), 723–732, 2006.)

vibrations (Wiberg et al. 2004; Mort and Autschbach 2006, 2007). Therefore, the temperature-dependent SOR measurements can also provide additional molecular information.

Concentration dependence: A higher concentration sample will have a larger observed OR than dilute concentration samples. For chiral molecules that do not aggregate, form assemblies, or do not interact with solvent molecules, OR is expected to change linearly with concentration, and hence SOR is expected to be independent of concentration. However, for molecules that aggregate, form organized assemblies, or have interactions with solvent molecules, OR can vary nonlinearly with concentration. As a consequence, SOR need not be independent of concentration (Polavarapu et al. 2003; He et al. 2004). The measurement of OR at multiple concentrations, to determine the functional dependence of OR on concentration, should be considered a prerequisite. Figure 7.6 shows the concentration dependence of OR and SOR for (*S*)-(−)-3-butyn-2-ol in CCl_4. Note that 3-butyn-2-ol can behave differently at higher and lower concentrations due to the formation of intermolecular hydrogen bonds at higher concentrations. Nevertheless, OR should approach zero as the concentration approaches zero. However, SOR need not approach zero as the concentration approaches zero; instead SOR extrapolated to zero concentration is referred to as the intrinsic rotation (Eliel et al. 2001; Polavarapu et al. 2003) and designated with α placed in curly brackets. Intrinsic rotation provides the SR value of a solute molecule isolated in the solvent cage, so solute–solute interactions can be considered to be absent in the experimentally measured intrinsic rotation. When intrinsic rotations are measured in different solvents, the differences among experimental intrinsic rotation values in different solvents provide the magnitude of solvent cage influence. However, such studies require much patience. Special precautions to minimize the errors (see below) associated with the use of low concentrations are to be taken.

Errors in SOR measurements: Errors in typical polarimeter measurements are large, unless special precautions are taken (Dewey and Gladysz 1993).

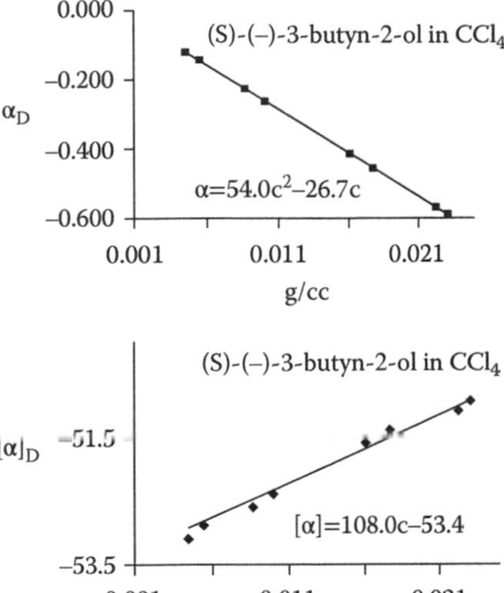

FIGURE 7.6
Optical rotation and specific optical rotation of (*S*)-(–)-3-butyn-2-ol in CCl_4 at sodium D line showing dependence on concentration. Equations indicated on the graphs are used to fit the experimental data. (From He, J., *J. Phys. Chem. A*, 108(10), 1671–1680, 2004.)

For example, consider a measurement that used a 1-mg sample in 2 mL solution and a 1 dm cell and yielded an observed OR of 0.01°. Then, c = 0.001 g/2 mL = 5 × 10^{-4} g/cc; ℓ = 1 dm; $[\alpha] = \alpha/c\ell = 0.01°/(5 \times 10^{-4} \times 1\text{ dm})$ = 20 deg cc/(g dm). Assuming an error of 0.0001 g in the weight of the sample, 10% error in volume, and 0.002° error in observed rotation (typical in commercial polarimeters), standard error propagation gives

$$\frac{S^2_{[\alpha]}}{[\alpha]^2} \sim \frac{S^2_{\text{weight}}}{[\text{weight}]^2} + \frac{S^2_{\text{volume}}}{[\text{volume}]^2} + \frac{S^2_{\text{obs rotation}}}{[\text{obs rotation}]^2} = \frac{(0.0001)^2}{(0.001)^2} + \frac{(0.2)^2}{(2)^2} + \frac{(0.002)^2}{(0.01)^2}$$

$$= 0.01 + 0.01 + 0.04$$

or $S_{[\alpha]} = 4.9$ and $[\alpha] = 20 \pm 4.9$, which amounts to ~25% error. Of course, another source of error would be when solutions are not made up volumetrically. From the example illustrated above, a large portion of the error originates from the measured OR with magnitude of 0.01 or less. Thus, when measurements are made with typical polarimeters with a stated accuracy of 0.002°, it is necessary to adjust the concentration and path length to yield measured OR magnitudes of 0.01° or greater.

The errors in SOR measurements can be reduced with the following precautions: (a) using a larger amount of sample minimizes the error in weight; (b) preparing a larger volume of solution minimizes the error in volume;

(c) choosing concentrations that yield observed rotation in the first decimal place (~0.1°) minimizes the error in OR measurement.

Caveats in OR measurements: If the polarization of light was rotated by the sample solution by α, then the null position can be obtained by rotating the analyzer by either +α, or −(180−α) (Eliel et al. 2001) (see Figure 7.2). If +α is correct, decreasing the concentration of the sample solution by one-half should yield a measurement of +α/2. In contrast, if −(180−α) is correct, then decreasing the concentration of the sample solution by one-half should yield −(180−α)/2. For example, suppose that the null position was obtained by rotating the analyzer by +10°. One could also have obtained the null position by rotating the analyzer by −(180−10)° = −170°. Then, α could have been either +10° or −170°. If +10° is correct, at one-half of the sample concentration, the null position should be +5°. If −170° is correct, then at one-half of the sample concentration, the null position should be −175°/2 = −87.5°. Typical samples, at typical concentrations, exhibit small magnitudes of rotations, so large magnitude rotations of the order of tens of degrees are unlikely. Nevertheless, a good experimental practice is to keep all possible options in consideration. The precautions mentioned in this paragraph apply to manual polarimeters, where the user rotates the analyzer manually. The research-grade commercial polarimeters, however, rotate the analyzer with electronic control and report the OR value without user intervention. The electronic control, for the rotation of the analyzer, is guided by a modulator that oscillates the input polarization using either a Faraday modulator or a mechanical modulator.

Recent advances in OR measurements of liquid solutions: A new method (Bougas et al. 2015) for measuring the OR for liquid solutions is the use of a chiral cavity ring down polarimetry (CCRDP) (see Section 7.2.2). Using this method, the ORs for three different sucrose/water solutions were reported at 800 nm.

7.2.2 OR of Gas-Phase Samples

Although the commercial instruments can be used to measure the OR of liquid or liquid solution samples, it is challenging to do the same for vapor-phase samples (especially for samples with low vapor pressures), because the number of molecules present in a short path length gas cell may not be enough to register the OR. In principle, one can device an instrument, like the one used for liquid-phase measurements, to accommodate very long gas cells (Lowry 1964). The need for such long cells is avoided by the recent technological advances in cavity ring down techniques utilizing a two-mirror cavity. Vaccaro and coworkers (Müller et al. 2000) developed cavity ring down polarimetry (CRDP) using a two-mirror cavity. CRDP led to reliable vapor-phase measurements and is playing an important role in understanding the role of solvation that is gleaned through the differences between vapor-phase and liquid-phase SORs (Lahiri et al. 2014). A detailed description of CRDP with a two-mirror cavity is beyond the scope of this book, and the interested reader is referred to the original articles (Müller et al. 2000).

A more recent development (Sofikitis et al. 2014; Bougas et al. 2015) is the use of a bowtie ring cavity with two distinct counter-propagating laser beams and the utilization of reversible longitudinal magnetic field to suppress the artifacts. This method was abbreviated as CCRDP. The measurements of OR for α-pinene, 2-butanol, and α-phellandrene vapor samples at 800 nm were demonstrated.

7.2.3 Time-Resolved ORD Measurements

The methods for ORD measurements with nanosecond time resolution (Shapiro et al. 1995) and femtosecond resolution (Rhee, Eom, and Cho 2011) have been developed, and these measurements are useful for protein folding studies. The time-resolved ORD measurements were also coupled with the T-jump method for studying the fast reaction mechanisms (Chen et al. 2010).

7.2.4 Evanescent Wave OR Measurements

The OR measurement method using a bowtie ring cavity with two distinct counter-propagating laser beams, described in Section 7.2.2, has been extended to evanescent wave OR measurements (Sofikitis et al. 2014; Bougas et al. 2015). The liquid solution samples were held in a flow-through prism cell, and the ORs of maltodextrin and fructose solutions in the evanescent wave were measured (Sofikitis et al. 2014).

7.3 Circular Dichroism Measurements

When light in the visible or ultraviolet region is passed through a chemical sample of interest, the absorption of that light results in molecules going from the ground electronic state to an excited electronic state. Differential absorption of left and right circularly polarized visible/ultraviolet light is then referred to as ECD, because this differential absorption results from molecular electronic transitions. The ECD spectral region is divided into two subregions, namely vacuum ultraviolet (VUV, <~190 nm) and ultraviolet–visible (UV–vis, ~190–700 nm) regions. Measurements in the VUV regions are not accessible with most of the commercial CD spectrometers, and such measurements are limited to specialized laboratories or to those with access to synchrotron radiation sources. Measurements in the UV–vis region are accessible with commercial CD spectrometers, and therefore most widely studied ECD spectra are in the UV–vis region.

ECD spectroscopy has been in use for several decades as an established method (Berova et al. 2000; Eliel et al. 2001; Harada and Nakanishi 1983; Lightner and Gurst 2000; Woody 1996, 1995; Johnson 1990; Gray et al. 1992;

Sreerama and Woody 2000; Berova et al. 2007; Superchi et al. 2004; Berova et al. 2010; Berova et al. 2000). The popular use of ECD spectroscopy in the past has resulted from (a) the low sample concentrations that can be employed for the experimental studies and (b) relating the observed ECD spectral patterns to molecular stereochemistry using interpretational methods that included empirical sector rules (Lightner and Gurst 2000), exciton coupling model (Harada and Nakanishi 1983) and Devoe's polarizability model (Superchi et al. 2004). However, as the number of applications of ECD for determining stereochemistry has increased, the limitations of empirical sector rules, as well as questions regarding the reliability of interpretational methods, began to emerge. Resurgence of ECD spectroscopy in recent years as a valuable technique for chiral structural analysis resulted from the advances made in quantum chemical predictions of ECD spectra. This latest development brought a new outlook for ECD spectroscopy.

The absorption of infrared (IR) light at appropriate wavelength results in a vibrational transition (molecules going from the ground vibrational state to an excited vibrational state, both states belonging to the same ground electronic state). When left and right circularly polarized IR light is passed through a sample, the resulting CD is then referred to as IR VCD (Holzwarth et al. 1974). The IR region is divided into three subregions, namely near-IR (~1–2.5 μm), mid-IR (~2.5–25 μm), and far-IR (~25–100 μm). In the near-IR region, vibrational transitions are mostly classified as overtones and combinations, besides O–H, N–H, and C–H stretching vibrations. Overtone transitions result when the change in quantum number for a vibrational mode is greater than 1. Combination transitions result when the quantum numbers for more than one vibrational mode change simultaneously. The interpretation of VCD associated with overtones and combination bands is not as straightforward (Polavarapu 1996; Abbate et al. 2000). VCD in the mid-IR region arises mostly from the fundamental vibrational transitions that can be associated with functional groups and can be predicted reliably using quantum chemical methods. VCD measurements have been attempted in the far-IR region (Polavarapu and Deng 1996), although they are not yet routine. Thus, most of the VCD measurements have been carried out in the mid-IR region, with a low-frequency cutoff at ~800 cm^{-1} due to the restricted transmission of the ZnSe photoelastic modulator (PEM) that is used for generating circularly polarizing the IR light.

The successful emergence of VCD in the mid-IR as a viable technique for chiral structural analysis results from the advances made both in instrumentation and in quantum chemical predictions of VCD spectra. These two developments together brought this area to the research forefront. The advantages resulting from shifting the region of measurement of CD from the visible to the mid-IR region are as follows: (a) Since every nonlinear molecule has $(3N - 6)$ vibrations, where N is the number of atoms in the molecule, that many vibrational transitions, in principle, are available for VCD investigations. The number of accessible transitions in the mid-IR region is much greater than that for accessible electronic transitions in the UV–vis range, even though all of the (3*N*–6) vibrational transitions may not be accessible in practice. The process of analyzing and interpreting a larger number of transitions introduces more stringent requirements. Therefore, the conclusions resulting from analyzing a large number of transitions have a lower probability of going wrong. (b) The vibrational absorption (VA) and VCD spectra in the mid-IR region can be reliably predicted using quantum chemical calculations. The established reliability in the quantum chemical predictions of mid-IR VCD has been a crucial factor for successful chiral structural analysis using the mid-IR VCD. However, there are also some disadvantages, as with any other method, associated with mid-IR VCD measurements: (a) Longer wavelengths associated with the mid IR region probe the molecular chirality less efficiently compared with shorter wavelengths associated with the visible region. As a result, CD signals in the mid-IR region are much smaller (by about two to three orders of magnitude) than those in the visible. (b) Detectors used for mid-IR spectral detection are inherently less sensitive than the corresponding detectors used for visible spectral detection. As a result, the signal-to-noise ratio associated with CD in the mid-IR region is lower than that in the visible region. These disadvantages translate into the need for longer data collection times and larger sample concentrations. (c) Another disadvantage, not often mentioned, is the "(3*N* – 6) disadvantage" for larger molecules. As the size of the molecule increases (e.g., large size natural products), the number of vibrational transitions that appear in the mid-IR region increases, resulting in the overlap of multiple vibrational transitions, poorer resolution, and higher uncertainty in understanding and interpreting them.

It should be emphasized that neither ECD nor VCD is uniquely favored for all types of chemical compounds. Therefore, one should investigate both techniques simultaneously, when possible, to evaluate the merits of each of these two methods in the specific context under consideration.

Irrespective of the wavelength region (visible vs. IR) used for CD measurements, the principles of CD measurement are the same. The differences for different regions arise from the corresponding differences in light sources, appropriate transmission properties of the materials (polarizing devices, lenses), and light detectors. There are two types of instruments commercially available for measuring CD: (a) dispersive CD spectrometers that utilize a monochromator (Appendix 6), and (b) Fourier transform (FT) CD spectrometers that utilize an interferometer (Appendix 6). The dispersive instruments are the only ones commercially available for ECD measurements while no commercial dispersive instrument is available for VCD measurements. One typical drawback of dispersive instruments is attributed to a significant light loss at the entrance and exit slits of the monochromator. This throughput disadvantage is avoided in FT spectrometers. Furthermore, in FT spectrometers, the nature of interference phenomenon permits the detection of light at all wavelengths simultaneously. These two advantages have led to the adoption of FT spectrometers for the IR region and for VCD measurements (Nafie et al. 1979). Commercially available VCD instruments are based on modified Fourier transform infrared (FTIR) spectrometers.

7.3.1 Dispersive Spectrometers

Historically, dispersive spectrometers have been the workhorses for spectroscopic measurements. While dispersive spectrometers for ECD measurements in the visible region have been around for a long time, those for CD measurements in the IR region were developed in the mid-1970s. The early instruments were developed for measuring CD associated with low-lying electronic transitions of metal complexes in the near-IR region (Chabay et al. 1972; Osborne et al. 1973). The first reference to CD in the vibrational transitions, and hence the discovery of VCD, was made by Holzwarth and coworkers (Holzwarth et al. 1974) in 1974. VCD measurements with better signal quality were reported in the following year by Stephens and coworkers (Nafie et al. 1975). Following these developments, the dispersive VCD spectrometers began to be built in individual laboratories. The most recent work of Keiderling and coworkers with a dispersive VCD spectrometer (Lakhani et al. 2009) can be consulted for further information.

A schematic of dispersive CD instrument is shown in Figure 7.7. Instrumental components suitable for different spectral regions are summarized in Table 7.1. Light from an appropriate source is processed to achieve monochromatic light (light component at a particular wavelength) in the region of interest. Commercial ECD spectrometers and the in-house developed early-stage VCD spectrometers used monochromators for wavelength selection (see Appendix 6 for details on wavelength dispersion achieved by monochromators). The light component of the desired wavelength exiting the monochromator is passed through a linear polarizer to achieve linearly polarized light. Calcite polarizers and wire grid polarizers (metal wires deposited on IR transmitting substrate, such as BaF_2) are used, respectively, for visible and IR regions. The linear polarization is then converted and modulated between left and right circular polarizations using a PEM, at a frequency that is characteristic of the optical component used for constructing the PEM. The alternating left and right circularly polarized light components are passed through a chiral sample and then on to an appropriate

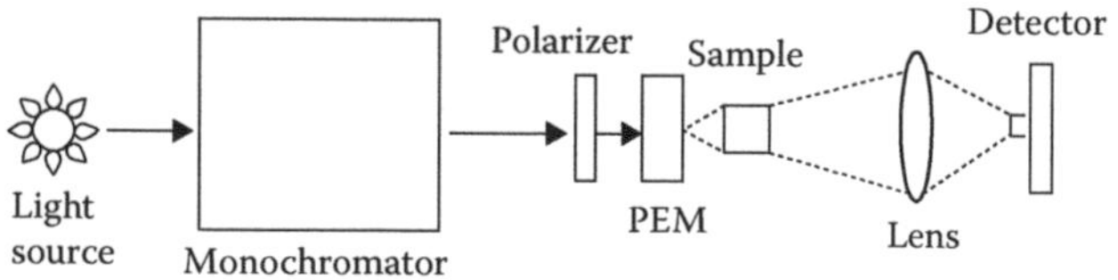

FIGURE 7.7
Schematic of a dispersive circular dichroism instrument. PEM, photoelastic modulator.

TABLE 7.1

Instrumental Components Suitable for CD Measurements in Different Spectral Regions

		Infrared Region		
	Visible Region	**Near-IR**	**Mid-IR**	**Far-IR**
Light source	Xenon lamp		Carbon rod	
Diffraction	Prisms, Gratings		Gratings	
Interferometer			Michelson, polarizing	
Polarizers	Calcite	Wire grids on BaF_2 substrate		Wire grids on Mylar
Optical element for PEM	Si, CaF_2	ZnSe, CaF_2		NA
Detector	PMT, CCD	InSb	HgCdTe	Bolometer

PMT, photomultiplier tube; CCD, charge-coupled device.

light detector. When the chiral sample absorbs left and right circularly polarized components differently, the signal at the detector varies synchronously with polarization modulation. This oscillating detector signal is processed electronically to extract the CD signal. The principles associated with polarization modulation and procedures to extract CD signals can be found in Appendix 7.

7.3.2 FT Spectrometers

The central component of FT spectrometers is the interferometer. Although several types of interferometers have been developed, the Michelson interferometer is widely used in commercial FT instruments. FT spectrometry is widely used in the IR region but not in the UV–visible region. Therefore, the discussion in this section applies to FTIR spectrometers based on the Michelson interferometer and their use for VCD measurements. Two types of FT-VCD instruments are commercially available: FT-VCD instruments with (a) single polarization modulation and (b) dual polarization modulation. The former method discussed in this section relies on reducing the artifacts through well aligned optics, while the latter method, discussed in Section 7.3.3, eliminates some of the artifacts using an additional PEM and associated processing electronics. The principles of polarization modulation are given in Appendix 7.

The schematic of FT-VCD spectrometers using a Michelson interferometer and single polarization modulation is shown in Figure 7.8a. Light from a polychromatic mid-IR source is passed through a Michelson interferometer, which contains a beam splitter (BS), a fixed mirror, and a movable mirror. The functioning of a Michelson interferometer is described in Appendix 6. The light intensity exiting the interferometer is modulated at different frequencies for

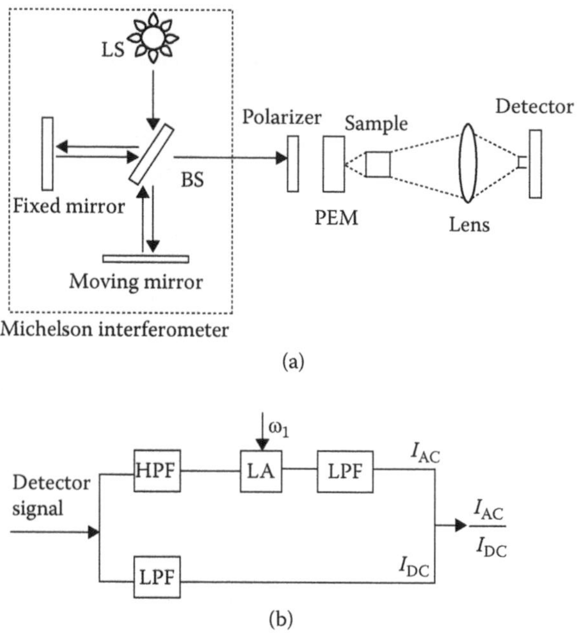

FIGURE 7.8
(a) Schematic of an FT-VCD spectrometer with single polarization modulation. LS, light source; BS, beam splitter; PEM, photoelastic modulator. (b) Electronic signal processing. HPF, high-pass filter; LA, lock-in amplifier (tuned to the frequency of PEM); LPF, low-pass filter; I_{AC}, Fourier-transformed ac signal; I_{DC}, Fourier-transformed dc signal.

different wavelengths, due to interference of the light waves from the two arms of the interferometer. These frequencies, called Fourier frequencies, depend on the velocity of the moving mirror in the interferometer. The light exiting the interferometer is circularly polarized using a linear polarizer and PEM. The PEM converts the linearly polarized light into alternating left and right circularly polarized light components (see Appendix 7). The modulation between circularly polarized components takes place at a frequency that is characteristic of the optical element used in PEM (usually ~37 kHz for ZnSe optical element). This alternating circularly polarized light is passed through the sample of interest and then focused onto a liquid nitrogen cooled detector using a lens. The use of lenses, to manipulate the light after it was circularly polarized, has been a key factor in reducing the artifact signals (Malon and Keiderling 1988; Chen et al. 1994; Tsankov et al. 1995).

The outline of electronics used to process the detector signal is shown in Figure 7.8b. The electronic signal at the detector contains a high-frequency (referred to as ac) component riding on top of a low-frequency (referred to as dc) component. The high-frequency component arises from differential absorption of left and right circularly polarized light components, and the low-frequency component comes from intensity modulation in the interferometer (see Appendix 7). These two components are separated using a high-pass electronic filter and another low-pass electronic filter (LPF). The high-frequency component is processed by a lock-in amplifier (LA), which is tuned to the frequency of PEM, further filtered with an LPF and then Fourier transformed to extract the signal I_{AC}. The low-frequency component, isolated using an LPF, is Fourier transformed separately to extract the signal I_{DC}. The CD signal is derived from the ratio of the Fourier transformed ac component to the Fourier transformed dc component, that is, I_{AC}/I_{DC}. During the measurement of FTIR-VCD, both VA and CD spectra are measured for each sample. The recent availability of dedicated commercial instruments (from BioTools, Jasco and Bruker) for measuring VCD, based around this FT method, has been a major factor for the increased adaptation of this technique.

Enhanced signal quality using an FT spectrometer with cube corner mirrors and dual light sources has also been reported (Nafie et al. 2004).

An alternative to CD measurements using the FT method is the use of polarizing interferometers (Dignam and Baker 1981). Although VCD measurements were reported using polarizing interferometers (Ragunathan et al. 1990; Polavarapu et al. 1994; Polavarapu and Deng 1996), they have not been routinely adopted. More details on polarizing interferometers can be found in Appendix 6 and in the literature (Polavarapu 1997).

7.3.3 Potential Artifacts

The magnitudes of CD signals are usually small. The CD signals (normalized to corresponding absorption signals) generally range from 10^{-2} to 10^{-3} in ECD spectra and from 10^{-4} to 10^{-6} in VCD spectra. Because of these smaller magnitudes (more so in the case of VCD), the real signals can often be dominated by artifacts resulting from linear birefringence and linear dichroism of the samples and of optical components. As a result, the practitioners of chiroptical spectroscopy need to exercise extreme caution for experimental measurements. As a rule of thumb, the expected mirror image CD features for the enantiomers should be verified, whenever both enantiomers are available. Artifacts in the ECD spectra of liquid solutions are rare (except for large aggregates whose sizes are larger than the wavelength of light; vide infra), but the same cannot be guaranteed for the VCD spectra of liquid solutions, unless special precautions are taken and implemented. Artifacts can appear for both ECD and VCD spectra of solid-state samples, and literature should be surveyed to understand the sources of artifacts before attempting to measure the solid-state CD spectra.

ECD spectra of large-size biological aggregates are known to contain contributions from circular intensity differential scattering (CIDS), when the size of the aggregate is larger than the wavelength of probing light. The telltale sign of these contributions in ECD spectra is the presence of long wavelength tails in the ECD spectra where there is no corresponding absorption. Tinoco and coworkers have measured (Maestre et al. 1982) the CIDS contributions for a 43-μm-long sperm head of the octopus *Eledone cirrhosa* using a

He–Cd laser at 442 nm (0.442 μm). Expressions for the interpretation of CIDS have been reported (Bustamante et al.1985), and methods to eliminate the CIDS contributions from ECD spectra have been developed (Reich et al. 1980; Maestre and Reich 1980).

The artifacts in solution-state VCD spectra generally appear as baseline and absorption-dependent artifacts and can arise from different sources.

Baseline artifacts: For randomly oriented achiral molecules in the absence of external fields, the CD signal should be zero. In practice, exact zero is never realized, so whatever CD magnitude obtained for achiral samples may be considered as the baseline. When both enantiomers are available, CD measurements for both enantiomers are to be recorded, and the expected mirror image features verified. One-half of the difference between these measurements represents the CD signal, and the sum of these measurements represents the baseline. Simultaneous CD measurements are recommended for both enantiomers.

When both enantiomers are not available for liquid solution samples, the baseline is determined by measuring the CD spectrum for the solvent that was used for dissolving the sample. In an ideal instrument, achiral solvents used for CD measurements should yield a flat line with zero intensity as a function of wavelength. In practice, one would see a curved/slanting line (instead of a flat line) with nonzero intensity for achiral samples. The CD signal obtained for achiral solvents is subtracted from that of corresponding sample solutions. For gas-phase samples, the CD spectrum of a blank cell serves to determine the baseline. For film samples deposited on a transparent widow, the blank window serves to determine the baseline.

Absorption-dependent artifacts: The procedure outlined under baseline artifacts is applicable only when no absorption-dependent artifacts are generated by the instrument. An absorption-dependent artifact can be recognized whenever there is a CD signal associated with absorption bands that should not have accompanying CD. For example, the absorption bands of achiral solvents should not exhibit CD. If the baseline obtained for achiral solvents shows CD signals at the locations of absorption bands, the instrument used should be considered to generate absorption-dependent artifacts. In such cases, subtraction of solvent CD spectrum from that of the sample solution does not guarantee the complete elimination of artifacts. In such cases, one must use one-half of the difference between the CD spectra of enantiomers to reduce artifacts.

Minimization of artifacts: The sources of artifacts are better understood now than they were in the early stages of CD measurements. Different instrumental modifications are known to reduce artifacts: (a) use of lenses, instead of mirrors, for steering the light in post-PEM optics; (b) careful alignment of optics; and (c) scrambling the polarization of light before reaching the detector. Polarization scrambling was first introduced for dispersive VCD spectrometers (Cheng et al. 1975) by incorporating a second PEM between the sample and the detector (see Appendix 7). This concept extended to FT-VCD spectrometers, accompanied with an analysis of the sources of artifact signals (Nafie 2000), and is referred to as the dual polarization modulation method. The schematic of FT-VCD spectrometer with dual polarization modulation and associated electronics signal processing is shown in Figure 7.9. Although the difference between single polarization modulation and dual polarization modulation methods is only the introduction of a second PEM in the optical train after the sample, the electronic signal processing of detector signal is more involved in the latter method (compare Figures 7.8b and 7.9b). In the dual polarization method, the detector signal is processed by two LAs—LA1 and LA2—where the signal detected by LA1 is demodulated at the frequency of PEM1 and the signal detected by LA2 is demodulated at the frequency of PEM2. The latter signal is electronically subtracted from the former, which accomplishes the reduction of certain artifact signals. The most recent developments in FT-VCD instrumentation can be found in the recent article by Nafie (2012).

Even after introducing the second PEM in an FT-VCD spectrometer, the signal processing can still be undertaken with a single LA as shown in Figure 7.8b. This approach amounts to polarization scrambling and may help improve the baseline over that obtained with FT-VCD spectrometers using single polarization modulation. A comparison of the VCD spectra

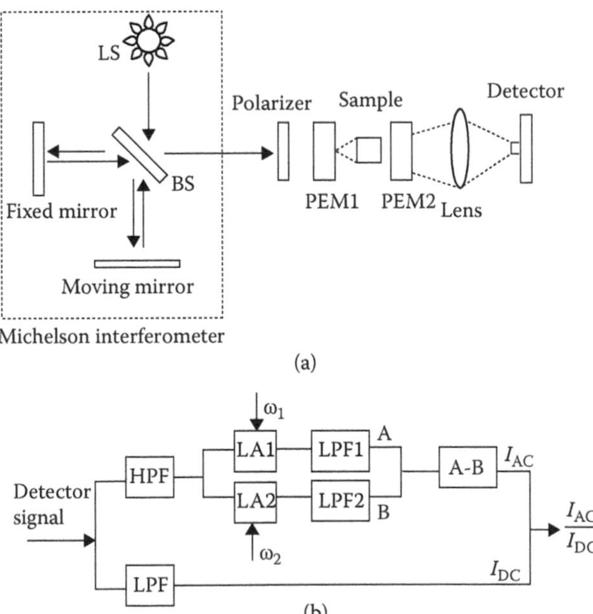

FIGURE 7.9
(a) Schematic of an FT-VCD spectrometer with dual polarization modulation. LS, source; BS, beam splitter; PEM, photoelastic modulators. (b) Electronic signal processing. HPF, high-pass filter; LA1, LA tuned to the frequency of PEM1; LA2, LA tuned to the frequency of PEM2; LPF, low-pass filters; I_{AC}, Fourier-transformed ac signal; I_{DC}, Fourier-transformed dc signal. Signal from LA1: A, signal from LA2: B, and difference signal: A–B.

obtained in different polarization modulation methods has been presented (Polavarapu and Shanmugam 2011).

There is no guarantee, even with the dual polarization method, that all artifact signals are eliminated. To eliminate the remaining artifact signals, the incorporation of a rotating quarter wave (QW) plate between the second PEM and the detector (Cao et al. 2008) and the rotation of the sample around the light propagating axis are recommended (Nafie 2012). Dual polarization modulation adaptation is not available in all commercial FT-VCD spectrometers but can be implemented in individual laboratories relatively easily.

7.3.4 Practical Considerations

Light sources: Polychromatic light sources needed for ECD measurements in the visible region and VCD measurements in the IR region are different. A xenon lamp is often used as the visible light source, while a hot carbon rod is often used as the IR light source. A deuterium lamp covers the ~180–350 nm region while a tungsten halogen lamp covers the ~320–1,000 nm region. For covering shorter wavelengths (up to 160 nm), a xenon lamp (controlled by a high stability, constant current, DC power supply) is used. Synchrotron light sources (Miles and Wallace 2006) are often used for operation in the vacuum UV region. Standard ECD spectrometers operate in the UV–vis (190–800 nm) region and require purging with nitrogen gas. Spectrometers that can operate in the vacuum UV region (below 190 nm) need to be evacuated to prevent UV absorption by air and require special optics.

Calibration of CD spectrometers: The users can calibrate their own instruments by measuring the calibration curves using a multiple quarter-wave plate as described in Appendix 7. The manufacturers of commercial CD spectrometers normally provide guidelines to verify the calibration of their spectrometers.

Verification of correct functioning of spectrometers: Malfunction of instrumental components can cause poor or incorrect performance of CD spectrometers. To avoid finding out such problems after completing lengthy scheduled experiments, one may run some standards before scheduled experiments are carried out. For this purpose, it is sometimes recommended to measure the ECD spectrum of the aqueous solution of ammonium-D-camphor-10-sulfonate (see Figure 7.10) and VCD spectrum of neat liquid, or CCl_4 solution of α-pinene (see Figure 7.11), to verify the correct functioning of respective instruments. In this process, preparing and disposing involved chemicals can be inconvenient and expensive. Such problems can be avoided by preparing a nondegradable freestanding collagen film and using this film for routine verification of CD spectrometers (Shanmugam and Polavarapu 2005). Collagen has very large VCD signals that are characteristic of PPII conformation, and it takes less than a minute to obtain the VCD spectrum. The VCD spectrum of a collagen film is shown in Figure 7.12. The option to use collagen film for routine testing is much more convenient and has been used

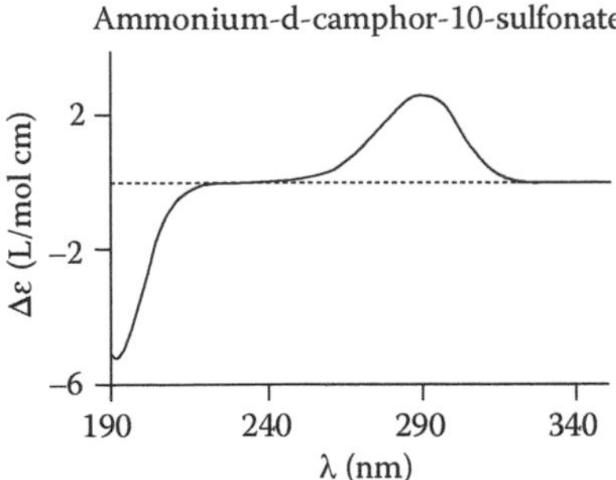

FIGURE 7.10
ECD spectrum of ammonium-d-camphor-10-sulfonate. (From Polavarapu, P. L., *Chirality*, 18(9), 723–732, 2006.)

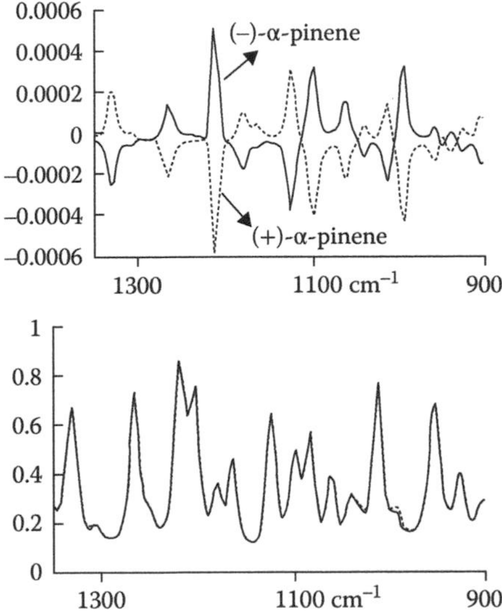

FIGURE 7.11
Raw VCD spectra (top frame) and corresponding absorption spectra (bottom frame) of (+) and (−)-α-pinene enantiomers measured at 8 cm^{-1} resolution in CCl_4 in the mid-infrared region (concentration: 75% by volume). VCD spectra were calibrated with a CdSe wave plate (see Appendix 7).

in the author's laboratory since 2004. No noticeable degradation of collagen film prepared in 2004 is evident to date. The use of collagen film for VCD measurements is analogous to using polystyrene film standard for IR measurements in commercial FTIR spectrometers. However, until a procedure to prepare collagen films of reproducible thickness can be developed, the film prepared in a laboratory remains as a standard just for that laboratory alone for routine verification of the performance of the instrument in that laboratory.

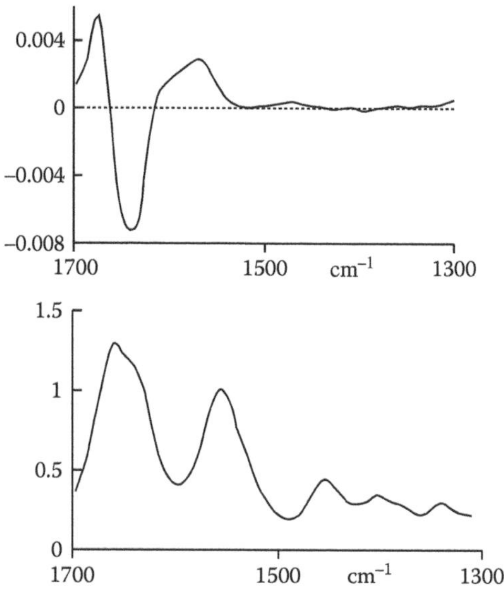

FIGURE 7.12
Vibrational absorption (bottom) and VCD (top) spectra of freestanding collagen film measured at 8 cm^{-1} resolution in the mid-infrared region obtained using 2-min data collection.

Amount of sample needed for CD measurements: The required amount of sample is dictated by the intensities of the sample absorption bands. As a rule of thumb, one needs to keep the absorbance of a band of interest to be less than ~1 (because at this level the amount of light reaching the detector would only be 10% of that entering the sample), with optimal absorbance being ~0.4. In any case, the measured CD signal needs to be above the noise level. For ECD studies, sample concentrations required in the liquid solution phase vary from ~μg/mL to ~mg/mL. The sample volumes required are ~10 μL with 0.1-mm cells, but several milliliters would be required when larger path length cells are used. Typical sample concentrations required for VCD measurements are ~4–100 mg/mL for the liquid-phase studies. The sample volumes required for VCD measurements are ~10 μL when cells of <50 μm path length are used, ~100–200 μL for fixed path length cells of 50–200 μm, and of the order of a milliliter or more when larger path length cells (especially variable path length cells) are used.

A majority of the reported CD measurements to date have been done for liquid solutions. Only a limited number of CD measurements have been carried out for the gas-phase samples, because most chiral samples do not have enough vapor pressure to undertake the vapor-phase CD measurements. Gas-phase ECD (Gedanken et al. 1984) and VCD (Polavarapu 1989; Cianciosi et al. 1989) measurements have been reported for a few samples that have high enough vapor pressure at room temperature.

Measurements in solid state: ECD and VCD measurements for solid-state samples are not routine and require special instrumentation in both cases. When commercial instruments are used for solid-state measurements, care should be exercised in recognizing and eliminating spurious signals (artifacts) that arise from linear birefringence. ECD measurements on solid samples have been successfully measured by Kuroda and coworkers using an in-house designed instrument (Kuroda et al. 2001). VCD measurements have been successfully made for the film samples (Sen and Keiderling 1984). For use with FT-VCD instruments, the films were prepared by evaporating the solution using either the drop-cast method (Shanmugam and Polavarapu 2004) or spray method (Lombardi et al. 2004) on an IR transparent plate. When the film samples are used, it is necessary to check for consistency in the VCD spectra when the film is rotated around the light beam axis and also around the axis perpendicular to the light beam axis (Buffeteau et al. 2005; Merten et al. 2008). VCD measurements have also been reported for solid samples ground as oil-mulls (Nafie 2004, 2012).

Sample cells: For measurements in the UV–vis region (~190–800 nm), fused quartz cells are used. Cells with fixed path lengths varying from 0.01 to 10 cm are commercially available. For measurements in the IR region, a variety of cells with IR transparent windows (BaF_2, CaF_2, KBr) are commonly used. Demountable, fixed path length and variable path length cells with IR transparent windows are commercially available. Since a proper combination of sample concentration and path length is needed to obtain the desired absorbance, variable path length cells are convenient to vary the path length at a given concentration. Variable path length cells permit the use of large path lengths, but too large a path length may result in interference from dominating solvent absorption bands. Variable path length cells are also expensive, and care must be exercised in the maintenance of these cells. Demountable cells are economical, but assembling the cell with a proper spacer, and ensuring that the sample solution does not leak (or evaporate), is a time-consuming process and needs some practice. Spacers with thicknesses from 6 to 500 μm are available for demountable cells. Note that for aqueous solutions, it is necessary to use 6-μm spacer (so that water absorption is minimized), and at the same time, a high sample concentration should be used. Fixed path length cells for a specified path length are also commercially available. For any of the IR cells mentioned earlier, it is preferable to have the cell windows made of BaF_2 for the mid-IR region. One may also use CaF_2 windows, but the transmission of CaF_2 does not permit measurements below ~1,100 cm^{-1}. Although KBr or KCl windows can also be used, their hygroscopic nature renders them inconvenient for routine use. The higher refractive index associated with ZnSe and KRS-5 windows may result in loss of throughput and in artifact signals. For the gas-phase VCD studies (Polavarapu and Michalska 1983; Polavarapu 1989), a simple single-pass gas cell is to be preferred and multiple-pass cells with reflective mirrors are not recommended.

Solvent considerations: From a practical point of view, the solvent should be one that dissolves the sample and does not have absorption interference in the region of measurement. For the UV–vis region different solvents can be used, but their absorbance of UV–vis radiation limits the range in which they can be used. Commonly used solvents and their optimal range are as follows: acetonitrile (>190 nm), chloroform (>245 nm), cyclohexane (>185 nm), dimethylsulfoxide (>300 nm), dioxane (>232 nm), methanol or ethanol (>205 nm), and water (>190 nm).

For the IR region, if a solvent has a rich mid-IR spectrum in the region of interest, then that solvent should be avoided. Inert solvents such as CCl_4 and CS_2 are ideal choices for the mid-IR region. This is because CCl_4 can be used in the 1,700–900 cm^{-1} region, and CS_2 can be used in the 1,350–900 cm^{-1} region, with up to 100 μm path length without its absorption bands overwhelming the sample absorption bands (see Figure 7.13). If a 1,700–1,500 cm^{-1} region is not important, then CCl_4 may be used up to a 200 μm path length. However, CCl_4 and CS_2 solvents, being inert, need not be suitable for dissolving several chemical compounds. In such cases, polar yet nonhydrogen bonding solvents such as $CHCl_3$, CH_2Cl_2 may also be used for measurements in the restricted regions where each of these solvents do not absorb. Deuterated analogs, $CDCl_3$, CD_2Cl_2, can be used to study the regions where parent solvents have interfering absorption bands. The IR absorption spectra of some of these solvents are shown in Figure 7.13. Since CH_2Cl_2 has interfering absorption bands at ~1,420 and 1,265 cm^{-1}, the usable region with this solvent is restricted to ~1,700–1,500

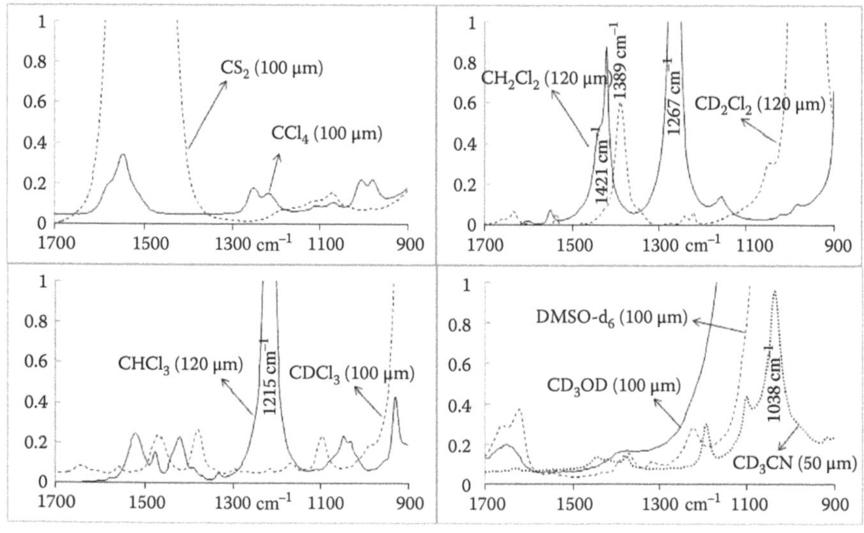

FIGURE 7.13
Infrared absorption spectra of different solvents identifying the transparent regions suitable for VCD measurements.

and ~1,200–950 cm^{-1} region. In contrast, its deuterated analog, CD_2Cl_2, has interfering absorption bands at ~1,390 and 970 cm^{-1} so the usable regions with this solvent are ~1,700–1,450 and ~1,300–1,000 cm^{-1}. $CHCl_3$ has an interfering absorption band at 1,215 cm^{-1}, so a portion of the 1,300–1,150 cm^{-1} region may not be accessible with this solvent. $CDCl_3$ is more appropriate for covering the ~1,700–1,000 cm^{-1} region, as the strong absorption band of this solvent occurs below ~950 cm^{-1} (see Figure 7.13). Note that the usable regions suggested here can vary somewhat depending on the exact path length used for VCD measurements.

Polar and hydrogen bonding solvents, CH_3OH, $(CH_3)_2SO$, and CH_3CN, are not very useful due to interfering absorption bands associated with methyl group vibrations in these molecules. However, the corresponding deuterated solvents, CD_3OD, $(CD_3)_2SO$, and CD_3CN, are useful for the mid-IR region, as strong absorption bands of these solvents appear at lower frequencies. The usable regions with CD_3OD, $(CD_3)_2SO$, and CD_3CN solvents are, respectively, ~1,700–1,250, 1,700–1,150, and 1,700–1,100 cm^{-1} (see Figure 7.13).

The IR absorption spectra of H_2O and D_2O are shown in Figure 7.14. When H_2O is used as the solvent, the path length needs to be kept small at ~6–10 µm due to the strong water absorption band at 1,643 cm^{-1}, which necessitates higher sample concentration and restricts the usable region to ~1,500–950 cm^{-1}. One useful rule of thumb is that the absorbance of the H_2O band at 1,643 cm^{-1} is approximately the path length in µm × 0.1. Therefore, a ~10-µm path length yields an absorbance of ~1.0 for the H_2O band at 1,643 cm^{-1}. D_2O can be used up to a 50-µm path length to cover the 1,700–1,300 cm^{-1} region. The strong absorption band of D_2O at ~1,200 cm^{-1} swamps out the lower frequency region. One should keep in mind the H/D exchange that takes place with D_2O and CD_3OD solvents when these solvents are not used in airtight IR cells (avoiding exposure to atmospheric water). The conversion of D_2O to HDO in a fixed path length IR cell is shown in Figure 7.14.

It should also be noted that some of the solvents (H_2O, CD_3CN, $(CD_3)_2SO$, CD_3OD, and D_2O) may form hydrogen bonds with the sample, which can result in the formation of solute–solvent molecular complexes, or solute–solvent clusters, that lead to additional complexities in the spectral interpretations.

Temperature dependence: When multiple conformations or aggregates are anticipated for a given sample in the solution state, it is useful to undertake temperature and concentration-dependent CD studies. In the harmonic approximation, the fundamental vibrational spectral intensities of a fixed conformer molecule are temperature independent (Wilson et al. 1980). Thus, in the absence of anharmonic effects, any temperature-dependent variations in the VCD spectrum may be attributed to the presence of multiple conformers and to the shift in conformer equilibrium as a function of temperature. The temperature-dependent CD would be useful to study the change in the population of conformers. For increasing the temperature, cells embedded with heating cartridges can be purchased. For lowering the temperature,

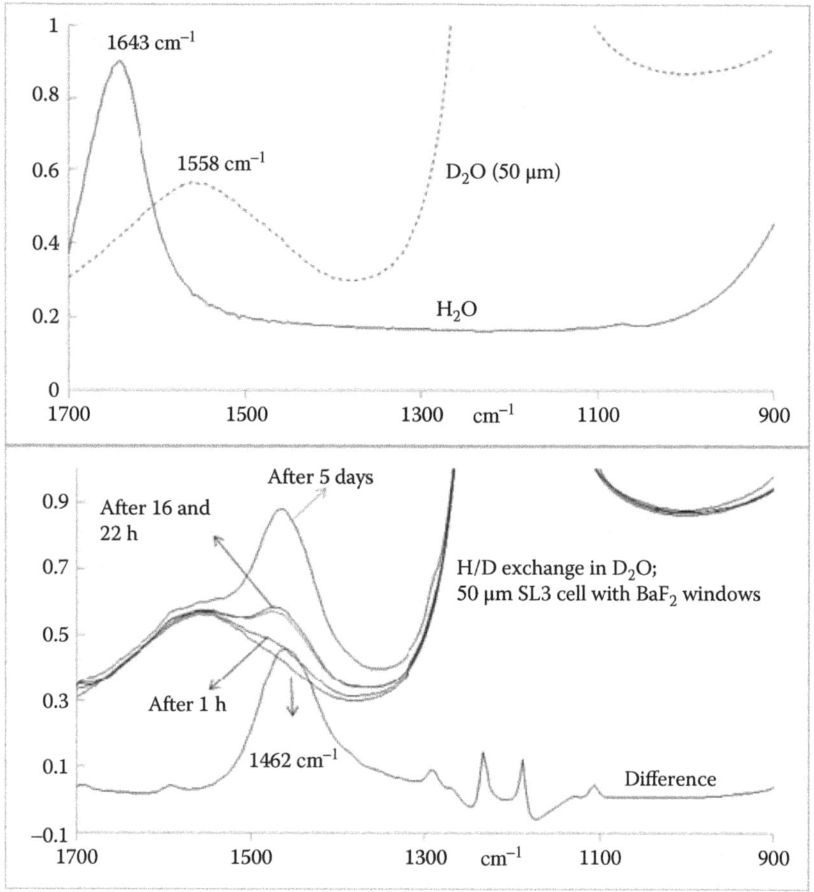

FIGURE 7.14
Infrared absorption spectra of H_2O and D_2O (top) to identify the transparent regions suitable for VCD measurements. The H/D exchange that occurs when D_2O is not used in airtight infrared cells is displayed in the bottom panel.

jacketed cells (for circulating a coolant) are available, but to avoid water condensation on the windows, an evacuated outer jacket is needed.

Concentration dependence: Concentrations that lead to sample aggregation (intra- and intermolecular hydrogen bonding, molecular complexes, etc.) lead to complexities in interpreting the spectra resulting from such effects. Thus, unless the goal is to study aggregation effects, dilute solutions are to be preferred. However, a balance needs to be found between concentration of the sample and path length of the cell to maintain the desired absorbance of bands and to avoid the interference from solvent bands. These demands may restrict the CD studies for certain classes of molecules.

For example, to eliminate intermolecular hydrogen bonding in carboxylic acids, one has to use extremely dilute solutions, but such low concentrations require larger path lengths where solvent absorption interference precludes VCD measurements. One approach to overcome this limitation is to chemically convert the carboxylic acid to corresponding methyl ester (He et al. 2005) and perform the VCD measurements and calculations for this ester.

7.3.5 Tunable Laser-Based Spectrometers

The dispersive/FT spectrometers utilize a broadband thermal light source and the intensities at individual wavelengths are detected using either the dispersion (monochromator) method or interference (interferometry) method. The tunable laser sources avoid the need for dispersion/interference processes, because the laser source wavelength is tunable. The second advantage of using tunable laser sources is that light flux delivered by these sources is very high and hence strongly absorbing samples or solutions in longer path lengths can be studied. It is necessary to choose a combination of concentration and path length that does not exceed maximum absorbance of ~1.0 in dispersive and FT spectrometers, but this restriction can be relaxed in laser-based spectrometers, and measurements with a maximum absorbance of ~3.0 have been reported (Lüdeke et al. 2011). Another advantage is that tunable lasers allows spectra to be measured at a higher resolution and hence closely spaced transitions can be better resolved and better comparisons realized with theoretical calculations. However, the limited wavelength region accessible with these lasers restricts the measurements to a limited portion of the entire mid-IR region.

7.3.6 Time-Resolved Circular Dichroism

Conventional CD instruments use a continuously emitting light source. Pulsed laser light sources differ from these conventional sources by emitting light pulses at certain frequency. Spectral changes that are time resolved on an ultrafast time (picosecond to femtosecond) scale require pulsed laser sources. While these measurements for ECD have been well developed (Lewis et al. 1992; Kliger et al. 2012; Rhee, Eom, and Cho 2011), (Meyer-Ilse et al. 2012), those for VCD are relatively new. There are two objectives in using these pulsed laser sources for VCD measurements. First, it is necessary to demonstrate that VCD spectra can be measured with pulsed light sources. This objective has been achieved via VCD measurements with picosecond (Bonmarin and Helbing 2008, 2009; Helbing and Bonmarin 2009) and femtosecond light pulses (Rhee, June, Lee, et al. 2009; Rhee, Eom, and Cho 2011; Rhee, June, Kim, et al. 2009). Second, it is necessary to demonstrate that time-resolved VCD spectral changes can be measured as well. Although picosecond time-resolved photoinduced changes in the cobalt–sparteine complex have been reported, they have not yet been firmly established.

7.3.7 Nonlinear Spectroscopic Methods

There are several nonlinear spectroscopic methods that include sum frequency generation, second harmonic generation, and two photon circular dichroism measurements. These methods are beyond the scope of this book, and the reader is recommended to consult the reviews (Fischer 2012; Wang 2012) and original articles (Jarrett et al. 2015; De Boni et al. 2008).

7.4 Interconversion of CD and ORD

ORD and CD are related via the Kramers–Kronig (KK) transform (Tinoco and Cantor 1970). Thus, if one of these two properties is measured as a function of wavelength, in a wide wavelength region, then the second can be obtained, at least in principle, via KK transform. Although such transformation between experimental CD and ORD has been undertaken rarely, in practice, it is possible to convert the experimental ECD into ORD spectrum and vice versa, when only one of them is measurable experimentally (as for highly absorbing colored samples).

The details of the KK transform (Polavarapu 2005) and its applications can be found in Appendix 2. The important points that can be learned through KK transform are on the interdependence of ORD and CD. OR at a given wavelength has contributions from all ECD bands. Thus, these two methods are not really independent, but they can serve complementary roles: (a) In the UV–visible spectral region, an experimental ECD spectrum may show only a limited number of bands. Since an experimentally observed ORD pattern in the visible wavelength region depends also on the CD associated with electronic transitions in the short wavelength region, combined experimental investigation of ECD and ORD spectra can yield, in favorable cases, information on ECD bands present in the experimentally inaccessible short wavelength region (Polavarapu et al. 2006). (b) Conformational changes in a molecule may result in the corresponding changes in ECD bands that appear in vacuum UV and are not routinely measurable. Since these short-wavelength ECD bands can influence the ORD in the visible region, the above-mentioned conformational changes may lead to changes in ORD in the visible region. Thus, some conformational changes may not be "seen" through ECD but may be "seen" through ORD. In an opposite situation, conformational changes may lead to corresponding changes in ECD bands in the UV–vis region but may not lead to corresponding changes in ORD in the visible region, because contribution to ORD in the UV–vis region may have been dominated by ECD bands that are not sensitive to conformational changes at short wavelengths. Thus, it will be useful, in general, to undertake both ECD and ORD measurements simultaneously for a given sample.

7.5 VROA Measurements

7.5.1 Normal VROA Measurements

Unlike in ECD and VCD phenomena, where differential absorption contains the stereochemical information, in VROA phenomenon, the stereochemical information is contained in differential Raman scattering from chiral molecules. The molecular parameters responsible for VCD and VROA phenomena are different, although the vibrational Raman scattering involves vibrational transitions, just as in VA spectra. Barron and Buckingham developed the theoretical formalism that predicted the existence of VROA phenomenon (Barron and Buckingham 1971). The experimental efforts were subsequently made by different research groups to observe this phenomenon. Reliable and reproducible VROA signals were measured (Barron et al. 1973) in the 90° scattering geometry (see Chapter 3, Figure 3.3), and these measurements were later confirmed (Hug et al. 1975). The instrumentation in that early period used monochromators with photomultiplier tube (PMT) detection methods and was not as advanced as it is today; as a consequence, the VROA measurements required ~24 h or more of measuring time to record a short width of spectrum (just 300 cm^{-1}). This can be appreciated by noting that the VROA signals are four to five orders of magnitude smaller than Raman signals. Also in that early period, a clear understanding of the factors responsible for artifacts was lacking, so some early measurements were later considered to contain spurious signals. However, stunning progress has been made in recent years, both in instrumentation for, and quantum chemical predictions of VROA spectra to the point of turning VROA spectroscopy into a practical technique.

During this journey, a significant advance in the experimental VROA measurements was the introduction of optical multichannel analyzer detectors, in combination with a dispersive spectrograph (SG), which led Brocki et al. (1980) to a 10-fold increase in speed over the earlier instruments that used monochromators with PMT. The introduction of self-scanning silicon photodiode array detectors (Hug and Surbeck 1979) demonstrated further increase in the speed for VROA measurements. Incorporating these developments into a dedicated instrument, a large collection of VROA spectra were generated (Barron, et al.1987). The decreased measurement time for VROA spectra also facilitated the analysis of the origin of artifacts (Hug 1981; Barron and Vrbancich 1984; Escribano 1985; Hecht et al. 1987; Hecht and Barron 1990; Che and Nafie 1993) and of new experimental designs (Barron et al. 1989; Hecht and Barron 1989; Hug 1982; Hecht et al. 1989; Spencer et al. 1988; Nafie and Freedman 1989). Until about 1989, most of the VROA measurements were made in the 90° scattering geometry with the incident light modulated between left and right circularly polarizations and collecting the scattered polarization parallel to the scattering plane (see Chapter 3, Figure 3.4), as this arrangement is less susceptible to artifacts. The spectra obtained in this configuration are referred to as the depolarized VROA spectra. When the incident light is circularly polarized and the scattered light (in 90° scattering geometry) with polarization perpendicular to the scattering plane is collected, the corresponding measurement is referred to as the polarized VROA. This measurement was not easy because of the large artifacts associated with the polarized Raman bands. However, it became possible, with the increased understanding of artifacts and their control, to measure (Barron et al. 1987, Barron et al. 1989) polarized VROA. Another form of VROA, known as magic angle VROA (Hecht and Barron 1989), was also measured. Most useful measurement, however, was recognized to be in the backscattering geometry (see Chapter 3, Figure 3.3) where VROA intensity is four times larger, and conventional Raman intensity is twofold larger, and artifacts are reduced over those in 90° scattering geometry (Hecht et al. 1989). Thus, backscattering VROA measurements have been used for achieving higher signal-to-noise and reduced artifacts.

VROA measurements that utilized the circularly polarized incident light are referred to as the incident circular polarization (ICP) measurements. In the 90° scattering geometry, they are ICP-depolarized VROA (where scattered light with polarization parallel to the scattering plane is collected) and ICP-polarized VROA (where scattered light with polarization perpendicular to the scattering plane is collected). In the 180° backscattering geometry, the measurement is ICP-backscattering VROA.

The next development was the realization that, instead of circularly polarizing the incident light, one can analyze the Raman light scattered by chiral molecules for circular polarizations (see Chapter 3, Figure 3.4). These measurements are referred to as scattered circular polarization (SCP) measurements (Spencer et al. 1988); for 90° scattering geometry, they are called SCP-VROA in 90° geometry; for backscattering geometry, they are called SCP-VROA in backscattering geometry. For SCP-VROA in 90° geometry, the incident light is linearly polarized with incident light polarization in the scattering plane, being less susceptible for artifacts. For SCP-VROA in the backscattering geometry, the incident light is unpolarized.

Another variation is to use circularly polarized incident light and analyze the Raman light scattered by chiral molecules for circular polarizations. These measurements are referred (Nafie and Freedman 1989) as to dual circular polarization (DCP)-VROA (see Chapter 3, Figure 3.4). There are two versions of DCP measurements: one where incident and SCPs are in-phase (DCP-I) and another where they are out-of-phase (DCP-II).

The introduction of charge-coupled devices (CCDs) as Raman detectors has been the most recent advance for VROA spectroscopy. The higher quantum efficiency, low read-out noise, and low dark current at cryogenic temperatures associated with CCDs resulted in significantly improved signal to noise (Hecht et al. 1992). Incorporating the CCD detectors into an ICP-backscattering instrument, VROA spectra of biological molecules could be measured (Hecht et al. 1992). The demonstration of the feasibility of application of VROA to biological molecules has widened the applications of VROA spectroscopy. The CCD detectors have been incorporated into SCP and DCP-VROA (Hecht et al. 1991; Che et al. 1991; Vargek et al. 1997) and ICP-depolarized VROA (Hecht and Barron 1994) instruments.

All these past and some new developments were incorporated (Hug and Hangartner 1999) into a unique new design (see Figure 7.15) with high throughput for SCP-VROA measurements. In this design, both right and left circularly polarized scattered light components are detected simultaneously using a dual channel design (vide infra) with distinct advantages: (1) Any laser intensity fluctuations will affect the two channels similarly, so the effect of laser flicker noise on VROA is eliminated. The laser intensity fluctuations used to be a major source of problems in all previous designs due to long data collection times involved. (2). Scattering artifacts arising from particulate matter (such as dust, etc.) are also eliminated in the simultaneous detection of right and left circularly polarized scattered light. (3). Thermal Schlieren effects (localized differences in optical path length) are also eliminated. These advantages were demonstrated by constructing (Hug and Hangartner 1999) a backscattering VROA instrument and performing the VROA measurements on (-)-(M)-σ-[4] Helicene (Hug et al. 2001). Typical data collection times and laser power used were 40 min and 115 mW, respectively.

The commercial VROA instrument, marketed by BioTools Inc., is based on these developments (Hug and Hangartner 1999), with backscattering geometry

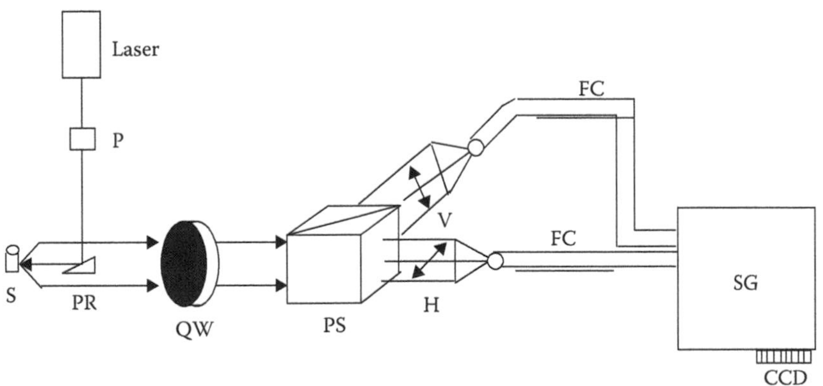

FIGURE 7.15
Optical layout of VROA spectrometer in backscattering geometry (Hug and Hangartner 1999). P, polarization conditioning optics; PR, right angle prism, S, sample; QW, quarter wave plate; PS, polarizing cube beam splitter; H, horizontal polarization; V, vertical polarization; FC, fiber optic cable; SG, spectrograph; CCD, charge-coupled detector. (Reprinted from Polavarapu, P. L., Determination of molecular stereochemistry using optical rotatory dispersion, vibrational circular dichroism and vibrational Raman optical activity. In K. W. Busch and M. A. Busch [eds.], *Chiral Analysis*, New York, NY, Elsevier Science, Copyright 2011, with permission from Elsevier.)

and unpolarized incident light, and digitizes the two circularly polarized Raman scattered components simultaneously, as shown in Figure 7.15.

A concise description of the instrument is as follows: The polarization of monochromatic laser light is rendered unpolarized, on time average, using polarization condition optics (P) that contain a linear polarizer and rotating half-wave plate, and directed to the sample (S) using a small right angle prism. The backscattered light from the sample is collected and passed through a circular analyzer, which is composed of a QW plate and linear polarizer. The linear polarizer here is a polarizing cube beam splitter (PS), which directs the orthogonal linear polarizations in two different directions that are at 90° to each other. These two orthogonal directions are referred to as two channels. Light in each channel is collected through a fiber optic cable (FC), and these two FCs are brought to the entrance of an SG and juxtaposed. The Raman light dispersed by this spectrograph (SG) is then detected by a CCD detector. The CCD detector pixels are binned so that in each column of CCD detector pixels, the top half of the pixels are grouped to give one signal and the bottom half of the pixels are grouped separately to give another signal. The image of the light is processed such that the light coming from one channel is detected by the top half of the pixels and that from the second channel is detected by the bottom half of the pixels. Thus, the two circular polarization components scattered by a chiral sample are digitized simultaneously in two different channels by the CCD detector. The difference between, and the sum of, these two channels can be generated using appropriate software. As a representative example of modern ROA experimental capabilities, the Raman and ROA spectra of α-pinene measured on commercial ChiralRaman spectrometer are shown in Figure 7.16. While the commercial ROA spectrometer is configured for SCP-ROA measurements, with appropriate modifications, all four forms of ROA (namely, ICP, SCP, DCP-I, and DCP-II) could be measured for α-pinene (Li and Nafie 2012).

Sample cells: Most of the reported VROA measurements were done for either neat liquids or liquid solutions. A quartz cuvette or capillary is normally used for holding the samples in VROA measurements. One gas-phase ROA measurement was reported (Šebestík and Bouř 2011), and solid samples are not normally investigated.

Amount of sample needed for measurements: In principle, one would require only a small amount of a sample whose volume matches that of the laser beam diameter incident on the sample. Microgram amount of samples and liquid volumes of ~1 μL were used (Hug 2012). Typical concentrations used (Mutter et al. 2015) are ~50–200 mg/mL for small molecules, ~30–100 mg/mL for proteins and nucleic acids, and ~5–30 mg/mL for polysaccharides.

Solvent considerations: In most cases, neat liquid samples were used. In the case of biological samples, measurements were done for aqueous solutions. Water is an excellent solvent for Raman spectroscopy because water has a weak Raman spectrum, so solvent interference is not an issue. Organic solvents can certainly be used, but here one should be aware of interfering

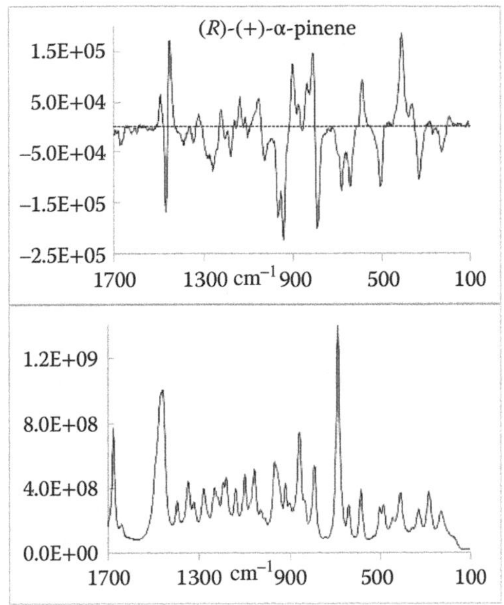

FIGURE 7.16
Vibrational Raman (bottom) and VROA (top) spectra of (+)-α-pinene sample as neat liquid. (Courtesy of Dr. Rina Dukor, BioTools.)

Raman bands of the solvent. Special precautions are needed to avoid (or suppress) interfering fluorescence from the impurities.

Excitation laser wavelength: Since scattering intensity is inversely proportional to the fourth power of incident wavelength, shorter excitation wavelength is to be preferred. However, when this wavelength is close to that of an electronic transition, resonance effect can lead to sample "burning." A laser wavelength in the region of ~400–700 nm is commonly used. An ROA spectrometer with near-IR excitation (785 nm) was used for studying bacteriorhodopsin (Unno et al. 2013). Measurements with deep ultraviolet excitation source (244 nm) have been reported recently (Kapitán et al. 2015).

7.5.2 FT-VROA Measurements

A Fourier transform technique using polarization division interferometer (Martin 1998) was suggested for measuring VROA (Polavarapu 1988). However, this proposition remained as a theoretical suggestion and was never practically implemented.

7.5.3 Resonance VROA Measurements

When the exciting laser wavenumber corresponds to the energy gap between excited and ground electronic states of chiral molecules, resonance effects come into play. Only a few experimental ROA measurements involving resonance phenomenon have been reported (Vargek et al. 1998; Merten, Li, and Nafie 2012).

7.5.4 Induced Resonance ROA Measurements

Here the resonance of probing laser light energy with electronic states of an achiral metal complex is exploited while optical activity is induced into the electronic states of the achiral metal complex by a chiral adduct. The adducts formed among europium complex, Eu^{III}-tris(1,1,1,2,2,3,3-heptafluoro-7,7-dimethyl-4,6-octanedionate-d_{27} and chiral organic molecules were reported to yield much larger ROA signals than those associated with normal VROA phenomenon (Yamamoto and Bouř 2012). However, the observed signals are actually associated with electronic transitions of $Eu^{3}+$ ion that appear in the 2,200–1,600 cm^{-1} region, and not vibrational transitions, and therefore the mechanism involved here should perhaps be labeled as electronic ROA. The induced ROA signals are found to be mirror images when enantiomers of adduct are used, so the chirality of adducts is revealed through these measurements.

7.5.5 Surface-Enhanced VROA Measurements

One of the drawbacks of VROA spectroscopy is the need for high concentration of compounds under investigation. This limitation may be overcome if surface-enhanced (SE) effect can be incorporated into VROA measurements. Therefore, surface-enhanced VROA (SEROA) measurements attracted much attention (Abdali and Blanch 2008; Brinson 2009; Osińska et al. 2010). However, the reliability of experimental SEROA measurements has not yet been firmly established, although much progress appears to have been made (Pour et al. 2011).

7.5.6 Coherent Anti-Stokes VROA Measurements

The first observation of VROA in coherent anti-Stokes Raman scattering has been reported recently (Hiramatsu et al. 2012). Using heterodyne detection methods and pulsed lasers, improved contrast from an achiral background has been reported.

7.5.7 Magnetic VROA Measurements

VROA exhibited by achiral samples in an external magnetic field is referred to as magnetic VROA. Barron and coworkers have demonstrated the advantages of these magnetic VROA measurements (Barron and Vrbancich 1984; Barron 1985; Barron et al. 1982). More recently, magnetic VROA of paramagnetic NO_2 gas was shown to be significantly larger and the advantages associated with measurements on paramagnetic substances were emphasized (Šebestík and Bouř 2014).

7.6 Related Methods

Recently some novel methods have been reported that aid in the separation, absolute configuration determination, and/or sub-femtosecond dynamics of chiral molecules. These methods utilize microwaves (Patterson et al. 2013), Fourier transform microwave spectroscopy (Lobsiger et al. 2015), chiral gratings (Cameron et al. 2014), high harmonic generation (Cireasa et al. 2015; Cho 2015), NMR (Buckingham 2014), and rotating fields (Schamel et al. 2013; Kibar et al. 2011). These methods are beyond the current scope of this book, and the reader is encouraged to read the original articles.

7.7 Summary

The procedures for the measurement of OR and ECD are well established. The instrumentation for these measurements in liquid-phase has been around for a long time, but that for gas-phase samples using cavity ring down methods is a new development. The procedures for the measurement of VCD and VROA have become established in recent years, and the instrumentation for these measurements is also well developed. The availability of commercial instruments for OR (except for CRDP), ECD, VCD, and VROA makes these methods attractive for practicing researchers. However, one should be aware of the possible artifacts in these measurements, and it is important to implement appropriate checks during the measurements.

References

Abbate, S., G. Longhi, and C. Santina. 2000. Theoretical and experimental studies for the interpretation of vibrational circular dichroism spectra in the CH-stretching overtone region. *Chirality* 12(4):180–190.

Abdali, S., and E. W. Blanch. 2008. Surface enhanced Raman optical activity (SEROA). *Chem. Soc. Rev.* 37(5):980–992.

Barron, L. D. 1985. Magnetic Raman optical activity and Raman electron paramagnetic resonance. *Pure Appl. Chem.* 57(2):215–223.

Barron, L. D., C. Meehan, and J. Vrbancich. 1982. Magnetic resonance-Raman optical activity of ferrocytochrome c. theory and experiment *J. Raman Spectrosc.* 12(3):251–261.

Barron, L. D., M. P. Bogaard, and A. D. Buckingham. 1973. Raman scattering of circularly polarized light by optically active molecules. *J. Am. Chem. Soc.* 95(2):603–605.

Barron, L. D., and A. D. Buckingham. 1971. Rayleigh and Raman scattering from optically active molecules. *Mol. Phys.* 20(6):1111–1119.

Barron, L. D., L. Hecht, and S. M. Blyth. 1989. Polarized Raman optical activity of menthol and related molecules. *Spectrochim. Acta Part A* 45(3):375–379.

Barron, L. D., J. F. Torrance, and D. J. Cutler. 1987. A new multichannel Raman optical activity instrument. *J. Raman Spectrosc.* 18(4):281–287.

Barron, L. D., and J. Vrbancich. 1984. On the theory of dominant artifacts in natural and magnetic Raman optical activity. *J. Raman Spectrosc.* 15(1):47–50.

Berova, N., L. Di Bari, and G. Pescitelli. 2007. Application of electronic circular dichroism in configurational and conformational analysis of organic compounds. *Chem. Soc. Rev.* 36:914–931.

Berova, N., K. Nakanishi, R. W. Woody, and Editors. 2000. *Circular Dichroism: Principles and Applications*. New York, NY: Wiley-VCH.

Berova, N., G. A. Ellestad, and N. Harada. 2010. Characterization by circular dichroism spectroscopy. In: L. Mander and H.-W. B. Liu (eds). *Comprehensive Natural Products II: Chemistry and Biology*. New York, NY: Elsevier.

Berova, N., N. Harada, and K. Nakanishi. 2000. Exciton coupling. In: J. C. Lindon, *Encyclopedia of Spectroscopy and Spectrometry*. New York, NY: Elsevier.

Bonmarin, M., and J. Helbing. 2008. A picosecond time-resolved vibrational circular dichroism spectrometer. *Opt. Lett.* 33(18):2086–2088.

Bonmarin, M., and J. Helbing. 2009. Polarization control of ultrashort mid-IR laser pulses for transient vibrational circular dichroism measurements. *Chirality* 21(1E):E298–E306.

Bougas, L., D. Sofikitis, G. E. Katsoprinakis, A. K. Spiliotis, P. Tzallas, B. Loppinet, and T. P. Rakitzis. 2015. Chiral cavity ring down polarimetry: Chirality and magnetometry measurements using signal reversals. *J. Chem. Phys.* 143(10):104202.

Brinson, B. E. 2009. *Nonresonant Surface Enhanced Raman Optical Activity, Chemistry*. Houston, TX: Rice University.

Brocki, T., M. Moskovits, and B. Bosnich. 1980. Vibrational optical activity—circular differential Raman scattering froma series of chiral terpenes *J. Am. Chem. Soc.* 102(2):495–500.

Buckingham, A. D. 2014. Communication: Permanent dipoles contribute to electric polarization in chiral NMR spectra. *J. Chem. Phys.* 140(1):011103.

Buffeteau, T., F. Lagugné-Labarthet, and C. Sourisseau. 2005. Vibrational circular dichroism in general anisotropic thin solid films: Measurement and theoretical approach. *Appl. Spectrosc.* 59(6):732–745.

Bustamante, C., M. F. Maestre, and D. Keller. 1985. Expressions for the interpretation of circular intensity differential scattering of chiral aggregates. *Biopolymers* 24(8):1595–1612.

Cameron, R. P., A. M. Yao, and S. M. Barnett. 2014. Diffraction gratings for chiral molecules and their applications. *J. Phys. Chem. A* 118(19):3472–3478.

Cao, X., R. K. Dukor, and L. A. Nafie. 2008. Reduction of linear birefringence in vibrational circular dichroism measurements: Use of rotating half wave plate. *Theor. Chem. Acc.* 119:69–79.

Chabay, I., E. C. Hsu, and G. Holzwarth. 1972. Infrared circular dichroism measurement between 2000 and 5000 cm^{-1}: Pr3+-tartrate complexes. *Chem. Phys. Lett.* 15(2):211–214.

Che, D. P., L. Hecht, and L. A. Nafie. 1991. Dual and incident circular polarization Raman optical activity backscattering of (−)-trans-pinane *Chem. Phys. Lett.* 180(3):182–190.

Che, D. P., and L. A. Nafie. 1993. Theory and reduction of artifacts in incident, scattered and dual circular polarization forms of Raman optical activity. *Appl. Spectrosc.* 47(5):544–555.

Chen, E., R. A. Goldbeck, and D. S. Kliger. 2010. Nanosecond time-resolved polarization spectroscopies: Tools for probing protein reaction mechanisms. *Methods* 52(1):3–11.

Chen, G.-C., P. L. Polavarapu, and S. Weibel. 1994. New design for Fourier transform infrared vibrational circular dichroism spectrometers. *Appl. Spectrosc.* 48(10):1218–1223.

Cheng, J. C., L. A. Nafie, and P. J. Stephens. 1975. Polarization scrambling using a photoelastic modulator. Application to circular dichroism measurement. *J. Opt. Soc. Am.* 65:1031–1035.

Cho, M. 2015. High-harmonic generation: Drive round the twist. *Nat Phys* advance online publication.

Cianciosi, S. J., K. M. Spencer, T. B. Freedman, L. A. Nafie, and J. E. Baldwin. 1989. Synthesis and gas-phase vibrational circular dichroism of (+)-(S,S)-cyclopropane-1,2-2H2. *J. Am. Chem. Soc.* 111(5):1913–1915.

Cireasa, R., A. E. Boguslavskiy, B. Pons, M. C. H. Wong, D. Descamps, S. Petit, H. Ruf, N. Thire, A. Ferre, J. Suarez, J. Higuet, B. E. Schmidt, A. F. Alharbi, F. Legare, V. Blanchet, B. Fabre, S. Patchkovskii, O. Smirnova, Y. Mairesse, and V. R. Bhardwaj. 2015. Probing molecular chirality on a sub-femtosecond timescale. *Nat Phys* advance online publication.

Crabbé, P. 1965. *Optical Rotatory Dispersion and Circular Dichroism in Organic Chemistry, Holden-Day Series in Physical Techniques in Chemistry*. San Francisco, CA: Holden-Day.

De Boni, L., C. Toro, and F. E. Hernández. 2008. Synchronized double L-scan technique for the simultaneous measurement of polarization-dependent two-photon absorption in chiral molecules. *Opt. Lett.* 33(24):2958–2960.

Dewey, M. A., and J. A. Gladysz. 1993. Optical rotation measurements of organometallic compoounds: Caveats and recommended procedures. *Organometallics* 12(6):2390–2392.

Dignam, M. J., and M. D. Baker. 1981. Analysis of a polarizing Michelson interferometer for dual beam Fourier-transform infrared, circular dichroism infrared, and reflectance ellipsometric infrared spectroscopies. *Appl. Spectrosc.* 35(2):186–193.

Djerassi, C. 1960. *Optical Rotatory Dispersion; Applications to Organic Chemistry. McGraw-Hill Series in Advanced Chemistry*. New York, NY: McGraw-Hill.

Eliel, E. L., S. H. Wilen, and M. P. Doyle. 2001. *Basic Organic Stereochemistry*. New York, NY: Wiley.

Escribano, J. R. 1985. The influence of finite collection angle on Rayleigh and Raman optical activity. *Chem. Phys. Lett.* 121(3):191–193.

Fischer, P. 2012. Nonlinear optical spectroscopy of chiral molecules. In: N. Berova, P. L. Polavarapu, K. Nakanishi, and R. W. Woody (eds). *Comprehensive Chiroptical Spectroscopy*, Vol. 1. New York, NY: John Wiley.

Gedanken, A., K. Hintzer, and V. Schurig. 1984. Chirality rule for substituted oxiranes. *J. Chem. Soc. Chem. Commun.* (23):1615–1616.

Gray, D. M., R. L. Ratliff, and M. R. Vaughan. 1992. Circular dichroism spectroscopy of DNA. *Methods Enzymol.* 211:389–406.

Harada, N., and K. Nakanishi. 1983. *Circular Dichroic Spectroscopy. Exciton Coupling in Organic Stereochemistry*. Herndon, VA: University Science Books.

He, J., A. Petrovich, and P. L. Polavarapu. 2004. Quantitative determination of conformer populations: Assessment of specific rotation, vibrational absorption, and vibrational circular dichroism in substituted butynes. *J. Phys. Chem. A* 108(10):1671–1680.

He, J., F. Wang, and P. L. Polavarapu. 2005. Absolute configurations of chiral herbicides determined from vibrational circular dichroism. *Chirality* 17(S1):S1–S8.

Hecht, L., and L. D. Barron. 1989. Magic angle Raman optical activity: βeta pinene and nopinone. *Spectrochim. Acta Part A* 45(6):671–674.

Hecht, L., and L. D. Barron. 1990. An analysis of modulation experiments for Raman optical activity. *Appl. Spectrosc.* 44(3):483–491.

Hecht, L., and L. D. Barron. 1994. Instrument for natural and magnetic Raman optical activity studies in right angle scattering. *J. Raman Spectrosc.* 25(7–8):443–451.

Hecht, L., L. D. Barron, A. R. Gargaro, Z. Q. Wen, and W. Hug. 1992. Raman optical instrument for biochemical studies. *J. Raman Spectrosc.* 23(7):401–411.

Hecht, L., L. D. Barron, and W. Hug. 1989. Vibrational Raman optical activity in backscattering. *Chem. Phys. Lett.* 158(5):341–344.

Hecht, L., D. Che, and L. A. Nafie. 1991. A new scattered circular polarization Raman optical activity instrument equipped with a charge coupled device detector. *Appl. Spectrosc.* 45(1):18–25.

Hecht, L., B. Jordanov, and B. Schrader. 1987. Mueller-Stokes treatment of artifacts in natiral Raman optical activity. *Appl. Spectrosc.* 41(2):295–307.

Helbing, J., and M. Bonmarin. 2009. Time-resolved chiral vibrational spectroscopy. *CHIMIA Int. J. Chem.* 63(3):128–133.

Hiramatsu, K., M. Okuno, H. Kano, P. Leproux, V. Couderc, and H. Hamaguchi. 2012. Observation of Raman optical activity by heterodyne-detected polarization-resolved coherent anti-stokes Raman scattering. *Phys. Rev. Lett.* 109(8):083901.

Holzwarth, G., E. C. Hsu, H. S. Mosher, T. R. Faulkner, and A. Moscowitz. 1974. Infrared circular dichroism of carbon-hydrogen and carbon-deuterium stretching modes. Observations. *J. Am. Chem. Soc.* 96(1):251–252.

Hug, W. 1981. Optical artifacts and their control in Raman circular difference scattering measurements. *Appl. Spectrosc.* 35(1):115–124.

Hug, W. 1982. Instrumental and theoretical advances in Raman optical activity. In: J. Lascombe and P. V. Huong (eds). *Raman Spectroscopy, Linear and Nonlinear: Proceedings of the Eighth International Conference on Raman Spectroscopy*. Bordeaux, France, 6–11 September 1982. Hoboken, NJ: John Wiley & Sons.

Hug, W. 2012. Measurement of Raman optical activity. In: N. Berova, P. L. Polavarapu, K. Nakanishi and R. W. Woody (eds). *Comprehensive Chiroptical Spectroscopy*. Vol. 1. Hoboken, NJ: John Wiley & Sons.

Hug, W., and G. Hangartner. 1999. A novel high-throughput Raman spectrometer for polarization difference measurements. *J. Raman Spectrosc.* 30(9):841–852.

Hug, W., S. Kint, G. F. Bailey, and J. R. Scherer. 1975. Raman circular intensity differential spectroscopy. Spectra of (−)-alpha-pinene and (+)-slpha-phenylethylamine. *J. Am. Chem. Soc.* 97(19):5589–5590.

Hug, W., and H. Surbeck. 1979. Vibrational Raman optical activity spectra recorded in perpendicular polarization. *Chem. Phys. Lett.* 60(2):186–192.

Hug, W., G. Zuber, A. de Meijere, A. F. Khlebnikov, and H. J. Hansen. 2001. Raman optical activity of a purely σ-bonded helical chromophore: (−)-(M)-σ-[4] Helicene. *Helv. Chim. Acta* 84(1):1–21.

Jarrett, J. W., X. Liu, P. F. Nealey, R. A. Vaia, G. Cerullo, and K. L. Knappenberger. 2015. Communication: SHG-detected circular dichroism imaging using orthogonal phase-locked laser pulses. *J. Chem. Phys.* 142(15):151101.

Johnson, W. C. 1990. Protein secondary structure and circular dichroism. A practical guide. *Proteins Struct. Funct. Genet.* 7(3):205–214.

Kapitán, J., L. D. Barron, and L. Hecht. 2015. A novel Raman optical activity instrument operating in the deep ultraviolet spectral region. *J. Raman Spectrosc.*:n/a-n/a.

Kibar, O., M. Chachisvilis, E. Tu, and T. H. Marsilje. 2011. Separation and manipulation of a chiral object. USA: Dynamic Connections LLC. US Patent 20140209464 A1.

Kliger, D. S., E. Chen, and R. A. Goldbeck. 2012. Probing kinetic mechanisms of protein function and folding with time-resolved natural and magnetic chiroptical spectroscopies. *Int. J. Mol. Sci.* 13(1):683–697.

Kuroda, R., T. Harada, and Y. Shindo. 2001. A solid-state dedicated circular dichroism spectrophotometer: Development and application. *Rev. Sci. Instrum.* 72(10):3802–3810.

Lahiri, P., K. B. Wiberg, P. H. Vaccaro, M. Caricato, and T. D. Crawford. 2014. Large solvation effect in the optical rotatory dispersion of norbornenone. *Angew. Chem.* 126(5):1410–1413.

Lakhani, A., P. Malon, and T. A. Keiderling. 2009. Comparison of vibrational circular dichroism instruments: Development of a new dispersive VCD. *Appl. Spectrosc.* 63(7):775–785.

Lakhtakia, A. 1990. *Selected Papers on Natural Optical Activity*. Bellingham WA: SPIE Optical Engineering Press.

Lewis, J. W., R. A. Goldbeck, D. S. Kliger, X. Xie, R. C. Dunn, and J. D. Simon. 1992. Time-resolved circular dichroism spectroscopy: Experiment, theory, and applications to biological systems. *J. Phys. Chem.* 96(13):5243–5254.

Li, H., and L. A. Nafie. 2012. Simultaneous acquisition of all four forms of circular polarization Raman optical activity: Results for α-pinene and lysozyme. *J. Raman Spectrosc.* 43(1):89–94.

Lightner, D. A., and J. E. Gurst. 2000. *Organic Conformational Analysis and Stereochemistry from Circular Dichroism Spectroscopy*. New York, NY: Wiley-VCH.

Lobsiger, S., C. Perez, L. Evangelisti, K. K. Lehmann, and B. H. Pate. 2015. Molecular structure and chirality detection by Fourier transform microwave spectroscopy. *J. Phys. Chem. Lett.* 6(1):196–200.

Lombardi, R. A., X. Cao, and L. A. Nafie. 2004. Detection of chirality in solid-state samples using Fourier transform vibrational circular dichroism. In: *16th International Symposium on Chirality*. New York.

Lowry, T. M. 1964. *Optical Rotatory Power*. New York, NY: Dover Publications.

Lüdeke, S., M. Pfeifer, and P. Fischer. 2011. Quantum-cascade laser-based vibrational circular dichroism. *J. Am. Chem. Soc.* 133(15):5704–5707.

Maestre, M. F., C. Bustamante, T. L. Hayes, J. A. Subirana, and I. Tinoco. 1982. Differential scattering of circularly polarized light by the helical sperm head from the octopus *Eledone cirrhosa*. *Nature* 298(5876):773–774.

Maestre, M. F., and C. Reich. 1980. Contribution of light scattering to the circular dichroism of deoxyribonucleic acid films, deoxyribonucleic acid-polylysine complexes, and deoxyribonucleic acid particles in ethanolic buffers. *Biochemistry* 19(23):5214–5223.

Malon, P., and T. A. Keiderling. 1988. A solution to the artifact problem in Fourier transform vibrational circular dichroism. *Appl. Spectrosc.* 42(1):32–38.

Martin, D. H. 1998. The principles of polarization-division interferometric spectrometry. In: P. L. Polavarapu (ed). *Principles and Applications of Polarization-Division Interferometry*. New York, NY: Wiley.

Merten, C., T. Kowalik, and A. Hartwig. 2008. Vibrational circular dichroism spectroscopy of solid polymer films: Effects of sample orientation. *Appl. Spectrosc.* 62(8):901–905.

Merten, C., H. Li, and L. A. Nafie. 2012. Simultaneous resonance Raman optical activity involving two electronic states. *J. Phys. Chem. A* 116(27):7329–7336.

Meyer-Ilse, J., D. Akimov, and B. Dietzek. 2012. Ultrafast circular dichroism study of the ring opening of 7-Dehydrocholesterol. *J. Phys. Chem. Lett.* 3(2):182–185.

Miles, A. J., and B. A. Wallace. 2006. Synchrotron radiation circular dichroism spectroscopy of proteins and applications in structural and functional genomics. *Chem. Soc. Rev.* 35(1):39–51.

Mort, B. C., and J. Autschbach. 2006. Temperature dependence of the optical rotation of fenchone calculated by vibrational averaging. *J. Phys. Chem. A* 110(40):11381–11383.

Mort, B. C., and J. Autschbach. 2007. Temperature dependence of the optical rotation in six bicyclic organic molecules calculated by vibrational averaging. *Chem Phys Chem* 8(4):605–616.

Müller, T., K. B. Wiberg, and P. H. Vaccaro. 2000. Cavity ring-down polarimetry (CRDP): A new scheme for probing circular birefringence and circular dichroism in the gas phase. *J. Phys. Chem. A* 104(25):5959–5968.

Mutter, S. T., F. Zielinski, P. L. A. Popelier, and E. W. Blanch. 2015. Calculation of Raman optical activity spectra for vibrational analysis. *Analyst.* 140(9):2944–2956.

Nafie, L. A. 2004. Advanced applications of mid-IR and near-IR vibrational circular dichroism: New spectral regions, reaction monitoring and quality control. In: *16th International Symposium on Chirality*. New York.

Nafie, L. A. 2012. Infrared vibrational optical activity: Measurement and instrumentation. In: N. Berova, P. L. Polavarapu, K. Nakanishi and R. W. Woody (eds). *Comprehensive Chiroptical Spectroscopy.* Vol. 1. Hoboken, NJ: John Wiley & Sons.

Nafie, L. A., J. C. Cheng, and P. J. Stephens. 1975. Vibrational circular dichroism of 2,2,2-trifluoro-1-phenylethanol. *J. Am. Chem. Soc.* 97 (Copyright (C) 2012 American Chemical Society (ACS). All Rights Reserved.):3842–3843.

Nafie, L. A., and T. B. Freedman. 1989. Dual circular polarization Raman optical activity. *Chem. Phys. Lett.* 154(3):260–266.

Nafie, L. A. 2000. Dual polarization modulation: A real-time, spectral-multiplex separation of circular dichroism from linear birefringence spectral intensities. *Appl. Spectrosc.* 54(11):1634–1645.

Nafie, L. A., H. Buijs, A. Rilling, X. Cao, and R. K. Dukor. 2004. Dual source Fourier transform polarization modulation spectroscopy: An improved method for the measurement of circular and linear dichroism. *Appl. Spectrosc.* 58(6):647–654.

Nafie, L. A., M. Diem, and D. Warren Vidrine. 1979. Fourier transform infrared vibrational circular dichroism. *J. Am. Chem. Soc.* 101(2):496–498.

Osborne, G. A., J. C. Cheng, and P. J. Stephens. 1973. Near-infrared circular dichroism and magnetic circular dichroism instrument. *Rev. Sci. Instrum.* 44:10–15.

Osińska, K., M. Pecul, and A. Kudelski. 2010. Circularly polarized component in surface-enhanced Raman spectra. *Chem. Phys. Lett.* 496(1–3):86–90.

Patterson, D., M. Schnell, and J. M. Doyle. 2013. Enantiomer-specific detection of chiral molecules via microwave spectroscopy. *Nature* 497(7450):475–477.

Polavarapu, P. 1997. *Principles and Applications of Polarization-Division Interferometry*. New York, NY: John Wiley & Sons.

Polavarapu, P. L. 1988. Fourier transform Raman optical activity. *Chem. Phys. Lett.* 148(1):21–25.

Polavarapu, P. L. 1989. Rotational–vibrational circular dichroism. *Chem. Phys. Lett.* 161(6):485–490.

Polavarapu, P. L. 1996. Vibrational optical activity of anharmonic oscillator. *Mol. Phys.* 89(5):1503–1510.

Polavarapu, P. L., G. C. Chen, and S. Weibel. 1994. Development, justification, and applications of a midinfrared polarization-division Interferometer. *Appl. Spectrosc.* 48(10):1224–1235.

Polavarapu, P. L., and D. F. Michalska. 1983. Vibrational circular dichroism in (S)-(-)-epoxypropane. Measurement in vapor phase and verification of the perturbed degenerate mode theory. *J. Am. Chem. Soc.* 105(19):6190–6191.

Polavarapu, P. L., A. Petrovic, and F. Wang. 2003. Intrinsic rotation and molecular structure. *Chirality* 15:S143–S149.

Polavarapu, P. L. 2005. Kramers–Kronig transformation for optical rotatory dispersion studies. *J. Phys. Chem. A* 109(32):7013–7023.

Polavarapu, P. L. 2011. Determination of molecular stereochemistry using optical rotatory dispersion, vibrational circular dichroism and vibrational Raman optical activity. In: K. W. Busch and M. A. Busch (eds). *Chiral Analysis*. New York, NY: Elsevier Science.

Polavarapu, P. L., and Z. Deng. 1996. Measurement of vibrational circular dichroism below ~600 cm^{-1}: Progress towards meeting the challenge. *Appl. Spectrosc.* 50(5):686–692.

Polavarapu, P. L., A. G. Petrovic, and P. Zhang. 2006. Kramers–Kronig transformation of experimental electronic circular dichroism: Application to the analysis of optical rotatory dispersion in dimethyl-L-tartrate. *Chirality* 18(9):723–732.

Polavarapu, P. L., and G. Shanmugam. 2011. Comparison of mid-infrared Fourier transform vibrational circular dichroism measurements with single and dual polarization modulations. *Chirality* 23(9):801–807.

Pour, S. O., S. E. J. Bell, and E. W. Blanch. 2011. Use of a hydrogel polymer for reproducible surface enhanced Raman optical activity (SEROA). *Chem. Commun.* 47(16):4754–4756.

Ragunathan, N., N. S. Lee, T. B. Freedman, L. A. Nafie, C. Tripp, and H. Buijs. 1990. Measurement of vibrational circular dichroism using a polarizing Michelson interferometer. *Appl. Spectrosc.* 44(1):5–7.

Reich, C., M. F. Maestre, S. Edmondson, and D. M. Gray. 1980. Circular dichroism and fluorescence-detected circular dichroism of deoxyribonucleic acid and poly[d(A-C).cntdot.d(G-T)] in ethanolic solutions: A new method for estimating circular intensity differential scattering. *Biochemistry* 19(23):5208–5213.

Rhee, H., I. Eom, and M. Cho. 2011. Ultrafast chiroptical spectroscopy: Monitoring optical activity in quick time. *J. Anal. Sci. Technol.* 2(3):103–107.

Rhee, H., Y. G. June, Z. H. Kim, S. J. Jeon, and M. Cho. 2009. Phase sensitive detection of vibrational optical activity free-induction-decay: Vibrational CD and ORD. *J. Opt. Soc. Am. B* 26(5):1008–1017.

Rhee, H., Y. G. June, J. S. Lee, K. K. Lee, J. H. Ha, Z. H. Kim, S. J. Jeon, and M. Cho. 2009. Femtosecond characterization of vibrational optical activity of chiral molecules. *Nature* 458(7236):310–313.

Schamel, D., M. Pfeifer, J. G. Gibbs, B. Miksch, A. G. Mark, and P. Fischer. 2013. Chiral colloidal molecules and observation of the propeller effect. *J. Am. Chem. Soc.* 135(33):12353–12359.

Šebestík, J., and P. Bouř. 2014. Observation of paramagnetic Raman optical activity of nitrogen dioxide. *Angew. Chem. Int. Ed.* 53(35):9236–9239.

Šebestík, J., and P. Bouř. 2011. Raman optical activity of methyloxirane gas and liquid. *J. Phys. Chem. Lett.* 2(5):498–502.

Sen, A. C., and T. A. Keiderling. 1984. Vibrational circular dichroism of polypeptides. III. Film studies of several α-helical and β-sheet polypeptides. *Biopolymers* 23(8):1533–1545.

Shanmugam, G., and P. L. Polavarapu. 2004. Vibrational circular dichroism of protein films. *J. Am. Chem. Soc.* 126(33):10292–10295.

Shanmugam, G., and P. L. Polavarapu. 2005. Film techniques for vibrational circular dichroism measurements. *Appl. Spectrosc.* 59(5):673–681.

Shapiro, D. B., R. A. Goldbeck, D. Che, R. M. Esquerra, S. J. Paquette, and D. S. Kliger. 1995. Nanosecond optical rotatory dispersion spectroscopy: Application to photolyzed hemoglobin-CO kinetics. *Biophys. J.* 68(1):326–334.

Sofikitis, D., L. Bougas, G. E. Katsoprinakis, A. K. Spiliotis, B. Loppinet, and T. Peter Rakitzis. 2014. Evanescent-wave and ambient chiral sensing by signal-reversing cavity ringdown polarimetry. *Nature* 514(7520):76–79.

Spencer, K. M., T. B. Freedman, and L. A. Nafie. 1988. Scattered circular polarization Raman optical activity. *Chem. Phys. Lett.* 149(4):367–374.

Sreerama, N., and R. W. Woody. 2000. Estimation of protein secondary structure from circular dichroism spectra: Comparison of CONTIN, SELCON, and CDSSTR methods with an expanded reference set. *Anal. Biochem.* 287(2):252–260.

Superchi, S., E. Giorgio, and C. Rosini. 2004. Structural determinations by circular dichroism spectra analysis using coupled oscillator methods: An update of the applications of the DeVoe polarizability model. *Chirality* 16(7):422–451.

Tinoco, I., Jr., and C. R. Cantor. 1970. Application of optical rotatory dispersion and circular dichroism in biochemical analysis. In: D. Glick (ed). *Methods Biochem. Anal.* Vol. 18. New York, NY: John Wiley & Sons.

Tsankov, D., T. Eggimann, and H. Wieser. 1995. Alternative design for improved FT-IR/VCD capabilities. *Appl. Spectrosc.* 49(1):132–138.

Unno, M., T. Kikukawa, M. Kumauchi, and N. Kamo. 2013. Exploring the active site structure of a photoreceptor protein by Raman optical activity. *J. Phys. Chem. B* 117(5):1321–1325.

Vargek, M., T. B. Freedman, and L. A. Nafie. 1997. Improved backscattering dual circular polarization Raman optical activity spectrometer with enhanced performance for biomolecular applications. *J. Raman Spectrosc.* 28(8):627–633.

Vargek, M., T. B. Freedman, E. Lee, and L. A. Nafie. 1998. Experimental observation of resonance Raman optical activity. *Chem. Phys. Lett.* 287(3–4):359–364.

Wang, H. F. 2012. In situ measurement of chirality of molecules and molecular assemblies with surface nonlinear spectroscopy. In: N. Berova, P. L. Polavarapu, K. Nakanishi, and R. W. Woody (eds). *Comprehensive Chiroptical Spectroscopy*. Vol. 1. New York, NY: John Wiley.

Wiberg, K. B., Y. G. Wang, M. J. Murphy, and P. H. Vaccaro. 2004. Temperature dependence of optical rotation: α-pinene, β-pinene pinane, camphene, camphor and fenchone. *J. Phys. Chem. A* 108(26):5559–5563.

Wilson, E. B., J. C. Decius, and P. C. Cross. 1980. *Molecular Vibrations: The Theory of Infrared and Raman Vibrational Spectra*. New York, NY: Dover Publications.

Woody, R. W. 1995. Circular dichroism. *Biochem. Spectrosc.* 246:34–71.

Woody, R. W. 1996. Circular dichroism and the conformational analysis of biomolecules. In: G. D. Fasman (ed). *Circular Dichroism and the Conformational Analysis of Biomolecules*. New York, NY: Plenum.

Yamamoto, S., and P. Bouř. 2012. Detection of molecular chirality by induced resonance Raman optical activity in Europium complexes. *Angew. Chem.* 124(44):11220–11223.

8

Comparison of Experimental and Calculated Spectra

The central theme in deriving the molecular structural information from chiroptical spectra rests in the comparison of experimental spectra with those predicted using reliable quantum chemical calculations. If the chiroptical spectrum predicted for a molecule of a given configuration (weighted for possible conformations) matches the corresponding experimental spectrum measured for the sample, then the configuration used for the predictions is assigned to the molecules constituting the experimental sample. The word "match" is important, because it is possible to evaluate the comparison between experimental and predicted spectra either qualitatively or quantitatively. These different levels of interpretations are summarized in Sections 8.1 and 8.2.

8.1 Comparison of Experimental and Calculated Spectra

Most measurements and predictions of optical rotatory dispersion (ORD) are carried out at discrete wavelengths, unlike in electronic circular dichroism (ECD), vibrational circular dichroism (VCD), and vibrational Raman optical activity (VROA), and there are no bands to analyze for ORD. Therefore, the discussion in the following sections pertains to ECD, VCD, and VROA. Depending on the spectral region investigated, one can encounter a large number of bands or sometimes only a handful of bands to analyze in that region. Regardless of the number of bands available for analysis, it is possible to undertake the analysis either qualitatively or quantitatively. These approaches are discussed in Sections 8.1, 8.2, and 8.3. In the case of ECD and VCD, respective electronic absorption (EA) and vibrational absorption (VA) spectra are to be measured and predicted simultaneously. While circular dichroism (CD) is determined by the product of electric and magnetic dipole transition moments, the corresponding absorption is determined by the square of the electric dipole transition moment (see Chapter 4, Equations 4.127 and 4.129). Therefore, electric dipole transition moment contributes to both absorption and CD spectra. In the case of VROA, vibrational Raman spectra are measured and calculated along with corresponding VROA spectra. VROA intensity is determined by the interference between the normal coordinate derivatives of electric dipole, magnetic dipole, and electric quadrupole polarizabilities, and the corresponding vibrational Raman intensity is determined by the mean and/or anisotropy of the normal coordinate derivatives of the electric dipole polarizability tensor (see Chapter 3, Equations 3.50 through 3.53 and Equations 3.57 through 3.61). Therefore, normal coordinate derivatives of the electric dipole polarizability tensor contribute to both vibrational Raman and VROA spectra. A successful comparison between experimental and predicted CD spectra should therefore accompany a corresponding successful comparison between the corresponding experimental and predicted absorption spectra. Similarly, a successful comparison between experimental and calculated VROA spectra should accompany a corresponding successful comparison between the corresponding experimental and predicted vibrational Raman spectra. For these reasons, the comparison of experimental and predicted ECD/VCD/VROA spectra should also report the comparison of corresponding EA/VA/vibrational Raman spectra. As mentioned earlier, since ORD is not included in the discussion in this chapter, the chiroptical spectra terminology in this chapter will pertain to ECD, VCD, and VROA spectra only.

8.1.1 Qualitative Comparisons

A qualitative analysis of comparison between experimental and predicted spectra focuses on finding a correspondence between experimental and calculated bands. In analyzing the chiroptical spectra, bisignate couplets are often encountered. A positive bisignate couplet (or simply positive couplet) is one with the positive part of the couplet appearing on the lower energy (or wavenumber) side and negative part of the couplet appearing on the higher energy (or wavenumber) side of the couplet. The opposite situation refers to the negative bisignate couplet (or, simply, the negative couplet). For instance, if the experimental chiroptical spectrum of interest contains one positive bisignate couplet, then the qualitative analysis seeks to find that positive bisignate couplet in the calculated chiroptical spectrum and queries whether these couplets match in their energy (wavenumber) or wavelength order and appearance. Extending this example, when the experimental chiroptical spectrum in a given spectral region of interest contains a series of bands with a given order of signs and intensities, the qualitative analysis looks for the same series of bands and order of signs and intensities in the corresponding calculated chiroptical spectrum. For this purpose, the experimental and calculated chiroptical spectra are placed one above the other, and correlating lines are drawn between the experimental and calculated bands. A simultaneous analysis is carried out for the corresponding absorption (or Raman) spectra. As an example, the experimentally observed bands in VCD and VA spectra are correlated to those obtained at the B3LYP/aug-cc-pVTZ level of theory for (+)-(*R*)-3-chloro-1-butyne (He et al. 2004) in Figure 8.1. The wavenumber or wavelength positions of bands in the calculated spectra are often shifted from those in the

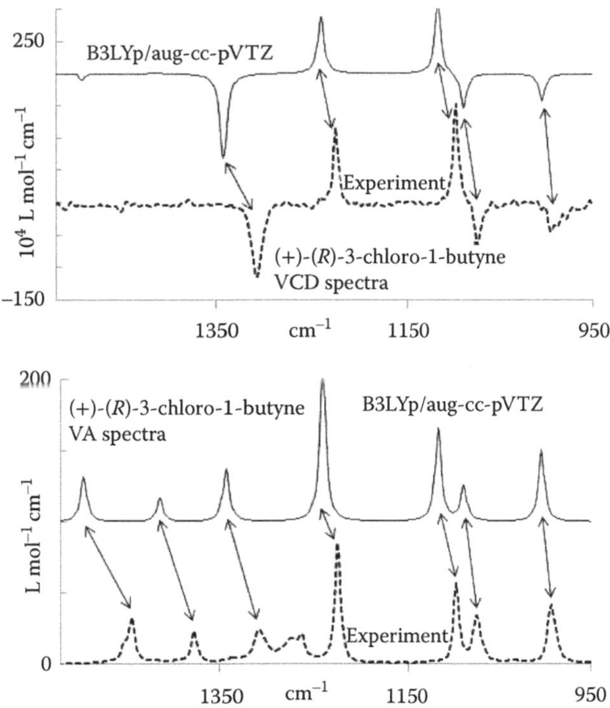

FIGURE 8.1
Experimental (dotted line) and calculated (solid line) vibrational absorption (VA) (bottom) and vibrational circular dichroism (VCD) spectra (top) of (+)-(R)-3-chloro-1-butyne. (Data from He, J. et al., *J. Phys. Chem. A*, 108, 1671–1680, 2004.)

experimental spectra due to the deficiencies in the theoretical model used. For these reasons, the lines correlating to the bands appear slanted. This problem is often circumvented by scaling the predicted transition wavenumbers, or wavelengths, with a constant σ, referred to as the *x*-axis scale factor.

Although one can see a one-to-one correlation for the vibrational bands of chlorobutyne, it is possible for other cases that some experimental bands may not be seen to have corresponding calculated bands. In such cases, the criterion used is that the number of bands that can be correlated between experimental and calculated chiroptical spectra should be much greater than the number of bands that cannot be correlated. Sometimes it is also possible to find a correlation between experimental and calculated bands when some bands in the calculated spectra are considered to be interchanged in their positions. If the user interchanges the wavenumber or wavelength order of band positions to find correlated signs and intensities, some independent evidence to support the proposed interchange is needed. Otherwise, the reliability of such analyses becomes uncertain. Furthermore, qualitative analyses focus only on the qualitative appearance of the experimental and calculated spectra, and no emphasis is placed on the agreement between experimental and calculated absolute intensities. These practices, which were commonly employed in the early stages of development of chiroptical spectroscopy, can lead to inadvertent user bias in determining the level of agreement between experimental and calculated spectra. This uncertainty necessitated the need for some type of quantification of the agreement between experimental and calculated spectra.

In this process, the concept of robustness was introduced for VCD (Nicu and Baerends 2009). Here, the angle between electric and dipole transition moments was calculated for each vibrational transition and only those transitions for which the angle is significantly different from 90° are considered robust and therefore emphasized in the comparison between calculated and experimental VCD spectra. The same concept was introduced later for VROA spectra (Tommasini et al. 2014). Unfortunately the values of these angles are origin dependent (Gobi and Magyarfalvi 2011; Nicu and Baerends 2011). Instead of using the origin-dependent angle between electric and dipole transition moments, the magnitude of the ratio of rotational strength to dipole strength of *k*th vibration transition, $\zeta_k = R_k/D_k$, was suggested to be used as the robust criterion. The transitions with ζ_k >10 ppm were suggested to be considered as robust (Gobi and Magyarfalvi 2011; Góbi et al. 2015). The robustness criterion was concluded to be of limited use for conformationally flexible molecules (Gussem et al. 2012).

The measures of agreement between experimental and calculated spectra should preferably be done with similarity measures as discussed in Section 8.1.3.

8.1.2 Quantitative Comparisons

Just as calculated rotational and dipole strengths can be converted to experimentally measured CD and absorption bands, the reverse transformation from experimentally measured CD and absorption bands to corresponding rotational and dipole strengths can be achieved using Equations 8.1 and 8.2 (He et al. 2004):

$$D_k = \frac{0.92 \times 10^{-38}}{\bar{\nu}_k^o} \int \varepsilon(\bar{\nu})\, d\bar{\nu} \tag{8.1}$$

$$R_k = \frac{0.23 \times 10^{-38}}{\bar{\nu}_k^o} \int \Delta\varepsilon(\bar{\nu})\, d\bar{\nu} \tag{8.2}$$

(see Appendix 8 for details on these conversions). In Equation 8.1, the molar extinction coefficient ε (in units of $L{\cdot}mol^{-1}{\cdot}cm^{-1}$) is obtained from the experimental absorbance by dividing it with the sample concentration (*C*) and pathlength (*l*) used to measure the experimental data (i.e., $\varepsilon = A/Cl$). Similarly, in Equation 8.2, $\Delta\varepsilon = \Delta A/Cl$. By plotting the rotational strengths obtained in the calculations ($R_{k,calc}$) against the corresponding rotational strengths obtained for bands in the experimental spectrum ($R_{k,expt}$), one can determine the correlation coefficient from these plots (Devlin et al. 1997). For ideal situations, the correlation coefficient will be ±1; +1 indicates that the absolute configuration (AC) used in the calculations should be same as that of the molecules of the sample used for experimental measurements; –1 indicates that the AC used in the calculations should be opposite to that of the molecules of the sample used for experimental measurements. A similar plot of the dipole strengths obtained in the calculations ($D_{k,calc}$) against the corresponding dipole strengths of bands in the experimental spectrum ($D_{k,expt}$) can be used to determine the correlation coefficient for dipole strengths. For ideal situations, this correlation coefficient will be +1, independent of the AC used for calculations. A good correlation for rotational strengths should be verified by a correspondingly good correlation coefficient for dipole strengths. The correlation plot for the experimental and calculated strengths of (+)-(*R*)-3-chloro-1-butyne (He et al. 2004) is shown in Figure 8.2. The VCD spectrum of (+)-(*R*)-3-chloro-1-butyne shows five major bands, but there are more bands in the VA spectra that could be correlated. The correlation coefficient for experimental and predicted rotational strengths is 0.99, and the corresponding correlation coefficient for dipole strengths is 0.90. These correlation coefficients indicate that the assignment of (*R*) configuration to (+)-3-chloro-1-butyne has very little uncertainty.

This correlation plot procedure, as seen in Figure 8.2, is sound but fraught with practical difficulties. The identity of individual transitions in the

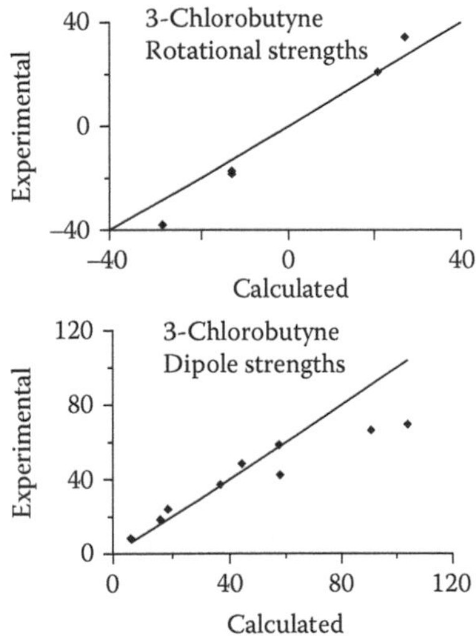

FIGURE 8.2
Correlation plots of calculated and experimental rotational strengths (top) and dipole strengths (bottom) for (+)-(*R*)-3-chloro-1-butyne. Correlation coefficients are 0.99 and 0.90, respectively, for rotational and dipole strengths. (Data from He, J. et al., *J. Phys. Chem. A*, 108, 1671–1680, 2004.)

calculated spectra is known, because the respective normal mode displacements are also obtained in the calculations. The identity of individual bands in the experimental spectra is generally not known (except for those associated with well-behaved functional groups such as C=O stretching vibration). Therefore, one does not know a priori which calculated transition corresponds to which of the bands in the experimental spectrum. The usual practice is to arrange the calculated transitions in a decreasing order of energy (or wavenumber) and align them with the experimentally observed bands (also arranged in the same order). However, the experimentally observed bands are broad and may have several transitions within the observed band envelope. It often becomes necessary to deconvolute the experimental bands and determine the integrated intensities (or rotational/dipole strengths) of individual band components, so they can be compared with the corresponding calculated counterparts. This process is not only laborious but also introduces some uncertainties.

A modification of this approach is to calculate root mean square percent (RMSP) deviations (He et al. 2004) given as

$$\mathrm{RMSP} = \sqrt{\sum_{k=1}^{m} \frac{\left[100\left(P_{k,\mathrm{calc}} - P_{k,\mathrm{expt}}\right) / P_{k,\mathrm{expt}}\right]^2}{m}} \tag{8.3}$$

where m is the number of resolved bands in the experimental spectrum, and P_k equals either R_k or D_k, depending on the choice of analyzing CD or absorption. However, this modification inherits all of the problems mentioned earlier in the context of correlation plots between $R_{k,\mathrm{calc}}$ versus $R_{k,\mathrm{expt}}$, and $D_{k,\mathrm{calc}}$ versus $D_{k,\mathrm{expt}}$.

8.1.3 Numerical Measures of Similarity

Since both experimental and calculated spectra are generally available in the digital format (as x and y pairs), one can easily calculate the similarity between experimental and calculated spectra using spectral overlap integrals. Different similarity measures have been introduced in the last several years.

8.1.3.1 Carbó-Similarity Index (SI)

SI is calculated from the similarity overlap integral as a dimensionless number (Kuppens et al. 2003):

$$\mathrm{SI} = \frac{\int f(x)g(x)\,dx}{\sqrt{\int f^2(x)\,dx \int g^2(x)\,dx}} \tag{8.4}$$

In Equation 8.4, $f(x)$ represents the spectrum derived from quantum chemical predictions (a scale factor σ multiplying the transition wavenumbers/wavelengths may be used to match calculated and experimental x-values); $g(x)$ is the experimental spectrum; both represent intensity as a function of x (frequency/wavelength).

The value of SI can range from 0 to 1 for absorption/Raman spectra (where all intensities are positive) and between –1 and 1 for CD/VROA spectra where intensities can be positive or negative. An ideal value of +1 corresponds to the situation when the AC used for calculations is the same as that of molecules constituting the sample used for experimental measurements and –1 for the opposite situation. It is implied in Equation 8.4 that the experimental and calculated spectral are normalized. That is,

$$N_f^2 \int \left(f(x)\right)^2 dx = 1 \quad \text{and} \quad N_g^2 \int \left(g(x)\right)^2 dx = 1 \tag{8.5}$$

Because of this implied normalization, one would not see any difference in the SI values even when $f(x) = \alpha g(x)$, where α is a positive constant that multiplies the intensities (y-axis values). This insensitivity to the intensity scale factor can be seen as a positive for some situations and as a negative for other situations. When the sample concentration and pathlength used for experimental measurements are uncertain, such uncertainties are avoided by the above-mentioned normalization. When the quantitative accuracy of calculated spectra intensities is sought by comparing them with the experimental spectral intensities, such information cannot be gained from SI.

8.1.3.2 Neighborhood Similarity

To include the influence of neighborhood at any spectral point, a triangular weighing function is introduced (Kuppens et al. 2005). The expression for neighborhood similarity (NS) is obtained by modifying that for SI as follows:

$$\text{NS} = \frac{\int w_{fg}(r) c_{fg}(r)\, dr}{\sqrt{\int w_{ff}(r) c_{ff}(r)\, dr \int w_{gg}(r) c_{gg}(r)\, dr}} \tag{8.6}$$

where

$$c_{fg}(r) = \int f(x) g(x+r) dx; \quad c_{ff}(r) = \int f(x) f(x+r) dx \tag{8.7}$$

$$w_{fg}(r) = 1 - \frac{|r|}{l} \quad \text{for } |r| < l \quad \text{and 0 for } |r| \geq l \tag{8.8}$$

The width of triangular function, l, can vary depending on the bandwidth. Values for l ranging from 40 to 10 cm^{-1} have been used.

8.1.3.3 Enantiomeric Similarity Index (ESI)

A somewhat more complicated measure of similarity, which is applicable only for the spectra of enantiomers, is given as ESI (Debie et al. 2011):

$$\mathrm{ESI} = \left|\Sigma_{fg} - \Sigma_{\bar{f}g}\right| \tag{8.9}$$

where

$$\Sigma_{fg} = \frac{\Phi^{++}S_{fg}^{++} + \Phi^{--}S_{fg}^{--}}{\Phi^{++} + \Phi^{--}} \tag{8.10}$$

$$\Sigma_{\bar{f}g} = \frac{\Phi^{-+}S_{fg}^{-+} + \Phi^{+-}S_{fg}^{+-}}{\Phi^{-+} + \Phi^{+-}} \tag{8.11}$$

where Φ^{++}, Φ^{--}, Φ^{+-}, and Φ^{-+} are all defined as positive quantities:

$$\Phi^{++} = \int_{f(x)>0} f(x)dx + \int_{g(x)>0} g(x)dx \tag{8.12}$$

$$\Phi^{+-} = \int_{f(x)>0} f(x)dx + \left|\int_{g(x)<0} g(x)dx\right| \tag{8.13}$$

$$\Phi^{-+} = \left|\int_{f(x)<0} f(x)dx\right| + \int_{g(x)>0} g(x)dx \tag{8.14}$$

$$\Phi^{--} = \left|\int_{f(x)<0} f(x)dx\right| + \left|\int_{g(x)<0} g(x)dx\right| \tag{8.15}$$

Similarly, $S_{fg}^{++}, S_{fg}^{--}, S_{fg}^{+-}, S_{fg}^{-+}$ are defined as positive quantities; for example,

$$S_{fg}^{+-} = \int_{\substack{f(x)>0,\\ g(x)<0}} f(x)\left|g(x)\right| dx \tag{8.16}$$

The spectra $f(x)$ and $g(x)$ used in Equations 8.12 through 8.16 are normalized (see Equation 8.5). Σ_{fg} measures the similarity by separately taking the weighted overlaps of positive and negative parts of the spectra. The mismatch between positive and negative parts of the spectra is taken into account by calculating Σ_{fg} for the opposite enantiomer (labeled with a bar on the letter f, as $\Sigma_{\bar{f}g}$). The range for ESI is 0 to 1. When the spectrum computed for one of the enantiomers has perfect agreement with the experimentally measured spectrum, the ESI value becomes 1. ESI has been used for analyzing the similarity among calculated and experimental VCD (Debie et al. 2011), ECD (Bruhn et al. 2013), as well as VROA (Polavarapu and Covington 2014) spectra.

8.1.3.4 Spectral Similarity (Sim)

A different measure of similarity, in analogy to the Tanimoto coefficient of chemical finger print similarity, for VA and VCD spectra was given as follows (Shen et al. 2010):

$$Sim\text{VA} = \frac{I_{fg}}{I_{ff} + I_{gg} - I_{fg}} \tag{8.17}$$

$$Sim\text{VCD} = \frac{I_{fg}}{I_{ff} + I_{gg} - \left|I_{fg}\right|} \tag{8.18}$$

where

$$I_{fg} = \int f(x)\, g(x)\, dx \tag{8.19}$$

The spectra $f(x)$ and $g(x)$ for Equations 8.17 through 8.18 can be normalized (as in Equation 8. 3) or left non-normalized. If they are not normalized, *Sim* values are labeled as *Sim*_NN. For $f(x) = \alpha g(x)$, *Sim*_NN values will vary as $\alpha/(1+\alpha^2-\alpha)$. For $\alpha = 1$ and 2, *Sim*_NN values will change from 1 to 0.66. Therefore, *Sim*_NN values can reflect any quantitative differences between experimental and calculated intensities. A comparison of *Sim* and *Sim*_NN values is a convenient way to judge the quantitative differences between calculated and experimental spectra. The *Sim* method has been used to analyze EA and ECD, VA and VCD, and Raman and VROA spectra (Covington and Polavarapu 2013; Polavarapu and Covington 2014).

8.1.3.5 Square Root of the Squared Difference

For analyzing the similarity among the calculated VROA spectra for diastereomers, a different measure of similarity, Δ^2, was used (Simmen et al. 2012):

$$\Delta^2 = \sqrt{\int_a^b \left(f(x) - h(x)\right)^2 dx} \tag{8.20}$$

In Equation 8.20, both $f(x)$ and $h(x)$ are the calculated spectra for diastereomers. However, Equation 8.20 can also be extended to evaluate the similarity between experimental and calculated absorption, Raman, or CD spectra; in that case, $h(x)$ is replaced by $g(x)$, representing the experimental spectrum.

In all of the similarity measures mentioned above, the plot of similarity measure (SI, NS, ESI, *Sim*, or Δ^2) versus x-axis scale factor (σ) for theoretical spectra yields a similarity plot that provides the information on the maximum similarity obtainable when the calculated transition wavenumbers/wavelengths are scaled for possible errors in their predictions. For ideal cases, such plots should exhibit a single well-defined maximum, and the x-axis scale factor σ should be close to 1. In practice, a double maxima can be seen, which indicates that when calculated spectra are translated in relation to experimental spectra, one may find some overlapping bands at one point and other overlapping bands at another point during translation.

8.2 Comparison of Experimental and Calculated Ratio Spectra

The comparison of experimental and calculated spectra is normally done separately for EA and ECD, VA and VCD, and Raman and VROA spectra. However, the experimental and calculated chiroptical spectra can be found to have good similarities for both "right" reasons and "wrong" reasons.

For example, if the calculated magnetic dipole transition moment is overestimated and the corresponding electric dipole transition moment is underestimated by the same amount, then the predicted magnitude of a CD band can agree with the corresponding experimental magnitude. Such situations yield a good agreement for the wrong reasons. To ensure that a good agreement is obtained for the right reasons, a good agreement between predicted and experimental magnitudes of a CD band should also be accompanied by a good agreement between experimental and calculated magnitudes of the corresponding absorption band. In other words, for each band in the CD spectrum, it is necessary to verify that not only CD intensities match in the experimental and calculated CD spectra but also the corresponding absorption intensities match in the experimental and calculated absorption spectra. This requirement, however, is not met by the separate similarity overlap comparisons for CD and absorption spectra.

The same arguments apply for VROA and vibrational Raman spectra because the normal coordinate derivative of the electric dipole polarizability tensor is common for predicting both Raman and VROA intensities. Therefore, for each VROA band, the experimental and calculated magnitudes of the corresponding Raman band intensity should match as well. In other words, for each band in the VROA spectrum, it is necessary to verify that not only VROA intensities match in the experimental and calculated VROA spectra but also the corresponding vibrational Raman intensities match in the experimental and calculated Raman spectra. This requirement, however, is not met by the separate similarity overlap comparisons for VROA and Raman spectra. Also, a practical problem in vibrational Raman and VROA spectra is that the measured experimental spectral intensities are not on an absolute scale but are obtained as photon counts. Therefore, in comparing the experimental and predicted vibrational Raman spectra, the experimental spectral intensities are multiplied by a constant to bring their magnitudes to the same level as that in the calculated spectra. The same constant should be used for comparing the experimental and predicted VROA spectral intensities.

The correct predictions of intensities for CD and absorption (or VROA and Raman) for individual bands can be verified by focusing on the ratio of a CD spectrum to the corresponding absorption spectrum (or the ratio of a VROA spectrum to the corresponding vibrational Raman spectrum) (Covington and Polavarapu 2013; Polavarapu and Covington 2014). The ratio of the CD spectrum to the corresponding absorption spectrum is referred to as the dimensionless dissymmetry factor (DF) spectrum, and the ratio of the VROA spectrum to the corresponding vibrational Raman spectrum as the dimensionless circular intensity differential (CID) spectrum:

$$g(\bar{\nu}) = \frac{\Delta\varepsilon(\bar{\nu})}{\varepsilon(\bar{\nu})} = \frac{\Delta A(\bar{\nu})}{A(\bar{\nu})} = \frac{4R(\bar{\nu})}{D(\bar{\nu})} \tag{8.21}$$

$$\Delta(\bar{\nu}) = \frac{I_{\alpha}^{\gamma}(\bar{\nu}) - I_{\beta}^{\delta}(\bar{\nu})}{I_{\alpha}^{\gamma}(\bar{\nu}) + I_{\beta}^{\delta}(\bar{\nu})} \tag{8.22}$$

The DF spectrum is labeled as electronic DF (EDF) for electronic transitions and as vibrational DF (VDF) for vibrational transitions. There are several advantages to analyzing the DF and CID spectra. Experimentally, the regions with larger DF or CID values are considered to be more reliably measured (robust) while those with smaller DF or CID values are less robust. This is because DF represents the strength of the CD signal per unit absorption intensity, and CID represents the strength of the VROA signal per unit Raman intensity. Regions with larger DF or CID values are expected to carry better signal-to-noise ratio, while regions with smaller DF or CID values, especially those within a certain threshold, may fall within the noise level of the measurement. One can specify a robustness threshold, as determined by the quality of the instrument used to measure the experimental spectra, and consider only those spectral regions with signals that are above the robustness threshold for similarity analysis. This process ensures that the analysis is carried out for robust regions (with favorable signal-to-noise ratio), and signals with poorer signal-to-noise ratio are not relied upon. It is possible that a given region has a larger intensity in both absorption and CD (or vibrational Raman and VROA), but their ratio can be smaller than that for other regions. The separate similarity analyses of absorption and CD, or vibrational Raman and VROA, spectra emphasize on the regions with larger intensities therein, but these regions may become less important in the ratio spectra. Therefore, the analysis of DF/CID spectra focusing on the robust regions gives a different, and possibly better, perspective than that obtained from separate similarity analyses of absorption and CD or vibrational Raman and VROA.

Earlier, in Section 8.1.1, we mentioned the use of the origin-dependent angle between electric and magnetic dipole transition moment vectors for robust criterion (Nicu and Baerends 2009, 2011). More recently, it was found (Longhi et al. 2015) that this angle can always be made robust ($\theta = 0°$ or $180°$) by a suitable translation of the origin of the coordinate system. However, normal modes differ in the rate at which this angle changes with further translations from this point. The changes in this angle are small for normal modes with larger DFs and vice versa. Therefore, the use of DF as a criterion is also favored from the point of angle between electric and magnetic dipole transition moment vectors.

In deriving the experimental DF/CID spectra, however, one has to take some precautions (Covington and Polavarapu 2013; Polavarapu and Covington 2014). In the experimental absorption and vibrational Raman spectra, there will be baseline regions (regions between the bands and regions where there are no bands) with intensity approaching zero. The corresponding regions in CD and VROA spectra will represent the respective baselines, which is essentially the noise. In those regions the ratio spectrum may incur amplified noise, because the baseline noise signal from the numerator, when divided by the denominator that approaches zero, will result in amplified noise in the ratio spectrum. The introduction of a baseline tolerance parameter for absorption or vibrational Raman spectra avoids such unpleasant effects. The baseline tolerance parameter defines the minimum baseline intensity value usable for absorption or vibrational Raman spectra in taking the ratio. The regions with absorption/Raman intensities below that the baseline tolerance value will be excluded from the ratio spectra. Because of these exclusions, the DF/CID spectra need not have continuous spectral distribution, and the band shapes will not follow conventional (Lorentzian/Gaussian) band shapes. For an isolated transition, the absorption (or vibrational Raman) and CD (or VROA) spectra will have conventional (Lorentzian/Gaussian) band shapes, but the baseline tolerance mentioned above will impart a boxcar shape (Covington and Polavarapu 2013) for that transition in DF/CID spectra.

A computer program that provides the similarity overlap analyses for absorption, CD, and DF as well as Raman, VROA, and CID spectra has been developed in the author's laboratory and is made freely available (Covington and Polavarapu 2015). The baseline tolerance parameters for experimental spectra are user defined and depend on the type of chiroptical spectrum being considered. In the experimental VA spectra, the user-defined baseline tolerance is usually of the order of ~10 $L{\cdot}mol^{-1}{\cdot}cm^{-1}$, while most peak band intensities are usually of the order of >200 $L{\cdot}mol^{-1}{\cdot}cm^{-1}$ units. In the case of experimental Raman spectra, the baseline tolerance can be ~10^6 photon counts, while most peak band intensities are greater than 10^8 photon counts. The robustness threshold parameter, predefined in the CDSpecTech program (Covington and Polavarapu 2015), is 4×10^{-5} for vibrational DF and CID spectra and zero for electronic DF spectra (since normal ECD and EDF magnitudes are quite large compared with vibrational counterparts). These preset values can be changed by the user as desired.

During the similarity analysis of DF (or CID) spectra, the analysis for all three (absorption, CD, and DF or vibrational Raman, VROA, and CID) spectra is carried out simultaneously, and the user will have all three similarity plots at hand. A good criterion for the confident assignment of the AC is to require that similarity measure values are large (preferably closer to 1), and that x-axis scale factors corresponding to maximum similarity overlaps do not deviate significantly from each other, among three spectra.

The experimental and B3LYP/aug-cc-pVTZ predicted VDF spectra for (+)-(*R*)-3-chloro-1-butyne, derived from the absorption and VCD spectra shown in Figure 8.1, are displayed in Figure 8.3. The absorption baseline

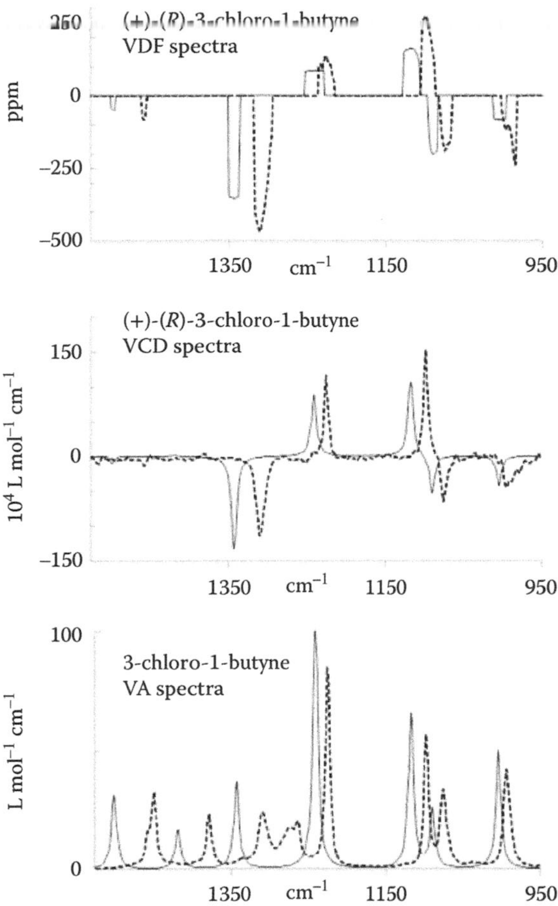

FIGURE 8.3
B3LYP/aug-cc-pVTZ predicted (solid traces) and experimental (dashed traces) vibrational absorption (bottom), VCD (middle), and VDF (top) spectra for (+)-(*R*)-3-chloro-1-butyne. A baseline tolerance of 10 L·mol^{-1}·cm^{-1} and robustness of 40 ppm were used for obtaining VDF. The displayed predicted spectra are without applying the frequency scale factor. (Data from Covington, C. L. and Polavarapu, P. L., *J. Phys. Chem. A*, 117, 3377–3386, 2013.)

tolerance of 10 $L \cdot mol^{-1} \cdot cm^{-1}$ and robustness threshold of 40 ppm are used in deriving the VDF spectra.

The *Sim*VA, *Sim*VCD, and *Sim*VDF plots for (+)-(*R*)-3-chloro-1-butyne (Covington and Polavarapu 2013) are shown in Figure 8.4. The maximum values obtained for *Sim*VA, *Sim*VCD, and *Sim*VDF are, respectively, 0.65, 0.51, and 0.57, and the vibrational frequency scale factors for these maxima are 0.99, 0.98, and 0.97, respectively. It is common to encounter larger similarity values for absorption (or vibrational Raman) than for CD (or VROA) and DF (or CID).

The similarity overlap analysis is particularly suited for analyzing the chiroptical spectra of diastereomers. A revealing application can be described for (2*S*,3*R*)-tetrahydro-3-hydroxy-5-oxo-2,3-furandicarboxylic acid, referred to as hibiscus acid, and its diastereomer, (2*S*,3*S*)-tetrahydro-3-hydroxy-5-oxo-2,3-furandicarboxylic acid, referred to as garcinia acid. To avoid hydrogen bonding effects associated with carboxylic acids, the corresponding esters, hibiscus acid dimethyl ester (HADE) and garcinia acid dimethyl ester (GADE) (see Figure 8.5), have been investigated in a CD_2Cl_2 solvent. The

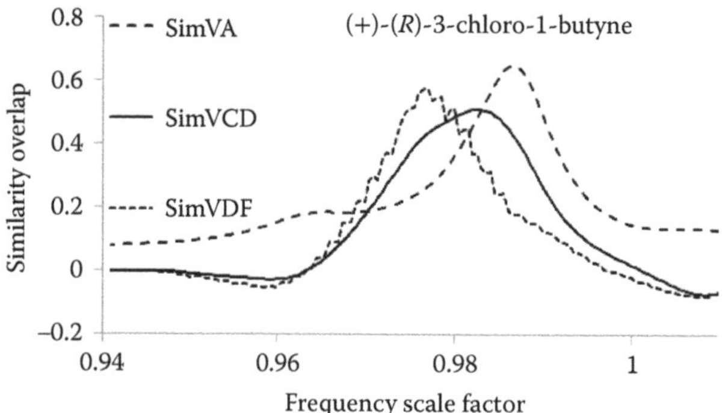

FIGURE 8.4
Similarity overlap plots of VA, VCD, and VDF spectra of (+)-(*R*)-3-chloro-1-butyne. (Data from Covington, C. L. and Polavarapu, P. L., *J. Phys. Chem. A*, 117, 3377–3386, 2013.)

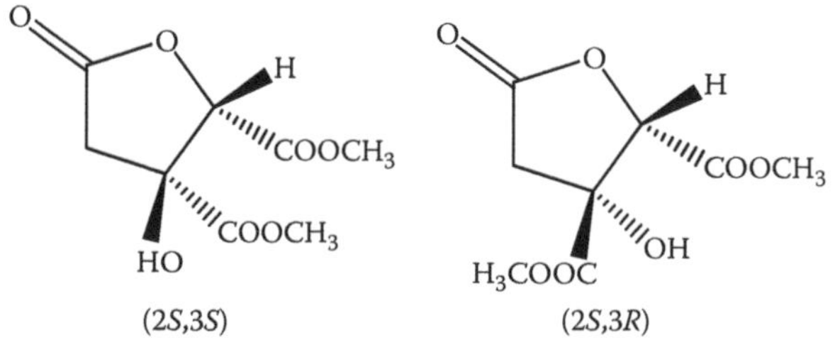

(2*S*,3*S*) (2*S*,3*R*)

FIGURE 8.5
Structures of dimethyl esters of (2*S*,3*S*)-tetrahydro-3-hydroxy-5-oxo-2,3-furandicarboxylic acid and (2*S*,3*R*)-tetrahydro-3-hydroxy-5-oxo-2,3-furandicarboxylic acid diastereomers.

experimental VA and VCD spectra of these two esters are compared with those calculated for four diasteremoers, (2*S*,3*S*), (2*S*,3*R*), (2*R*,3*S*), and (2*R*,3*R*) in Figure 8.6. It is difficult to visually correlate the observed experimental VCD bands with those of individual diastereomers. Nevertheless, similarity overlap plots clearly reveal the diastereomer that is favored for reproducing the experimental VCD and VDF spectra. In the case of (+)-GADE (see Figure 8.7, top panel), (2*S*,3*S*) diastereomer gives a similarity overlap of 0.79 and 0.71, respectively, for VCD and VDF, at a frequency scale factor of 0.99. There is also a negative overlap of −0.16 and −0.12, respectively, for VCD and VDF, at a frequency scale factor of 0.96. This negative overlap indicates that the antipode,

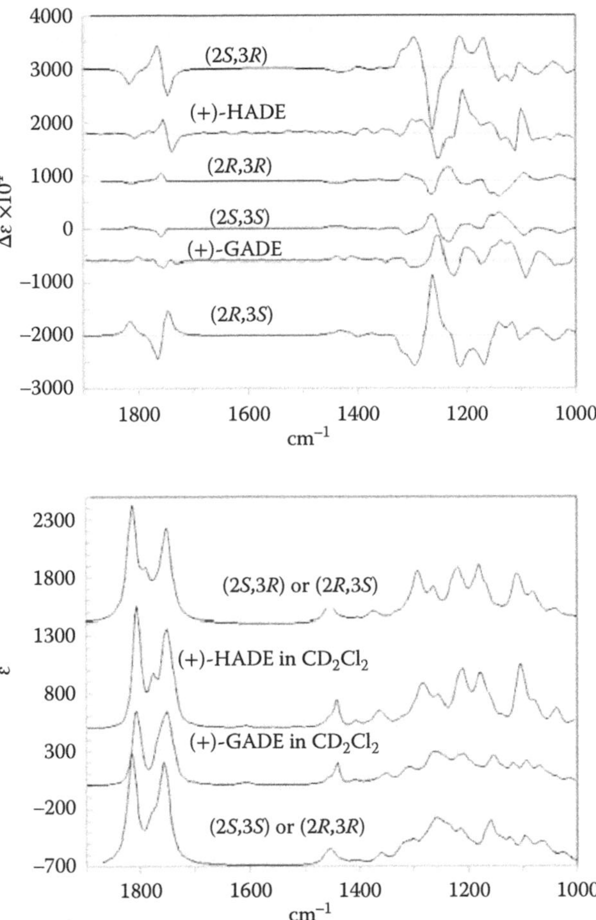

FIGURE 8.6
VA and VCD spectra of (+)-garcinia acid dimethyl ester (GADE) and (+)-hibiscus acid dimethyl ester (HADE) and their comparison to calculated spectra for four diastereomers. Reprinted with permission from Polavarapu, P. L. et al., *J. Phys. Chem.* A, 115, 5665–5673. Copyright 2011 American Chemical Society.

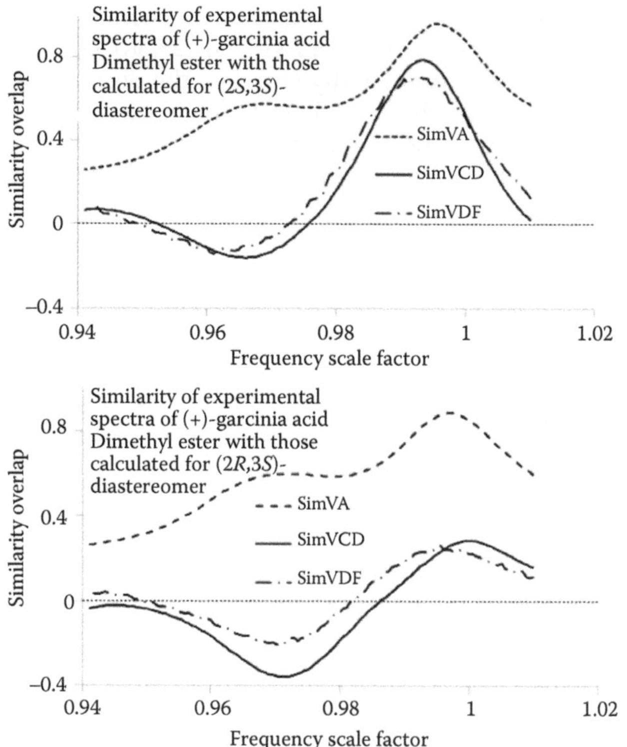

FIGURE 8.7
Similarity overlap of VA, VCD, and VDF spectra of (+)-GADE with those calculated for (2*S*,3*S*) (top) and (2*R*,3*S*) (bottom) diastereomers.

(2*R*,3*R*) diastereomer, also gives a similarity overlap of 0.16 and 0.12, respectively, for VCD and VDF of (+)-GADE. The bottom panel of Figure 8.7 shows a similarity overlap of (+)-GADE spectra with those for (2*R*,3*S*) diastereomer. Here similarity overlaps of 0.29 and 0.26 are seen for VCD and VDF, respectively, at a frequency scale factor of ~1. There is also a negative overlap of −0.36 and −0.21, respectively, for VCD and VDF, at a frequency scale factor of 0.97. This negative overlap indicates that the antipode, (2*S*,3*R*) diastereomer, also gives a similarity overlap 0.36 and 0.21, respectively, for VCD and VDF of (+)-GADE. However, since the similarity overlap with (2*S*,3*S*) diastereomer is the highest, the AC of GADE can be assigned as (2*S*,3*S*).

Similarly, the experimental VCD and VDF spectra of (+)-HADE have similarity overlaps with different diasteromers as follows (see Figure 8.8): 0.68 and 0.47 for (2*S*,3*R*), 0.22 and 0.12 for (2*R*,3*S*), 0.22 and 0.12 for (2*R*,3*R*), and 0.17 and 0.14 for (2*S*,3*S*). Since the similarity overlap with (2*S*,3*R*) diastereomer is the highest, the AC of HADE can be assigned as (2*S*,3*R*).

Even though the experimental VCD spectrum of (+)-GADE has the highest overlap with that predicted for (2*S*,3*S*) diastereomer and the experimental

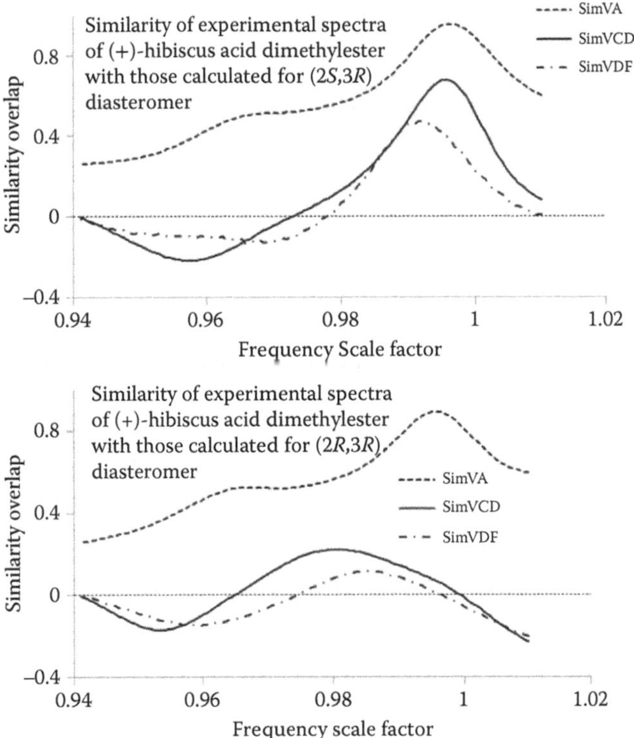

FIGURE 8.8
Similarity overlap of VA, VCD, and VDF spectra of (+)-HADE with those calculated for (2*S*,3*R*) (top) and (2*R*,3*R*) (bottom) diastereomers.

VCD spectrum of (+)-HADE has the highest overlap with that predicted for (2*S*,3*R*) diastereomer, the overlap with remaining diastereomers is nonzero. These examples indicate that significant magnitudes for similarity overlap in *Sim*VCD and *Sim*VDF are needed to be confident of the similarity overlap analyses for the AC assignment.

The overlap magnitudes obtained with SI and ESI criteria will be different, usually higher, from those obtained with *Sim*. However, one has to use the same criterion (*Sim* or SI or ESI) in a comparative analysis for a given set of compounds.

The examples presented earlier pertain to the VCD spectral analyses, but similar analyses were recommended for ECD (Derewacz et al. 2014; Junior et al. 2014) and VROA (Polavarapu and Covington 2014) spectra as well (Polavarapu 2016). In some cases more information could be gained from EDF analyses than from separate ECD and EA spectral analyses (Covington and Polavarapu 2016). A freely available computer program (Covington and Polavarapu 2015) for carrying out these similarity analyses facilitates general applications.

There are some caveats in the practical use of similarity analysis. When the experimental vibrational spectra are measured in hydrogen bonding solvents, such as CH_3OH, $(CH_3)_2SO$, and solute molecules that participate in hydrogen bonding interactions with the solvent, it is possible that some vibrational bands of the solute can be shifted due to hydrogen bonding effects, while others are not (Zhang et al. 2016). The use of a constant x-axis scaling factor may only bring some calculated bands to coincide with experimental bands, while other calculated bands may not match the experimental band positions. In such cases, it is likely to obtain low similarity overlap values. This will reflect a problem due to the deficiencies in theoretical calculations to account for the influence of hydrogen bonding interactions in vibrational spectra. In such situations, the spectra can be separated into two regions: one region with hydrogen bonding influence and another without hydrogen bonding influence. Carrying out separate similarity analysis for the two separate regions (Covington et al. 2016) is likely to improve the similarity measure values. In such cases, the x-axis scale factor corresponding to the maximum similarity overlap will be different for the two regions.

8.3 Significance of Similarity Measure Values

It is possible that the calculated spectrum may reproduce the experimental spectrum better, in the spectral region of interest, when a fewer number of bands are present and are analyzed. If the spectral region of interest has a larger number of bands, then the calculated spectrum is more likely to show larger deviations from the experimental spectrum. As a result, the similarity measure values for the former case can be larger than that for the latter. The number of available bands in the spectral region of interest can vary from molecule to molecule. Moreover, electronic spectra usually have a smaller number of bands in the accessible electronic spectral region, compared with the number of bands present in the accessible vibrational spectral region. For these reasons, the magnitude of similarity measure value becomes dependent on the molecule as well as on the type of chiroptical spectrum being considered. Therefore, similarity measure values are most useful, and less ambiguous, when comparing the experimental spectrum with those calculated for diastereomers of a given chiral molecule. In such cases, similarity measures obtained can confidently indicate the spectrum of a diastereomer that closely matches the experimental spectrum.

To associate significance with any given similarity measure value, confidence level (Debie et al. 2011) and statistical robustness (Vandenbussche et al. 2013), criterion have been introduced. In the former case, one compares the similarity measure value obtained for a given molecule with those obtained for molecules in a dataset and estimates the percent likelihood of obtaining the value being analyzed (which need not be a true reflection of the correct AC assignment). This procedure requires the availability of similarity measure values for a large set of molecules and for the type of chiroptical spectrum being analyzed. Such a requirement is avoided in determining the statistical robustness. In the latter method, a large number of randomly generated chiroptical spectra are obtained, and their similarity measures with the corresponding experimental spectrum and with the calculated spectrum (at a reliable quantum theoretical level) are calculated. The plot of similarity measures of randomly generated spectra with the experimental spectrum versus that with the predicted spectrum gives a straight line if the similarity between experimental and calculated spectra is statistically sound. What this means is that none of the randomly generated spectra can yield as much similarity as that obtained from the calculated spectrum at a reliable quantum theoretical level.

However, it is not clear if the comparison with a set of random spectra can provide a real test. This is because vibrational spectra depend on the characteristic group frequencies and therefore have nonrandom nature. Therefore, it may prove to be more realistic and advantageous to compare the similarity measure of a given predicted chiroptical spectrum with the experimental chiroptical spectra of compounds that are in the same chemical category as that of compound of interest.

8.4 Summary

In any comparison of the experimental and calculated spectra, the first level of analysis should involve the visual comparison of these spectra. However, discerning the level of agreement via visual judgment is subject to individual discretion and can be subject to inadvertent user bias. Also it would not be possible to report on the quantification of the agreement. The procedure for determining the experimental dipole and rotational strengths and comparing them with corresponding calculated strengths is definitely valuable, but the labor involved and associated practical difficulties make this approach generally difficult to implement. The numerical measure of similarity is one way to estimate the unbiased level of agreement between experimental and calculated spectra. The choice among different similarity measures (namely SI, ESI, or *Sim*) is a mere personal preference as all of them will lead to similar conclusions. The analysis of all three spectra associated with a given chiroptical spectrum, namely absorption, CD and DF or vibrational Raman, VROA and CID, and presentation of corresponding similarity plots, is certain to provide a better insight and is strongly recommended.

References

Bruhn, T., A. Schaumlöffel, Y. Hemberger, and G. Bringmann. 2013. SpecDis: Quantifying the comparison of calculated and experimental electronic circular dichroism spectra. *Chirality* 25(4):243–249.

Covington, C. L., F. M. S. Junior, J. H. S. Silva, R. M. Kuster, M. B. de Amorim, and P. L. Polavarapu. 2016. Atropoisomerism in biflavones: The absolute configuration of agathisflavone determined from chiroptical spectroscopy. *J Nat. Products* (submitted).

Covington, C. L., and P. L. Polavarapu. 2013. Similarity in dissymmetry factor spectra: A quantitative measure of comparison between experimental and predicted vibrational circular dichroism. *J. Phys. Chem. A* 117(16):3377–3386.

Covington, C. L., and P. L. Polavarapu. 2015. CDSpecTech: Computer programs for calculating similarity measures for experimental and calculated dissymmetry factors and circular intensity differentials. https://sites.google.com/site/cdspectech1/ (accessed May 4, 2016).

Covington, C. L., and P. L. Polavarapu. 2016. Solvation dependence observed in the electronic dissymmetry factor spectra: How much information are we missing by analyzing the circular dichroism spectra alone? *Phys. Chem. Chem. Phys.* 18(20):13912–13917.

Debie, E., E. D. Gussem, R. K. Dukor, W. Herrebout, L. A. Nafie, and P. Bultinck. 2011. A confidence level algorithm for the determination of absolute configuration using vibrational circular dichroism or Raman optical activity. *ChemPhysChem* 12(8):1542–1549.

Derewacz, D. K., C. R. McNees, G. Scalmani, C. L. Covington, G. Shanmugam, L. J. Marnett, P. L. Polavarapu, and B. O. Bachmann. 2014. Structure and stereochemical determination of hypogeamicins from a cave-derived actinomycete. *J. Nat. Prod.* 77(8):1759–1763.

Devlin, F. J., P. J. Stephens, J. R. Cheeseman, and M. J. Frisch. 1997. Ab initio prediction of vibrational absorption and circular dichroism spectra of chiral natural products using density functional theory: α-Pinene. *J. Phys. Chem. A* 101(51):9912–9924.

Góbi, S., and G. Magyarfalvi. 2011. Reliability of computed signs and intensities for vibrational circular dichroism spectra. *Phys. Chem. Chem. Phys.* 13(36):16130–16133.

Góbi, S., G. Magyarfalvi, and G. Tarczay. 2015. VCD robustness of the amide-I and amide-II vibrational modes of small peptide models. *Chirality*:27(9):625–634.

Gussem, E. D., P. Bultinck, M. Feledziak, J. Marchand-Brynaert, C. V. Stevens, and W. Herrebout. 2012. Vibrational circular dichroism versus optical rotation dispersion and electronic circular dichroism for diastereomers: The stereochemistry of 3-(1[prime or minute]-hydroxyethyl)-1-(3[prime or minute]-phenylpropanoyl)-azetidin-2-one. *Phys. Chem. Chem. Phys.* 14(24):8562–8571.

He, J., A. Petrovich, and P. L. Polavarapu. 2004. Quantitative determination of conformer populations: Assessment of specific rotation, vibrational absorption, and vibrational circular dichroism in substituted butynes. *J. Phys. Chem. A* 108(10):1671–1680.

Junior, F. M. S., C. L. Covington, M. B. de Amorim, L. S. M. Velozo, M. A. C. Kaplan, and P. L. Polavarapu. 2014. Absolute configuration of a rare sesquiterpene: (+)-3-Ishwarone. *J. Nat. Prod.* 77(8):1881–1886.

Kuppens, T., W. Langenaeker, J. P. Tollenaere, and P. Bultinck. 2003. Determination of the stereochemistry of 3-hydroxymethyl-2,3-dihydro-[1,4]dioxino[2,3-b]-pyridine by vibrational circular dichroism and the effect of DFT integration grids. *J. Phys. Chem. A* 107(4):542–553.

Kuppens, T., K. Vandyck, J. Van der Eycken, W. Herrebout, B. J. van der Veken, and P. Bultinck. 2005. Determination of the absolute configuration of three as-hydrindacene compounds by vibrational circular dichroism. *J. Org. Chem.* 70(23):9103–9114.

Longhi, G., M. Tommasini, S. Abbate, and P. L Polavarapu. 2015. The connection between robustness angles and dissymmetry factors in vibrational circular dichroism spectra. *Chem. Phys. Lett.* 639:320–325.

Nicu, V. P., and E. J. Baerends. 2009. Robust normal modes in vibrational circular dichroism spectra. *Phys. Chem. Chem. Phys.* 11(29):6107–6118.

Nicu, V. P., and E. J. Baerends. 2011. On the origin dependence of the angle made by the electric and magnetic vibrational transition dipole moment vectors. *Phys. Chem. Chem. Phys.* 13(36):16126–16129.

Polavarapu, P. L. 2016. Structural analysis using chiroptical spectroscopy: Insights and cautions. Chirality doi: 10.1002/chir.22601.

Polavarapu, P. L., and C. L. Covington. 2014. Comparison of experimental and calculated chiroptical spectra for chiral molecular structure determination. *Chirality* 26:179–192.

Polavarapu, P. L., E. A. Donahue, G. Shanmugam, G. Scalmani, E. K. Hawkins, C. Rizzo, I. Ibnusaud, G. Thomas, D. Habel, and D. Sebastian. 2011. A single chiroptical spectroscopic method may not be able to establish the absolute configurations of diastereomers: Dimethylesters of hibiscus and garcinia acids. *J. Phys. Chem. A* 115(22):5665–5673.

Shen, J., C. Zhu, S. Reiling, and R. Vaz. 2010. A novel computational method for comparing vibrational circular dichroism spectra. *Spectrochim. Acta A Mol. Biomol. Spectrosc.* 76(3–4):418–422.

Simmen, B., T. Weymuth, and M. Reiher. 2012. How many chiral centers can Raman optical activity spectroscopy distinguish in a molecule? *J. Phys. Chem. A* 116(22):5410–5419.

Tommasini, M., G. Longhi, G. Mazzeo, S. Abbate, B. Nieto-Ortega, F. J. Ramírez, J. Casado, and J. T. López Navarrete. 2014. Mode robustness in Raman optical activity. *J. Chem. Theory Comput.* 10(12):5520–5527.

Vandenbussche, J., P. Bultinck, A. K. Przybył, and W. A. Herrebout. 2013. Statistical validation of absolute configuration assignment in vibrational optical activity. *J. Chem. Theory Comput.* 9(12):5504–5512.

Zhang, Y., M. R. Poopari, X. Cai, A. Savin, Z. Dezhahang, J. Cheramy, and Y. Xu. 2016. IR and vibrational circular dichroism spectroscopy of matrine- and artemisinin-type herbal products: Stereochemical characterization and solvent effects. *J. Nat. Products*, DOI: 10.1021/acs.jnatprod.5b01082.

9

Chiral Aggregation, Equilibria, and the Horeau Effect

In the presence of solute–solute and solute–solvent intermolecular interactions, it is known that specific optical rotation (SOR) may no longer be independent of the concentration of solution. In 1968, it was demonstrated that $[\alpha]_D$ for (α-methyl-α-ethyl)succinic acid changes from being positive at low concentrations to negative at higher concentrations (Krow and Hill 1968) (see Figure 9.1). Such drastic concentration dependence, attributed to sample aggregation at higher concentrations, has consequences in the way SOR can be used. For example, the enantiomeric composition (i.e., composition of enantiomers) of a synthesized sample used to be estimated in the literature by comparing its $[\alpha]_D$ with that of enantiomerically pure substance and designated as optical purity (op); op is the ratio of SOR measured for an enantiomeric mixture to that of pure enantiomer. In the presence of aggregation effects, op determined from the experimental SORs need not match the enantiomeric excess (ee) (determined by mixing known amounts of enantiomers or from chiral chromatography). The plot of %op against %ee generated from the data reported (Horeau 1969) for (+)-(α-methyl-α-ethyl) succinic acid is shown in Figure 9.2. From this plot, it can be seen that for a sample with 50%ee, the observed op is ~36%, which amounts to a deviation of −28%. These deviations are referred to as the Horeau effect. Chiral aggregation is considered to be responsible for such observed deviations (Eliel and Wilen 1994).

The formation of dimers represents the simplest form of aggregation. When both monomers and dimers of a chemical substance exist in equilibrium, it is necessary to consider the associated equilibrium for analyzing the measured experimental data. This type of situations can occur for molecules containing, carboxylic, amine, and/or hydroxyl groups when those substances are dissolved in inert solvents, such as CCl_4, CH_2Cl_2, and $CHCl_3$. These inert solvents are not expected to participate in solute–solvent hydrogen bonding interactions, thereby promoting the aggregation of solute molecules.

The monomer–dimer equilibrium in achiral substances is well known (Chen et al. 2004). When a single enantiomer is used for the experimental measurements, the resulting homochiral monomer–dimer equilibrium (Polavarapu and Covington 2015) can be treated similar to monomer–dimer equilibrium for achiral molecules. However, if a mixture of enantiomers is used for the experimental measurements, then both homochiral and heterochiral

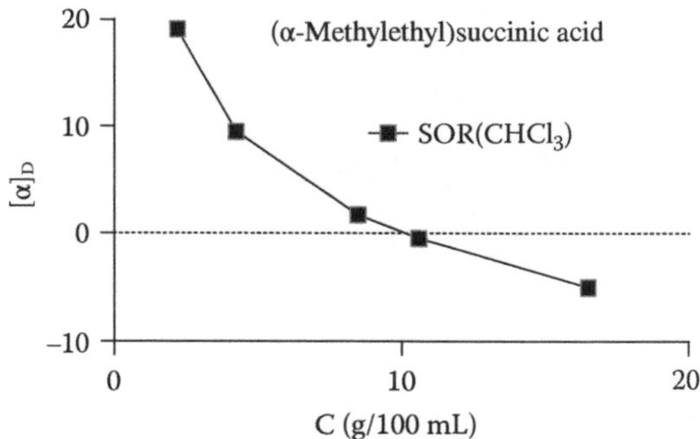

FIGURE 9.1
Concentration dependence of $[\alpha]_D$ for (S)-(−)-α-methyl-α-ethyl succinic acid. (Data from Krow, G. and Hill, R. K., *Chem. Comm.*, (8), 430–431, 1968.)

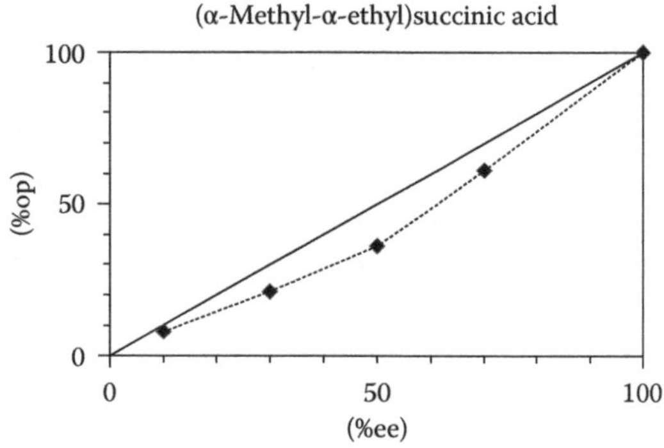

FIGURE 9.2
Comparison of % optical purity (%op) with % enantiomeric excess (%ee) of α-methyl-α-ethyl succinic acid. The diagonal line represents the situation for %op= %ee. The data points (square symbols) connected by the dotted line are the observed %op values for known %ee. (Data from Horeau, A., *Tetrahedron Lett.*, 10, 3121–3124, 1969.)

monomer–dimer equilibria need to be considered (Baciocchi et al. 2002, 2004; Covington and Polavarapu 2015a; Polavarapu and Covington 2015).

The pertinent expressions for analyzing the experimental chiroptical spectroscopic data for pure enantiomers and enantiomeric mixtures, in the presence of monomer–dimer equilibrium, are presented in this chapter.

9.1 Homochiral Monomer–Dimer Equilibrium

Homochiral monomer–dimer equilibrium pertains to a single enantiomer exhibiting monomer–dimer equilibrium. The monomeric form of single enantiomeric molecules of a chiral substance is designated as M_A and dimeric form of the molecules of same substance as D_{AA}. The monomeric form includes all possible conformers of the monomer, and the dimeric form includes all possible conformers of the dimer. Here A can represent either of the two enantiomers of a chiral substance. The monomer–dimer equilibrium is represented by the expression

$$2M_A \rightleftharpoons D_{AA} \tag{9.1}$$

Since there is only one enantiomer involved in homochiral monomer–dimer equilibrium, we will not specifically identify the enantiomer. However, it is implied that the monomer and dimer discussed in this section belong to a specified enantiomer. Then, the homochiral equilibrium constant K can be written as

$$K = \frac{C_d}{C_m^2} \tag{9.2}$$

where C_m and C_d are, respectively, the concentrations of monomeric and dimeric forms of the enantiomeric substance (in mol L^{-1}). An enantiomeric substance is one that is made up of single enantiomer molecules. In Equation 9.2, it is implied that $C_m = \sum_i C_{m,i} = \sum_i f_{m,i} C_m$ and $C_d = \sum_i C_{d,i} = \sum_i f_{d,i} C_d$, where $C_{m,i}$ is the concentration and $f_{m,i}$ is the fractional population of ith conformer of monomer and $\sum_i f_{m,i} = 1$ and $C_{d,i}$ is the concentration and $f_{d,i}$ is the fractional population of ith conformer of dimer and $\sum_i f_{d,i} = 1$. The material balance requires that the concentration of dissolved sample, C_o, is given as

$$C_o = C_m + 2C_d \tag{9.3}$$

Substituting Equation 9.2 in 9.3, one obtains

$$C_o = C_m + 2KC_m^2 \tag{9.4}$$

Upon rearrangement, Equation 9.4 can be written as

$$C_m^2 + \frac{C_m}{2K} - \frac{C_o}{2K} = 0 \tag{9.5}$$

Solving the quadratic equation in C_m,

$$C_m = \frac{-\frac{1}{2K} \pm \sqrt{\left(\frac{1}{2K}\right)^2 + \frac{4C_o}{2K}}}{2} = \frac{-1 \pm \sqrt{1 + 8KC_o}}{4K} \tag{9.6}$$

The minus combination is not allowed because concentrations cannot be negative. Then, the solution to Equations 9.4 and 9.5 becomes

$$C_m = \frac{-1+\sqrt{1+8KC_o}}{4K} \tag{9.7}$$

To simplify Equation 9.7, multiply and divide with $1+\sqrt{1+8KC_o}$. Then,

$$C_m = \frac{-1+\sqrt{1+8KC_o}}{4K} \times \frac{1+\sqrt{1+8KC_o}}{1+\sqrt{1+8KC_o}} = \frac{8KC_o}{4K\left(1+\sqrt{1+8KC_o}\right)} \tag{9.8}$$

Upon cancelling the common terms in numerator and denominator, one obtains

$$C_m = P_m C_o \tag{9.9}$$

where P_m is given as (Polavarapu and Covington 2015)

$$P_m = \frac{2}{\left(1+\sqrt{1+8KC_o}\right)} \tag{9.10}$$

On substituting Equation 9.9 into Equation 9.3, the expression for C_d can be obtained as

$$C_d = P_d C_o \tag{9.11}$$

where P_d is given as (Polavarapu and Covington 2015)

$$P_d = \frac{1}{2}\frac{\left(-1+\sqrt{1+8KC_o}\right)}{\left(1+\sqrt{1+8KC_o}\right)} \tag{9.12}$$

Note that P_m and P_d do not represent the mole fractions because they do not add up to one. However, they can be related to the mole fractions, x_m and x_d, respectively, of monomer and dimer, using the relation

$$P_m + 2P_d = 1 \tag{9.13}$$

as

$$x_m = \frac{C_m}{C_m + C_d} = \frac{P_m}{P_m + P_d} = \frac{2P_m}{1+P_m} = \frac{P_m}{1-P_d} \tag{9.14}$$

$$x_{\mathrm{d}} = \frac{C_{\mathrm{d}}}{C_{\mathrm{m}} + C_{\mathrm{d}}} = \frac{P_{\mathrm{d}}}{P_{\mathrm{m}} + P_{\mathrm{d}}} = \frac{1 - P_{\mathrm{m}}}{1 + P_{\mathrm{m}}} = \frac{P_{\mathrm{d}}}{1 - P_{\mathrm{d}}} = \frac{2P_{\mathrm{d}}}{1 + P_{\mathrm{m}}} \tag{9.15}$$

By measuring the spectral properties as a function of starting sample concentration, C_o, one can determine the equilibrium constant and hence the concentrations of monomers and dimers. The concentrations of chiral monomer and its homochiral dimer determine the circular dichroism (CD), vibrational Raman optical activity (VROA), and the corresponding absorption/vibrational Raman spectra properties of a given enantiomer. The concentrations of chiral monomer and its homochiral dimer will also determine the optical rotation (OR) property, even though the analysis of OR data will differ from that of absorption/CD/vibrational Raman/VROA spectral data. This is because, in the absorption/CD/vibrational Raman/VROA spectra, the bands due to the monomer and dimer may be separated enough in their positions (more so in vibrational spectra than in electronic spectra) to monitor them individually, while there is no such separation possible in the OR, which is a composite property.

9.1.1 Analysis of SOR Data

The measured OR, a composite of contributions from monomer and dimer species, as a function of concentration can be used to determine the equilibrium constants (Baciocchi et al. 2002). The emphasis in this section, however, will be to determine the SORs of monomer and dimeric species (Polavarapu and Covington 2015) involved in the equilibrium.

Assuming that SORs of monomer and dimer are independent of concentrations, the observed OR, α, for an enantiomeric substance in homochiral monomer–dimer equilibrium, can be written as follows:

$$\alpha = [\alpha]_{\mathrm{m}}\, c_{\mathrm{m}} l + [\alpha]_{\mathrm{d}}\, c_{\mathrm{d}} l = [\alpha]_{\mathrm{m}} \frac{C_{\mathrm{m}} M_{\mathrm{m}}}{1000} l + [\alpha]_{\mathrm{d}} \frac{C_{\mathrm{d}} M_{\mathrm{d}}}{1000} l \tag{9.16}$$

where $[\alpha]_{\mathrm{m}}$ and $[\alpha]_{\mathrm{d}}$ are the SORs, respectively, of monomer and dimer species; l is the path length of the cell used for OR measurement; c_{m} is concentration of monomer in g/cc; c_{d} is concentration of dimer in g/cc; M_{m} is the molar mass of monomer; and M_{d} is the molar mass of dimer. Note that upper case letter "C" is used for concentrations in mol L^{-1}, and lower case letter "c" is used for those in g/cc.

As mentioned already, since there is only one enantiomer involved in homochiral monomer–dimer equilibrium; we will not specifically identify the enantiomer. However, it is implied that the monomer and dimer discussed in this section belong to a specified enantiomer. For example, α, $[\alpha]_{\mathrm{m}}$, $[\alpha]_{\mathrm{d}}$, c_{m}, and c_{d} in Equation 9.16 belong to the enantiomer of choice and are implied to be, respectively, α_A, $[\alpha]_{m,A}$, $[\alpha]_{d,A}$, $c_{m,A}$, and $c_{d,A}$ with A representing the enantiomer under consideration.

One can simplify Equation 9.16 in two different ways:

1. Simplification of Equation 9.16 using Equation 9.2: Substituting Equation 9.2 and $M_d = 2M_m$ in Equation 9.16 yields the following:

$$\alpha = [\alpha]_m \frac{C_m M_m}{1000} l + [\alpha]_d \frac{2KC_m^2 M_m}{1000} l \tag{9.17}$$

Using $C_m = P_m C_o$ (see Equation 9.9), this equation modifies to

$$\alpha = \left\{[\alpha]_m P_m C_o + 2[\alpha]_d K P_m^2 C_o^2\right\} \frac{M_m}{1000} l \tag{9.18}$$

This equation clearly shows that observed rotation is a nonlinear function of the concentration of enantiomeric substance when monomer–dimer equilibrium is involved.

Writing the starting concentration, c_o in g/cc of enantiomeric substance as

$$c_o = \frac{C_o M_m}{1000} \tag{9.19}$$

and substituting it in Equation 9.18, the SOR of solution

$$[\alpha] = \frac{\alpha}{c_o l} \tag{9.20}$$

can be obtained from Equation 9.18 as

$$[\alpha] = \left\{[\alpha]_m + 2[\alpha]_d K P_m C_o\right\} P_m \tag{9.21}$$

In the presence of monomer–dimer equilibrium, the SOR of an enantiomeric substance need not be independent of its concentration, as is evident from Equation 9.21. However, this equation is not as transparent as one would like (vide infra), so let us simplify Equation 9.16 in another way.

2. Simplification of Equation 9.16 using Equations 9.9 and 9.11: A more compact expression can be obtained when equations $C_m = P_m C_o$ and $C_d = P_d C_o$ (see Equations 9.8 and 9.9 through 9.11) and $M_d = 2M_m$ are substituted in Equation 9.16. Then,

$$\alpha = \left\{[\alpha]_m P_m + 2P_d [\alpha]_d\right\} \frac{C_o M_m}{1000} l \tag{9.22}$$

Using Equations 9.19, 9.20, and 9.13, the corresponding expression for SOR becomes (Polavarapu and Covington 2015)

$$[\alpha]=\left\{[\alpha]_m P_m+(1-P_m)[\alpha]_d\right\} \tag{9.23}$$

Equation 9.23 is more transparent than Equation 9.21: (1) As C_o approaches zero, P_m approaches 1 (see Equation 9.10), so [α] in Equation 9.23 becomes $[\alpha]_m$. (2) As C_o approaches infinity, P_m approaches 0, and [α] in Equation 9.23 becomes $[\alpha]_d$. (3) When $[\alpha]_m = [\alpha]_d$, [α] in Equation 9.23 becomes independent of the concentration of the enantiomeric substance, as it should be, because if the SORs of the monomer and dimer are equal, then the monomer–dimer equilibrium should not influence the observed SOR. For these reasons, Equation 9.23 may be considered as the fundamental equation governing SOR for monomer–dimer equilibrium of an enantiomeric substance.

When equilibrium constant is known, SORs of monomer and dimer can be determined by plotting the experimentally determined [α] against the concentration of enantiomeric substance, C_o, and fitting this data to Equation 9.23 (Polavarapu and Covington 2015). When equilibrium constant is not known, enough experimental data points need to be gathered for determining the equilibrium constant along with SORs of monomer and dimer by fitting the experimental data to Equation 9.23. Nevertheless, since the experimentally measured values are α of the solution and Equation 9.23 uses [α] of the solution, one should use weighted nonlinear least square (WNLS) method. The weights can be determined using error propagation for Equation 9.20. Straightforward error propagation yields $\frac{S_{[\alpha]}^2}{[\alpha]^2}=\frac{S_\alpha^2}{\alpha^2}+\frac{S_{c_o}^2}{c_o^2}$. In the approximation that a major portion of the relative error may come from that in the measurement of OR, and relative error in concentration is negligible (see the example discussed in Chapter 7), $S_{[\alpha]}^2 \sim [\alpha]^2\frac{S_\alpha^2}{\alpha^2}=\frac{S_\alpha^2}{c_o^2 l^2}$. Taking S_α as the instrumental reproducibility (0.002° for typical commercial polarimeters), weights for the residuals of data points become $w_{[\alpha]}=\frac{1}{S_{[\alpha]}^2}=\frac{c_o^2 l^2}{(0.002)^2}\propto c_o^2$. That means, for the WNLS fitting with c_o^2 as weights, the residuals from higher concentration data points are weighted more than those at lower concentrations.

Here we digress to comment on a literature article for the analysis SORs of monomer–dimer equilibria. It was suggested that the observed OR, α, for monomer–dimer equilibrium system (Goldsmith et al. 2003) can be written in terms of ORs of monomer and dimer (α_m and α_d, respectively), multiplied with mole fractions of monomer and dimer, χ_m and χ_d, respectively, as

$$\alpha=\alpha_m\chi_m+\alpha_d\chi_d \tag{9.24}$$

and

$$[\alpha]=[\alpha]_m\chi_m^2+[\alpha]_d\chi_d^2 \tag{9.25}$$

Note that the definitions for mole fractions used in Equations 9.24 and 9.25 were not same as those defined in Equations 9.14 and 9.15. Nevertheless,

Equation 9.25 does not have the correct limiting values, that is, [α] does not become independent of the concentration of enantiomeric substance when $[\alpha]_m = [\alpha]_d$, because the sum $\chi_m^2 + \chi_d^2$ cannot be equal to 1.

9.1.2 Analysis of Absorption Spectral Data

Absorption spectra are useful for determining the homochiral equilibrium constants. Here we assume that the absorption bands due to monomer and dimer species appear at well-separated *x*-axis positions in the spectra, especially in vibrational spectra. Then, the integrated intensities of spectral bands belonging to monomer species (or dimer species) can be measured to determine the monomer–dimer equilibrium constant (Chen et al. 2004; Kuppens et al. 2006; Nakao et al. 1985). The integrated absorption intensity of *i*th band of monomer is given as

$$A_{m,i} = C_m l \int \varepsilon_{m,i}(\nu)d\nu = P_m C_o l \varepsilon_{m,i} = \frac{2}{\left(1+\sqrt{1+8KC_o}\right)} C_o l \varepsilon_{m,i} \tag{9.26}$$

where $\varepsilon_{m,i} = \int \varepsilon_{m,i}(\nu)d\nu$ represents the integrated molar absorption coefficient of *i*th band of monomer, and l is the path length used. In lieu of integrated intensity, the peak intensity of spectral bands associated with the monomer may be used if the band under consideration is isolated from the neighboring bands.

Nonlinear least squares fitting of the integrated intensities of individual bands of monomer, $A_{m,i}$, as a function of C_o provides the equilibrium constant K and hence the population of monomer at a given concentration C_o. The commercial programs, such as PeakFit (2015), may be used to find the integrated band areas, and nonlinear least squares fitting can be undertaken using commercial programs such as KaleidaGraph or a freely available open source program (Gnuplot).

Equation 9.26 can be rearranged to convert it to the form of a straight line equation as follows:

$$\left(1+\sqrt{1+8KC_o}\right) = \frac{2}{A_{m,i}} C_o l \varepsilon_{m,i} \tag{9.27}$$

Squaring this equation and rearranging it gives

$$1+8KC_o = \left(\frac{2}{A_{m,i}} C_o l \varepsilon_{m,i} - 1\right)^2 = \left(\frac{2}{A_{m,i}} C_o l \varepsilon_{m,i}\right)^2 + 1 - \frac{4}{A_{m,i}} C_o l \varepsilon_{m,i} \tag{9.28}$$

Further rearrangement of Equation 9.28 leads to

$$\frac{4}{A_{m,i}^2} C_o^2 l^2 \varepsilon_{m,i}^2 = \frac{4}{A_{m,i}} C_o l \varepsilon_{m,i} + 8KC_o \tag{9.29}$$

Canceling C_o in Equation 9.29 and rearranging it yields

$$\frac{C_o}{A_{m,i}^2} = \frac{1}{l\varepsilon_{m,i}}\frac{1}{A_{m,i}} + \frac{2K}{l^2\varepsilon_{m,i}^2} \tag{9.30}$$

Equation 9.30 is in the familiar form of a straight line equation $y = mx + c$. A plot of $\left(C_o/A_{m,i}^2\right)$ versus $1/A_{m,i}$ will be a straight line with slope of $(1/l\varepsilon_{m,i})$ and intercept of $(2K/l^2\varepsilon^2_{m,i})$.

Alternately, Equation 9.30 can be written as a quadratic equation in the form of $y = ax + bx^2$, with $y = C_o$ and $x = A_{m,i}$,

$$C_o = \frac{1}{l\varepsilon_{m,i}}A_{m,i} + \frac{2K}{l^2\varepsilon_{m,i}^2}A_{m,i}^2 \tag{9.31}$$

and use least squares fit of C_o versus $A_{m,i}$ values. To obtain the same results (especially the estimated errors) with Equations 9.30 and 9.31, however, a WNLS fit is needed for using Equation 9.30. Writing the left-hand side of Equation 9.30 as $z = y/A_{m,i}^2$, where $y = C_o$, error propagation assuming $A_{m,i}$ is error free gives, the standard deviation as $S_z = \frac{1}{A_{m,i}^2}S_y$. Then, WNLS fit of $C_o/A_{m,i}^2$ versus $1/A_{m,i}$ requires weighting the residuals with $A_{m,i}^4$.

Note that the results obtained from nonlinear least square fitting of Equation 9.26, need not be same as those obtained from linear least square fitting of Equation 9.30. This is because, in the least square fitting procedures provided in most commercial programs, x-values are assumed to be error free and y-values to have distributed errors. Then in using Equation 9.26, C_o are assumed to be error free and errors are assumed to be present in $A_{m,i}$. On the contrary, in using Equation 9.30 or Equation 9.31, $A_{m,i}$ are assumed to be error free, and errors present in C_o. The assumptions associated with analyzing Equations 9.26 and 9.30 (or 9.31) are thus different.

In the modern computer era, there is no need to linearize a nonlinear equation and Equation 9.26 can be used as such with nonlinear least squares fitting algorithms.

There is no reason to restrict the analysis only to the monomer bands. If dimer bands can be identified, then the equation analogous to Equation 9.26 for a dimer band becomes (Chen et al. 2004)

$$A_{d,i} = C_d l\int \varepsilon_{d,i}(\nu)d\nu = P_d C_o l\varepsilon_{d,i} = \frac{1}{2}\frac{\left(-1+\sqrt{1+8KC_o}\right)}{\left(1+\sqrt{1+8KC_o}\right)}C_o l\varepsilon_{d,i} \tag{9.32}$$

In Equation 9.32, $\varepsilon_{d,i} = \int \varepsilon_{d,i}(\nu)d\nu$. Nonlinear least squares fitting of the integrated intensities of individual bands of dimer, $A_{d,i}$, as a function of starting sample concentration C_o provides the equilibrium constant K.

Even though it is not necessary to modify a nonlinear equation to the form of a linear equation, it is sometimes convenient to use the linear form of the equation to estimate initial values of the parameters needed during the nonlinear least squares fitting procedure. For this purpose, the nonlinear form of Equation 9.32 can be recast (Chen et al. 2004) in the form of a straight line equation, as follows:

$$\frac{2A_{\mathrm{d},i}}{C_{\mathrm{o}}l\varepsilon_{\mathrm{d},i}}=\frac{\left(-1+\sqrt{1+8KC_{\mathrm{o}}}\right)}{\left(1+\sqrt{1+8KC_{\mathrm{o}}}\right)} \tag{9.33}$$

Since, $\frac{a}{b}=\frac{c}{d}$ is same as $\left(\frac{a+b}{a-b}\right)=\left(\frac{c+d}{c-d}\right)$, Equation 9.33 can be rewritten as

$$\frac{2A_{\mathrm{d},i}+C_{\mathrm{o}}l\varepsilon_{\mathrm{d},i}}{2A_{\mathrm{d},i}-C_{\mathrm{o}}l\varepsilon_{\mathrm{d},i}}=\frac{\left(-1+\sqrt{1+8KC_{\mathrm{o}}}\right)+\left(1+\sqrt{1+8KC_{\mathrm{o}}}\right)}{\left(-1+\sqrt{1+8KC_{\mathrm{o}}}\right)-\left(1+\sqrt{1+8KC_{\mathrm{o}}}\right)}=-\sqrt{1+8KC_{\mathrm{o}}} \tag{9.34}$$

Taking square of this equation gives

$$\frac{\left(C_{\mathrm{o}}l\varepsilon_{\mathrm{d},i}+2A_{\mathrm{d},i}\right)^2}{\left(C_{\mathrm{o}}l\varepsilon_{\mathrm{d},i}-2A_{\mathrm{d},i}\right)^2}=1+8KC_{\mathrm{o}} \tag{9.35}$$

Since, $\frac{a}{b}=\frac{c}{d}$ is same as $\left(\frac{a-b}{b}\right)=\left(\frac{c-d}{d}\right)$, Equation 9.35 can be rewritten as

$$\frac{\left(C_{\mathrm{o}}l\varepsilon_{\mathrm{d},i}+2A_{\mathrm{d},i}\right)^2-\left(C_{\mathrm{o}}l\varepsilon_{\mathrm{d},i}-2A_{\mathrm{d},i}\right)^2}{\left(C_{\mathrm{o}}l\varepsilon_{\mathrm{d},i}-2A_{\mathrm{d},i}\right)^2}=1+8KC_{\mathrm{o}}-1=8KC_{\mathrm{o}} \tag{9.36}$$

Using the relation that $(a+b)^2-(a-b)^2 = 4ab$, Equation 9.36 becomes

$$\frac{A_{\mathrm{d},i}C_{\mathrm{o}}l\varepsilon_{\mathrm{d},i}}{\left(C_{\mathrm{o}}l\varepsilon_{\mathrm{d},i}-2A_{\mathrm{d},i}\right)^2}=KC_{\mathrm{o}} \tag{9.37}$$

Rearrangement of Equation 9.37 gives

$$\sqrt{\left(\frac{A_{\mathrm{d},i}l\varepsilon_{\mathrm{d},i}}{K}\right)}=\left(C_{\mathrm{o}}l\varepsilon_{\mathrm{d},i}-2A_{\mathrm{d},i}\right) \tag{9.38}$$

which can be written in the $y = mx + c$ form as

$$\frac{2A_{\mathrm{d},i}}{C_{\mathrm{o}}}=l\varepsilon_{\mathrm{d},i}-\sqrt{\frac{l\varepsilon_{\mathrm{d},i}}{K}}\times\frac{\sqrt{A_{\mathrm{d},i}}}{C_{\mathrm{o}}} \tag{9.39}$$

A plot of $2A_{\mathrm{d},i}/C_{\mathrm{o}}$ versus $(A_{\mathrm{d},i})^{1/2}/C_{\mathrm{o}}$ will be a straight line yielding a slope of $-(l\varepsilon_{\mathrm{d},i}/K)^{1/2}$ and intercept $l\varepsilon_{\mathrm{d},i}$. As mentioned earlier, the nonlinear least square fitting (of Equation 9.32) and least square fitting (of Equation 9.39) need not give the same results (especially the estimated errors). It is recommended to use Equation 9.32 with the WNLS method to determine K. Then, Equation 9.39 can be used to estimate the K and $\varepsilon_{d,i}$ values and transfer them as initial estimates for use with Equation 9.32 in WNLS fitting.

When overlapping bands are involved, the individual bands have to be resolved using spectra fitting procedures, and this process can introduce some error in the derived integrated band areas. If all of the overlapping bands under consideration belong to only monomer (or dimer) species, then one can avoid the spectral fitting process and consider the area of a larger region that contains these overlapping monomer (or dimer) bands.

9.1.3 Analysis of CD Spectral Data

To use CD spectra (either vibrational CD or electronic CD), for the analysis mentioned earlier, integrated absorption, A_i, and molar absorption coefficient, $\varepsilon_{i,}$ are to be replaced, respectively, by difference quantities ΔA_i and $\Delta\varepsilon_i$ in Equations 9.26 through 9.39. Practical applications of CD spectra to determine the monomer–dimer equilibrium constants do not appear to have been explored.

9.1.4 Analysis of Vibrational Raman and VROA Spectral Data

Since vibrational Raman and VROA scattering intensities depend on exciting laser intensity, relative intensities may be determined using an appropriate internal standard, but absolute intensities are not generally reported. Raman scattering intensities are proportional to the sample concentration, so in analogy to the absorption intensities, the Raman band intensity may be written as

$$R_{m,i} = k_{R,m} C_m \int I_{m,i}(v)dv = k_{R,m} P_m C_o I_{m,i} = \frac{2}{\left(1+\sqrt{1+8KC_o}\right)} C_o k_{R,m} I_{m,i} \quad (9.40)$$

where $k_{R,m}$ is a constant and $I_{m,i} = \int I_{m,i}(v)dv$ is the integrated vibrational Raman band intensity per mole of the substance or the molar intensity. Therefore, the equations presented in the previous sections can be applied to Raman spectral bands by replacing $A_{m,i}$ and $A_{d,i}$ with Raman band intensities $R_{m,i}$ and $R_{d,i}$ and the products $l\varepsilon_{m,i}$ and $l\varepsilon_{d,i}$ with different constants $k_{R,m}$ and $k_{R,d}$ in Equations 9.26 through 9.39. For VROA spectral data, vibrational Raman band intensities are replaced with VROA band intensities, ΔR. It is not apparent that the practical applications of vibrational Raman/VROA spectra to determine the monomer–dimer equilibrium constants have been explored.

9.2 Heterochiral Monomer–Dimer Equilibrium

When arbitrary amounts of both enantiomers are mixed together, instead of a single equilibrium involved for homochiral monomer–dimer equilibrium, three equilibria are to be considered (Baciocchi et al. 2002; Covington and Polavarapu 2015a) as follows:

$$2M_R \rightleftharpoons D_{RR} \tag{9.41}$$

$$2M_S \rightleftharpoons D_{SS} \tag{9.42}$$

$$M_R + M_S \rightleftharpoons D_{RS} \tag{9.43}$$

where M_R and M_S represent, respectively, the monomers of (*R*) and (*S*) enantiomers; D_{RR} and D_{SS} represent, respectively, the homochiral dimers of these two enantiomers; and D_{RS} represents the heterochiral dimer formed between the enantiomers. Because of the presence of both enantiomers, they are specifically identified with (*R*) and (*S*) subscripts, unlike in the case of homochiral equilibrium (see Section 9.1). The concentrations of dissolved enantiomers, C_R and C_S, respectively, are governed by the relations

$$C_R = C_{m,R} + 2C_{d,RR} + C_{d,RS} \tag{9.44}$$

$$C_S = C_{m,S} + 2C_{d,SS} + C_{d,RS} \tag{9.45}$$

where $C_{m,R}$ is the concentration of the monomer of (*R*) enantiomer; $C_{m,S}$ is the concentration of the monomer of (*S*) enantiomer; $C_{d,RR}$ is the concentration of the homochiral dimer of (*R*) enantiomer; $C_{d,SS}$ is the concentration of the homochiral dimer of (*S*) enantiomer; and $C_{d,RS}$ is the concentration of the heterochiral dimer of (*R*) and (*S*) enantiomers. The two dimer concentrations, namely $C_{d,RR}$ and $C_{d,SS}$, can also be written for short, without loss of generality, as $C_{d,R}$ and $C_{d,S}$, respectively. We will designate the homochiral equilibrium constant as K (as in Section 9.1) and heterochiral equilibrium constant as K_{ht}:

$$K = \frac{C_{d,RR}}{C_{m,R}^2} = \frac{C_{d,SS}}{C_{m,S}^2} \tag{9.46}$$

$$K_{ht} = \frac{C_{d,RS}}{C_{m,R}C_{m,S}} \tag{9.47}$$

Then, the concentrations satisfy the relations (Baciocchi et al. 2002):

$$C_R = C_{m,R} + 2KC_{m,R}^2 + K_{ht}C_{m,R}C_{m,S} \tag{9.48}$$

$$C_S = C_{m,S} + 2KC_{m,S}^2 + K_{ht}C_{m,R}C_{m,S} \tag{9.49}$$

Closed expressions for the concentrations of monomers can be obtained from these equations when $K_{ht} = 2K$ (see Section 9.2.1). Iterative solutions to Equations 9.48 and 9.49 are needed when K_{ht} is not equal to $2K$ (see Section 9.2.2). A computer program for these iterative solutions is freely available (Covington and Polavarapu 2015b).

9.2.1 Equilibria with $K_{ht} = 2K$

In this situation, the combination of Equations 9.48 and 9.49 gives

$$2K(C_{m,R} + C_{m,S})^2 + (C_{m,R} + C_{m,S}) - (C_R + C_S) = 0 \tag{9.50}$$

Designating the total concentration of monomers as C_m,

$$C_m = (C_{m,R} + C_{m,S}) \tag{9.51}$$

and the total concentration of starting mixture as C_o,

$$C_o = (C_R + C_S) \tag{9.52}$$

the abovementioned quadratic equation becomes

$$2KC_m^2 + C_m - C_o = 0 \tag{9.53}$$

This quadratic equation in C_m is similar to that for homochiral equilibrium, Equation 9.4, except for the contextual differences in the definitions of C_m and C_o. In the context of homochrial equilibrium, C_m and C_o represent, respectively, the concentrations of monomer and starting substance for a given enantiomer. In the context of heterochrial equilibrium, C_m represents the sum of the concentrations of monomers of both enantiomers, and C_o represents the concentration of starting enantiomeric mixture.

Equation 9.53 can be solved as was done for homochiral equilibrium in Section 9.1 (see Equations 9.4 through 9.9), with the result given below:

$$C_m = \frac{2}{\left(1+\sqrt{1+8KC_o}\right)}C_o = P_m C_o \tag{9.54}$$

This equation is the same as Equation 9.9, except for the difference in definitions of C_m and C_o as noted earlier. Equation 9.54 can be separated (Baciocchi et al. 2002) into two, one each for the enantiomers, as follows:

$$C_{m,R} = \frac{2}{\left(1+\sqrt{1+8KC_o}\right)}C_R \tag{9.55}$$

$$C_{m,S} = \frac{2}{\left(1+\sqrt{1+8KC_o}\right)} C_S \tag{9.56}$$

The following relations among the concentrations of monomers can be noted (Covington and Polavarapu 2015a):

$$\frac{(C_{m,R} - C_{m,S})}{(C_R - C_S)} = \frac{(C_{m,R} + C_{m,S})}{(C_R + C_S)} = \frac{C_m}{C_o} = \frac{2}{\left(1+\sqrt{1+8KC_o}\right)} = P_m \tag{9.57}$$

and

$$\frac{(C_{m,R} - C_{m,S})}{(C_{m,R} + C_{m,S})} = \frac{(C_R - C_S)}{(C_R + C_S)} = \text{ee} \tag{9.58}$$

where "ee," as stated early on, stands for enantiomeric excess.

The difference in concentrations of dimers can be written in two different ways. One way is to use Eqaution 9.46 and obtain

$$(C_{d,RR} - C_{d,SS}) = K\left(C_{m,R}^2 - C_{m,S}^2\right) \tag{9.59}$$

The second way is to subtract Equations 9.44 and 9.45, yielding

$$2(C_{d,RR} - C_{d,SS}) = (C_R - C_S) - (C_{m,R} - C_{m,S}) \tag{9.60}$$

Then, substituting Equation 9.58 for the second term on the right-hand side of 9.60 gives

$$2(C_{d,RR} - C_{d,SS}) = (C_R - C_S) - \frac{(C_R - C_S)}{(C_R + C_S)} C_m = \frac{(C_R - C_S)}{(C_R + C_S)} C_o - \frac{(C_R - C_S)}{(C_R + C_S)} C_m \tag{9.61}$$

Using $C_m = P_m C_o$ from Equation 9.57 and the definition of ee from Equation 9.58, the difference in concentrations of dimers simplifies to (Covington and Polavarapu 2015a)

$$2(C_{d,RR} - C_{d,SS}) = \text{ee} \times (C_o - C_m) = \text{ee} \times C_o(1 - P_m) \tag{9.62}$$

This equation will be used in the next section.

9.2.1.1 Analysis of SOR Data

Assuming that SORs of monomer and dimer are independent of their concentrations, and that the heterochiral dimer, D_{RS}, does not contribute to OR (because of possible inversion symmetry for *RS* dimer or mutual cancelation from *SR* and *RS* dimers), the OR for an enantiomeric mixture (em), α_{em}, can be written for heterochiral monomer–dimer equilibrium system, as follows:

$$\alpha_{em} = \left([\alpha]_{m,R}\, c_{m,R} + [\alpha]_{m,S}\, c_{m,S} + [\alpha]_{d,RR}\, c_{d,RR} + [\alpha]_{d,SS}\, c_{d,SS}\right) l \tag{9.63}$$

In this equation, $[\alpha]_{m,R}$, $[\alpha]_{m,S}$, $[\alpha]_{d,RR}$, and $[\alpha]_{d,SS}$ are, respectively, the SORs of monomers of enantiomers and of their homochiral dimers; $c_{m,R}$, $c_{m,S}$, $c_{d,RR}$, and $c_{d,SS}$ are, respectively, the concentrations (all in g/cc) of monomers of enantiomers and of their homochiral dimers. The SORs of enantiomers are opposite to each other, and therefore, the above-mentioned equation becomes

$$\alpha_{em} = [\alpha]_{m,R}\left(c_{m,R} - c_{m,S}\right) l + [\alpha]_{d,RR}\left(c_{d,RR} - c_{d,SS}\right) l \tag{9.64}$$

Since molar masses, $M_R = M_S$, $M_{RR} = M_{SS}$, and $M_d = 2M_m$, the abovementioned equation becomes

$$\alpha_{em} = \left\{[\alpha]_{m,R}\left(C_{m,R} - C_{m,S}\right) + 2[\alpha]_{d,RR}\left(C_{d,RR} - C_{d,SS}\right)\right\}\frac{M_m l}{1000} \tag{9.65}$$

Equation 9.65 can be simplified in two different ways.

1. Simplification of Equation 9.65 using Equation 9.59: Substitution of Equation 9.59 modifies Equation 9.65 to

$$\begin{aligned}\alpha_{em} &= \left\{[\alpha]_{m,R}\left(C_{m,R} - C_{m,S}\right) + 2K[\alpha]_{d,RR}\left(C_{m,R}^2 - C_{m,S}^2\right)\right\}\frac{M_m l}{1000} \\ &= \left(C_{m,R} - C_{m,S}\right)\left\{[\alpha]_{m,R} + 2K[\alpha]_{d,RR}\left(C_{m,R} + C_{m,S}\right)\right\}\frac{M_m l}{1000}\end{aligned} \tag{9.66}$$

Dividing this equation with $c_o l = \dfrac{C_o M_m l}{1000}$, where C_o is the molar concentration of the mixture of enantiomers, the corresponding SOR for a mixture of enantiomers is obtained as

$$[\alpha_{em}] = \frac{\left(C_{m,R} - C_{m,S}\right)}{C_o}\left\{[\alpha]_{m,R} + 2K[\alpha]_{d,RR}\left(C_{m,R} + C_{m,S}\right)\right\} \tag{9.67}$$

Substitution of Equation 9.58 and further use of Equations 9.51 and 9.54 leads to

$$[\alpha_{em}] = \text{ee} \times \left\{[\alpha]_{m,R} + 2K[\alpha]_{d,RR}\, C_o P_m\right\} P_m \tag{9.68}$$

This equation is not as transparent as one would like (vide infra), so let us use a second way for simplifying Equation 9.65.

2. Simplification of Equation 9.65 using Equations 9.58 and 9.62: A more compact expression can be obtained by substituting Equations 9.58 and 9.62 in Equation 9.65. Then,

$$\alpha_{em} = ee \times \left\{[\alpha]_{m,R} C_m + [\alpha]_{d,RR} (1 - P_m) C_o\right\} \frac{M_m l}{1000} \tag{9.69}$$

Dividing this equation with $c_o l = \frac{C_o M_m l}{1000}$, and use of Equation 9.57, yields the simplified SOR expression for a mixture of enantiomers as follows (Covington and Polavarapu 2015a):

$$[\alpha_{em}] = ee \times \left\{[\alpha]_{m,R} P_m + [\alpha]_{d,RR} (1 - P_m)\right\} \tag{9.70}$$

Note that Equations 9.68 and 9.70 differ from the corresponding equations for homochiral equilibrium, Equations 9.21 and 9.23, in two respects: (1) the presence of ee in the former equations; and (2) the P_m term reflects the fractional concentration of monomers of both enantiomers in the former equations, while that in latter equations reflects the fractional concentration of monomer of single enantiomer. However, at a given concentration, C_o, these two P_m values will be the same (compare Equations 9.9 and 9.54). Thus, Equations 9.68 and 9.70 can be written as

$$[\alpha_{em}] = ee \times [\alpha_A] \tag{9.71}$$

where $[\alpha_A]$ is the SOR of pure enantiomer A (see Equations 9.21 and 9.23) at the same total concentration as that for mixture of enantiomers. Then, the op, which is the ratio of SOR of enantiomerc mixture to that of pure eanantiomer, at a given total concentration, becomes

$$op = \frac{[\alpha_{em}]}{[\alpha_A]} = ee \tag{9.72}$$

That means no distinction can be made between op and ee, under the following conditions: (1) $K_{ht} = 2K$ and (2) heterochiral dimer does not contribute to the observed OR. Then, the Horeau effect (Horeau 1969) is not expected to be evident under these conditions, even though aggregation occurs in the form of dimer formation.

For pantolactone (Nakao et al. 1985), binaphthol (Baciocchi et al. 2002), omeprazole, and Pirkle's alcohol (Baciocchi et al. 2004), the heterochiral monomer–dimer equilibrium constant was reported to be approximately twice that of homochiral monomer–dimer equilibrium constant. As a result, these compounds are not expected to exhibit the Horeau effect. The experimental measurements on pantolactone confirmed the absence of the Horeau effect (Covington and Polavarapu 2015a).

9.2.1.2 Analysis of Absorption Spectral Data

As in Section 9.1.2, absorption spectral data are useful for determining the homochiral equilibrium constant. Here also, we assume that the absorption bands due to monomer and dimer species appear at well-separated x-axis positions in the spectra (especially in the case of vibrational spectra). For homochiral dimer bands, the bands due to enantiomers appear at the same wavenumber position in the spectra. The integrated intensities of spectral bands belonging to the monomer species can be measured and processed individually, and the analysis is similar to that in Section 9.1.2, with the following differences: C_m represents the sum $C_{m,R} + C_{m,S}$; C_o represents the sum $C_R + C_S$; and concentrations of the species are governed by Equations 9.54 through 9.56.

For analyzing the homochiral dimer bands, the concentrations of homochiral dimers can be determined from (see Equation 9.46) $\left(C_{d,RR} + C_{d,SS}\right) = K\left(C_{m,R}^2 + C_{m,S}^2\right)$. Then, the integrated absorption intensity for a homochiral dimer band "i" is given by the relation

$$A_{d,i} = \left(C_{d,RR} + C_{d,SS}\right) l \int \varepsilon_{d,i}(\nu) d\nu = l\varepsilon_{d,i} K\left(C_{m,R}^2 + C_{m,S}^2\right) \tag{9.73}$$

where $\varepsilon_{d,i} = \int \varepsilon_{d,i}(\nu) d\nu$.

The bands due to heterochiral dimer can appear at different wavenumber positions than those from homochiral dimer. For heterochiral dimer bands, if they can be clearly identified, the concentrations of heterochiral dimers can be determined from $C_{d,RS} = 2KC_{m,R}C_{m,S}$. Then, the integrated absorption intensity for hetrochiral band "i" is given by the relation

$$A_{d,RS,i} = C_{d,RS} l \int \varepsilon_{d,RS,i}(\nu) d\nu = 2KC_{m,R}C_{m,S} l\varepsilon_{d,RS,i} \tag{9.74}$$

Substituting the expressions for monomer concentrations (Equations 9.55 and 9.56), into Equations 9.73 and 9.74, one can carry out the nonlinear least square fitting of resulting equations to determine K and ε.

When the monomer and dimer bands cannot be separated along the x-axis positions (especially in electronic spectra), one can sum the contributions from the monomer and dimer species to a given band and use the composite equation containing the monomer and dimer band intensities (Baciocchi et al. 2002).

9.2.1.3 Analysis of CD Spectral Data

The analysis of CD spectra follows that mentioned above for absorption, by replacing the integrated absorption, A_i, and molar absorption coefficient, ε_i, with difference in quantities ΔA_i and $\Delta\varepsilon_i$, respectively. However, it is to be noted that the CD signs belonging to enantiomers are opposite to each other.

The integrated CD intensity for a homochiral dimer band is given, using Equation 9.60 through 9.62, by the relation

$$\Delta A_{d,i} = (C_{d,RR} - C_{d,SS})l\int \Delta\varepsilon_{d,RR,i}(\nu)d\nu = \frac{ee \times C_o(1-P_m)}{2} \times l\Delta\varepsilon_{d.RR,i} \quad (9.75)$$

The heterochiral dimer bands may not appear in the CD spectra due to possible inversion symmetry for heterochiral dimer or to mutual cancelation from the *SR* and *RS* dimers. Practical applications of Equation 9.75 have not yet appeared.

9.2.1.4 Analysis of Vibrational Raman and VROA Spectral Data

The vibrational Raman spectra can be analyzed (see Section 9.1.4) in the same way by replacing integrated absorption band intensities (see Equations 9.73 and 9.74) with integrated Raman band intensities and the products of molar extinction coefficient and path length with corresponding Raman constants.

For VROA spectral data, Equation 9.75 can be used by replacing $\Delta A_{d,i}$ with integrated VROA band intensities, $\Delta R_{d,i}$, and $l\varepsilon_{d,i}$ with a corresponding Raman constant. Practical applications of vibrational Raman/VROA spectra for studying the heterochiral monomer–dimer equilibrium have not yet appeared.

9.2.2 Equilibria with K_{ht} Not Equal to $2K$

In situations where the equilibrium constant for heterochiral dimerization, K_{ht}, is not equal to twice that for homochiral dimerization, closed expressions for concentrations of monomers and dimers are not possible. Nevertheless, the two interdependent quadratic equations (Equations 9.48 and 9.49) can be solved through the iterative procedure for determining the concentrations of monomers. As a result, general expressions for this situation cannot be written. Nevertheless, for a given set of K and K_{ht} values, one can simulate the chiroptical properties using the concentrations of monomers derived from iterative solutions to Equations 9.48 and 9.49 and of dimers using Equations 9.46 and 9.47.

9.2.2.1 Analysis of SOR Data

Equations 9.50 through 9.58 and Equation 9.60 through 9.62 are not applicable here. We start with Equation 9.63 and proceed through Equation 9.65 for the OR of an enantiomeric mixture. Then divide Equation 9.65 with $c_o l = \frac{C_o M_m l}{1000}$, and rewrite it with additional annotation as

$$[\alpha_{em}]_{K_{ht}\neq 2K} = \frac{1}{C_o}\left\{[\alpha]_{m,R}(C_{m,R} - C_{m,S})_{K_{ht}\neq 2K} + 2[\alpha]_{d,RR}(C_{d,RR} - C_{d,SS})_{K_{ht}\neq 2K}\right\} \quad (9.76)$$

where the subscript $K_{ht} \neq 2K$ indicates the conditions under which concentrations (of monomers, dimers) and SOR of solution are applicable. The dependence on heterochiral dimerization constant, K_{ht}, in these equations comes through the dependence of $C_{m,R}$, $C_{m,S}$, $C_{d,RR}$, and $C_{d,SS}$ on K_{ht} (see Equations 9.48 and 9.49).

Simulations of $[\alpha_{em}]_{K_{ht}\neq 2K}$ with different values for K and K_{ht} and for $[\alpha]_{m,R}$ and $[\alpha]_{d,RR}$ indicate (Covington and Polavarapu 2015a) that one can anticipate both positive and negative Horeau effects as well as no Horeau effect. A positive Horeau effect refers to the op of an enantiomeric mixture being greater than the enantiomeric purity of the sample. A negative Horeau effect refers to the op of an enantiomeric mixture being less than the enantiomeric purity of the sample. These situations are depicted in Figure 9.3. Since these predictions depend on the values assumed for sample concentration, K, K_{ht}, $[\alpha]_{m,R}$, and $[\alpha]_{d,RR}$, the magnitude and direction of Horeau effect can vary from sample to sample. The Horeau effect simulations for systems exhibiting monomer–dimer equilibrium can be performed using a freely available program (Covington and Polavarapu 2015b).

Earlier we mentioned that the Horeau effect will not be seen when $K_{ht} = 2K$ (see Equation 9.72). Besides this situation, regardless of the values for K and K_{ht}, the Horeau effect will also not be seen when the SORs for monomer and dimer are equal, that is, $[\alpha]_{m,R} = [\alpha]_{d,RR}$. This can be seen from Equation 9.76 as follows: Substituting $[\alpha]_{m,R} = [\alpha]_{d,RR} = [\alpha]_R$ in Equation 9.76, the following equation can be obtained:

$$
\begin{aligned}
[\alpha_{em}]_{K_{ht}\neq 2K} &= [\alpha]_R \times \frac{1}{C_o}\Big\{\left(C_{m,R} + 2C_{d,RR} + C_{d,RS}\right)_{K_{ht}\neq 2K} \\
&\quad - \left(C_{m,S} + 2C_{d,SS} + C_{d,RS}\right)_{K_{ht}\neq 2K}\Big\} \\
&= [\alpha]_R \times \frac{(C_R - C_S)}{C_o} = [\alpha]_R \times \text{ee}
\end{aligned}
\tag{9.77}
$$

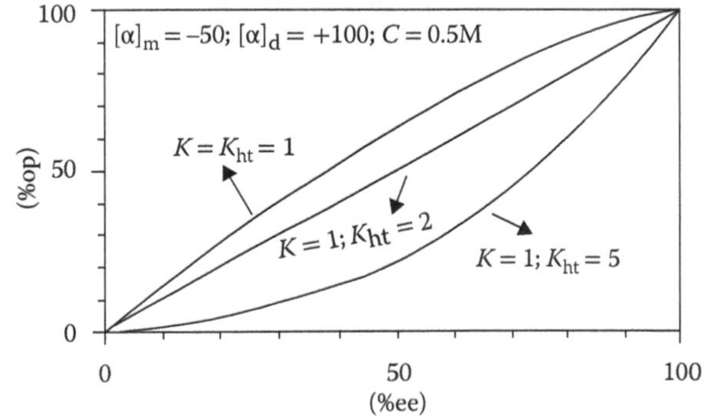

FIGURE 9.3
Simulated optical purity (op) as a function of enantiomeric excess (ee) for the mixtures of enantiomers, with assumed K, K_{ht}, $[\alpha]_m$, and $[\alpha]_d$. A positive Horeau effect is expected when $K = K_{ht} = 1$ and a negative Horeau effect when $K = 1$ and $K_{ht} = 5$, for $[\alpha]_m = -50$ and $[\alpha]_d = +100$ at a concentration of 0.5 M. Horeau effect is not expected to be seen when $K_{ht} = 2K$.

In Equation 9.77, the term $C_{d,RS}$ has been added and subtracted such that $C_{m,R} + 2C_{d,RR} + C_{d,RS}$ can be used for the total concentration of (*R*) enantiomer, and $C_{m,S} + 2C_{d,SS} + C_{d,SR}$ can be used for the total concentration of (*S*) enantiomer in a mixture of enantiomers. Thus,

$$\text{op} = \frac{[\alpha_{\text{em}}]_{K_{\text{ht}} \neq 2K}}{[\alpha]_R} = \text{ee} \tag{9.78}$$

Equations 9.77 and 9.78 predict that op becomes equal to ee when $[\alpha]_{m,R} = [\alpha]_{d,RR}$, regardless of the relative magnitudes for *K* and K_{ht}. This prediction has been verified (Covington and Polavarapu 2015a) through actual simulations as shown in Figure 9.4.

There are a limited number of molecules whose SORs were reported to exhibit the Horeau effect: (1) α-ethyl-α-methylsuccinic acid (Horeau 1969; Krow and Hill 1968); (2) α-isopropyl-α-methylsuccinic acid (Horeau and Guetté 1974); (3) 6-Phenyl-2,3,5,6-tetrahydroimidazo[2,1-b]thiazole (Acs 1990), also known as tetramisole; and (4) hydroxypinanone (Solladié-Cavallo and Andriamiadanarivo 1997). The homochiral and heterochiral monomer–dimer equilibrium constants for these systems are not available. However, the reported Horeau effect in hydroxypinanone (Solladié-Cavallo and Andriamiadanarivo 1997) is in contradiction to the prediction of Equation 9.77 because the SOR of hydroxypinanone solution is independent of concentration (Covington and Polavarapu 2015a), which occurs when $[\alpha]_m = [\alpha]_d$. Under this condition, the Horeau effect is not observable as stated earlier.

9.2.2.2 Analysis of Absorption/Raman Spectral Data for Racemic Mixtures

The absorption/Raman spectral data of a racemic mixture (Nakao et al. 1985) are useful for determining the heterochiral equilibrium constant, K_{ht}. For a

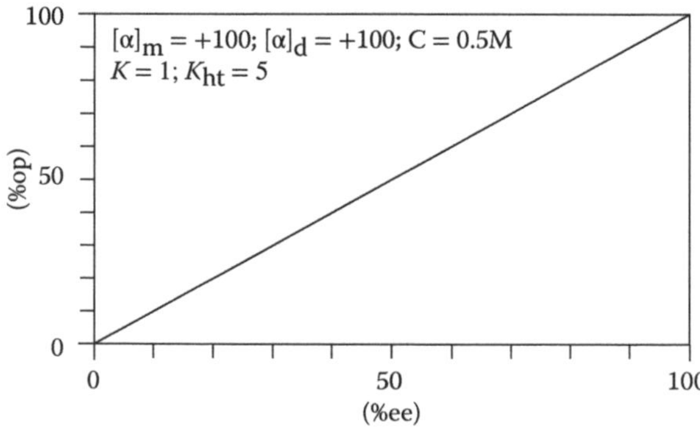

FIGURE 9.4
Simulated optical purity (op) as a function of enantiomeric excess (ee) for the mixtures of enantiomers, with assumed *K*, K_{ht}, $[\alpha]_m$, and $[\alpha]_d$. Horeau effect is not expected to be seen when the specific optical rotation (SOR) of monomer, $[\alpha]_m$, is equal to the SOR of dimer $[\alpha]_d$.

racemic mixture, $C_R = C_S$ and $C_{m,R} = C_{m,S} = C_m/2$, where $C_m = C_{m,R} + C_{m,S}$, the concentration of total monomer. Designating $C_R + C_S = C_o$ as the total concentration of the racemic mixture, the combination of Equations 9.48 and 9.49 yields

$$C_o = C_m + (K + 0.5K_{ht})C_m^2 \tag{9.79}$$

Note that when $K_{ht} = 2K$, Equation 9.79 will reduce to that for homochiral equilibrium (see Equation 9.3 and 9.4) as follows:

$$C_o = C_m + (K + 0.5K_{ht})C_m^2 = C_m + 2KC_m^2 = C_m + 2C_d \tag{9.80}$$

Expressing C_m in terms of absorption/Raman band intensity of monomer band and noting that the bands due to monomer will contain contributions from both enantiomers, one can use Equation 9.79 for C_o versus monomer band intensity plot to determine ($K + 0.5K_{ht}$). Then, one can determine K_{ht} using the K value determined from the measurements on pure enantiomer.

9.3 Equilibrium Between Secondary Structures of Biological Systems

Well-known examples of this category are the helix-coil transitions in polypeptides with a widely studied molecule being poly-γ-benzyl-L-glutamate. In the terminology of helix-coil transition, "coil" conformer is a misnomer, and it is well established now that "coil" conformer actually contains local regions of polyproline-II (PP-II)-type conformer (Shi et al. 2002). The expressions for secondary structure equilibria are straightforward. The equilibrium is written as

$$H_A \rightleftharpoons P_A \tag{9.81}$$

where H_A is the helix conformer of enantiomer A and P_A is the coil/PP-II conformer of enantiomer A. The equilibrium constant is $K = C_P/C_H$, and material balance requires $C_o = C_H + C_P$, where C_o is the molar concentration of dissolved sample, and C_H and C_P are, respectively, the molar concentrations at equilibrium of helix and PP-II conformers. Since the molar masses of helix and PPII-conformers are the same, $K = C_P/C_H = c_P/c_H = x_P/x_H$, where lower case "*c*" represents concentration in g/cc, and "*x*" represents the mole fraction.

Since OR method is widely used for studying the conformational transitions of polypeptides, we will restrict this section only to this method. The OR of dissolved sample is written as

$$\alpha = [\alpha]_H c_H l + [\alpha]_P c_P l = [\alpha]_H \frac{C_H M_H}{1000} l + [\alpha]_P \frac{C_P M_P}{1000} l \tag{9.82}$$

Substituting $C_P = KC_H$ and noting that molar masses of helix and PP-II conformers, M_P and M_H, are the same, the above-mentioned equation can be written as

$$\alpha = \{[\alpha]_H + K[\alpha]_P\} \frac{C_H M_H}{1000} l = \{[\alpha]_H + K[\alpha]_P\} c_H l \tag{9.83}$$

Dividing this equation with $c_o l$, the equation for SOR is obtained as

$$[\alpha] = \{[\alpha]_H + K[\alpha]_P\} \frac{c_H}{c_o} \tag{9.84}$$

which simplifies, using $Kc_H = c_P, x_H = \frac{c_H}{c_H + c_P} = \frac{c_H}{c_o}$ and $x_P = \frac{c_P}{c_o}$, to

$$[\alpha] = x_H [\alpha]_H + x_P [\alpha]_P \tag{9.85}$$

Equation 9.85 is widely quoted in the physical chemistry text books (Garland et al. 2009) for equilibrium between secondary structures of polypeptides.

9.4 Summary

When chiral aggregates are formed via dimer formation, the presence of monomer–dimer equilibrium will have profound influence on the measured molecular chiroptical properties. With regard to SOR of systems exhibiting monomer–dimer equilibrium, the following points are important to note: (1) If the SOR of monomer is equal to that of dimer, then the SOR of solution should be independent of the concentration of solution. (2) If the SOR of monomer is equal to that of dimer, then op and ee are equivalent, and the Horeau effect will not be observed. (3) If the heterochiral equilibrium constant is equal to twice that of homochiral equilibrium constant, then op and ee are equivalent and the Horeau effect will not be observed. (4) When the Horeau effect is present, the sign and magnitude of the effect will depend on the concentration of the solution, the relative magnitudes of homochiral and heterochiral equilibrium constants, and the relative magnitudes of the SORs of monomer and dimer species.

With regard to CD/Raman optical activity, these spectra do not appear to have been used for investigating the monomer–dimer equilibria. Although the corresponding absorption/Raman spectral data are useful for determining the homochiral and heterochiral equilibrium constants, their utilization appears to be limited.

References

Acs, M. 1990. Chiral recognition in the light of molecular associations. In: M. Simonyi (ed), *Problems and Wonders of Chiral Molecules*. Budapest, Hungary: Akademiai Kiado.

Baciocchi, R., M. Juza, J. Classen, M. Mazzotti, and M. Morbidelli. 2004. Determination of the dimerization equilibrium constants of omeprazole and Pirkle's alcohol through optical-rotation measurements. *Helv. Chim. Acta* 87:1917–1926.

Baciocchi, R., G. Zenoni, M. Valentini, M. Mazzotti, and M. Morbidelli. 2002. Measurement of the dimerization equilibrium constants of enantiomers. *J. Phys. Chem. A* 106:10461–10469.

Chen, J.-S., C.-C. Wu, and D.-Y. Kao. 2004. New approach to IR study of monomer–dimer self-association: 2,2-dimethyl-3-ethyl-3-pentanol in tetrachloroethylene as an example. *Spectrochim. Acta A Mol. Biomol. Spectrosc.* 60(10):2287–2293.

Covington, C. L., and P. L. Polavarapu. 2015a. Specific optical rotations and the Horeau effect. *Chirality* 28(3):181–185.

Covington, C. L., and P. L. Polavarapu. 2015b. A computer program to predict horeau effect for systems exhibiting monomer-dimer equilibrium. *git clone*. https://github.com/polavarapu-lab/public.git polavarapu_lab (accessed November 2015).

Eliel, E. L., and S. H. Wilen. 1994. *Stereochemistry of Organic Compounds*. New York, NY: John Wiley & Sons.

Garland, C. W, J. W. Nibler, and D. P. Shoemaker. 2009. *Experiments in Physical Chemistry*, 8th ed. Boston, MA: McGraw-Hill Higher Education.

Gnuplot. 2016. GNUPLOT 4.2—A brief manual and tutorial. http://people.duke.edu/~hpgavin/gnuplot.html (accessed May 4, 2016).

Goldsmith, M.-R., N. Jayasuriya, D. N. Beratan, and P. Wipf. 2003. Optical rotation of noncovalent aggregates. *J. Am. Chem. Soc.* 125:15696–15697.

Horeau, A. 1969. Interactions d'enantiomeres en solution; influence sur le pouvoir rotatoire: Purete optique et purete enantiomerique. *Tetrahedron Lett.* 10:3121–3124.

Horeau, A., and J. P. Guetté. 1974. Interactions diastereoisomeres d'antipodes en phase liquide. *Tetrahedron* 30:1923–1931.

KaleidaGraph. 2013. Version 4.1. Reading, PA: Synergy Software.

Krow, G., and R. K. Hill. 1968. Absolute configuration of a dissymmetric spiran. *Chem. Comm.* (8):430–431.

Kuppens, T., W. Herrebout, B. van der Veken, and P. Bultinck. 2006. Intermolecular association of tetrahydrofuran-2-carboxylic acid in solution: A vibrational circular dichroism study. *J. Phys. Chem. A* 110(34):10191–10200.

Nakao, Y., H. Sugeta, and Y. Kyogoku. 1985. Intermolecular hydrogen bonding of enantiomers of pantolactone studied by infrared and ^{1}H-NMR spectroscopy. *Bull. Chem. Soc. Jpn.* 58:1767–1771.

PeakFit. 2015. *Software for Automated Peak Separation and Analysis*. San Jose, CA: Systat Software, Inc.

Polavarapu, P. L., and C. L. Covington. 2015. Wavelength resolved specific optical rotations and homochiral equilibria. *Phys. Chem. Chem. Phys.* 17:21630–21633.

Shi, Z., R. W. Woody, and N. R. Kallenbach. 2002. Is polyproline II a major backbone conformation in unfolded proteins? *Adv. Protein Chem.* 62:163–240.

Solladié-Cavallo, A., and R. Andriamiadanarivo. 1997. Hydroxypinanone: Solute/solute interactions and non-linear chiroptical properties. *Tetrahedron Lett.* 38:5851–5852.

10

Applications of Chiroptical Spectra

10.1 Introduction

The absolute configurations (ACs) are most commonly determined using stereoselective chemical synthesis, X-ray crystallography, and/or nuclear magnetic resonance (NMR) spectroscopy. Stereoselective chemical syntheses require the knowledge of the ACs of the synthetic precursors and of the stereochemical course in the reaction pathway. The X-ray diffraction methods for determining the ACs require quality single crystals of the compounds (or their derivatives) being studied. The NMR methods for determining the ACs require chiral shift reagents and empirical correlations associated with interpreting the chemical shifts. Chiroptical spectroscopic methods are not limited by these requirements and the experimental measurements can be conducted for native chiral compounds in a liquid solution phase. Due to these distinct advantages, many nonspecialists were attracted, in recent years, to adopt the chiroptical spectroscopic methods for AC determinations. In practical terms, these methods are facilitated by the readily available instrumentation for experimental chiroptical spectroscopic measurements, and the computer algorithms for modern quantum chemical predictions of the corresponding properties.

However, the prerequisite for applying chiroptical spectroscopy for structural analysis is a definite knowledge of the chemical structure (chemical groups and their connectivity in a molecule). Unlike X-ray diffraction and NMR spectroscopy, where atom/group connectivity can be ascertained, chiroptical spectroscopy does not address the structural information at the level of atomic resolution. Therefore, one should have definitive information on what chemical groups a molecule has and how these groups are connected. Given that information, one can address the three-dimensional arrangement of groups, that is, ACs and relative orientations of different groups, that is, conformations, using chiroptical spectroscopy. Section 10.4.2.7 is informative for recognizing the dangers involved in applying chiroptical spectroscopy without a definite knowledge of the chemical structure.

In the applications of chiroptical methods based on electronic transitions, namely electronic circular dichroism (ECD) and optical rotatory dispersion

(ORD), the molecular orbitals (MOs) form the basic units of interpretations. However, canonical MOs are delocalized over the entire molecule and are hard to visualize. Therefore, localization of MOs (Lehtola and Jónsson 2013) was the preferred approach for easy interpretations and clear understanding of the involved electronic transitions. Recently, graphical interfaces (Gaussian 09 2013) to display the MOs have helped visualize the electronic transitions, and therefore localization of MOs is hardly used in interpreting the electronic transitions.

Similarly, in the applications of chiroptical methods based on vibrational spectra, namely vibrational circular dichroism (VCD) and vibrational Raman optical activity (VROA), one of the nontransparent aspects for nonspecialists is the theory and understanding of molecular vibrations. Before the later portion of twentieth century, vibrational spectral analyses were dependent on calculation of the vibrational frequencies with Wilson's GF matrix method (Wilson et al. 1980), use of assumed force constants, and explaining the nature of molecular vibrations using potential energy distributions (PEDs). With the development of quantum mechanical methods for computing Cartesian force constants and vibrational properties, Wilson's GF matrix method has become one of historical interest and is hardly used nowadays. The use of PEDs has been replaced with the display of normal modes of vibrations using commercial graphical interfaces (Gaussian 09 2013) or CYLView freeware (Legault 2009). Thus, currently, a nonspecialist does not ever have to worry about the theory and nature of molecular vibrations, which is one reason for not including a chapter on the theory of molecular vibrations (Polavarapu 1998) in this book. Normal vibrations even when displayed with graphical displays, nevertheless, represent depictions of the displacements of all of the atoms in a molecule and do not convey transparent details for the local origin of a chosen vibrational property. A better perspective can be obtained for localized vibrations (such as those localized in a given bond or group in the molecule). For large molecules that contain repeating units, combinations of these local modes, with different phases, from different repeating units will have to be understood. In general, visualizing different segments of a molecule that contribute to a given vibrational band in terms of localized vibrations (and not as delocalized normal vibrations) will have interpretational and pedagogical value. For such interpretations, localization of normal modes (Jacob and Reiher 2009) can be used.

Even though the abovementioned localization methods help in the visualization of appropriate transitions involved, the comparison of experimental and calculated chiroptical spectra (as discussed in Chapter 8) needed to arrive at the preferred structural information does not require the use of these localization methods.

Historically, the predominant literature applications of ECD spectroscopy have been in the areas of AC determination using the exciton coupling (EC) [also referred to as exciton chirality or coupled oscillator (CO)] model (see Appendix 4) and in the determination of the secondary structures of biological molecules. The use of EC/CO model requires the presence of two (preferably identical) interacting chromophores. For chiral substrates with a single chromophore, a zinc porphyrin tweezer (ZPT) method was used (Huang et al. 2000). In the ZPT method, the mono chromophoric chiral substrate is conjugated with an achiral carrier that contains two amine groups. The binding of this conjugate with ZPT, through coordination of N atoms of a carrier with Zn atoms, produces a chiral macrocyclic host–guest complex. This complex has twisted porphyrin units with the nature of the twist determined by the chirality of the substrate. The macrocyclic complex exhibits exciton-coupled CD spectra reflecting the AC of the chiral substrate. Chiral diamines (Huang et al. 1998), however, can be directly used with ZPT, without the need for conjugation with an achiral carrier. This ZPT method is a popular approach, even currently, for determining the ACs using experimental ECD spectra.

The experimental ORD and ECD spectra are also being widely used as probes of dynamic stereochemistry, that is, for monitoring changes in chirality induced by various external triggers (Canary and Dai 2012; Dai et al. 2012; Mammana et al. 2012).

In the last two decades, ECD, ORD, and VCD and VROA spectra have all been considered as competent and convenient approaches for determining the ACs. The goal for measuring the experimental chiroptical spectra, in most cases, is to extract the three-dimensional molecular structure information from them. For this purpose, one may choose to adopt one or more of three different approaches: (1) spectra–structure correlations, (2) spectral predictions using approximate models, and (3) spectral predictions using quantum chemical calculations. The last option is definitely the most preferred, but there can be situations where one has to resort to one of the other two options.

10.2 Spectra–Structure Correlations

The correlation of specific optical rotation (SOR) with structure has found widespread applications, notably for carbohydrates (Shallenberger 1982), natural products (Djerassi 1960), and organic molecules (Lightner and Gurst 2000), in the older literature, although such applications are now coming under greater scrutiny. The most useful applications of correlating the ECD, VCD, and/or VROA spectra with molecular structure are for the determination of secondary structures biological molecules.

10.2.1 ORD Spectra

The hypothesis that the SOR of a molecule can be visualized as a sum of the contributions from individual chiral centers originated from the work of van't Hoff (van't Hoff 1875) and is referred to as van't Hoff's principle of optical superposition. Historically, the measurement and interpretation has been practiced for SOR measured at 589 nm, that is, $[\alpha]_D$, where the subscript D stands for sodium D line (589 nm). The van't Hoff's concept was extended to suggest that the $[\alpha]_D$ of a molecule can be viewed as a sum of those associated with the fragments that make up the molecule. The widely known applications of the optical superposition rule are in interpreting the $[\alpha]_D$ of carbohydrates (Shallenberger 1982), molar rotation of steroids (Eliel et al. 2001), and natural products (Kondru et al. 1997, 1998, 2000). There is also a large amount of ORD, that is, $[\alpha]$ versus wavelength, data for peptides and proteins with an eye for distinguishing between protein secondary structures (Yang and Doty 1957). Different secondary structures were considered to give different ORD patterns. A large amount of ORD data correlating the structures of chemical compounds, especially steroids, are available in the literature (Djerassi 1960).

A common assumption in the older literature for analyzing the SOR data for molecular structure determination was that the compounds with structurally similar molecules may exhibit $[\alpha]_D$ of the same sign or ORD with similar patterns. Recently, these assumptions are coming under greater scrutiny. One of the limitations in relating the $[\alpha]_D$ to molecular structures is that what one considers to be structurally similar molecules may not turn out to be so, due to the differences in dominant conformations, and such differences can influence the sign and magnitudes of SORs. The sign of SOR often depends on the conformation as well as environment in which sample molecules reside. As a consequence, although one may find attractive applications of SOR for limited cases, universally valid spectra–structure correlations are often difficult to find for SOR.

Also, the concept of van't Hoff's optical superposition clearly fails when strong interactions, such as those imparted by intermolecular hydrogen bonds, are present. Such interactions are considered to be the source of the Horeau effect (Horeau 1969) (See Chapter 9).

The large amount of literature on spectra–structure correlations using SOR prevents a detailed discussion here. A limited discussion on Clough–Lutz–Jirgensons (CLJ) rule for amino acids and α-hydroxy acids is presented here.

Amino acids: At neutral pH, amino acids are expected to be in the zwitter ionic form (i.e., contain NH_3^+ and COO^- groups), while at lower pH amino acids are expected to contain NH_3^+ and COOH groups. Experimental studies indicated that the $[\alpha]_D$ becomes more positive at lower pH, compared to neutral solutions, for naturally occurring L-amino acids. These observations formed the basis for CLJ rule as follows: *If the molecular rotation of an optically active amino acid is shifted toward a more positive direction upon the addition of acid to its aqueous solution, the amino acid is of L-configuration; a negative direction of shift, however, is characteristic of a D-amino acid* (Greenstein and Winitz 1961). Autschbach et al. undertook quantum theoretical studies of molar rotations at 589 nm for different amino acids to investigate why the molar rotation of an L-amino acid becomes more positive upon the addition of acid (Kundrat and Autschbach 2008). They found that the molar rotation of L-amino acid molecules with a carboxylic acid functional group is more positive than that of the corresponding carboxylate anion. This study also indicated that if $[\alpha]_{589}$ nm is dominated by the contribution from a low-lying n-π* C=O transition, then in principle the same effect is observable in the ECD spectra of carboxylic acid–carboxylate anion pair. That means, for molecules containing carboxylate anion, ECD associated with n-π* transition is expected to exhibit a less positive band than the corresponding parent molecule containing the carboxylic group. It was also noted that the opposite CLJ effect may be seen for molecules where the electronic transition of another chromophore occurs near the n-π* transition of carboxylic acid chromophore (as in amino acids containing aromatic chromophore). CLJ effect is not expected to be seen for molecules that do not have a carboxylic acid functional group. To assign the ACs based on theoretical predictions, Autschbach et al. also concluded (Kundrat and Autschbach 2008) that the difference in calculated molar rotations of carboxylic acid–carboxylate anion pair can be more reliable than the calculated molar rotations of individual species.

α-Hydroxy acids: The assignment of AC, from the experimental $[\alpha]_D$, for α-hydroxy acids, was based on the empirical rule of Clough, which was stated in Greenstein and Winitz (1961) as follows: *A monoasymmetric α-hydroxy acid isomer which becomes more dextrorotatory upon the addition of acid to an aqueous solution of its metal salt is of the L-configuration, where as a shift of optical rotation in the levorotatory direction is characteristic of the D-isomer.* Autschbach et al. have also investigated different nonamino acid molecules containing COOH groups using quantum chemical calculations and found behavior similar to that seen for amino acids (Nitsch-Velasquez and Autschbach 2010). The rule of Clough can be tested using the pH-dependent ORD data (Polavarapu and Hammer 2010, unpublished data) shown in Figure 10.1 and the $[\alpha]_D$ summarized in Table 10.1, for different α-hydroxy carboxylic acids. The observed pattern for change in pH-dependent variation of $[\alpha]_D$ is in line with the rule of Clough; that is, $[\alpha]_D$ becomes more positive at acidic pH (compared to that at basic pH) for (*S*) configuration. The same observation is noted for tartaric acid, even though it has two chiral centers, and for an alkyl mono carboxylic acid, even though it does not have an α-hydroxy group. Garcinia and hibiscus acids (Haleema et al. 2012), the former with (2*S*,3*S*)-configuration and the latter with (2*S*,3*R*), also displayed the same behavior. Then the rule of Clough does not appear to be specific for α-hydroxy monocarboxylic acids and may well apply to molecules with a carboxylic acid functional group. However, since ORD and ECD are related through the Kramers-Kronig relation (see Appendix 2), the pH-dependent behavior of $[\alpha]_D$ has to be understood by simultaneously analyzing the changes in ORD and the corresponding changes in ECD spectra.

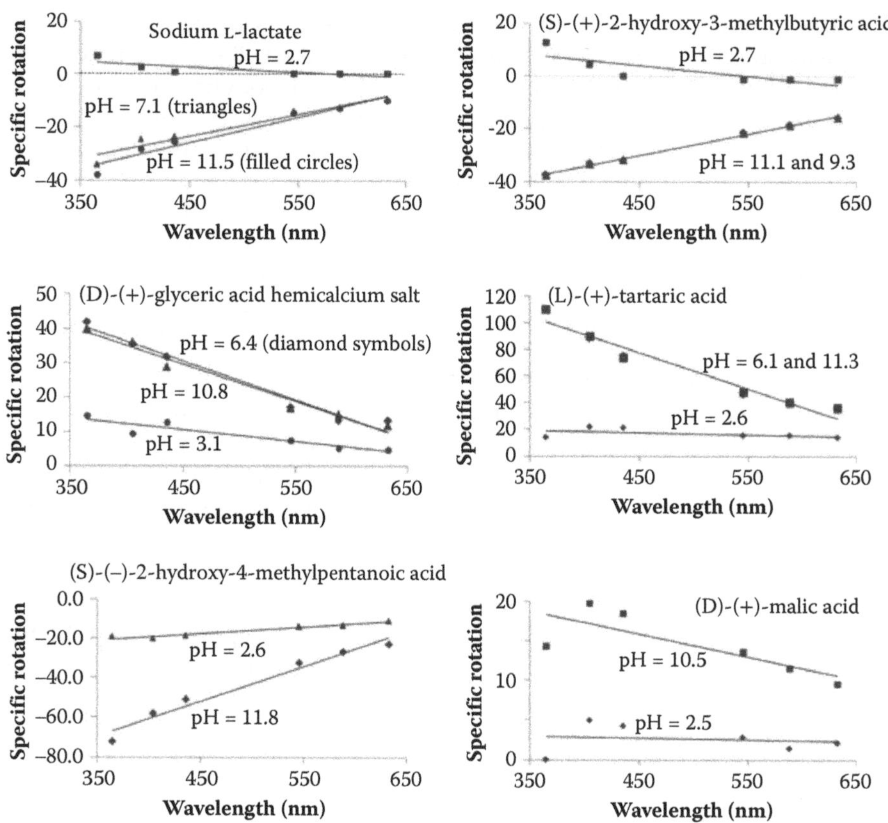

FIGURE 10.1
The pH-dependent optical rotatory dispersion spectra of some α-hydroxy carboxylic acids. The straight lines in each panel represent linear trend lines when the data were to fit to a straight line. Significant deviation from linearity is noted for malic acid. (Data from Polavarapu and Hammer, 2010, unpublished data.)

The pH-dependent ECD and corresponding electronic absorption (EA) spectra of some α-hydroxy carboxylic acids (Polavarapu and Hammer 2010, unpublished data) are shown in Figure 10.2 and the peak ECD intensities are summarized in Table 10.1. ECD spectrum changes from being monosignate at acidic pH to bisignate at basic pH, for sodium lactate, glyceric acid, hydroxyl-3-methylbutyricacid, and 2-methylbutyric acid (not shown). However, monosignate ECD spectrum at acidic pH remains monosignate at basic pH, for 2-hydroxyl-4-methylpentanoic acid, tartaric acid, and malic acid. If $[\alpha]_{589}$ is dominated by the contribution from low-lying n-π* C=O transition (i.e., ECD band intensity at ~212 nm), then one should expect the ECD intensity of D-(+)-malic acid with (*R*)-configuration to be more negative in an acidic medium than in a basic medium. But the opposite is

TABLE 10.1

pH Dependent $[\alpha]_D$ and ECD Peak Intensities for Some α-Hydroxy Carboxylic Acids

		Circular Dichroism, $\Delta\varepsilon_{max}$				Specific Rotation, $[\alpha]_{589}$		
	Configuration	λ_{max} (nm) at acidic pH	Acidic	Basic	Change (acidic–basic)	Acidic	Basic	Change (acidic–basic)
α-Hydroxy-monocarboxylic acids								
Sodium L-(−)-lactate	(*S*)	210	0.86	0.19	0.67	0	−12.8	12.8
D-(+)-glyceric acid	(*R*)	210	−0.77	−0.11	−0.66	5.3	15.1	−9.8
L-(−)-2-Hydroxy-4-methylpentanoic acid	(*S*)	206	1.6	0.71	0.89	−13.3	−26.6	13.3
S-(+)-2-hydroxy-3-methylbutyric acid	(S)	210	1.2	−0.08	1.28	−1.3	−18.8	17.5
α-Hydroxy-dicarboxylic acids								
L-(+)-tartaric acid	(2*R*,3*R*)	216	−4.45	−2	−2.45	15.3	39.1	−23.8
D-(+)-malic acid	(*R*)	212	−0.93	−2.1	1.17	1.42	11.6	−10.18
Alkyl-monocarboxylic acid								
S-(+)-2 -methylbutyric acid	(*S*)	210	0.24	−0.01	0.25	19.4	9.8	9.6

observed. The ORD spectra for lactic acid, glyceric acid, 2-hydroxyl-4-methylpentanoic acid, 2-hydroxy-3-methylbutyric acid, and tartaric acid are nearly linear in wavelength. However, the ORD spectrum for D-(+)-malic acid displays a significant deviation from linearity as a function of wavelength and indicates a possible switchover of optical rotation sign from positive to negative at ~365 nm. Such switchover can occur when contributions to optical rotation from two different short wavelength electronic transitions counteracts each other (Polavarapu 2005). In such instances, the sign and magnitude of $[\alpha]_D$ is subject to the relative positions of the electronic transitions as well as to the relative ECD intensities associated with those transitions, which can depend on the predominant conformations and solvent environment. Then it is likely that one has to take into account

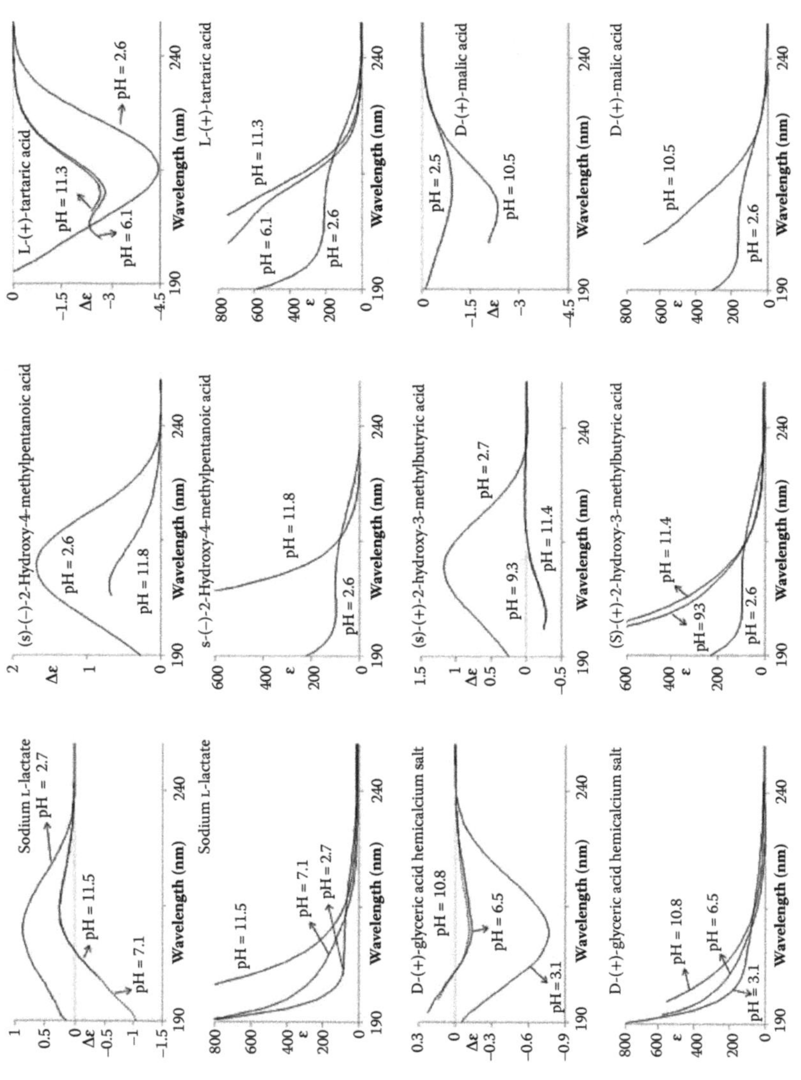

FIGURE 10.2
The pH-dependent electronic circular dichroism (ECD) and EA spectra of some α-hydroxy carboxylic acids. (Data from Polavarapu and Hammer, 2010, unpublished data.)

the different conformations that each molecule may prefer and the influence of the solvent environment. But such criteria are not embedded in the simple rules such as the rule of Clough. As a result, even though the rule of Clough may appear to be applicable, it may be so for wrong reasons and unlikely to hold, in general.

10.2.2 ECD Spectra

The secondary structures of peptides and proteins are often characterized using ECD spectral signatures obtained for those with known structures (Figure 10.3). ECD spectrum with two negative bands in the

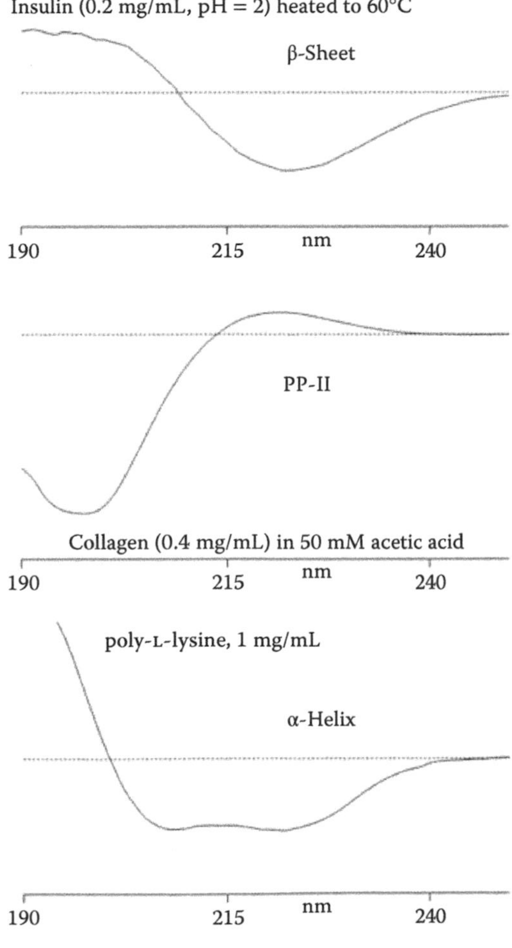

FIGURE 10.3
ECD spectra characteristic of α-helix, polyproline II (PP-II), and β-sheet type protein secondary structures. Data measured in the author's laboratory.

200–230 nm region and a strong positive ECD band at ~190 nm is considered to be characteristic of right-handed α-helical structures. An ECD spectrum with a weak positive band at ~220 nm and a strong negative ECD band at ~200 nm is considered (Woody 1992) to be characteristic of left-handed polyproline II (PPII)-type structure. Note that what used to be considered as a random coil structure is now associated with a structure with local left-handed helical stretches or PPII structure (Shi et al. 2002; Woody 1992). A negative ECD band at ~220 nm followed by a positive ECD band in the 190–200 nm region is considered to be a characteristic of β-sheet conformation (Terzi et al. 1994). A β-turn structure on the other hand is characterized by the presence of a weak negative band at a longer wavelength (~230 nm) and an intense positive band at ~210 nm (Tamburro et al. 1990). When a mixture of secondary structures is present, the quantitative estimates of different secondary structures can be obtained from the experimental ECD data using data analysis programs (Sreerama and Woody 2000). Online determination of protein secondary structure content from ECD spectra is also available (http://dichroweb.cryst.bbk.ac.uk/html/references.shtml).

10.2.3 VCD Spectra

The secondary structures of peptides and proteins are often characterized using VCD spectral signatures in the amide I vibrational region (Keiderling 1986). Two oppositely signed VCD bands appearing next to each other are referred to as a VCD couplet. Just as for ECD couplets, a positive VCD couplet has a positive band on the lower frequency (i.e., longer wavelength) side and negative band on the higher frequency (i.e., shorter wavelength) side of the couplet. A negative VCD couplet has a negative band on the lower frequency side and positive band on the higher frequency side of the couplet. A positive VCD couplet centered at ~1650 cm^{-1}, as seen for hemoglobin in the amide I region, is considered to be characteristic of a right-handed α-helix structure (Figure 10.4). A negative VCD couplet, as seen for poly-L-lysine in the amide I region, centered at ~1660 cm^{-1} is characteristic of a left-handed PP II structure (Dukor and Keiderling 1991). A negative VCD band at ~1630 cm^{-1}, as seen for chymotrypsin in the amide I region, is characteristic of β-sheet structure. Antiparallel β-sheet structure also exhibits a negative VCD band at ~1680 cm^{-1}. A positive–negative–positive series of overlapping bands, as seen in the amide I region for cytochrome c, represents a mixture of α-helical and β-sheet structures.

When a mixture of different secondary structures is present, the quantitative estimates of individual secondary structures can be obtained from a "trained set" of experimental VCD data using principle component analysis (Keiderling 2000).

VCD spectral measurements were also used for structural characterization of viruses (Shanmugam et al. 2005b).

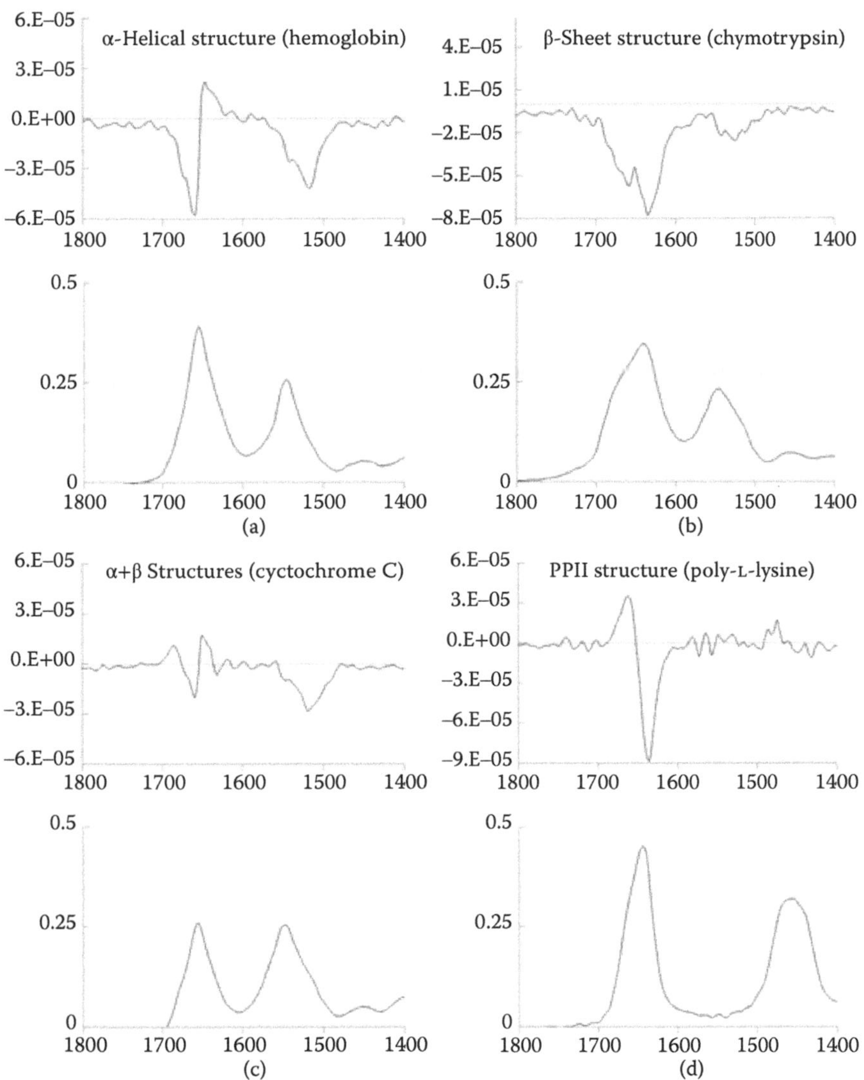

FIGURE 10.4
Vibrational absorption (bottom) and vibrational circular dichroism(top) spectra of (a) α-helix, (b) β-sheet, (c) mixture of α + β, and (d) PP-II type protein secondary structures. (Data from Polavarapu, P. L. and Zhao, C., *Fresenius J. Anal. Chem.*, 366(6–7), 727–734, 2000.)

10.2.4 VROA Spectra

Distinct differences in VROA spectra for different protein secondary structures have been reported indicating that VROA is useful for assessing the secondary structures of peptides and proteins (Barron et al. 2000). A VROA couplet in the amide I region is present for α-helix (negative at 1638 and positive at 1662 cm^{-1} for human serum albumin), and for β-sheet (negative at 1648 and positive at 1677 cm^{-1} for human immunoglobulin) structures. The couplet found for β-sheet structures is shifted to higher frequencies, compared to that for α-helix, by ~10 cm^{-1}. In the extended amide III region, α-helix exhibits a positive VROA band at 1345 cm^{-1} and β-sheet structures exhibit a negative VROA band at 1247 cm^{-1}. The characteristic VROA bands of disordered proteins (whose major conformational element is considered to be the PPII structure) include a positive amide I VROA band at 1673 cm^{-1} and strong positive extended amide III VROA band at 1318 cm^{-1}. In addition, VROA spectra have been used to distinguish between hydrated and unhydrated α-helices (McColl et al. 2004). A stronger sharp positive VROA band at 1340 cm^{-1} has been associated with a hydrated α-helix, while unhydrated α-helix structures are associated with a weaker positive VROA band at 1300 cm^{-1}. More information on structure determination of proteins, nucleic acids, carbohydrates, glycoproteins, and viruses using VROA can be found widely documented work in the literature (Barron et al. 2000, 2007; Blanch et al. 2002; Zhu et al. 2006a, 2006b).

The unknown structure of a protein can be determined from its VROA spectrum using nonlinear mapping, which is a multivariate analysis method that clusters the proteins into different types of secondary structures, as mentioned above (Zhu et al. 2006b).

10.3 Spectral Predictions Using Approximate Models

With the development of quantum chemical calculations for chiroptical properties, the interest in the applications of approximate models is waning. Therefore, this section will be very brief.

10.3.1 ORD Spectra

The Kirkwood polarizability model, Whiffen's model, and the atom–dipole interaction (ADI) model of Applequist are used to calculate SOR (see Chapter 5). But the predictions of SOR obtained with the use of these models are not considered to be reliable.

10.3.2 ECD Spectra

Exciton chirality/EC/CO model (see Appendix 4) has been widely practiced for applications of ECD, especially to biaryl systems. For biaryl compounds, the spectroscopic notation of Platt (Klevens and Platt 1949; Platt 1949) is used for the analysis of ECD spectra. In the case of phenanthrene, the five lowest energy electronic transitions, namely ^1L_b at 353 nm, ^1L_a at 303 nm, ^1B_b at 254 nm, ^1C_b at 212 nm, and ^1B_a at 187 nm have been identified in the visible spectrum. Of these the $^{1}B_{b}$ transition is the most intense and is polarized along the long-axis of phenanthrene. Thus to apply the EC/CO method for biphenanthryl compounds one looks for the $^{1}B_{b}$ absorption band and analyzes the associated bisignate ECD band. A positive bisignate ECD couplet associated with a $^{1}B_{b}$ band indicates that the long axes of two phenanthryl groups make a right-handed screw sense; a negative bisignate ECD couplet indicates left-handed screw sense (Harada and Nakanishi 1983). There are several articles reviewing these applications.

The exciton CD effects were also stated to be applicable for long-range interactions, as observed for a brevetoxin derivative, and explained using density functional studies (Moore and Autschbach 2012).

The interpretation of ECD spectra of a variety of molecules, including proteins, nucleic acids, carbohydrates, natural products, and self-associating aggregates, in the earlier years depended heavily on models such as EC/CO or spectra–structure correlations (Allenmark 2000).

The EC/CO method provided easier assignment of AC in some cases. For 4,4′ biphenanthrene-3,3′-diol, and its thiol analogue, a simple negative ECD couplet was found to be associated with the $^{1}B_{b}$ absorption band and this couplet did correlate with the AC (Gottarelli et al. 1996). Nevertheless, it should be remembered that this method is prone to give incorrect AC when (1) observed ECD is complicated (because the $^{1}B_{b}$ band overlapped with other electronic transitions); (2) it is not clear which electronic transition is responsible for which ECD band; and (3) the predominant conformation (dihedral angle 3-4-4′-3′ in the case of 4,4′-biphenanthryl) is not known. The identification of the $^{1}B_{b}$ band, for practical applications, might not be easy when this band overlaps with those from other electronic transitions appearing in the same wavelength range. As an illustration of the difficulties in assigning the AC of phenathryl compounds using EC/CO method a few examples noted in the literature are summarized here: (1) (+)-(a*R*) configuration was assigned for 1,1′-biphenanthryl-2,2′-diol (Hattori et al. 1998), based on the formation mechanism of cyclic diester with 1,1′-binaphthalene-2,2′-dicarbonyldichloride. The (a*R*) configuration of 1,1′-biphenanthryl-2,2′-diol has two long axes oriented with negative chirality, so the authors expected a negative ECD couplet for $^{1}B_{b}$ transition, but a positive ECD couplet was found for (+)-1,1′-biphenanthryl-2,2′-diol and its methoxy derivative. This discrepancy was thought to arise from the overlap with a second transition polarized along the short axis. Thus biphenanthryl molecules are considered to be difficult to analyze with the EC/CO method. As a result, the EC/CO method is known to have limited applicability for some compounds containing phenanthrene chromophores. (2) Based on ^{1}H NMR studies, the AC of 1-(9-phenanthryl)-2-naphthoic acid and 1-(9-phenanthryl)-2-naphthalene methanol were assigned as (–)-(a*S*). The EC/CO method required several assumptions because the observed ECD spectra of these compounds gave complex ECD patterns and it was not clear which ECD band corresponds to which electronic transition. Based on chemical correlation and X-ray results, the configurational assignment was reversed (Harada et al. 1993) as (+)-(a*S*).

(3) For assigning the AC of 1-(9-phenanthryl)ethylamine, a bichromophoric derivative obtained by complexing 1-(9-phenanthryl)ethylamine with naphthimido chromophore was synthesized. The EC/CO method suggested the configuration of 1-(9-phenanthryl)ethylamine to be (+)-(a*S*). But this assignment had to be reversed (Solladié-Cavallo et al. 2002) based on (a) the stereochemical outcome of catalytic hydrogenation, (b) VCD studies on 1-(9-phenanthryl) ethylamine, and (c) ECD studies on Zn–porphyryn complex, where all three methods suggested (+)-(a*R*) assignment. Thus the EC/CO method could not be used for determining the AC of 1-(9-phenanthryl)ethylamine.

In some paracyclophanes also the EC/CO model was found to be inadequate to explain the experimental ECD spectra (see Sections 10.4.1.10.1 through 10.4.1.10.4).

Although several successful applications of EC/CO model have appeared, there are instances of failure, both in the older (as mentioned above) and in recent literature (Aamouche et al. 1999; Mori et al. 2007). Most of the failures of the EC/CO method originate from two sources: (1) failure of the consideration of all possible conformers and (2) the overlap of different electronic transitions with the chromophoric transitions.

The EC/CO method can also fail, as described in Appendix 4, when the interaction between the electric dipole and magnetic dipole transition moments of the interacting groups is not considered (Bruhn et al. 2014; Jurinovich et al. 2015). The electric dipole transition moment coupling, normally considered to determine exciton chirality, was suggested in some cases to be overwhelmed by the coupling between electric dipole and magnetic dipole transition moments of interacting groups. The failure of the EC/CO model in explaining the observed bisignate ECD couplets in some 1,1′-biphenanthryl compounds, as stated above, was recently attributed (Jurinovich et al. 2015) to a significant contribution from the interaction between the electric dipole and magnetic dipole transition moments that was not considered in the previous literature reports.

From quantum chemical (QC) calculations on stacked chiral (twisted) ethylenes, it was found that the exciton contribution are intertwined with those from chiral units, so the indiscriminate use of the EC model can be perilous (Norman and Linares 2014). These examples suggest that one needs to exercise caution in placing excessive confidence on the EC/CO model predictions.

10.3.3 VCD Spectra

The models for VCD based on atomic charge and/or bond moment concepts (See Chapter 5), are approximate in nature and are not widely used. The EC/CO model of VCD has been used in several instances (Birke et al. 1992; Schweitzer-Stenner 2004; Su and Keiderling 1980), but was not pursued widely due to the ever increasing applicability of reliable quantum chemical calculations. In recent years, renewed interest has appeared for applying the EC/CO method, with its use for C=O stretching motions in a number of molecules containing carbonyl groups (Asai et al. 2013; Komori et al. 2014; Taniguchi and Monde 2012a; Taniguchi et al. 2015; Wu and You 2012). It is tempting to utilize the dipole interaction based EC/CO model for isolated vibrational transitions, but one has to be careful (Abbate et al. 2015; Bour and Keiderling 1992; Covington et al. 2015) even for such cases as the carbonyl transitions. This is because the dipole interaction mechanism may not be appropriate for vibrational transitions where coupling arises predominantly from interaction force constants. One should also be aware that for VCD interpretations, the EC/CO model can result in the right answers for the wrong reasons or in a wrong conclusion altogether (Covington et al. 2015).

10.3.4 VROA Spectra

The models based atomic polarizability and/or bond polarizability concepts (see Chapter 5) require parameterization and are approximate in nature. The two-group model of VROA, the analog of the CO model for VCD, has been instrumental in understanding the underlining concepts, but its applications are limited.

10.4 Spectral Predictions Using Quantum Chemical Calculations

Interpretations of experimental ORD, ECD, VCD, and VROA spectra using quantum chemical calculations represent the state-of-the-art in the current times. However, as quantum chemical calculations require considerable computational resources, the applications to large size molecules can be difficult. With the availability of both experimental and computational resources, it has been recognized (Stephens et al. 2007; Polavarapu 2008) that the simultaneous use of more than one of the chiroptical methods leads to higher confidence for the molecular structure determination. As a result, many researchers are adapting multiple chiroptical spectroscopic methods for determining the structure of a given chiral molecule. Instead of discussing the applications of individual chiroptical methods, the intention here is to present applications to individual, or classes of, molecules. However, because of an exponential increase in the number of literature papers on this topic, and limitations of space, it is not possible to reference all of the reported studies. For this reason, only selected applications are presented.

10.4.1 Determination of the Molecular Structures

In a broader sense, there are three kinds of applications for molecular structure determination using chiroptical spectroscopy: (1) confirmation of the known ACs, thereby validating the use of chiroptical spectroscopy, and obtaining additional information regarding various possible conformers; (2) determination of the unknown ACs for the first time; and (3) correcting the literature ACs that are found to be incorrect during the course of modern chiroptical spectroscopic analyses. All these applications represent the triumphs of chiroptical spectroscopy. An exhaustive coverage of all reported studies will, however, be overwhelming. Therefore, the representatives examples of (1) and (2) are described in this section, while the representative examples of (3) are described separately in Section 10.4.2.

10.4.1.1 Fluorinated Ethers

A few chiral haloethers, namely isoflurane [CF_2OH)$CHClCF_3$], desflurane [(CF_2OH)$CHFCF_3$)], 1,2,2,2-tetrafluotoethylmethylether [(CH_3O)$CHFCF_3$)], enflurane [(CF_2OH)CF_2CHClF)] along with achiral sevoflurane ($CH_2FOCH(CF_3)_2$), and its chiral degradation product, 1,1,1,3,3,-pentafluoro-2-(fluoromethoxy)-3-methoxypropane, were evaluated for anesthetic action. These haloethers attracted much attention in the early nineties because of their medicinal use. The ACs of some of these chiral haloethers were determined for the first time, and confirmed for the others, using chiroptical spectroscopy.

10.4.1.1.1 Isoflurane [(CF_2OH)$CHClCF_3$]

The AC of isoflurane was not known until it was determined using VCD spectroscopy. The VCD spectra of (+)-isoflurane, measured in the ~1500–800 cm^{-1} region, were compared to those predicted using localized MO (LMO) theory (see Chapter 5) and a 6-31G* basis set. The VCD spectrum predicted as the average of those for the two lowest energy conformers of (*S*)-isoflurane was found to reproduce those measured for (+)-isoflurane in CS_2 and CCl_4 solvents (Polavarapu et al. 1992). Based on this observation, the AC of isoflurane was determined to be (+)-(*S*). This was the first application of VCD spectroscopy to determine an unknown AC and this AC assignment was later found to be consistent with the X-ray structure (Schurig et al. 1996), as well as with the comparison of experimental and predicted SOR (Polavarapu 1997a; Polavarapu and Zhao 1998) and VROA (Polavarapu 2011) of isoflurane.

10.4.1.1.2 Desflurane [(CF_2OH)$CHFCF_3$)]

Investigations in the same manner as that for isoflurane initially led to an (–)-(*S*) assignment for desflurane (Polavarapu et al. 1993). However, this assignment led to a controversy for the stereoselective reactions involving isoflurane and desflurane. The X-ray structure suggested that the correct absolution configuration of desflurane should be (+)-(*S*) (Schurig et al. 1996). It was later found that the original incorrect assignment resulted from a mislabeling of the enantiomer samples used for VCD measurements and the original assignment was latter corrected (Polavarapu et al. 1997a). The reinvestigation of the experimental VCD spectra and theoretical analysis

(Polavarapu et al. 1999a) concurred with the correct (+)-(*S*) assignment. The comparison of experimental and predicted SOR values (Polavarapu 1997a; Polavarapu and Zhao 1998) also agreed with the (+)-(*S*) assignment.

10.4.1.1.3 1,2,2,2-Tetrafluoroethyl Methyl Ether ([(CH_3O)$CHFCF_3$)] (TFEME)

The experimental VCD spectra of (–)-TFEME measured in CCl_4 were reproduced by the predicted spectra for the (*R*) configuration at two different theoretical levels, B3LYP/6-31G* and B3PW91/6-311G(2d) (Polavarapu et al. 1999b). Among the three possible conformers of this molecule only one conformer was found to be dominant. The (–)-(*R*) assignment for TFEME is found to be consistent with stereoselective reactions involving isoflurane, desflurane and TFEME.

10.4.1.1.4 Enflurane [(CF_2OH)CF_2CHClF)]

Unlike the previously discussed three haloethers, the AC of enflurane was known previously from the synthetic route used in its preparation as (+)-(*S*). The quantum chemical predictions of SOR (Polavarapu 1997b; Polavarapu and Zhao 1998) supported this conclusion. An independent confirmation of the AC assignment was obtained from VCD spectral analysis (Zhao et al. 2000). The predicted VCD sign pattern for (*S*)-enflurane, at B3LYP/6-31G* and B3PW91/6-311G(2d) levels, matched the experimental VCD spectrum of (+)-enflurane measured in CCl_4. Four different dominant conformations of enflurane were considered to be responsible for the observed VCD spectra.

10.4.1.1.5 1,1,1,3,3,-Pentafluoro-2-(fluoromethoxy)-3-methoxypropane (PFFMMP)

PFFMMP is the chiral product generated during the degradation of sevoflurane. The AC of PFFFMMP, known from X-ray structure to be (–)-(*R*), was confirmed using VCD spectroscopy (Wang et al. 2002b). The experimental VCD spectra were measured for $CDCl_3$ solutions in the 1500–900 cm^{-1} region and compared to those predicted at B3LYP/6-31G* level. The spectra measured for (+)-PFFMMP were found to be in excellent agreement with the conformer population weighted spectra for (*S*)-PFFMME. Among the 81 conformers investigated, six conformers were found to be dominant.

10.4.1.2 Halomethanes

10.4.1.2.1 Bromochlorofluoromethane

One of the simplest chiral molecules that has been used in many organic chemistry textbooks to introduce chirality for undergraduate students is bromochlorofluoromethane. Nevertheless chemists had a difficult time establishing its AC for two reasons: (1) resolution, or isolation, of the enantiomers of bromochlorofluoromethane turned out to be a challenge; and (2) even with successful resolution of the enantiomers, growing a single crystal of the enantiomer was not possible to deduce its stereochemistry using X-ray diffraction. The deduction of AC from the observed SOR of bromochlorofluoromethane was not possible because the approximate model predictions of

SOR yielded conflicting results. These difficulties lead to the statement, "The experimental determination of the AC of bromochlorofluoromethane remains a challenge," even as late as 1985 (Wilen et al. 1985). The emergence of modern quantum chemical methods for predicting SOR (Polavarapu 1997a) and VROA (Polavarapu 1990) and of experimental VROA measurements (Barron et al. 1973) overcame this challenge. By measuring the experimental VROA spectrum of (–)-bromochlorofluoromethane and reproducing this spectrum using the quantum chemical prediction for (*R*)-bromochlorofluoromethane, and comparing the experimental $[\alpha]_D$ of –1.78 with the quantum chemical prediction of $[\alpha]_D$ of –8.0, the AC was inferred as (*R*)-(–) and (*S*)-(+) (Costante et al. 1997). Although the early quantum chemical predictions were made with modest size basis sets, the conclusions remained unaltered when quantum chemical predictions were obtained with larger size basis sets and SOR at different wavelengths, that is, ORD, were compared (Polavarapu 2002) (see Figure 10.5). Thus, the AC of bromochlorofluoromethane has been established with little ambiguity for the first time using VROA and ORD spectroscopic methods.

10.4.1.2.2 Chlorofluoroiodomethane

By comparing the experimental and quantum chemically predicted ORD, which are displayed in Figure 10.5, the AC of chlorofluoroiodomethane was also established as (*S*)-(+) (Crassous et al. 2004).

10.4.1.3 Molecules with Chirality Imparted by Isotopic Substitution

Molecules that become chiral due to isotopic substitution will have the same achiral electronic distribution as that of their parent analogue. Some perturbation to that achiral electronic distribution can result from vibrational–electronic coupling, and such effects can introduce only small influence on the electronic properties. As a result, for chiral molecules whose chirality originates from isotopic substitution, the associated SOR is usually very small and ECD spectral signatures are weak, but are nevertheless measurable

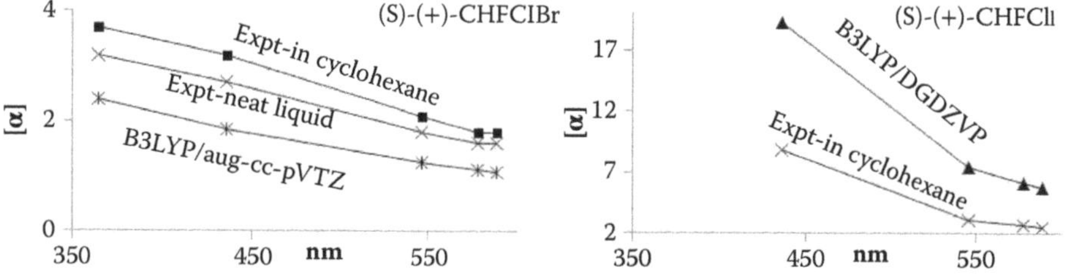

FIGURE 10.5
Specific optical rotation of (*S*)-(+)-bromochlorofluoromethane (left) and of (*S*)-(+)-chlorofluoroiodomethane (right). (Data from Crassous, J. et al., *Tetrahedron Asymmetry*, 15(13), 1995–2001, 2004; Costante, J., et al., *Angew. Chem. Int. Ed. Engl.*, 36(8), 885–887, 1997; Polavarapu, P. L., *Angew. Chem. Int. Ed. Engl.*, 41(23), 4544–4546, 2002.)

(Barth and Djerassi 1981; Canceill et al. 1984) and used for structure determination (Lin et al. 2012).

However, the vibrational motions of chiral molecules whose chirality originates from isotopic substitution are inherently chiral and therefore their VCD and VROA spectra offer greater potential and versatility for molecular structure determination. The VCD spectra for [α-^{2}H]-neopentyl chloride (Holzwarth et al. 1974), [α-^{2}H]-cyclohexanone (Devlin and Stephens 1999; Polavarapu et al. 1981), [2,3-$^2H_2$]oxirane (Freedman et al. 1987; Freedman et al. 1991b; Stephens et al. 1993), cyclopropane-1,2-$^2H_2$ (Cianciosi et al. 1989, 1990; Lowe et al. 1986), cyclopropane-1-^{13}C,^{2}H-2,3-$^2H_2$ (Freedman et al. 1991a), cyclopropane-1-^{13}C-1,2,3-d_3 (Cianciosi et al. 1991), cyclobutane-1,2-$^2H_2$ (Annamalai et al. 1985), trans-succinic-d_2 anhydride (Bouř et al. 1996), 3,4-dideuterio cyclobutane-1,2-dione (Malon et al. 1991), 2,3-dideuteriocyclobutane-1,2-dione (Malon et al. 1992), allene-1,3-$^2H_2$ (Annamalai et al. 1990), perdeuteriophenyl-phenyl-sulfoxide (Drabowicz et al. 2008), and VROA spectra for [α-^{2}H]-benzyl alcohol (Barron 1977), 4,4-dideuterioadamantanone (Barron et al. 1978) demonstrated that VCD and VROA are uniquely optimal methods for probing the chirality of molecules whose chirality originates from isotopic substitution.

Among the molecules that owe their chirality to isotopic substitution, (*R*)-[$^2H_1$,$^2H_2$,$^2H_3$]-neopentane posed by far the greatest challenge, not only for its synthesis, but also for its structure determination (Haesler et al. 2007), where the difficulty was compounded by the presence of nine rotamers for this molecule. The SOR and ECD associated with this molecule were perceived to be small and so were not measured. Being a gas at room temperature, [$^2H_1$,$^2H_2$,$^2H_3$]-neopentane had to be condensed and sealed in a glass capillary for VROA measurements (~5 mg sample) and in glass tubes for NMR measurements (~25 mg sample). Using an in-house built state-of-the art VROA spectrometer operating in backscattering geometry, Hug and coworkers succeeded not only in measuring the VROA spectrum of [$^2H_1$,$^2H_2$,$^2H_3$]-neopentane, but also in reproducing the experimentally measured VROA spectrum using quantum chemical calculations. Based on an excellent agreement between experimental VROA spectrum of the synthesized [$^2H_1$,$^2H_2$,$^2H_3$]-neopentane and quantum chemically predicted VROA spectrum for the (*R*)-enantiomer, the AC of synthesized compound was assigned as (*R*)-[$^2H_1$,$^2H_2$,$^2H_3$]-neopentane (Haesler et al. 2007). [$^2H_1$,$^2H_2$,$^2H_3$]-neopentane represents the second molecule, after CHFClBr, whose AC was established for the first time using VROA spectroscopy.

10.4.1.4 Metal Wires

A linear arrangement of three nickel atoms surrounded by four dipyridylamine ligands represents a miniaturized metal wire (Armstrong et al. 2007). The chirality of these metal wires arises from the helical orientation of the pyridyl groups of the dipyridylamine ligand. The inability to grow single crystals of these metal wire samples prevented the use of X-ray crystallography for determining the absolution configuration. Solution-phase studies using chiroptical spectroscopic methods alleviated this situation and the AC of the $Ni_3[(C_6H_5N)_2N]_4Cl$ metal wire was determined to be $(-)_{589}$-*P*- $Ni_3[(C_6H_5N)_2N]_4Cl$. This conclusion was reached from a comparison of experimental and quantum chemical predictions of three different chiroptical spectroscopic properties, namely ECD, ORD, and VCD (Armstrong et al. 2007).

10.4.1.5 Fullerenes

The ability to separate the pure enantiomers of chiral fullerenes (Crassous et al. 1999; Kessinger et al. 1998) opened up the possibility to determine their ACs using ECD spectroscopy. Unlike for typical organic molecules, special configurational descriptors, (^{f}A) and (^{f}C), are used for fullerenes (Thilgen et al. 1997). SOR measurements are not available for chiral fullerenes (probably because fullerenes are colored samples and strongly absorb the visible light). As a result the connection between the sign of SOR and configurational descriptors (^{f}A) and (^{f}C) is not usually provided. VCD and VROA measurements were also not undertaken for chiral fullerenes.

10.4.1.5.1 D_2 Symmetric C_{84}

Although the enantiomers of C_{84} with D_2 symmetry were separated and ECD spectra reported (Crassous et al. 1999), the connection between ECD spectra and AC was not made until quantum chemical calculations of ECD were undertaken (Furche and Ahlrichs 2002). The enantiomer with negative ECD bands in the 500–650 nm region, positive ECD bands in the ~400–475 nm region, and negative ECD bands in the ~350–400 nm region is found to belong to the (^{f}A) configuration.

10.4.1.5.2 D_2 Symmetric C_{76}

The experimental ECD spectra of D_2 symmetric C_{76} (Kessinger et al. 1998) were analyzed with the SCF-CI-DV MO method to determine the AC (Goto et al. 1998). The AC was determined as [CD(+)281]-(^{f}C)-C_{76}, where [CD(+)281] indicates the enantiomer that exhibits positive ECD band at 281 nm. In the interest of making connection between the configurational descriptors and the sign of SOR, the experimental ORD was derived from the ECD spectrum using the Kramers–Kronig transformation (see Appendix 2) and compared to the corresponding ORD calculated with the density functional theory (Polavarapu et al. 2005). A good agreement was found between the experimental ORD derived from the ECD spectrum of the enantiomer that exhibited positive CD at 281 nm and that predicted for the (^{f}C)-enantiomer. Although the ORD has both positive and negative values, it is an accepted practice to relate the AC to the sign of SOR at 589 nm. Thus, based on this analysis, the AC of C_{76} is given as $(+)_{589}$-(^{f}C)-C_{76}.

10.4.1.5.3 Fullerene Adducts

There are several adducts of fullerenes that are chiral (whether or not the fullerene core itself is chiral). The ACs of such adducts are also determined from ECD spectroscopy (Kuwahara et al. 2005), but the literature on such compounds is not covered here.

10.4.1.6 Helicenes

Helicenes can be broadly categorized as π-[*n*]helicenes and σ-[*n*]helicenes. The former helicenes contain "*n*" number of ortho fused aromatic rings with extensive π-delocalization and are referred to simply as helicenes without reference to π-character. At least five aromatic rings are required for obtaining stable enantiomers, so pentahelicene is the first member of this series that can be used for chiroptical spectroscopy studies. The π-[*n*]helicenes exhibit rich ECD and VCD spectra and large SOR. The σ-[*n*]helicenes contain orthogonal cyclopropane rings (with the orthogonality occurring at one commonly shared carbon) and contain σ-bonding network. The smallest member of these spiro helicenes is the one with $n = 4$. These compounds, also referred to as [*n*]triangulanes, do not have any electronic bands in the accessible UV–Vis region (>200 nm), so no ECD spectra are available for these compounds. However, these compounds exhibit large SORs and vibrational optical activity signals. A different type of σ-helicene is that formed from the fused adamantane cages, the first member of which is called [123] Tetramantane (Schreiner et al. 2009). These compounds also do not exhibit ECD in the accessible UV–Vis (>190 nm) region, but they do have SOR that increases in magnitude toward shorter wavelengths.

10.4.1.6.1 π-[n]Helicenes

The AC of 2-bromohexahelicene was determined to be (–)-(*M*) from X-ray crystal structure, and since (–)-hexahelicene could be obtained from (–)-2-bromohexahelicene by treatment with *n*-butyllithium in ether, the AC of hexahelicene was also determined to be (–)-(*M*) (Lightner et al. 1972). The quantum chemical calculations of SOR (Srebro et al. 2011) and ECD (Furche et al. 2000; Grimme and Neese 2007; Nakai et al. 2012b) indicate that the ACs of carbon helicenes is (–)-(*M*). The same AC applies for heteroatom-substituted helicenes (Rajca et al. 2009). VCD (Abbate et al. 2009b) and VROA (Johannessen et al. 2013) spectral analyses of 2-bromohexahelicene also concurred with (–)-(*M*) configuration. The experimental VCD spectrum of (+)-heptahelicene was also reproduced well (Burgi et al. 2004) by that calculated with B3LYP and B3PW91 functionals and 6-31G(d,p) and cc-pVDZ basis sets for (*P*)-heptahelicene. The experimental and quantum chemical ECD spectra for several disubstituted dimethyl- and diaza[6]helicenes (Nakai et al. 2012a) and mono- and diazonia[6]helicenes (Nakai et al. 2013) are also consistent with a (+)-(*P*) configuration. Thus all helicenes, with all carbon as well as heterohelicenes, generally belong to the (–)-(*M*)/(+)-(*P*) configuration. More recently, the VCD spectra of 5-aza-hexahelicene and the ECD spectra for hexahelicene, 2-methylhexahelicene, and 2-bromohexahelicene were measured and analyzed (Abbate et al. 2013) with corresponding quantum chemical predictions. In these molecules, the features invariant to different substituents were identified as H-features and those that depended on the substituents as S-features.

Metallated helicenes were also investigated by Crassous and coworkers using ECD and ORD spectroscopies and these studies are referenced under transition metal complexes (see Appendix 9).

Quantum chemical predictions of VROA spectra for hexahelicene and their comparison to those predicted for tetrathia-[7]-helicene, and its pyrrole and furan analogs, provided an understanding of how much change one can expect with a change in heteroatom (Liégeois and Champagne 2009). In the forward scattered VROA spectrum, left-handed helicenes are predicted to display positive VROA signals for most of the vibrations.

10.4.1.6.2 σ-[n]Helicenes

The enantiomerically pure σ-[4]helicene (de Meijere et al. 1999), σ-[5]helicene (de Meijere et al. 2002), and their analogs have been reported. The ACs of these compounds are known from their synthetic route as (*M*)-(–)-σ-[n]helicenes. The (*M*)-(–)-σ-[4]helicene was found to exhibit a large SOR with increasing magnitude toward shorter wavelengths, which was successfully reproduced by the calculated SORs, as the KK transform of ECD calculated at the B3LYP/TZVP level (de Meijere et al. 1999), and also by coupled cluster theoretical calculations of SOR (Crawford et al. 2005). The experimental VROA spectrum for (*M*)-(–)-σ-[4]helicene (Hug et al. 2001) was satisfactorily reproduced by the quantum chemical calculations. The (*M*)-(–)-σ-[5]helicene was also found to exhibit large SOR with increasing magnitude toward shorter wavelengths, and these observations were also successfully reproduced by the calculated SORs, as the KK transform of ECD calculated at the B3LYP/TZVP level (de Meijere et al. 2002).

10.4.1.6.3 Cationic Helicenes

Dimethoxyquinacridinum cation adopts a twisted helical conformation, like that of helicene derivatives. By comparing the experimental VCD spectrum of the dimethoxyquinacridinum cation, stabilized as hexafluoride salt, to that predicted for a cation with (*P*) configuration, its AC was assigned as (+)-(*P*) (Herse et al. 2003). The experimental ECD spectrum of the salt was also measured, but due to the complex nature of this spectrum, no attempt was made to confirm the AC assignment independently using ECD.

Cationic azaoxa[4]helicenes were also investigated using chiroptical spectroscopy. Both enantiomers of dimethoxychromenoacridinium ions, as BF_4 salts, were prepared and their ACs determined using VCD spectral studies. The (+) and (–) enantiomers were assigned (*P*) and (*M*) configurations, respectively. The temperature-dependent ECD spectra were used to determine the barrier of racemization as ~33 Kcal/mol (Gouin et al. 2014).

10.4.1.6.4 [123]Tetramantane

The AC determination was achieved from crystal structure of (+)-7-bromo-[123]Tetramantane as (*P*) (Schreiner et al. 2009). The structure of the hydrocarbon moiety does not change upon bromination, so the structure of [123]Tetramantane was deduced as (*P*)-(+). To confirm this assignment, ORD and VCD spectral studies were undertaken. The experimental ORD curve for (–)-[123]Tetramantane was well reproduced by that calculated for (*M*)-[123]Tetramantane at B3LYP/6-31G(d,p) level. The experimental VCD spectrum of (–)-[123]Tetramantane in the 1600–1000 cm^{-1} region was also reproduced very well by that calculated for (*M*)-[123]Tetramantane, confirming that the AC of [123]Tetramantane is (*M*)-(–)/(*P*)-(+).

10.4.1.6.5 Diazaoxatricornan Derivative

These molecules differ somewhat from helicenes and have ring-closed polycyclic structures with a bowl shape, and chirality arises from dissymmetry of the central core. The experimental VCD spectrum of (–)-enantiomer in CD_2Cl_2 was found to be in agreement with those calculated for (*S*)-configuration at B3PW91/6-31G(d,p) level (Mobian et al. 2008). Although ECD spectra were also measured, these were not used for independent assignment of AC. This study was stated to represent the first report of a closed-capped chiral bowl molecule.

10.4.1.7 Metal Nanoparticles

A characteristic of gold nanoparticles of ~10 nm diameter is the localized surface plasmon resonance, LSPR, (which represents the collective oscillation frequency of conduction electrons) that gives rise to a visible electronic transition at ~530 nm. When the size of nanoparticles is less than ~2 nm, nanoparticles containing 10–200 atoms are referred to as nanoclusters (NCs). Well-resolved metal-based electronic transitions (MBETs), unlike a single LSPR for metal nanoparticles, are seen for metal NCs. The wavelengths of these transitions are size dependent. Chirality can be probed using ECD associated with an LSPR transition or with MBETs. ECD associated with LSPR was observed for gold nanoparticles functionalized with peptides (Slocik et al. 2011), grown on peptide nanotubes (George and Thomas 2010), and grown on twisted fibers as nanorods (Guerrero-Martínez et al. 2011). ECD bands associated with resolved MBETs were observed when organic chiral ligands are attached to metal NCs (Schaaff et al. 1998) and a pencillamine ligand is attached to silver NCs (Yao et al. 2010). ECD associated with metal electronic transitions can arise from (Gautier and Bürgi 2009) (1) a chiral metal core, (2) chirality induced into the metal core from the chiral ligands, or (3) chiral induction onto the metal surface by the attached ligands. Mid-infrared VCD spectral measurements, and simultaneous quantum chemical predictions of VCD, have been used (Gautier and Bürgi 2005, 2006, 2009, 2010; Yao et al. 2010) for deducing the conformational preferences of the attached chiral ligand. The plasmonic and excitonic resonances of achiral inorganic NCs capped and/or formed with chiral molecules were reviewed by Markovich and coworkers (Ben-Moshe et al. 2013). High-level quantum chemical ECD spectral predictions were shown to provide theoretical support for understanding the measurement in nanostructures (Noguez and Hidalgo 2014).

10.4.1.8 Organic Nanoparticles

(R)-(+)-1,1′-binaphthol dimethylether (BNDE) nanoparticles of varying size (25–100 nm) were prepared and their ECD spectra measured. ECD spectra of dilute solutions in acetonitrile were found to be opposite to those of nanoparticles dispersed in aqueous solution. Using quantum chemical calculations for both monomer and dimer of BNDE, it was concluded that the intermolecular EC between adjacent BNDE molecules is responsible for the inversion of ECD spectra (Xiao et al. 2004).

10.4.1.9 Cryptophanes

Cryptophanes are made up of two cyclotriveratrylene (CTV) bowls, assembled such that these appear as the top and bottom outward pointing caps of a round soccer ball and are connected via three identical bridges that form the sides (Collet 1996). Small guest molecules or atoms can be entrapped in these cage molecules. C_3-symmetric CTVs themselves are chiral with *P* or *M* helicity and can be combined to give optically active cryptophanes of D_3 symmetry. The experimental ECD spectra of six different D_3-symmetric cryptophanes were explained using the EC/CO model, and indirect information on the conformation of cryptophanes in solution was obtained (Canceill et al. 1987). Cryptophane-A, with chemical formula $C_{54}H_{54}O_{12}$, contains aliphatic $-O(CH_2)_2O-$ bridges and has D_3 symmetry (Collet 1996). Such large molecules may not have been amenable for quantum chemical calculations of chiroptical properties not too long ago, but with modern computer processors, VCD spectral predictions became possible. The experimental VCD spectra for cryptophane-A enantiomers were reproduced by quantum chemical predictions at the B3PW91/6-31G* level for cryptophane-A (Brotin et al. 2006). Excellent agreement obtained between experimental and calculated VCD spectra revealed the utility of VCD spectroscopy for structural determination of large organic molecules. More recently cryptophanes of C_1 symmetry were also synthesized and their ECD, ORD, and VCD spectra investigated (Cavagnat et al. 2007). Quantum chemical calculations of VCD spectra at the B3PW91/6-31G* level and their comparison with experimental VCD spectra revealed the preferential conformation of the aliphatic linkers of free cryptophane as well as of guest-encapsulated cryptophane (Brotin et al. 2008). The ACs and conformations of nona- and dodecamethoxy cryptophanes were also inferred from ECD, ORD, and VCD spectral investigations (Brotin et al. 2014).

The AC of cryptophane-A determined from VCD studies was also confirmed using VROA spectral studies (Daugey et al. 2014). The appearance of VROA spectra was not influenced by a change in symmetry of cryptophanes from D_3 to C_1, but was influenced by the addition of methoxy substituents to the rings of CTV.

The enantiomers of cryptophane derivatives containing ethylenedioxy and propylenedioxy links connecting the two cyclotribenzylene units have been separated, their ACs deduced from crystal structures, and the chiroptical spectroscopic properties (SOR ECD, VCD, VROA) determined (Brotin et al. 2015). Large differences were found in SOR values for enantiomers with ethylenedioxy and propylenedioxy links, but corresponding differences in ECD, VCD, and VROA spectra are small. DFT computations reproduced the experimental spectra well.

Four different CTV-C_{60} conjugates were also investigated using ECD and VCD spectroscopies to determine the constitution of fullerene addition pattern, their ACs, and the AC of CTV moiety (Kraszewska et al. 2010).

10.4.1.10 Cyclophanes

A cyclophane contains benzene (or another aromatic) ring and an aliphatic chain that makes a loop between two nonadjacent positions of the benzene ring. Paracyclophanes contain the loop between para positions of the benzene ring, and the number of methylene groups, *n*, involved in the loop are identified by a prefix, as [*n*]paracyclophanes. [*n.n*]paracyclophanes contain two benzene rings both connected through their para positions. [2.2]paracyclophanes are the most commonly used compounds for chiroptical investigations. The chirality associated with cyclophanes is represented by planar chirality designations (see Chapter 2). A recent review summarizes the ECD spectral investigations on planar chiral cyclophanes (Mori and Inoue 2011). The chiroptical properties of several allenic meta- and para-cyclophanes bearing anthracene and pyridine rings as spacers have been summarized recently (Lahoz et al. 2014).

10.4.1.10.1 4,7-Dicyano-12,15-dimethoxy[2.2]paracyclophanes (DCDMPC)

The experimental VCD spectra (in CD_2Cl_2) in the 1800–1000 cm^{-1} region and ECD spectra (in acetonitrile) in the 200–500 nm region were measured for both enantiomers of two different DCDMPC compounds to determine their ACs. One is staggered-DCDMPC and another is eclipsed-DCDMPC (Furo et al. 2005). The AC determination was found to be straightforward with VCD analyses. The predicted VCD spectra at the B3LYP/6-31G(d) level for ($4S_p$,12*Sp*)-staggered-DCDMPC were considered to match those measured for (–)-staggered-DCDMPC; and ($4R_p$,12*Sp*)-eclipsed-DPPC were considered to match those measured for (–)-eclipsed-DCDMPC. In a later investigation (Mori et al. 2007), the differences between some of the experimental and calculated VCD bands were highlighted, and it was concluded that, in situations where conformational variability is present, the simultaneous use of other methods is necessary to confirm the determined AC. The interpretation of ECD spectra using the EC/CO model was found to give the incorrect AC for eclipsed-DDPC, due to the presence of intramolecular charge transfer (CT) bands (Furo et al. 2005). Nevertheless, ECD calculations at BHLYP/TZV2p level were found to satisfactorily reproduce the experimentally observed ECD spectra and hence to determine the AC (Mori et al. 2007). From the analysis of excited states, it was concluded that the experimental ECD spectrum does not represent EC mechanism and that EC/CO model can give erroneous conclusions. The observed SOR sign for these compounds could be reproduced with quantum chemical calculations performed with the BHLYP functional, but not with the B3LYP functional as the excitation energies of CT transitions were underestimated with the latter. It was suggested (Mori et al. 2007) that to determine the ACs of cyclophanes, the comparison of experimental ECD spectra with corresponding high-level quantum chemical calculations is more reliable than corresponding analysis with SOR or VCD.

10.4.1.10.2 4,7-Bismethoxycarbonyl-12,15dimethoxy[2.2]paracyclophane (BMCDPC)

The experimental VCD spectra (in the 1800–1000 cm^{-1} region) for both enantiomers of staggered-BMCDPC and eclipsed-BMCDPC were measured in CD_2Cl_2 solvent, to determine their ACs (Furo et al. 2006). Quantum chemical calculations at the B3LYP/6-31G(d) level indicated the presence of six conformers for these molecules. The conformationally averaged VCD spectrum of ($4S_p$,$12S_p$)-staggered-BMCDPC was found to match the experimental VCD spectrum of (+)-staggered-BMCDPC. Similarly, the conformationally averaged VCD spectrum of ($4S_p$,$12R_p$)-eclipsed-BMCDPC was found to match the experimental VCD spectrum of (+)-eclipsed-BMCDPC. It was concluded that, despite the presence of strong intramolecular CT interactions, the combination of experimental and quantum chemical VCD spectral analysis can be used to determine the ACs of donor/acceptor-substituted [2.2]paracylophanes.

10.4.1.10.3 4,7-Dicyano-12,13.15,16-tetramethyl[2.2]paracyclophane (DCTMPC)

The AC of this compound was confidently assigned as (+)-($4S_p$)-DCTMPC by comparing the experimental ECD spectrum with that predicted at the BHLYP/TZV2P level (Mori et al. 2007).

10.4.1.10.4 4,7,12,15-Tetramethyl[2.2]paracyclophane (TMPC)

This compound differed from the previous three compounds in that there are no donor–acceptor interactions here. Seven conformers were found for this molecule and by comparing the experimental ECD spectrum with that predicted at BHLYP/TZV2P level (Mori et al. 2007), the AC was assigned as (−)-($4S_p$,$12S_p$)-TMPC. The same conclusion was obtained from the analysis of VCD spectra. From the analysis of excited states, it was concluded that the experimental ECD spectrum does not represent the EC/CO mechanism. Therefore, even for molecules that do not have CT transitions, it was suggested that EC/CO model can give erroneous conclusions.

10.4.1.10.5 [2.2]Paracyclophan-4-yl-2,2,2-trifluoroethanol (PCTFE)

There are two sources of chirality in this type of compounds: the planar chirality arising from cyclophane and the central chirality from the carbon atom that is attached to CF_3, H, and OH groups. Four diasteremoers, (R_p, *R*), (R_p, *S*), (S_p, *R*), and (S_p, *S*), are possible for PCTFE. The ECD and VCD spectra of (+)- and (–)-PCTFE were measured to determine their ACs (Abbate et al. 2011). While ECD spectra of the enantiomers are mirror images of each other, the diastereomers can have identical ECD spectra. Therefore ECD spectra were not used to identify the diastereomers. However, the experimental VCD spectra of (–)-PCTFE in the 1700–1000 cm^{-1} region, could be reproduced satisfactorily by those calculated for (R_p, *R*)-PCTFE, so the AC of (–)-PCTFE was concluded from VCD analysis to be (R_p, *R*). This conclusion was verified by the X-ray crystal structure of (–)-PCTFE.

10.4.1.10.6 4-X-[2.2]paracyclophanes

The ACs determined from the use of experimental VCD spectra, when analyzed with DFT calculations, were shown to be consistent with the known ACs of [2.2]paracylophanes with X = I, Cl, $COOCD_3$ (Abbate et al. 2007). For [2.2]paracylophanes with X = F,CH_2F, and $COCF_3$, experimental VCD as well as ECD spectra were analyzed with corresponding quantum chemical calculations and were found to be consistent with (+)-(*R*) for X = F, (–)-(*R*) for X = CH_2F and (+)-(*R*) for X = $COCF_3$ (Abbate et al. 2009a). Both VCD and ECD spectra were shown to contain information on the relative orientation of two phenyl rings and on the different conformers of the X group.

10.4.1.10.7 2,2′-Azobenzenophanes

These are novel compounds, differing from cyclophanes, obtained by the coupling of 2,2′-binaphthyl derivatives to azobenzene derivatives (Lu et al. 2015). The interesting property of these compounds is that cis–trans isomerization of azobenzenes can be driven by photo-irradiation with UV light and reversed by irradiation with visible light. The photoresponsive structural changes were monitored using ECD spectroscopy. The ECD spectra calculated for trans–trans, trans–cis, and cis–cis conformers at the B3LYP/6-31G(d) level were compared to the observed spectra, although higher level theoretical predictions would have been more comforting. The photoresponsive changes in experimental SORs were also reported.

10.4.1.11 Atropisomers

Stereoisomers resulting from hindered rotation about single bonds, with high enough energy barrier to rotation facilitating the isolation of the isomers, are referred to as atropisomers (see Chapter 2). Substituted biphenyl, binapthyl, and biphenanthryl compounds are among the commonly studied atropisomers. In most cases, the ACs of these compounds are obtained from X-ray studies. In cases where ACs are not known, the experimental ECD spectra were widely used with EC/CO model for suggesting the ACs. For cases where ACs are known, the chiroptical spectroscopic studies are used to determine if the known ACs can be supported by these studies and to discern how environmental and conformational variations influence the chiroptical spectra.

The AC of gossypol, a polyphenolic binaphthyl dialdehyde, was suggested as (–)-(*M*) based on the EC/CO model interpretation of its ECD spectrum. A definite assignment of its AC was obtained later by comparing the experimental VCD spectrum in $CDCl_3$ with that predicted at the B3LYP/6-31G(d) level (Freedman et al. 2003b). Three different conformers with intramolecular hydrogen bonding and different rotamers of the isopropyl group were inferred to contribute to the experimental VCD spectrum in $CDCl_3$. The AC of D_2-symmetric dimer of binaphthyl was also determined as (+)-(*R*) using VCD spectroscopy (Freedman et al. 2003c).

Chiroptical properties of the axially chiral biphenol ethers in polar and nonpolar solvents were measured and compared with the corresponding theoretical properties to determine their conformational behavior in solution (Mori et al. 2007). Calculated rotational strengths at the BHLYP/TZV2P level revealed that 6 of 18 conformers are crucial to reproduce the experimental ECD spectra and optical rotations.

The AC of an axially chiral pyridine-N-oxide derivative, containing two aryl chromophores, trans-2,6-diortho-tolyl-3,4,5-trimethylpyridine-N-oxide, was determined using VCD and ECD spectroscopies (Teodorescu et al. 2015). The experimental VCD spectra measured for CCl_4 solution were analyzed with those predicted at several theoretical levels (CAM-B3LYP/TZVP, CAM-B3LYP/6-311++Gdp, B3LYP/cc-PVDZ, and B3LYP/6-311++Gdp). The online HPLC-ECD spectra were analyzed with those predicted at the B3LYP/TZVP, BH&HLYP/TZVP, and CAM-B3LYP/TZVP theoretical levels. All these studies concurred with the AC assignment as (–)-(a*R*,a*R*).

VCD associated with the O-H stretching vibrations of 2,2′-dihydroxy-1,1′-binaphthyl was analyzed with the dynamic polarization model (Nakao et al. 1994). The potential of VCD spectra to reflect the detailed structure was investigated (Setnička et al. 2001) by measuring the experimental VCD spectra in the 1700–1000 cm^{-1} region, and analyzing them with B3PW91/6-31G* level calculations, for 1,1′-binaphthyl-2,2′-diol, 1,1′-binaphthyl-2,2′-diamine, 1,1′-binaphthol bis (trifluoromethane sulfonate), 1,1′-binaphthyl-2,2′-diol-3-carboxylic acid, and 1,1′-binaphthyl-2,2′-diyl hydrogen phosphate. The comparison of experimental VCD, ECD, and ORD spectra of 6,6′-dibromo-1,1′-binaphthol with those predicted for this molecule with three different orientations of the -O-H group indicated that VCD spectra depend sensitively on the conformation of the -O-H group, but such sensitivity was not reflected in the ECD and ORD spectra. Thus the analysis of VCD spectra is more appropriate for determining the conformation of -O-H groups (Polavarapu et al. 2009b). The elucidation of the structure of 3,3′-diphenyl-[2,2′-binaphthalene]-1,1′-diol, known as VANOL, using chiroptical spectroscopy is discussed in a later section (Polavarapu et al. 2009c). From the measurement and interpretation of VCD spectra of 1,1′-binaphthyl-2,2′-diol in different solvents, the importance of the consideration of solute–solvent complexes in acetonitrile and dimethylsulfoxide solvents was highlighted (Nicu et al. 2012).

The analyses of ECD spectra of chiral biphenanthryl compounds using the EC/CO model were ambiguous (see Section 10.3.2). However, the analyses of ECD spectra using quantum chemical calculations provided definitive conclusions on the ACs of biphenanthryl compounds. For example, the experimental VCD, ECD, and ORD spectra of 2,2′-diphenyl-[3,3′-biphenanthrene]-4,4′-diol, known as VAPOL, were analyzed with corresponding quantum chemical predictions to confirm the (+)-(*aS*) configuration (Petrovic et al. 2008b).

The solid-state VCD spectrum for a KBR pellet of a rigid bisetherketone macrocycle containing binaphthyl moieties were analyzed (Cao et al. 2005) with quantum chemical calculations at B3LYP/6-31G(d) level. But the details on artifact control in recording the solid-state experimental VCD spectra were not provided.

The conformational changes in photochromic receptors, such as azo benzenes, have been probed by attaching chiral binaphthalene as a sensitive probe. SOR and ECD were used to monitor the atropisomerism of binaphthalene group that takes place during the photoisomerization process (Takaishi et al. 2009, 2012). Numerous articles under the theme of chiroptical switches can be consulted for detailed information on the use of SOR and ECD spectroscopies in characterizing the molecular structural transitions involved in the photoswitching process (Canary 2009; van Delden et al. 2004).

The analysis of experimental VCD spectra of (*P*)-2′-[(4*S*)-4,5-dihydro-4-(1-methylethyl)oxazol-2-yl][1,1′-biphenyl]-2-methanol in $CDCl_3$ with B3LYP/6-31G* calculations indicated that two different H-bonded conformers (one with O-H…N- and another with O-H…O- hydrogen bonding) are present in solution (Freedman et al. 2003a). Thus rapid equilibration of the orientation of the dihydrooxazole group in solution was identified through the VCD spectroscopy. Such conformational variation could not be identified through ECD spectra, and the crystal structure indicates the presence of only one H-bonded conformer. Three additional biphenyls, with the isopropyl group replaced with the phenyl group and introducing methyl groups at 6,6′ positions, were studied similarly to identify the conformers in solution (Freedman et al. 2005).

The AC of atropisomeric dimers of curcuphenol was established using VCD spectral studies (Cichewicz et al. 2005). Here, truncated dimers were used as model molecules for predicting the contribution from axial chirality.

The AC of (+)-9,9′-dihydroxy-1,1′,3,3′,4,4′-hexamethoxy-6,6′,7,7′-tetrahydro-2,2′-bianthracene-8,8′(5H,5′H)-dione was determined as (*aR*) (Romaine et al. 2011) by comparing the experimental ECD spectrum in methanol with that calculated at the B3LYP/6-31G* level. By comparing the ECD spectrum of this (+)-(*aR*) compound with those of HMP-Y6, the central biaryl core of hibarimicins, the axial chirality of HMP-Y6 atropisomers was determined. Assuming that the oxidation to hibarimicin retains the configuration, the axial chirality of hibarimicin atropisomers was assigned as (*aS*).

When atropisomers possess central chirality, in addition to the axial chirality, the ECD spectra are often be dominated by the axial chirality, and the central chirality cannot be discerned from the ECD spectra. In such cases, VCD spectroscopy can be helpful (Polavarapu et al. 2009a) because vibrations originating from the atoms involved in the central chirality may be able to provide distinct signals. This situation was found for cephalochromin, a homodimeric naphthpyranone natural product, where VCD spectral analysis led to the AC assignment of cephalochromin as $(+)_{589}-\left[(-\mathrm{VCD})_{1038\,\mathrm{cm}^{-1}}\right]-(aS,2R,2'R)$ and $(+)_{589}-\left[(+\mathrm{VCD})_{1038\,\mathrm{cm}^{-1}}\right]-(aS,2S,2'S)$, indicating that the $(+)_{589}$ enantiomer with a negative VCD band at 1038 cm^{-1} has a (*aS*,2*R*,2′*R*) configuration, while the $(+)_{589}$ enantiomer with a positive VCD band at 1038 cm^{-1} has an (*aS*,2*S*,2′*S*) configuration.

Flavomannin A also has axial and central chiralities. However, in this case only the axial chirality could be determined using ECD and VCD spectra studies, and the central chirality could not be ascertained (Bara et al. 2013).

For a triply axial chiral binaphthyl based molecule, concentration-dependent changes were inferred for two of the axial chiralities, based on ECD, and ORD, and VCD studies, resulting in a change in the predominant diastereomer upon concentration change (Dezhahang et al. 2014).

The means of detecting local chirality in the presence of axial chirality was investigated using VROA spectroscopy (Herrmann et al. 2006). Using two diastereomers of right-handed helical deca-alanine, the (all-*S*) and the (*R*,*S*,*R*,*S*,*R*,*S*,*R*,*S*,*R*,*S*) form, the authors arrived at a positive conclusion.

10.4.1.12 Supramolecular Aggregates

The presence of functional groups that facilitate hydrogen bonding is a key factor in aggregate formation. The monomer–dimer equilibrium, ubiquitous in carboxylic acids, is the simplest form of aggregate formation. A description of the use of absorption, CD, Raman, and VROA spectra, as well as that of SOR, for studying the monomer–dimer equilibrium is summarized in Chapter 9. In addition to carboxylic acids, self-complimentary molecules that can form hydrogen bonding networks and molecules that favor stacking interactions (such as stacked aromatic systems) will have a propensity for supramolecular aggregation in the appropriate solvent media (Brunsveld et al. 2002; Hembury et al. 2007; Kudo et al. 2009). Chiral molecules that contain both hydrophilic and hydrophobic groups behave as surfactants and can form aggregate structures, such as micelles, reverse micelles, and vesicles, in suitable solvents. Another consequence of spontaneous aggregation of small molecules is the formation of gels with supramolecular architecture. Chiroptical spectroscopy is best suited for studying the supramolecular aggregates that possess chirality. Applications of ECD in the study of supramolecular systems have been reviewed (Pescitelli et al. 2014).

10.4.1.12.1 Dimeric Systems

As mentioned earlier, dimer formation is the simplest form of aggregation. The monomer–dimer equilibrium for both homochiral (pure enantiomers) and heterochiral (mixtures of enantiomers) systems was investigated using optical rotation at a single wavelength and equilibrium constants determined for binapthol (Baciocchi et al. 2002) and omeprazole (Baciocchi et al. 2004). The monomer–dimer equilibrium present for (*R*)-(−)-pantolactone in CCl_4 solutions was studied using SOR. The observed SOR is interpreted as the sum of SORs of monomer and dimer, each multiplied with a respective squared mole fraction (Goldsmith et al. 2003). The DFT-calculated SOR for monomer and dimer was used to explain the experimentally observed concentration-dependent SORs. But the analysis reported in this work is incorrect (see Chapter 9). The experimental determination of wavelength-resolved SORs of monomeric and dimeric forms of pantolactone was reported recently (Polavarapu and Covington 2015). The quantum chemical predictions of SORs for monomer and dimeric forms of pantolactone matched these experimental observations within a factor of 2.

The experimental vibrational absorption (VA) and VCD spectra of α-aryloxypropanoic acids could not be reproduced by those calculated for monomeric acids, but were satisfactorily reproduced by those calculated for dimeric acids (He and Polavarapu 2005b). Therefore, at the concentrations normally used for VCD measurements, it is necessary to consider the acids in their aggregated form and carry out the theoretical calculations accordingly.

The monomer–dimer equilibrium present for (*R*)-(+)- and (*S*)-(−)-tetrahydrofuran-2-carboxylic acid in $CDCl_3$ and CS_2 solvents has been studied using VA and VCD spectroscopies. The concentration-dependent absorbance of a C=O stretching vibrational band was used to estimate the populations of monomer and dimer. This information, combined with the B3LYP/aug-cc-pVTZ predictions of VA and VCD spectra for different structures, was used to interpret the experimental observations (Kuppens et al. 2006).

The monomeric and dimeric forms of glycidol were investigated by recording the VA, VCD, and SOR of glycidol as a function of concentration. It was suggested that the experimental data obtained at 0.2 M could be reproduced by the corresponding predicted data for a monomeric form, while that at 3.5 M could be reproduced by the corresponding predicted data for a dimeric form (Yang and Xu 2008).

The VCD and ECD spectra of chiral β-hydroxyesters, with long α-hydrocarbon chain substituents, have been measured for film samples (Poopari et al. 2015). The bisignate experimental VCD features observed for C=O stretching vibrations were reproduced by the predictions for a dimer model and not by those for a monomer model. However, since two of the four possible diastereomers were predicted to have the bisignate VCD features, further discrimination among these diastereomers was realized using ECD spectral analysis. Here also, the bisignate experimental ECD bands were reproduced by the corresponding predictions for a dimer model and not by those for a monomer model.

10.4.1.12.2 2,2′-Dimethyl-biphenyl-6,6′-dicarboxylic acid (DBDA)

The crystal structure of (*R*)-DBDA showed the presence of intermolecularly H-bonded tetramers. The presence of this tetramer in solution was verified using VCD spectroscopy (Urbanová et al. 2005). The experimental VA/VCD spectra measured for (+)-DBDA in $CDCl_3$ were found to be concentration dependent, indicating oligomerization at increased concentrations. The experimental VCD spectrum in the 1760–1100 cm^{-1} region for (+)-DBDA were reproduced by that calculated for tetramer, $[(S)\text{-DBDA}]_4$, with one of the six possible conformations. This study established that VCD spectroscopy can be used to determine the structures of supramolecular species.

10.4.1.12.3 Tris-[(R)-3,7-dimethyloctyl]benzene-1,3,5-tricarboxamide (TDOBTC)

The self-assembly of (*R*)-TDOBTC in a heptane solution at concentrations of 10^{-4}–10^{-5} M was investigated using ECD spectroscopy and a right-handed helical structure was inferred. VCD spectral measurements in a decahydronaphthalene solvent at different concentrations, 1.8 mM, 6.3 mM, and 10.2 mM, revealed concentration dependence. Totally different VCD spectra obtained at a higher concentration were attributed to possible spectral artefacts resulting from the gel formation and orientation. Using VCD calculations on model systems containing dimer, trimer, pentamer, and heptamer molecules, it was determined that pentamer represents the smallest structure to properly represent the VCD associated with helical H-bonded amide stacks (Smulders et al. 2008).

10.4.1.12.4 (4S,7R)-Campho[2,3-c]pyrazole (CP)

In CCl_4 solution CP exists as a monomer, but trimers are present in a crystalline state. The VCD spectra in solution are best reproduced by the calculated spectra for a monomer, while those in Fluorolube/Nujol mulls the experimental VCD spectra were reproduced by the calculated spectra for a trimer (Quesada-Moreno et al. 2014).

10.4.1.12.5 Surfactants

Above critical micelle concentrations (CMCs), surfactants form aggregates such as micelles, vesicles, and bilayers. The CMCs are fairly low for surfactants with larger hydrophobic alkyl chains. VCD and VROA spectroscopies may not be feasible at such low concentrations. Therefore VCD and VROA spectroscopies may not be useful for monitoring the process of aggregate formation and may only be useful at concentrations that are much above

CMCs. On the other hand, ECD and ORD spectroscopies can be undertaken at lower concentrations, so these represent practical tools for studying the surfactants. However, to derive any useful information on surfactants from ECD spectroscopy, chromophoric electronic transitions that appear in the visible region and influenced by the aggregate formation (Shinitzky and Haimovitz 1993) are needed. These criteria are not met by a large number of surfactants. On the other hand, ORD in the visible spectral region reflects the cumulative effect from all electronic transitions that appear in the visible as well as UV regions. Thus ORD spectroscopy can provide information that is not transmitted through ECD spectroscopy and represents by far the most sensitive method for studying the surfactant aggregation process.

The following advantages associated with ORD spectroscopy for studying the surfactants have been recognized recently: (1) chiral guest molecules can be used as probes for studying the aggregation of achiral surfactants (Polavarapu and Vijay 2012); (2) the SOR of some surfactants can increase dramatically beyond their CMCs, but ECD and VCD spectroscopies remain silent (Vijay and Polavarapu 2012); and (3) the SOR of surfactants increases with the size of spherical micelles (Vijay et al. 2013), indicating that the size of micelles may be determined using SOR measurements.

Even though some limitations appeared in utilizing VCD and ECD spectroscopies for studying the surfactants, these have been successfully used for deriving useful information on micelles (Shinitzky and Haimovitz 1993), reverse micelles (Colombo et al. 1991; Abbate et al. 2012, 2014), and bilayers (Nakashima et al. 1994).

Surfactants are also referred to as amphiphiles. When two monomeric amphiphiles are linked through a spacer, the resulting amphiphile is referred to as a gemini amphiphile. The spacer can be hydrophobic (aliphatic/aromatic), hydrophilic (polyether), rigid, or flexible. Usually the spacer is near the polar head groups and sometimes at the middle of the hydrophobic tail. However, if the spacer connects the ends of hydrophobic tails, then the amphiphile becomes a bola amphiphile. These amphiphiles are also designated as symmetric or asymmetric amphiphiles depending on whether the two monomeric units are the same or different. The symmetric gemini amphiphiles are represented as *m-s-m* amphiphiles, where *m* and *s* represent the number of carbon atoms in the hydrophobic and linker chains. The asymmetric amphiphiles are represented as *m-s-m'*. Oda and coworkers have studied an achiral 16-2-16 amphiphile with a chiral tartrate counter ion and found that the resulting complex forms twisted ribbons (Brizard et al. 2009). The degree of twist varies with the enantiomeric excess of the tartrate counter ion, with the racemate giving flat ribbons, while the L- and D-tartrates give oppositely twisted ribbons. The chiral structure of the achiral amphiphile was considered to be induced by the chiral tartrate counter ion. The inversion of chirality was noted and monitored using ECD and VROA spectra for helix suspension of 16-2-16 tartrate, when the tartaric acid solution of opposite enantiomer was added (Tamoto et al. 2015).

The conformational changes of two homopolypeptides, poly-L-lysine (PLL) and poly-L-arginine (PLAG), upon interaction with micelles composed of sodium dodecyl sulfate (SDS) and large unilamellar vesicles composed of phospholipids were studied using ECD and VCD spectroscopies (Novotná and Urbanová 2012). SDS was considered to represent a monolayer membrane and phospholipids to represent a bilayer membrane. The secondary structure of PLL and PLAG was found to be strongly influenced by the interaction with charged model membranes.

10.4.1.12.6 Organogels

Applications of chiroptical spectroscopy to organogels have appeared. ECD spectroscopy was utilized to determine the supramolecular helical structure adopted by 12-hydroxyoctadecanoic acid in a gel state (Tachibana et al. 1979), and by alkylamides derived from trans-1,2-diaminocyclohexane (Hanabusa et al. 1996).

The sensitivity and applications of VCD spectra to the gel formation was investigated using a brucine-appended-porphyrin gelator (Setnička et al. 2002). VCD spectral studies and structure characterizations were reported for gels formed from 12-hydroxyoctadecanoic acid (Sato et al. 2008; Sato et al. 2011a), N,N′-alkanoyl-1,2-diaminocyclohexane (Sato et al. 2012; Sato et al. 2014a), and N,N′-diperfluoroheptanoyl-1,2-diaminocyclohexane (Sato et al. 2011b; Kohno et al. 2012; Sato et al. 2014b). The AC and helical structure of 1-perfluoro-octyl-1-phenylmethanol in $CDCl_3$ solution, a related perfluoro compound though not a gel, was also determined using VCD (Monde et al. 2006).

The ROA spectrum of gel formed from FMOC-Leu-Leu-OH [where FMOC stands for N-(fluorenyl-9-methoxycarbonyl)] was reported and VROA peaks associated with FMOC group were suggested to indicate the stacking nature of these groups (Smith et al. 2009). The helical organization of oligo-p-phenylene-based organogelators was investigated using ECD, VCD, and VROA spectra. A gelator in this category containing two phenyl groups in the core was found (Aparicio et al. 2014) to invert the helicity through atropisomerism of the biphenyl core, and this inversion depended on the conditions (concentration, temperature, and time).

10.4.1.12.7 Langmuir–Blodgett Films

VCD spectra were reported for Langmuir–Blodgett films formed from both enantiomers of 12-hydroxyoctadecanoic acid and compared to those obtained for the gel form. The observed spectra were interpreted using predicted spectra for H-bonded dimer models (Sato et al. 2011a).

10.4.1.12.8 Solute–Solvent Clusters and Induced Optical Activity

The presence of hydrogen bonding interaction between solute and surrounding solvent molecules can be viewed as the formation of solute–solvent clusters and hence as another form of aggregation. The number of solvent molecules needed to correctly represent the solvent environment depends on the nature of solute. The situation can vary from consideration of a simple "solute–solvent molecular complex" model (Zhang and Polavarapu 2007) to explicit and implicit solvation models (Poopari et al. 2012a, 2012b, 2013), and to a "solute-embedded-in-solvent-bath" model requiring complicated molecular dynamics simulations (Cheeseman et al. 2011). In these cases, novel effects such as the induction of optical activity from chiral solute to achiral solvent (Debie et al. 2008; Losada and Xu 2007; Losada et al. 2008a, 2008b; Mukhopadhyay et al. 2007; Tarczay et al. 2009), and perturbation of solute molecular (electronic as well as vibrational) properties by the solvent molecules (Heshmat et al. 2014; Nicu et al. 2012) have been considered.

Chirality induction is referred to as chirality transfer in the literature, but this terminology is incorrect because "transfer" requires a loss at the source and corresponding gain at the target. In the chirality-induction process, a loss at the source need not be established or proven.

In cases where optical activity is induced in an achiral solvent, the induced CD in vibrational transitions of achiral solvent molecules can be weak and difficult to predict accurately (Nicu et al. 2008) and care should be exercised in interpreting this induced VCD. A review article on this topic can be consulted for additional information (Sadlej et al. 2010). Induced CD in the vibrations of an NH_4^+ counter ion in the presence of a chiral tris(tetrachlorobenzenediolato)-phosphate(V) anion was also reported (Bas et al. 2005).

A recent review article on solvation effects can be consulted for applications pertaining to vibrational optical activity spectroscopy (Perera et al 2016).

10.4.1.12.9 Avoidance of Aggregate Formation

The presence of OH and COOH groups leads to intramolecular or intermolecular hydrogen bonding, and consequential aggregate effects at higher concentrations normally used for vibrational spectroscopic measurements. The theoretical calculations needed to properly represent such aggregates can be challenging and computationally prohibitive. That type of theoretical challenge can be avoided by converting the acids into the corresponding esters (Devlin et al. 2005c; He and Polavarapu 2005a; He et al. 2005) and OH groups into acetates (Devlin et al. 2005a). For these derivatives, not only are the hydrogen bonding effects avoided, but the number of conformations one has to deal with are also reduced because of steric hindrance from the bulkiness of the ester and acetate groups. For this reason, such derivatization was referred to as conformational rigidification (Devlin et al. 2005b). The experimental measurements for the ester and acetate derivatives greatly simplify the theoretical analyses when compared to those needed for parent molecules containing COOH and/or OH groups.

The above-mentioned chemical modifications can be avoided if the spectral measurements can be made for molecules isolated in low temperature inert (argon, krypton, xenon, etc.) matrices. It is well known that isolated molecules can be investigated using matrix isolation (MI) methods, as long as the sample molecule to matrix gas atoms ratio is kept low (typically 1:300–1000). At that dilute concentrations, intermolecular association is avoided and one needs to be concerned only with intramolecular association. The MI-VCD measurements (Henderson and Polavarapu 1986; Schlosser et al. 1982) have been undertaken to exploit this advantage for a handful of systems, namely 2-amino-1-propanol (Tarczay et al. 2006), acetyl-N-methyl-L-alanine amide (Pohl et al. 2007), N-acetyl-L-proline amide (Pohl et al. 2008), N-acetyl-N′-methyl-L-alanine amide (Tarczay et al. 2009), 2-[(2S)-1-acetylpyrrolidin-2-yl]-N-methyl acetamide (Gobi et al. 2010), 2-chloropropionic acid (Gobi et al. 2011), and 3-butyn-2-ol (Merten and Xu 2013).

10.4.1.13 Pharmaceutical and Antimicrobial Compounds

Since the determination of ACs is a critical step before chiral compounds can be brought into the pharmaceutical and medicinal market, chiroptical spectroscopic methods are becoming popular techniques in the pharmaceutical industry. Several pharmaceutical companies have added the chiroptical spectroscopic methods as a part of their "tool boxes" (McConnell et al. 2007).

The fluorinated ethers and gossypol discussed earlier also belong to the class of pharmaceuticals. In addition, there are several other pharmaceutical compounds that were investigated using chiroptical spectroscopy for the purpose of determining their ACs and predominant conformations in a solution state. In the early stages of applications of chiroptical spectroscopy, a fragment method (Bouř and Keiderling 2005; Dunmire et al. 2005; Jiang et al. 2011; Yamamoto et al. 2012) was used to save required computational time, where a molecule of interest was considered to be made up of fragments and investigations were undertaken on separate fragments. With the generous availability of required computational resources, such approaches are discouraged and calculations are undertaken on whole molecules (Sherer et al. 2014), wherever possible. More recently, a molecule-in-molecule method was developed (Jose and Raghavachari 2016; Jose et al. 2015) to accurately predict the VA, VCD, Raman, and VROA spectra for large-sized molecules (see Chapter 5, Section 5.2.7).

10.4.1.13.1 Mirtazapine

The AC of mirtazapine is known from its synthetic route as (−)-(*R*). The experimental VCD spectrum of 0.10 M (−)-mirtazapine in $CDCl_3$ was found to compare well with that calculated for (*R*)-mirtazapine, thereby confirming the AC as (−)-(*R*)-mirtazapine. Only one predominant confirmation was identified for this molecule (Freedman et al. 2002).

10.4.1.13.2 Buagafuran

This (+)-isomer is stated to have antianxiety activity and is under phase II clinical trial in China. The AC of this compound is known from its synthetic route. Using a combination of experimental and theoretical (B3LYP functional and 6-31G(d), 6-31 + G(d,p) and aug-cc-pvdz basis sets) CD studies, the AC of buagafuran has been confirmed (Li et al. 2013) as (1*R*,6*S*,9*R*).

10.4.1.13.3 Benzodiazepine

The vasopressin receptor antagonists contain benzodiazepine as a key structural component (Dyatkin et al. 2002). When X-ray methods failed, the AC of benzodiazepine was determined using VCD spectroscopy. Experimental VCD spectra measured for $CDCl_3$ solutions in the 1700–800 cm^{-1} region were compared with calculated spectra at the B3LYP/6-31G* level for three different conformers to establish the AC as (+)-(*S*)-benzodiazepine.

10.4.1.13.4 McN-5652-X

(+)-McN-5652-X is known to have the potential for inhibiting serotonin reuptake in mammalian brain (Maryanoff et al. 2003). The enantiospecific synthesis of (+)-McN-5652-X leads to its AC assignment as (6*S*,10b*R*). The crystallographic results for (+)-McN-5652-X-$HClO_4$ provided trans relative configuration at carbon-6 and carbon-10b and indicated the AC to be (6*S*,10b*R*). The single crystals of (+)-McN-5652-X-(+)-(2*R*,3*R*)-tartaric acid also provided the same conclusions. The results from VCD analysis of (+)-McN-5652-X in $CDCl_3$ were found to be consistent with the AC determined from the other studies.

10.4.1.13.5 Calcium Channel Blockers

8-(4-Bromophenyl)-8-methoxy-5-methyl-8H-[1,4]thiazino-[3,4-c][1,2,4]oxadiazol-3-one was found to be most active myocardial calcium channel modulator (Carosati et al. 2006; Stephens et al. 2007). The AC of this molecule was determined from VCD, ECD, and ORD spectral analysis to be (*S*)-(+). Three different conformers, arising from the rotation of ethoxy moiety, were found to be dominant. The AC of 1-[(4-chlorophenyl)sulfonyl]-2-(2-thienyl)pyrrolidine, a cardiovascular L-type calcium channel blocker, was determined (Carosati et al. 2009) using experimental and B3PW91/TZ2p predicted VCD spectra to be (*R*)-(+)/(*S*)-(–).

10.4.1.13.6 Antiarrhythmic and Anti-Inflammatory Drugs

The VCD spectra of quinidine, flecainide, RAC 109, propranolol, ibuprofen, and naproxen in the O–H, N–H, and C–H stretching region were measured and analyzed with calculations on fragments of these drugs (Freedman et al. 1999). The dominant conformations were deduced from these analyses, and insight into stereo-specific binding of the antiarrhythmic drugs was obtained.

10.4.1.13.7 Lodine/Etodolac

The VA and VCD spectra of (+) and (–)-enantiomers of etodolac were measured in DMSO-d_6 and compared with those calculated for (*R*)-enantiomer (McConnell et al. 2007). The calculated spectrum for (*R*)-enantiomer closely resembled that measured for (–)-enantiomer, and differed from that for

(+)-enantiomer, confirming the previously known AC of etodolac as (*R*)-(–). In their article McConnell et al. (2007) also described how their parent company was able to develop a drug discovery toolbox for chiral molecular characterization.

10.4.1.13.8 Histamine H3 Receptor Antagonist, GT-2331

The X-ray diffraction studies provided conflicting data leaving questions on the AC of this drug substance that was used to regulate histamine (Minick et al. 2007). Using the experimental VCD spectra of (+)-enantiomer and calculations for reduced structure models, the AC was assigned as (+)-4-[(1*S*,2*S*)-2-(5,5-dimethyl-1-hexyn-1-yl)cyclopropyl]-1H-imidazole.

10.4.1.13.9 Anthracycline Antibiotics

Daunorubicin and doxorubicin, natural anthracyclin antibiotics, have effective anticancer activity, but have restricted usage due to cardiac toxicity and quickly developing drug resistance (Yang et al. 2010). Therefore anthracyclin analogues have been investigated for potential substitutes. Three different synthetic anthracycline analogues have been prepared and their ACs determined using VCD spectroscopy. The intramolecular H-bonded structures have been investigated and dominant conformations identified.

10.4.1.13.10 Ephedra Molecules

For six molecules in this class, namely, (1*S*,2*R*)-norephedrine, (1*S*,2*S*)-norpseudoephedrine, (1*S*,2*R*)-ephedrine, (1*S*,2*S*)-pseudoephedrine, (1*S*,2*R*)-N-methylephedrine, and (1*S*,2*S*)-N-methylpseudoephedrine, the experimental VCD spectra in the OH- and NH-stretching regions were analyzed with calculated spectra, at a rather low level of theory, and predominant conformations identified (Freedman et al. 1994). The VROA spectra of four of these compounds as hydrochloride salts in aqueous solutions were analyzed (Yu et al. 1993) and the marker bands for the configurations at C-1 and C-4 atoms were identified.

10.4.1.13.11 Otamixaban

The AC of otamixaban was known from a co-crystal structure, but to verify the AC of the compound derived from a different synthetic route, VCD spectral analysis was undertaken (Shen et al. 2014). In combination with the single crystal X-ray diffraction results for an intermediate and the proton NMR results, the AC of otamixaban was determined to be (*R*,*R*).

10.4.1.13.12 Thalidomide

The VA and VCD spectra of (*R*)-thalidomide when measured in DMSO-d_6 are found to be different from those in $CDCl_3$ (Izumi et al. 2006). The spectra observed in DMSO-d_6 were reproduced by the predicted spectra for (*R*)-thalidomide monomer, while those observed in $CDCl_3$ were reproduced by those predicted for thalidomide in a dimeric form, both calculations carried out at the B3LYP/6-31G(d) level. Two different conformers of a monomer and three different conformers of a dimer were used for the predicted spectra.

10.4.1.13.13 2-Pyrazolines

Some pharmaceutical compounds are chiral pyrazolines. The AC of 1,3,5-triphenyl-2-pyrazoline has been determined using chiroptical spectroscopy (Vanthuyne et al. 2011). The experimental VCD spectra of (+)-enantiomer in CD_2Cl_2 were reproduced by B3LYP/TZVP predicted VCD spectra for (*R*)-enantiomer. Also the experimental ECD spectra of (+)-enantiomer in hexane were reproduced by CAM-B3LYP/6-31++G(d,p) predicted ECD spectra for the (*R*)-enantiomer. Thus, using VCD and ECD spectroscopies, the AC of 1,3,5-triphenyl-2-pyrazoline was determined as (*R*)-(+).

10.4.1.13.14 Paclitaxel

The VCD and IR spectra of paclitaxel (referred to as taxol) and baccatin III, both in $CDCl_3$ solvent, were found to be similar. The predicted VCD spectra (calculated as population weighted average of minimum energy conformer VCD spectra) for the baccatin III ring and the experimental spectra of paclitaxel in solution were concluded to be similar (Izumi et al. 2008). The conformation of the baccatin III ring in a crystal structure was different from that deduced for solution state. This difference was attributed to a conformational change through the binding with β-tublin and the intermolecular interactions involving the hydroxyl and carbonyl of acetoxy groups. A new conformational code, augmenting IUPAC recommended nomenclature (see Chapter 2), was introduced to describe the conformations.

10.4.1.13.15 Ibuprofen

Interpretation of the experimental VCD spectra of (*S*)-ibuprofen using conformational analysis and predicted VCD spectra (Izumi et al. 2009a) for the low-energy conformers indicated that four conformers are dominant for both monomeric and dimeric forms. The monomer form of (*S*)-ibuprofen was suggested to exist in DMSO-d_6, and the dimer form in $CDCl_3$ solution. The dimer form of (*S*)-ibuprofen has the "U"-shape. The crystal structure of racemic ibuprofen was identified as an N-shaped structure. The conformational code recommended by IUPAC has been augmented with new definitions to accommodate additional conformations.

10.4.1.13.16 Pexiganan

Pexiganan, a 22-residue peptide with antimicrobial properties, is present in the skin of the African clawed frog. The structure of pexiganan in varying solvent environments has been investigated using ECD and VCD spectroscopies (Shanmugam et al. 2005a). The dominant structural component is α-helix in trifluoroethanol, a mixture of β-sheet and β-turn in methanol, β-turn in DMSO-d_6, and unordered in D_2O.

10.4.1.13.17 VP1 Peptide

This 11-residue peptide is in the domain V of m-calpain enzyme with antimicrobial activity. ECD and VCD spectral studies indicated that this peptide undergoes unusual structural transition forming an antiparallel β-sheet structure leading to aggregation in trifluoroethanol (Shanmugam et al. 2011). The TEM images of aged samples confirmed the presence of aggregates with fibril-like cross-assemblages.

10.4.1.13.18 Aeroplysinin-1

Extracted from the sponge *Aplysina cavernicola*, Aeroplysinin-1 is a candidate for the treatment of antiangiogenic diseases. The AC, and conformations in aqueous solution, of this compound were explored using VCD and VROA spectroscopies (Nieto-Ortega et al. 2011).

10.4.1.13.19 Hypogeamicin B

A cave-derived actinomycete, with a pyronaphthoquinone frame, hypogeamicin B has antimicrobial property against *Bacillus subtilis*. The absolute stereochemistry of hypogeamicin B was determined from combined experimental ORD, ECD, and VCD spectra by comparing them with corresponding quantum chemical calculations, after an extensive conformational search and including solute–solvent polarization effects (Derewacz et al. 2014).

10.4.1.13.20 γ-Aminobutyric acid (GABA) Modulators

The GABA modulator 9-Amino-2-cyclopropyl-5-(-2-fluoro-6-methoxyphenyl)-2,3-dihydro-pyrrolo[3, 4-b]quinolin-1-one has potential for use as an anxiolytic drug. This compound exists as atropisomers (atropisomer A and B) with chirality arising from restricted rotation around the C–C bond connecting the two phenyl groups. It is necessary to know the chirality of isolated atropisomers A and B for getting regulatory approvals for marketing them as drug candidates. Additionally, it is necessary to know the interconversion rate of the atropisomers for assessing their stability as drug candidates. The VCD spectral investigations at AstraZeneca Pharmaceuticals (Pivonka and Wesolowski 2013) indicated that the experimental VCD spectra of atropisomer A are in agreement with those calculated for the (*P*) enantiomer, and those of atropisomer B are in agreement with those calculated for the (*M*) enantiomer. In both cases, the calculations were undertaken using the B3PW91 functional and 6-311G** basis set. From the time-dependent VCD spectral measurements, the half-life for racemization was determined to be 31 days. The (*P*) enantiomer was stated to be 60 times more potent over the (*M*) enantiomer as anxiolytic drug.

10.4.1.14 Transport Antibiotics

Transport antibiotics are compounds that transport ions either through forming channels or through carrying ions across the membrane. Gramicidin D belongs to the former category, while valinomycin to the latter category.

Melittin forms voltage-gated channels in planar membranes, integrates into membranes causing cell lysis, but does not transport ions.

10.4.1.14.1 Gramicidin D

A hydrophobic 15-resdiue linear peptide, gramicidin D is an antibiotic, which when incorporated into phospholipid membranes serves as an ion channel for the transport of monovalent cations. ECD (Doyle and Wallace 1998) and VCD (Zhao and Polavarapu 2002; Zhao and Polavarapu 1999) spectroscopies have been used to infer the various conformations that gramicidin D adopts under different solvent environments as well as ionic environments. VCD spectra in the 1800–1300 cm^{-1} region of gramicidin D were correlated with its structures as follows: in the presence of Ca^{2+} ions, a parallel left-handed double helix structure is present in methanol-d_4, a mixture of parallel and antiparallel left-handed double helices in 1-propanol and high α and low β structures in $CHCl_3$-methanol mixture. In the presence of Cs^+ ions, a right-handed antiparallel double helix structure is present in methanol-d_4. In the presence of Li^+ ions, an unordered structure is present in methanol-d_4 and in 1-propanol. In ion-free methanol-d_4 solution, a mixture of different conformations are thought to exist. VCD spectra of gramicidin D in model membranes (SDS micelles, DODAC, and DMPC vesicles) are different from those in organic solvents and the presence of ions does not influence the membrane-bound conformation of gramicidin D (Zhao and Polavarapu 2001b).

10.4.1.14.2 Valinomycin

Chiroptical spectroscopies have been used to determine the conformations of valinomycin under different solvent environments and to determine the influence of binding to ions.

The ECD spectrum of valinomycin in *n*-hexane showed a single positive band at 218 nm and a negative ECD band in trifluoroethanol solvent (Grell and Funck 1973). The changes in ECD intensity as a function of solvent composition were characterized by a two-step conformational equilibrium. The ECD spectra of valinomycin complexed to K^+, Rb^+, and Cs^+ ions showed a positive ECD band at 210 nm, and a negative ECD band when complexed to Na^+ ions. Thus the ligand conformation was considered to depend on the bound cation.

The experimental VCD spectra obtained for the amide and ester carbonyl stretching vibrations in nonpolar solvents were reproduced satisfactorily by the corresponding quantum chemical calculations for bracelet-type conformation (Wang et al. 2004b). The VCD spectra of valinomycin in the presence of Li^+ and Na^+ ions are not different from those of free valinomycin in methanol-d_4 solution, but the presence of K^+ ions significantly influenced the spectra.

ROA spectroscopy has also been used (Yamamoto et al. 2010) to study the influence of K^+ ions on the conformation of valinomycin, and significant influences of complexation were evident. VROA spectroscopy was noted to be sensitive to the peptide side chain conformations that cannot be noticed through other chiroptical spectroscopic methods.

10.4.1.14.3 Melittin

A major component of honeybee venom, with antimicrobial, as well as therapeutic, activity, melittin is a 26-residue peptide that adopts different structures under different conditions. ECD and VCD spectroscopies have been used to investigate its conformational diversity.

ECD studies indicated that in aqueous solution at low pH melittin is unstructured. Addition of hexafluoroacetone hydrate at pH ≈ 2.0 induces a structural transition to a predominantly helical conformation. A similar structural transition is also observed in 2,2,2 trifluoroethanol (Bhattacharjya et al. 1999).

VCD spectra as a function of pH, KCl, and trifluoroethanol concentration indicated that melittin adopts a mixed structure at low pH, KCl, and TFE concentrations and α-helix structure at higher concentrations, but eventually aggregates at even higher concentrations (Wang and Polavarapu 2003).

10.4.1.15 Pesticides and Herbicides

10.4.1.15.1 Malathion

The dominant conformations of this pesticide were suggested (Izumi et al. 2009b) based on the comparison of experimental VCD spectra of (+)-malathion in CCl_4 solvent with those predicted for fragments of (*R*)-malathion structure.

10.4.1.15.1.2 Mecoprop and Dichloroprop

2-(4-Chloro-2-methylphenoxy) propanoic acid and 2-(2,4-dichlorophenoxy) propanoic acid are marketed as herbicides under the brand names of Mecoprop and Dichloroprop, respectively. The ACs of the enantiomers of these chiral herbicides were determined (He et al. 2005) from VCD spectral investigations on their methyl esters, as (+)-(*R*).

10.4.1.16 Thiophenes

The AC of (+)-3-(2-methylbutyl) thiophene was assigned as (*S*) by chemical correlation. The predicted VCD spectrum for the (*S*) configuration was found to match well with the experimental VCD spectrum of (+)-enantiomer, thereby confirming the AC derived from chemical correlation. Six different conformers were found to contribute to the spectra measured in $CDCl_3$ solvent. The AC of (+)-3,4-di(2-methylbutyl) thiophene was also determined as (*S*) in the same manner using VCD spectral analysis (Wang et al. 2002a).

The AC of heptathiophene was determined as (–)-(*R*), based on an excellent agreement between the experimental VCD spectrum of (–)-enantiomer in $CDCl_3$ solvent and calculated VCD spectrum for (*R*)-enantiomer at the B3LYP/6-31G(d) level (Freedman et al. 2003c).

10.4.1.17 Chiral Sulphur Compounds

The ACs of some sulfoxides were not known, while for some others the ACs were known from their synthetic routes, crystal structures of their derivatives, or from spectral correlations. The approximate methods of predicting ECD and ORD (Donnoli et al. 2003; Mislow et al. 1965; Rosini et al. 2001) were used in the early stages to establish the ACs of some sulfoxides. The purpose of modern chiroptical spectroscopic studies on these compounds is (1) to verify the suitability of these techniques for AC determination, (2) to determine/confirm the ACs independently, and/or (3) to determine the conformational preferences for the molecules involved.

10.4.1.17.1 t-Butyl Methyl Sulfoxide (TBMS)

The small size and conformational rigidity of this molecule was the attractive feature to test the quantum theoretical methods of VCD in confirming the known AC. The predicted VCD spectra for (*S*)-TBMS, using B3LYP and B3PW91 functionals and TZ2P and 6-31G*basis sets, reproduced the experimental VCD spectra of (+)-TBMS, thereby confirming the (+)-(*S*) configuration (Aamouche et al. 2000).

10.4.1.17.2 n-Butyl t-Butyl Sulfoxide (NBTBS)

The chemical correlation and crystal structure of complex with mercury chloride indicated the AC to be (+)-(*R*) or (–)-(*S*). The comparison of experimental and predicted VCD spectra also indicated the AC to be (+)-(*R*). Of the 27 possible conformations for this molecule, 10 conformers with significant populations were investigated and three of these conformers accounted for ~64% population (Drabowicz et al. 2001).

10.4.1.17.3 1-(2-Methylnaphthyl) Methyl Sulfoxide (MNMS)

Potential energy scan, as a function of O-S-C_{ring}-C_{ring}, at the B3PW91/6-31G* level indicated the presence of two energy minima corresponding to E and Z conformers. The energy difference between these two conformers is only ~0.5 Kcal/mol, with the Z conformer lower in energy, indicating the presence of a significant population of both conformers at room temperature. The population-weighted predicted VCD spectrum, at the B3PW91/TZ2P level, for (*S*)-MNMS reproduced the experimental VCD spectrum of (–)-MNMS in CCl_4 solvent. The level of agreement seen between these two spectra led to a confident assignment for the AC of MNMS as (–)-(*S*) (Stephens et al. 2001).

10.4.1.17.4 1-Thiochroman S-Oxide (TCSO)

The observation that the ECD spectrum of (–)-TCSO is similar to that of (–)-(*R*)-2,3-dihydrobenzo(b)thiophene-1-oxide led to the literature AC assignment of (–)-(*R*)-TCSO. However, as modern methods of analyses indicate, such correlations are not reliable because of the differences in conformational preferences of apparently similar compounds. To verify if the empirical AC assignment made in the literature is in fact correct, the experimental VCD spectrum of (+)-TCSO in CCl_4 solvent was compared to the quantum chemical VCD predictions. Three different conformers were found for this molecule (Devlin et al. 2002a). It was found that the experimental VCD spectrum of (+)-TCSO is satisfactorily reproduced by the predicted VCD spectrum for (*S*)-TCSO. Thus (+)-(*S*)-TCSO, or equivalently, (–)-(*R*)-TCSO has been determined to be the correct assignment (Devlin et al. 2001). The fact that the abovementioned empirical correlation suggested the same AC is fortuitous, because the ECD of (–)-TCSO is also similar to that of a different similar compound, (–)-(*S*)-phenylethylsulfoxide, a correlation that would suggest an opposite AC.

10.4.1.17.5 1-Thiochromanone S-Oxide (TCNSO)

The issues and goals associated with this study are similar to those for TCSO. Two different conformers were found for this molecule. From a comparison of experimental VCD of (+)-TCNSO in CCl_4 solvent with that predicted for (*S*)-TCNSO at the B3LYP/TZ2P level, as a population-weighted average of two conformers, the AC was determined to be (+)-(*S*)-TCNSO (Devlin et al. 2002b). The AC suggested from an empirical correlation of its ECD with related sulfoxides happened to be the same, but such empirical assignments depend on the sulfoxide chosen as reference.

10.4.1.17.6 t-Butane Sulfinamide (TBS)

The AC of TBS was known to be (+)-(*R*) from the synthetic route used in its preparation. Using three different chiroptical spectroscopies (ECD, ORD, and VCD) simultaneously (Petrovic and Polavarapu 2007), the AC that was known previously from the synthetic route was confirmed. Two stable staggered conformers were found, although only one of them has a large enough population to contribute to the spectra. TBS was also investigated using VROA spectroscopy (Qiu et al. 2012). The oppositely signed VROA bands associated with C-S stretching mode in (*R*)-TBS and (*R*)-t-butyl-t-butanethiosulfinate were suggested to reflect the inversion of configuration in these molecules.

10.4.1.17.7 N-α-phenylethyl-t-butylsulfinamide (PETBS)

Two diasteromers of PETBS, one with positive SOR and the other with negative SOR, were synthesized from the reaction of racemic t-butane sulfinyl chloride and (–)-(*S*)-α-phenylethylamine. The ACs of these diastereomers were determined (Petrovic et al. 2008a) from simultaneous investigations of theoretical and experimental ECD, ORD, and VCD spectra, despite the challenge to consider eight diastereomers, each with 81 possible conformers. The predicted spectra at B3LYP/6-31G* and B3LYP/aug-cc-pVDZ levels indicated that the experimental spectra for (+)-diastereomer of PETBS matched with those predicted for (*S,S,S*)-PETBS, and those for (–)-diastereomer of PETBS matched with those predicted for (*S,R,R*)-PETBS (the three-letter designation, in the order listed, identifies the ACs at C, N, and S centers, respectively). The

(–)-(*S*,*R*,*R*) assignment was independently confirmed with the X-ray structure of (–)-diastereomer of PETBS.

10.4.1.17.8 Brassicanal C

The AC of this cruciferous phytoalexin, a sulfinic natural product, isolated from cabbage or cauliflower in a racemic form, was not known (Taniguchi et al. 2008). To determine its AC, the (–)-enantiomer was subjected to experimental SOR, ECD, and VCD investigations and analyzed with corresponding quantum chemical predictions for the (*S*)-enantiomer. Four dominant conformations were considered. It was found that, in all three methods, the experimental observations for the (–)-enantiomer matched those predicted for the (*S*)-configuration and therefore the AC was assigned as (–)-(*S*). To gain confidence in this assignment, the authors had undertaken the SOR, ECD, and VCD investigations beforehand to verify the known AC of (+)-(*R*)-methyl p-toluenesulfinate.

10.4.1.17.9 Rubroflavin

The naturally occurring rubroflavin in the dried fruit bodies of *Calvatia rubro-flava* has a large negative SOR and complex ECD spectrum (Fugmann et al. 2001) in methanol. The AC of rubroflavin was determined to be (*S*) from a comparison of the experimental ECD spectrum in methanol with that calculated at the BP86/TZVP level. Thirty possible conformer structures were investigated and the ECD spectrum of the lowest energy quinoid conformer with (*S*) configuration was found to match better with the experimental spectrum of (–)-rubroflavin when the geometry-optimized solvent influence was used (Wang et al. 2003).

10.4.1.17.10 S-Oxides

The ACs of three S-oxides, namely 2H-naphtho[1,8-*bc*]thiophene 1-oxide, naphtho[1,8-*cd*]-1,2-dithiole 1-oxide, and 9-phenanthryl methyl sulfoxide, were successfully determined by comparing the experimental SORs with corresponding values predicted with quantum chemical calculations (Stephens et al. 2002). VCD spectroscopy was also used in an analogous manner to determine the AC of naphtho[1,8-*cd*]-1,2-dithiole 1-oxide (Holmén et al. 2003).

10.4.1.18 Chiral Phosphorous Compounds

The ACs of some of the phosphine oxides were known from their synthetic routes or chemical correlations, which depend on the assumption of the steric course of a given reaction. The ACs of phosphine oxides, whose enantiomers were separated by chromatography, were not known. The purpose of modern chiroptical spectroscopic studies on these compounds was (1) to verify the known ACs, (2) to determine the AC when not known, and/or (3) to determine the tautomeric structures and their conformational preferences for the molecules involved.

10.4.1.18.1 t-Butylphenylphosphineoxide (TBPPO)

The AC of TBPPO was assigned as (–)-(*S*) from chemical correlation. The experimental VCD spectra measured in $CDCl_3$ and $CHCl_3$ solvents in the 2000–900 cm^{-1} region were analyzed (Wang et al. 2000) with those predicted for pentavalent and trivalent structures of TBPPO. The pentavalent form has a much lower energy than the trivalent form. Only one tautomeric structure and one conformation were determined to be prevalent and the AC to be (–)-(*S*), thereby confirming the AC established by chemical correlation.

10.4.1.18.2 t-Butylphenylphosphinothioic Acid (TBPPTA)

The AC of TBPPTA was assigned as (+)-(*R*) from chemical correlation. This molecule can exist in two tautomeric structures, both pentavalent but differ in the presence of the O-H versus the S-H group. The comparison of experimental and predicted VCD spectra in the 2000–900 cm^{-1} region in CCl_4 solutions indicated that the AC is (–)-(*S*)-TBPPTA (Wang et al. 2001), confirming the conclusion deduced from chemical correlation. The tautomeric structure containing the O-H bond is more stable and only one confirmation is predominant.

10.4.1.18.3 t-Butyl-1-(2-methylnaphthyl)phosphine Oxide (TBMNPO)

The enantiomers of TBMNPO were separated using a HPLC column, so their ACs were not known, despite the availability of ECD spectra. Using the experimental VCD spectra measured in CH_2Cl_2 and CD_2Cl_2 solvents in the 2000–900 cm^{-1} region, and their analyses with quantum theoretical predictions, the AC was determined to be (+)-(*S*)-TBMNPO (Wang et al. 2002c). This assignment was consistent with the predicted SOR and chemical shifts observed in the ^1H NMR spectra (in the presence of a chiral solvating agent). Two conformers with populations in the ratio of ~2:1 were found for this molecule.

10.4.1.18.4 t-Butylphenylphosphinoselenoicacid (TBPPSA)

The experimental VCD spectra were measured in the 2000–900 cm^{-1} region in $CDCl_3$ solvent, and their comparisons to the spectra predicted for monomer and dimeric structures indicated the AC to be (–)-(*S*)-TBPPSA, and that TBPPSA exists in a monomeric form in $CDCl_3$ solutions and in one predominant conformation. The crystal structure, however, revealed the presence of intermolecularly H-bonded dimers. Thus one cannot assume the crystal structure to be retained in the solution state (Wang et al. 2004a).

10.4.1.18.5 Tert-butyl(dimethylamino)phenylphosphine-borane Complex (TBDAPPB)

Amino phosphines with a chiral P atom are not easily accessible. However, the stable enantiomers of tert-butyl(dimethylamino)phenylphosphine protected by borane could be resolved and their ACs determined using VCD spectroscopy (Naubron et al. 2006). The experimental VA and VCD spectra of both enantiomers of TBDAPPB were measured in CD_2Cl_2 and compared to those predicted at the B3LYP/6-31+G(d) level. It was found that three different conformers of TBDAPPB are in equilibrium and the population-weighted spectra obtained for (*S*)-TBDAPPB matched those measured for (+)-TBDAPPB. Thus the AC was assigned as (+)-(*S*)-TBDAPPB. The X-ray structure of enantiomerically pure TBDAPPB was found to be very close to the most stable conformer determined from B3LYP/6-31+G(d) calculations.

10.4.1.18.6 Spirophosphoranes

The structural characterization of phosphorous compounds is often done in the solid state using X-ray crystallography (Yang et al. 2010b). However, biologically relevant compounds often require structural information in the solution phase. The structures of phosphorous compounds with two simple amino acid residues, L(or D)-valine, L(or D)-leucine, were investigated in a solution phase using chiroptical spectroscopy. These compounds have three chiral centers, one at P and the other two at C atoms in the amino acid ligands. This study has revealed that the experimental observations, and quantum chemical predictions, of VCD spectra provide powerful and reliable approaches to determining the ACs and predominant conformers of phosphorous coordination complexes in solution.

10.4.1.19 Chiral Selenium Compounds

Hypervalent selenium compounds containing selenium as the central atom exhibit trigonal bipyramidal geometry and display chirality due to molecular dissymmetry. The AC of 3,3,3′,3′-tetramethyl-1,1′-spirobi[3H,2,1]benzoxaselenole was determined (Petrovic et al. 2005b) using experimental VCD, ECD, and ORD measurements, and their analyses with corresponding quantum chemical calculations as (+)-(*R*) and (−)-(*S*). All three methods independently led to the same conclusion.

10.4.1.20 Chiral Pheromones

10.4.1.20.1 Frontalin (1,5-dimethyl-6,8-dioxabicylco[3.2.1]octane)

1,5-Dimethyl-6,8-dioxabicylco[3.2.1]octane has been identified as a pheromone in the forntalis family. Its AC has been established through the synthesis of enantiomers as (+)-(1*R*,5*S*) and (−)-(1*S*,5*R*). The experimental VCD spectra of (+)-frontalin was reproduced by the predicted spectrum for (1*R*,5*S*)-frontalin using a B3LYP functional and 6-31G(d) basis set. Only one conformation, with the six-membered ring in a chair conformation and the seven-membered ring in a boat conformation, is predominant for this molecule (Ashvar et al. 1998).

10.4.1.20.2 1-Acetoxymethyl-2,3,4,4-tetra-methylcyclopentane

The experimental VCD spectra were measured for 0.90 M solutions of both enantiomers in $CDCl_3$ in the 1550–1300 cm^{-1} region. Conformational analysis, at the B3PW91/TZ2P level, identified the presence of 15 low-energy conformations and five of these conformations have populations greater than 10%. Comparison of the population-weighted VCD spectrum with the experimental VCD spectrum indicated that (+)-1-acetoxymethyl-2,3,4,4-tetra-methylcyclopentane has the (1*S*,2*S*,3*R*) configuration (Figadere et al. 2008). The (–)-(1*R*,2*R*,3*S*)-1-acetoxymethyl-2,3,4,4-tetra-methylcyclopentane is the female-produced sex pheromone of the obscure mealybug.

10.4.1.21 Fungal/Plant Metabolites

The compounds derived from plants and fungi can also be characterized as natural products (see Section 10.4.1.24). Nevertheless it might be useful to describe a few studies pertaining to the compounds derived from plants and fungi in a separate section. More studies on related compounds can be found in the review articles cited in Section 10.4.1.24.

10.4.1.21.1 Fungal and Plant Metabolites

Using the experimental measurements of ORD, ECD, and VCD spectra and their analyses using corresponding quantum chemical calculations, the ACs were determined for phyllostin, scytolide, and oxysporone as, respectively, as (–)-(3*S*,4a*R*,8*S*,8a*R*), (–)-(4*aR*,8*S*,8*aR*), and (+)-(4*S*,5R,6*R*) (Mazzeo et al. 2013). The AC of scytolide was derived from ECD and VCD spectral analyses, while the ORD was found to be not useful.

10.4.1.21.2 Cruciferous Phytoalexin-Related Metabolites

Using the experimental measurements of ECD and VCD spectra and their analyses with corresponding quantum chemical calculations, the ACs were determined for dioxibrassinin as (–)-(*S*) and for 3-cyanomethyl-3-hydroxyindole also as (–)-(*S*) (Monde et al. 2003).

10.4.1.21.3 Brevianamide B

The ECD and VCD spectroscopies were used to determine the AC of brevianamide B and to investigate if both monomeric and dimeric species are present. It was concluded that VCD is more reliable than ECD, as ECD predictions were found to be functionally dependent. Experimental and predicted VCD spectra in the limited carbonyl stretching region permitted the assignment of AC of brevianamide B and to confirm the absence of dimeric species (Bultinck et al. 2015).

10.4.1.21.4 Phytotoxins

The ACs of two phytotoxins, inuloxin A and seiricardine A, were investigated using ECD, ORD, and VCD spectroscopies (Santoro et al. 2015). Using the known relative configurations of these molecules, conformational analyses were conducted and the population-weighted spectra were compared to the corresponding experimental spectra. While the ECD and ORD spectra provided confidence in the assignment of (7*S*,8*S*,10*R*) configuration to

(-)-inuloxin A, the predicted VCD spectrum did not reveal a good correlation to the corresponding experimental data. Due to the absence of chromophores absorbing in the UV–Vis region, seiricardine A did not possess a usable experimental ECD spectrum. Also, the ORD and VCD spectra of seiricardine A were very weak. The ECD and ORD spectra of 2-O-p-bromobenzoate ester of seiricardine A and the corresponding predicted spectra provided the confidence for its AC assignment. But here also the predicted VCD spectrum did not reveal a good correlation to the corresponding experimental data.

10.4.1.22 Carbohydrates

Numerous SOR and ECD studies have been reported for carbohydrates in the literature. SOR values have been used in the past mostly for structural correlations (Shallenberger 1982). However, recent advances in quantum chemical calculations permitted the SOR calculations on glucose (da Silva et al. 2004). Eight different conformers of D-glucose, three of them being α-anomers and others β-anomers, were found from geometry optimizations at the B3LYP/6-311G++(2d,2p) level in a water solvent modeled with a polarizable continuum model (PCM). The population-weighted sum of the SORs of these conformers yielded a value that is close to the experimental SOR of D-glucose in water. This study clearly shows that reliable quantum chemical calculations of chiroptical properties of carbohydrates are feasible with modern-day computational resources.

As electronic transitions of most carbohydrates appear below 180 nm, ECD measurements on carbohydrates required special vacuum UV instruments. Carbohydrates that did not have visible electronic transitions could not be investigated using ECD. Since all carbohydrates have numerous vibrational transitions, the availability of VCD and VROA methods provided new means of deriving structural information for carbohydrates.

Early VCD measurements in the 1700–1100 cm^{-1} region for simple carbohydrates dissolved in DMSO-d_6 solvent attempted to correlate a significant VCD band at ~1150 cm^{-1} with the orientations of the O-H groups around the carbohydrate ring (Back and Polavarapu 1984; Tummalapalli et al. 1988). When hydroxyl groups were converted to acetates, enhanced VCD signals were found and the carbonyl group stretching vibrations appeared to serve as the probes of carbohydrate stereochemistry (Bose and Polavarapu 1999a).

When measured in D_2O solutions, VCD signals are rather weak and required high concentrations (Bose and Polavarapu 1999c), but this problem could be overcome by making measurements on film samples prepared from aqueous solutions (Petrovic et al. 2004). VCD measurements on oligosaccharides in D_2O solutions indicated that those with α-glycosidic linkage yield large VCD signals in the 1200–900 cm^{-1} region, while those with β-glycosidic linkage do not. Thus, VCD was indicated to be a sensitive probe of the glycosidic linkage (Bose and Polavarapu 1999b).

Later, VCD measurements on a series of methyl (or phenyl) glycosidic derivatives in DMSO-d_6 indicated (Monde et al. 2004) that only D-monosaccharides with α-glycosidic linkage exhibited negative VCD band at ~1145 cm^{-1}. There are only a limited number of VCD studies in the C-H stretching region (Marcott et al. 1978; Paterlini et al. 1986). Nevertheless, the VCD measurements on methyl D-glycosides in DMSO-d_6 indicated that a band at 2840 cm^{-1}, originating from the symmetric methyl stretching region, showed a positive VCD for α-anomers and a negative VCD for β-anomers (Taniguchi et al. 2004). Thus VCD spectroscopy was suggested to be useful to determine the anomeric configuration of carbohydrates.

A recent review discusses the applications of ECD, OR, and VCD for carbohydrates and glycoconjugates (Taniguchi and Monde 2012b).

Experimental (Bell et al. 1994b; Wen et al. 1993) and quantum chemical (Cheeseman et al. 2011; Macleod et al. 2006; Mutter et al. 2015a; Zielinski et al. 2015) VROA spectral analyses were also undertaken to establish the sensitivity of VROA spectroscopy for stereochemical analysis of carbohydrates. VROA spectra were suggested to reveal the configuration of glycosidic links in di- and polysaccharides (Bell et al. 1993). A VROA band reflecting the anomeric configuration was noted (Bell 1994a). Polysaccharides with different types of structures were shown to yield markedly different VROA spectra (Bell et al. 1995).

Recent reviews summarize the applications of VROA for carbohydrates and glycoproteins (Barron and Hecht 2012; Mutter et al. 2015b).

10.4.1.23 Transition Metal Complexes

The applications of ECD for transition metal complexes were reviewed by Kaizaki (2012). Applications of both ECD and ORD for transition metal complexes were reviewed by Autschbach and coworkers (Autschbach et al. 2011). Some of the early applications of VCD for transition metal complexes were mentioned in a review by Xu and coworkers (Yang and Xu 2011). More recently Wu et al. (2015) have reviewed the applications of ECD, VCD, and VROA spectroscopies for coordination compounds. The applications of VROA to transition metal complexes have been referenced in the most recent work on the ethylenediamine complex of rhodium (Humbert-Droz et al. 2014).

There are large number of references on chiroptical spectroscopic applications to transition metal complexes and it is difficult to summarize the work reported in individual references. For this reason, a summary of the references that used ECD, ORD, VCD, and/or VROA spectroscopies for studying the transition metal complexes is compiled in Appendix 9.

10.4.1.24 Natural Products

The applications of ECD and ORD for natural product structure determination have been around for a long time (Allenmark 2000). However, most of these applications were based on qualitatively defendable rules that, in some occasions, were noted to fail. VCD spectroscopy has found widespread applications for determining the structures of natural products. The current trend, to interpret the experimental ECD, ORD, or VCD spectra with corresponding quantum chemical predictions, is facilitating the reliable predictions of the ACs of natural products. Some natural products are not soluble in favorable solvents for VCD spectroscopy (Taniguchi et al. 2009) and some are not favorable for interpretations using ECD spectroscopy. Therefore one cannot claim one method to be generally superior over another method. Recent articles (Batista et al. 2011; Muñoz et al. 2014; Zhang et al. 2014) and reviews (Batista 2013; Batista Jr et al. 2015a; Batista Jr and Bolzani 2014; Bringmann et al. 2009; Joseph-Nathan and Gordillo-Romain 2015; Li et al. 2010; Nafie 2008; Nugroho and Morita 2014; Polavarapu 2012) on the applications of chiroptical spectroscopy can be consulted for variety of natural products investigated. (See also Sections 10.4.1.21 and 10.4.2.7.) On the contrary, the number of natural products investigated by VROA spectroscopy is limited. Nevertheless, the first determination of the AC of junionone, a natural product isolated from juniper berry oil, was achieved using VROA spectroscopy (Lovchik et al. 2008). The AC of highly flexible synoxazolidinone A was also explored using VROA spectroscopy (Hopmann et al. 2012). The AC of limonene, (+)-(E)-alpha-santalol (Sakamoto et al. 2012) and 5,7-dihydroxy-6-methoxy-3-(9-hydroxy-phenylmethyl)-chroman-4-one (Batista Jr et al. 2015b) was also investigated using VROA spectroscopy.

10.4.1.25 Polymers

One of the important questions related to helical polymers is the determination of helical handedness and pitch. ECD spectroscopy has been the workhorse for studying polymers in the early stages. The development of its vibrational analog, which is expected to offer more structural sensitivity, led to several recent studies with VCD spectroscopy. The studies of polymers with SOR and VROA spectroscopy are not as widespread as those with CD spectroscopies. Nevertheless VROA spectroscopy was used (Merten et al. 2011) for determining the backbone and side chain conformations of (+)-poly(trityl methacrylate).

Helical polymers can be broadly categorized into two classes: (1) polymerization of chiral monomers and (2) polymerization of achiral monomers.

10.4.1.25.1 Helical Polymers Derived from Chiral Monomers

10.4.1.25.1.1 Poly(menthylmethacrylate) The polymer with a higher isotactic content, made from anionic initiation, has a smaller SOR value than the one with lower isotactic content prepared from free radical initiation. The experimental VCD spectra of poly(menthylmethacrylate) polymerized using free radical and anionic initiators appeared to have minor changes between them, and these changes were interpreted as sensitive probes of stereoregularity (McCann et al. 1995).

10.4.1.25.1.2 Poly(vinylethers) with Chiral Pendants Poly(vinylethers) with (+)-menthol, (+)-neomenthol, and (+)-isomenthol as chiral pendants were studied using optical rotation (Liquori and Pispisa 1967). This investigation revealed significant enhancement of SOR from monomer to polymer, in the case of the first two pendants, but not for the third pendant. VCD spectroscopy studies on the same monomers and polymers revealed (McCann et al. 1997) similar trends. To understand the origins of different VCD bands in the experimental VCD spectrum of poly(menthyl vinyl ether), VCD calculations were performed for a five-repeat unit menthyl vinyl ether oligomer (Andrushchenko et al. 2000).

10.4.1.25.1.3 Polythiophenes These polymers with chiral substituents have been studied early on using ECD spectroscopy. Later, the polymers of (+)-3-(2-methylbutyl)thiophene and (+)-3,4-Di[(2-methylbutyl)]thiophene have been studied using VCD spectroscopy (Wang et al. 2002a). The repeating units in the polymers of these compounds have the same AC and conformations as those of respective monomers. A comparison of the VCD spectra of polymers with those of corresponding monomers indicated that the influence of polymerization is reflected in the vibrational bands associated with the backbone. The ECD spectra of these polymers in different solvents suggested the formation of different supramolecular arrangements (Lebon et al. 2002). For a different polythiophene, namely poly-3-[5-(mentholate)-pentyl] thiophene, ECD spectra are very weak, but VCD spectral signatures are strong enough to indicate the presence of interactions between thiophene rings and chiral substituents (Lebon et al. 2001).

10.4.1.25.1.4 Oligo(m-phenylurea)s Oligomers resulting from N,N′-dimethylated urea connected to benzene rings in the meta position have multilayered helical structures (Kudo et al. 2009). In these polymers, although N atoms represent chiral centers, the chirality of the system probed using ECD and VCD spectroscopies is that of a helical structure. The chiral polymers obtained by replacing one of the methyl groups at two urea nitrogen atoms with a chiral 2-(methoxyethoxyethoxy) propyl group were also investigated using ECD and VCD spectroscopies. These studies indicated that N-alkylated oligo(m-phenylurea)s have dynamic helical (rapid helical reversal) properties in solution. Oligomers with dynamic helical properties are referred to as foldamers.

10.4.1.25.1.5 Aromatic Foldamers The structure of a quinoline-derived (*R*)-chiral tetramer has been previously characterized using X-ray diffraction and ECD spectroscopy (Dolain et al. 2005). From these studies, this oligomer was known to have one right-handed and two left-handed conformers, with the dominant population being that of left-handed conformers. The chiral group attached at the end of the oligomer was considered to be responsible for a helical structure of this oligomer. The experimental VCD spectra of the tetramer in $CDCl_3$ were found to be reproduced by the population-weighed predicted VCD spectra of the tetramer at the B3LYP/6-31G(d) level. Model calculations performed for dimer and trimer were not considered to be as satisfactory as those predicted for the tetramer (Buffeteau et al. 2006; Ducasse et al. 2007).

10.4.1.25.2 Polymerization of Achiral Monomers

10.4.1.25.2.1 Poly(methylmethacrylate) The chiral polymers composed of complementary isotactic and syndiotactic poly(methylmethacrylate), it-PMMA and st-PMAA, respectively, were characterized (Kawauchi et al. 2008a). Helical it-PMMA could be incorporated into the cavity produced by the expulsion of C_{60} that was encapsulated in a chiral st-PMMA-C_{60} complex. Helical it-PMMA could also be incorporated into the cavity produced by the removal of chiral alcohol that was used to prepare the chiral st-PMMA–alcohol complex. The inclusion of the it-PMMA helix into the outer st-PMAA helix can generate assemblies with same or opposite-handed it- and st-PMMA helices. The calculated VA and VCD spectra for all possible combinations revealed that the calculated spectra for an assembly with the same it- and st-PMMA helicity better matched the experimental spectra. Thus it-PMMA encapsulated in st-PMAA was determined to have the same helicity as that of st-PMAA using VCD spectroscopy.

The helicity of encapsulated st-PMMA-C_{60} assembly generated in the presence of (*R*)-1-phenylethanol is opposite to that generated in the presence of (*S*)-1-phenylethanol. Even after the removal of chiral alcohol, the helicity is maintained by st-PMMA-C_{60}, (due to memory effect). This was evidenced by the experimentally observed oppositely signed ECD induced for encapsulated C_{60} (which by itself is optically inactive) and oppositely signed VCD for st-PMMA-C_{60}. The experimental VCD spectrum was also reproduced fairly well by the B3LYP/6-31G(d) calculations on st-PMMA (Kawauchi et al. 2008b).

10.4.1.25.2.2 Poly(4-carboxyphenylisocyanide) Achiral poly(4-carboxyphenylisocyanide) has been converted to the one with helical structure when poly(4-carboxyphenylisocyanide) was complexed with optically active amines in water or dimethyl sulfoxide (Hase et al. 2009). The helicity induced in water or aqueous organic solutions could be retained even after the complete removal of chiral amines. ECD and VCD, along with NMR spectroscopic measurements, indicated that isomerization around the C=N double bond is responsible for the helical induction process.

10.4.1.25.2.3 Polyguanidines Helical polyguanidines prepared from achiral carbodiimides, using chiral catalysts, have been studied using ECD, VCD, and specific rotation (Tang et al. 2007). Thermal and solvent-induced switching of chiroptical properties were also observed (Tang et al. 2005). The conformational reorientation that is responsible for the switching of chiroptical properties has been identified (Merten et al. 2014).

10.4.1.25.2.4 Poly(isocyanooctanes) The helical sense of poly (*R*)-2-isocyanooctane and poly (*S*)-2-isocyanooctane was determined (Schwartz et al. 2010) from the experimental VCD associated with the CN stretching band at ~1630 cm^{-1} using the CO model. The authors suggested that this method could be used for polyisocyanides.

10.4.1.25.2.5 Poly(trityl methacrylate) The helical conformation of poly(trityl methacrylate) is imparted by the initiator, n-butyl lithium/(–)-sparteine, which remains at the growing end of the polymer. The helical screw sense of (+)-poly(trityl methacrylate) was determined from a comparison of experimental and theoretical VCD spectra to be left-handed (Merten and Hartwig 2010). Experimental VCD spectra measured for solution and solid states were analyzed with calculated VCD spectra at the B3PW91/6-31G(d,p) level for short oligomers (3mer, 4mer, and 5mer). The similarity of SOR of poly(trityl methacrylate) in solution and suspension led to the belief that structures of polymer in solution and solid states are similar (Bartus and Vogl 1992).

10.4.1.25.2.6 Polypropylene The arrangement of a carbon backbone with consecutive 180° dihedral angles (referred to as TTTT structure) does not lead to helicity, but with consecutive 180° and 60° dihedral angles (referred to as TGTG structure), and 60° and 60° dihedral angles (referred to as GGGG structure) leads to overall helicity. To determine the characteristic VROA signature associated with these helical structures, quantum chemical calculations were carried out on 20 unit oligomers and analyzed using localized modes (Liégeois et al. 2010). It was found that a negative–positive VROA couplet at ~1100 cm^{-1} is characteristic of TGTG structure and a positive–negative–positive pattern is characteristic of GGGG structure. These features are retained even for smaller oligomers. Substitution of the TGTG polypropylene chain with a terminal chiral end group (Lamparska et al. 2006) also retained the characteristic VROA features of the TGTG chain.

10.4.1.25.2.7 Polysilane Quantum chemical predictions of VROA spectra for heptasilane, used as a model for polysilane, indicated that characteristic VROA signatures are present for helical TGTG, TGTG′, and GGGG structures (G′ refers to −60° dihedral angle) in two different regions: 440–510 cm^{-1} and 560–620 cm^{-1} (Liégeois et al. 2005).

10.4.1.25.2.8 Polyethylene Quantum chemical predictions of VROA spectra for heptane, C_7H_{16}, used as a model for polyethylene, indicated that characteristic VROA signatures are present for helical TGTG, TGTG′, and GGGG structures in three regions: rocking and wagging (680–765cm^{-1}), C–C stretching (980–1080 cm^{-1}), and CH stretching (2900–3000 cm^{-1}) (Liégeois et al. 2006).

10.4.1.26 Biomolecular Structures

Peptides, proteins, nucleic acids, carbohydrates, and viruses can all be broadly categorized as biomolecular systems. A comprehensive summary of applications to biomolecular systems is not possible due to the large number of investigations reported and space limitations. For this reason, it is best to refer to the recent individual reviews on these topics.

The applications of ECD for biomolecular structure determination are well known and have been around for a long time. ECD signatures associated with the n-π* and π-π* transitions associated with the peptides and the bases of nucleic acids are the main focus of the analyses. As mentioned in the spectra–structure correlation section (Section 10.2), ECD signatures associated with these transitions for known secondary structures are used to deduce the unknown structures of other peptides and nucleic acids. Several reviews (Corradini et al. 2012; Gray 2012; Kypr et al. 2012; Toniolo et al. 2012; Woody 2012) available on these topics provide the most recent updates. On the theoretical front, the DFT calculation of ECD for stacked base pairs was shown to provide a good representation for ECD spectra of DNA (Di Meo et al. 2015), and to provide insight into the origin of the hypochromic effect observed in going from B-DNA to N-DNA (Norman et al. 2015).

The applications of ORD to biomolecular systems are dated. Besides the well-known empirical correlations for carbohydrates (Shallenberger 1982) and older studies on proteins and nucleic acids (Gotoh et al. 1974; Steiner and Lowey 1966; Tinoco and Cantor 1970; Urray et al. 1967; Yang and Doty 1957), recent activity in this area is limited (da Silva et al. 2004).

The applications of VCD and VROA spectroscopies for biomolecular structure determination are relatively new and have added a wealth of new information. For amino acids and oligopeptides, where VCD signals are weak, intensities can be enhanced by up to two orders of magnitude by coupling them to a paramagnetic metal ion with low lying electronic states (He et al. 2001; Johannessen and Thulstrup 2007; Domingos et al. 2014a and 2014b). Although most applications rely on characteristic VCD/VROA band features noted for well-defined structures (e.g., α-helix, β-sheet, random coil structures of proteins, and B-DNA and Z-DNA type structures for nucleic acids; see Section 10.2 on spectra–structure correlations), quantum chemical calculations have also been undertaken for model systems, or using the concept of transferring parameters from smaller molecules to larger assemblies (Bouř et al. 1997; Yamamoto et al. 2012), or using a fragmentation approach (Choi et al. 2005). With the availability of large-scale computational facilities, calculations on larger biomolecular polymer systems are also becoming a reality (Keiderling and Lakhani 2012; Parchansky et al. 2014). New developments for treating large molecular systems accurately (Jose and Raghavachari 2016; Jose et al. 2015) have emerged recently (see Section 5.2.7).

Among the interesting applications of VCD to biomolecular systems is its sensitivity to protein fibril formation. Under the conditions of fibril formation, unusually large spectral intensities are found in the amide I and II regions for hen egg white lysozyme and bovine insulin (Ma et al. 2007). The origin of intensity enhancement was explained using an EC/CO model for stacked β-sheets (Measey and Schweitzer-Stenner 2010). It was also found that the transformation of insulin fibrils from one polymorph to another (Kurouski et al. 2012) was initiated by small changes in the pH. The fibrils formed by polyglutamic acid were investigated using VCD (Fulara et al. 2011), and the arrangement of side chains in these fibrils was inferred using MD simulations and quantum chemical calculations (Kessler et al. 2014). The supramolecular organization in polyglutamine fibril aggregates was interpreted to be distinct from that in amyloid fibrils (Kurouski et al. 2013, 2015).

In order to avoid the interference of water absorption in the infrared region, the use of dried film samples was also adopted for VCD studies on peptide and protein samples (Sen and Keiderling 1984; Shanmugam and Polavarapu 2004). However, one has to be careful in carrying out VCD studies on film samples when aggregating solutes are involved, because the nature of drying can influence the morphology of the films and spatial confinement can influence the aggregation (Vijay and Polavarapu 2013). As a consequence, artifacts can be introduced into the CD measurements. It is known that complex morphological patterns can be generated during the evaporation of droplets of aggregating peptides and proteins (Chen and Mohamed 2010; Sett et al. 2015).

By substituting the C-13 labels at selected carbonyl groups, VCD associated with $^{13}C{=}O$ vibration provides the local structure around that group. This concept is found to be a powerful way of determining the site-specific structural information of peptides (Silva et al. 2000). The packing pattern of fibrils formed by polyglutamic acid was inferred using the VCD spectra of ^{13}C substituted samples (Chi et al. 2013). Site-specific structural information was also obtained for amyloid peptides (Shanmugam and Polavarapu 2011, 2013). There are numerous studies on the applications of VCD to amyloid fibrils. Although it is difficult to describe all these studies due to space limitations, the latest reviews can be consulted for this information (Dzwolak 2014; Keiderling and Lakhani 2012).

Early VROA spectral investigations on biological systems depended on identifying qualitative spectral patterns for structural characterization. A number of reviews have appeared periodically documenting the progress (Barron 2012, 2013; Barron and Hecht 2012; Parchansky et al. 2014). But in recent years, quantum chemical calculations of VROA for large systems have become possible and are providing insights into structural interpretations using VROA. Quantum chemical calculations on 20-mers of alanine with α- and 3_{10}- helical structures revealed that α- and 3_{10}- helical structures can be identified using their characteristic VROA signals in the amide I region: the former gives a couplet with a negative band at a lower wavenumber, while the later gives an opposite couplet with a positive band at a lower wavenumber (Jacob et al. 2009). These differences arise because the coupling between nearest neighbor residues is significant for α-helix structure VROA bands, while that between second-nearest neighbor residues is significant for 3_{10}- helical structure VROA bands in the amide I region. Similar calculations on alanine and glycine oligomers with parallel and antiparallel β-sheet structures revealed that the negative–positive VROA couplet in the amide I region is a reliable marker of β-sheet structure (Weymuth and Reiher 2013). A sharp negative VROA band at 1350 cm^{-1} is also suggested as a robust characteristic band for β-sheet structures. Furthermore, a possibility for differentiating between parallel and antiparallel β-sheet structures using a VROA signature in the amide III region was pointed out. Calculations on β-turn structures revealed that type 1 β turns yield a positive VROA band in the amide I region (Weymuth et al. 2011). Furthermore, a negative VROA band in the 1200–1220 cm^{-1} region, positive VROA band in the 1260–1300 cm^{-1} region, and negative band in the 1340–1380 cm^{-1} region are found for all model peptides studied.

VROA spectroscopy was deemed useful to characterize the hydrated α-helical structures (McColl et al. 2004), as proteins with these structures exhibited a stronger positive VROA band at ~1340 cm^{-1} and a weaker positive VROA band at ~1300 cm^{-1}. The ratio of these band intensities is considered to reflect the extent of hydration. A theoretical examination of the solvated states of poly-L-alanine concurred with the experimental observations, and it was suggested that the dielectric constant of the solvent environment influences the intensities of these bands (Yamamoto et al. 2014).

10.4.1.27 Other Molecules

Some molecules whose ACs were corrected, but not discussed in Section 10.4.1, are discussed separately in Section 10.4.2. Still there are numerous other molecules that have been investigated using chiroptical spectroscopies, but limited space prevents a comprehensive coverage, unfortunately.

10.4.1.28 Role of Intramolecular CT

Vapor phase SORs of norbornenone are somewhat smaller than those of norbornenone in a liquid solution phase (Lahiri et al. 2014), but not withstanding this difference, the SORs of norbornenone are large compared to those for similar molecules lacking the C=O group. For example, the experimental-specific rotations of (1*S*,4*S*)-norbornenone in cyclohexane are five times larger (Caricato et al. 2014) than those of (1*S*,4*S*)-6-methylenenorbornene in isooctane. The investigations by Autschbach and coworkers (Moore et al. 2012) traced the source for the large SOR of norbornenone to delocalization and electronic coupling of π- orbitals of C=C and lone pair electrons of oxygen in the C=O bond. The investigation by Vaccaro and coworkers (Caricato et al. 2014) indicated that when C=O and C=C groups are in a plane, CT occurs among these groups in the first excited electronic state. If these two groups are on the opposite sides of the molecular frame, then the electric transition dipole moment and magnetic dipole transition moment are favorably oriented to yield a large rotational strength (which in turn translates into large specific rotation; see SOS expression, Equation 5.6, in Chapter 5). However, if these two groups are on the same side of the molecular plane, then the electric transition dipole moment orients orthogonal to the magnetic dipole transition moment, and leads to negligible specific rotation. It was suggested that these observations can be used to design systems with large rotational strength (leading to large SOR) by appropriately constructing two interacting groups.

10.4.2 Correcting the Incorrect or Ambiguous ACs

While adopting the chiroptical spectroscopy methods to independently determine their structures, some incorrect literature assignments of ACs were recognized. A few such cases are described here.

10.4.2.1 Troger's Base

The experimental ECD spectrum of Troger's base was interpreted using the EC/CO model to determine its AC as (+)-(5*R*,11*R*) (Mason et al. 1967). But the crystal structure of its diastereomeric salt revealed that the AC should be opposite; that is, (+)-(5*S*,11*S*) (Wilen et al. 1991). This discrepancy was revisited by undertaking the VCD spectral analysis. The experimental VCD spectrum of (–)-enantiomer in CCl_4 solution was reproduced by that calculated at the B3PW91/6-31G* level for (5*R*,11*R*) diastereomer. Therefore VCD spectral analysis suggested that the AC should be (–)-(5*R*,11*R*) (Aamouche et al. 1999), a conclusion that is opposite to that derived from ECD and in agreement with that determined from X-ray structure. This example illustrated the potential dangers involved in relying on the empirical models for determining the AC.

10.4.2.2 1,1-Dimethyl-2-Phenylethyl Phenyl Sulfoxide (DPPS)

The ACs of several chiral sulfoxides have been assigned in the literature using empirical spectra structure correlations and EC/CO calculations. The AC of (+)-DPPS was assigned as (*S*) (Berthod et al. 2002) [Note that in the original reference (Berthod et al. 2002), the chemical formula shown for DPPS in Table 1 contained an extra CH_2 group, and therefore was referred to as 1,1-dimethyl-3-phenylpropyl phenyl sulfoxide, but the structure displayed in Figure 6 was that of DPPS]. But the opposite conclusion was reached (Petrovic et al. 2005a) when chiroptical spectra of (+)-DPPS were analyzed.

The experimental ECD and VCD spectra as well as SOR of (+)-DPPS were measured and corresponding quantum chemical calculations were carried out for (*R*)-DPPS. Nine conformations of this molecule were investigated and four of them were found to have significant population. The population-weighted ECD spectra for (*R*)-DPPS obtained at B3LYP/6-31G*, B3LYP/6-31+G, and B3LYP/6-311G(2d,2p) levels matched the experimental ECD spectra of (+)-DPPS, suggesting that the AC should be (*R*)-(+). This prediction was confirmed by the positive SOR predicted for (*R*)-DPPS at 589 nm as well as 365 nm. The experimental VCD spectrum of (+)-DPPS showed weak signals. Nevertheless, the negative–positive VCD couplet observed in the 1470–1460 cm^{-1} region and the positive VCD band at ~1000 cm^{-1} are reproduced in the predicted VCD spectrum for (*R*)-DPPS at the B3LYP/6-31G* level. Thus, all three chiroptical spectroscopic methods used suggested that the AC of DPPS should be (*R*)-(+) (Petrovic et al. 2005a). Tracing back the origin for this discrepancy, it was concluded that the error in the original paper (Berthod et al. 2002) may have resulted from an incorrect application of the priority rules in assigning the AC.

10.4.2.3 2-(1-Hydroxyethyl)-Chromen-4-One (HECO)

The AC of (-)-HECO and (-)-6-bromo-2-(1-hydroxyethyl)-chromen-4-one (BrHECO)) were reported to be (*R*) and (*S*), respectively, using X-ray crystallography (Besse et al. 1999). Reversal of the AC by bromine substitution at the carbon-6 position on the benzene ring, a position far away from the chiral center, is unexpected. This unexpected conclusion was tested using VCD spectroscopy by preparing the acetate derivatives of HECO and BrHECO (Devlin et al. 2005). It was found that the (+)-enantiomers of both derivatives gave similar VCD spectra in the C=O stretching region, namely a bisignate-positive VCD couplet (positive VCD on the low frequency side and negative VCD on the high frequency side) at 1755 cm^{-1} and a positive VCD at 1665 cm^{-1}. Quantum chemical calculations of VCD spectra, at the B3LYP/TZ2P and B3PW91/TZ2P levels, for both derivatives yielded spectra that matched the experimental spectra, suggesting that the ACs of both acetate derivatives should be (*R*)-(+). Since derivatization of the O-H group to O-C(=O)CH_3 does not change the AC, the ACs of both parent compounds, (–)-HECO and (–)-BrHECO, should have been (*S*)-(–). Thus the AC of (–)-HECO determined from X-ray structure as (*R*) turned out to be incorrect.

10.4.2.4 3,3′-Diphenyl-[2,2′-Binaphthalene]-1,1′-Diol (Vanol)

From the X-ray structure of brucine binaphtholphosphate salt, the AC of Vanol was determined to be $(-)_{589}$-(*aS*) (Bao et al. 1996). During chiroptical spectroscopic investigations on Vanol, the reported X-ray structure (presented in the supplementary material) was noticed to actually belong to the (*aR*) configuration (Polavarapu et al. 2009c). This discrepancy raised the speculation if the AC of Vanol is $(-)_{589}$-(*aS*) as reported, or should have been $(-)_{589}$-(*aR*)?. To resolve this uncertainty, the experimental ECD, VCD, and ORD spectra were measured for $(+)_{589}$-Vanol and corresponding quantum chemical calculations were obtained at the B3LYP/6-31G* and B3LYP/6-311G(2d,2p) levels for (*aR*)-Vanol. The experimental chiroptical data for $(+)_{589}$-Vanol were found to be in very good agreement with those calculated for (*aR*)-Vanol, indicating that the original assignment of $(+)_{589}$-(*aR*)-Vanol, that is, $(-)_{589}$-(*aS*) was indeed correct (Polavarapu et al. 2009c). Then to resolve the uncertainty in the reported crystal structure, racemic Vanol hydrogen phosphate was reacted with brucine and the crystal structures of diastereomeric salts were reinvestigated. The source for error in the AC of the original crystal structure was attributed to either mislabeling of the crystallized salt or impure crystals (the crystallized salt might not have been pure enough that an incorrect crystal could have been picked for the analysis).

10.4.2.5 Chromanes

The AC of peperobtusin A and 3,4-dihydro-5-hydroxy-2,7-dimethyl-8-(3″-methyl-2″butenyl)-2-(4′-methyl-1′,3′-pentadienyl)-2H-1-benzopyran-6-carboxylic acid was assigned as (+)-(*S*) with *P*-helicity for a chromane ring (Batista et al. 2009). The VCD spectrum predicted for the (*R*) isomer, with *P*-helicity for a chromane ring, at the B3LYP/6-31G(d) level, was found to match well with that experimentally observed for the (+)-enantiomer, indicating that correct configuration should be (+)-(*R*) with *P*-helicity for a chromane ring (Batista Jr et al. 2010). The discrepancy was traced to the assumption, in the previous work, of isoprenyl group at C-2 in equatorial orientation, while this group is found to be energetically favored in the axial position. Additionally, quantum theoretical ECD calculations have predicted the ^1L_b ECD band to have the same sign for *P* and *M* helicities of a chromane ring with a given configuration at C-2, thereby questioning the validity of the empirical ECD helicity rule for chromanes.

10.4.2.6 Remisporine B

Remisporine A is derived from a marine fungus, but dimerizes in solution to remisporine B. Using the geometries optimized at the B3LYP/6-31G** level, ECD spectra were predicted for remisporine B using quantum chemical methods by varying the number of excited states from 20 to 100. It was found that a larger number of electronic states is necessary to satisfactorily reproduce the experimental ECD spectrum all the way to the shorter wavelength side. The predicted ECD spectra were used to assign the AC of remisporine B as *RRSSS* (Sherer et al. 2015), which is opposite to that previously assigned in the literature. The discrepancy was attributed to the lack of technology to accurately predict the ECD spectra in earlier years.

10.4.2.7 Natural Products

The AC of (–)-brevianamide M (a bioactive alkaloid), which was assigned in the literature as (2*S*,13*S*), was corrected using SOR, ECD, and VCD analyses to be (2*R*,13*R*) (Ren et al. 2013). The discrepancy was traced to the assumptions made in the mechanism for acid hydrolysis. The AC of (+)-schizandrin, a bioactive natural product, was also revised (He et al. 2014) from (7*S*, 8*S*) to (7*S*,8*R*). The discrepancy was traced to the use of the wrong axial chirality in the previous work. Desulfurization of epidithiodioxopiperazine fungal metabolites with triphenyl phosphine was stated in some literature papers to proceed through the inversion of the configuration at the bridgehead carbon atoms. But chiroptical spectroscopic studies, especially VCD and ORD, indicated this to be incorrect (Cherblanc et al. 2011, 2013) and the desulfurization process to proceed through the retention of configuration. This discrepancy was attributed to the correlation of the experimental ECD spectra of a given molecule to those of related molecules. Such correlations are not advised in the current literature.

The case of an antiparasitic natural product isolated from *Uvaria klaineana* illustrates that a definite knowledge of the chemical structure (chemical groups and their connectivity in a molecule) is a prerequisite for undertaking chiroptical spectroscopic analyses. The compound isolated from *U. klaineana* was originally identified as (+)-klaivanolide and its experimental VCD spectrum was reported to be satisfactorily reproduced by the DFT predictions at the B3PW91/TZ2P level for the (7*S*)-enantiomer (Devlin et al. 2009). A later investigation (Ferrié et al. 2014) indicated that the compound isolated from *U. klaineana* is actually a known compound, acetylmelodorinol, and that the experimental VCD spectrum could be reproduced by that calculated for its (7*S*)-enantiomer. The former compound contained a seven-membered ring, while the latter is a butyrolactone. The claim to reproduce the same experimental VCD spectrum by those calculated for two different chemical structures should serve as a wake-up call in using a single chiroptical spectroscopic method for structural determination.

The ACs of a few other natural products were also revised. These instances can be found in a recent review (Polavarapu 2012).

10.5 Analytical Applications

While analytical applications of ECD and ORD are numerous, only limited analytical applications involving the determination of the enantiomeric excess and reaction monitoring have been explored for VCD (Guo et al. 2004; Ma et al. 2010; Spencer et al. 1990; Urbanová et al. 2000; Zhao and Polavarapu 2001a) and VROA (Hecht et al. 1995; Rüther et al. 2014; Spencer et al. 1994, 1996). For determining the enantiomeric excess, chiroptical spectroscopic methods do not offer as much sensitivity as that offered by chiral chromatography.

The mechanisms involved in enantiomeric separations with chromatographic columns were investigated using VCD spectroscopy. The interaction of a chiral amine with sulfated β-cyclodextrin (Ma et al. 2009a), of α-substituted alanine esters with polysaccharide-based stationary phases (Ma et al. 2009b), and of aromatic amines with crown ether tetracarboxylic acid (Shen et al. 2009) were investigated using VCD spectral studies combined with chromatography and molecular mechanics calculations. The stereoselective interactions between the enantiomers of the derivatives of amino acids and chiral selectors (tert-butylcarbamoylquinine, tert-butylcarbamoylquinidine) have been studied using VCD and ECD spectroscopies (Julínek et al. 2009). The interaction between the tert-butylcarbamoylquinine selector and the (*S*)-3,5-dinitrobenzoyl alanine was also studied experimentally and theoretically using ECD and VCD spectroscopies (Julinek et al. 2010).

10.6 Chiroptical Spectroscopy in Undergraduate Teaching Laboratories

Considering the limited budgets available to acquire instruments for undergraduate teaching laboratories in most colleges and universities, the instruments for ECD, VCD, and VROA measurements will be out of reach for the purpose of exposing undergraduate students to chiroptical spectroscopy experiments. Inexpensive digital polarimeters, however, are available for designing the wet-laboratory experiments for undergraduate teaching laboratories. These wet-laboratory experiments can easily be complemented with corresponding computational experiments. When undergraduate students are exposed to the computational laboratory experiment on the same wet-laboratory experiment that they have done (or plan to do), it will help them to "see" the molecules involved in the experimental measurements. This combined wet-laboratory and computational laboratory experimental approach provides a better insight into the reactions involved than what they otherwise might imagine in the absence of accompanying computational experiments.

Different experiments that can be adopted for undergraduate student teaching laboratories are described here.

10.6.1 Inversion of Sucrose

A successful implementation of an undergraduate laboratory experiment to investigate the inversion of sucrose by combining the wet-laboratory experiment with quantum chemical calculations has been undertaken at Vanderbilt University. The experiments were designed to carry out the wet-laboratory experiment (Bettelheim 1971) in one laboratory period and the corresponding quantum chemical calculations in another laboratory period, with each student carrying out the computational experiment independently on a desktop computer.

The transparent name of sucrose, β-ᴅ-fructofuranosyl α-ᴅ-glucopyranoside, indicates that sucrose is made up of β-ᴅ-fructofuranose and α-ᴅ-glucopyranose. Upon acid hydrolysis, the parent sucrose molecule is cleaved into its constituent parts, namely β-ᴅ-fructofuranose and α-ᴅ-glucopyranose. The liberated α-ᴅ-glucopyranose undergoes anomeric equilibrium between α-ᴅ-glucopyranose and β-ᴅ-glucopyranose. Eventually equilibrium is established between α-ᴅ-and β-ᴅ-forms of glucopyranose. Similarly the liberated β-ᴅ-fructofuranose undergoes a complex conformational transition from furanose to pyranose form and equilibrium between α- and β-anomers. In aqueous solutions, fructose exists as an equilibrium mixture of fructopyranose and fructofuranose (along with minor components of noncyclic structures), with fructopyranose being the dominant form (~75%). At the beginning of sucrose hydrolysis, optical rotation is positive because sucrose exhibits positive optical rotation. At the end of hydrolysis, the sum of optical rotations of hydrolysis products becomes negative (because the negative rotation of equilibrated ᴅ-fructose dominates the positive rotation of equilibrated ᴅ-glucose). Therefore, the observed rotation changes from being positive (before hydrolysis) to negative (after hydrolysis).

Modern quantum chemical methods allow the undergraduate students to gain physical insight into this wet-laboratory experiment, from the visual display of the structures, and calculating the SORs, for sucrose, glucose, and fructose. Since the undergraduate computational exercises have to be completed in one laboratory period, certain simplifications had to be utilized.

The exploration of the conformational freedom of glucose, fructose, and sucrose has been avoided by providing the optimized structures of eight conformers (including both α- and β-anomers) of glucose (da Silva et al. 2004), β-ᴅ-fructopyranose (Cocinero et al. 2013), and a stable structure for sucrose (Yamabe et al. 2013) reported in the literature. Using these preselected structures, students undertook SOR calculations at the Sodium D line (589 nm) using a B3LYP functional, 6-31G(d) basis set and PCM for water solvent. The size of the basis set had to be limited to 6-31G(d), because the calculations on individual desktop computers (Dell Optiplex 960, with two processors), provided to each student, had to be completed in one laboratory period. The SOR of glucose was obtained as a population-weighted sum of the SORs of individual conformers. The populations reported at a higher level of theory (da Silva et al. 2004) were used for this purpose. This exercise yielded the predicted SORs of ᴅ-glucose as +49, β-ᴅ-fructopyranose as −84, and sucrose as +42. The corresponding experimental values (Shallenberger 1982) are, respectively, +53, −92, and +66, providing chemical insight into the chemical phenomenon behind the wet-laboratory experiment on inversion of sucrose.

The approximations implemented in this computational teaching experiment on chiroptical spectroscopy for undergraduate students are unavoidable, because the undergraduate teaching labs have a 3-hour time limit.

Nevertheless, this approach definitely served as an important step for introducing the undergraduate students to the importance of modern chiroptical spectroscopic methods.

10.6.2 Monomer–Dimer Equilibrium

A wet-laboratory experiment-based monomer–dimer equilibrium (see Chapter 9) in (R)-(–)-α-hydroxy-β,β-dimethyl γ-butyrolactone, known as (–)-pantolactone, can easily be adopted (Polavarapu and Covington 2015). Here students can prepare the solutions of a pantolactone enatiomer at different concentrations in CCl_4 (~0.002–0.04 M) and measure wavelength-resolved ORs. These ORs are then converted to SORs and fit to the monomer–dimer equilibrium equation (Equation 9.23) to determine the wavelength-resolved SORs of monomer, $[\alpha]_m$ and dimer, $[\alpha]_d$. The known equilibrium constant of 8.9 for monomer–dimer equilibrium in pantolactone can be used for this purpose. This wet-laboratory experiment can be carried out in one laboratory period (~3 hours). In the second laboratory period, $[\alpha]_m$ and $[\alpha]_d$ can be predicted using QC methods with the optimized geometries available for the monomer and dimer. Students can analyze the geometries and visually see the differences in monomer and dimer geometries. Different levels of theory can be used, if time permits, to give an idea to the students on the importance of using high-level theoretical methods. A comparison of experimentally determined $[\alpha]_m$ and $[\alpha]_d$ with corresponding QC-predicted values and submitting a report can be completed in the second laboratory period.

10.6.3 Other Experiments

A computational experiment for investigating the pH-dependence of the specific optical rotation of amino acids has been designed by Zurek and coworkers (Simpson et al. 2013). An experiment to investigate the ORD of alpha-pinene was suggested recently (Compton and Duncan 2016).

10.7 Summary

The applications of chiroptical spectroscopy for deriving three-dimensional molecular structural information have skyrocketed in recent years. Numerous applications identifying the agreement between measured chiroptical properties for a given enantiomer and corresponding predicted chiroptical properties for the correct AC have attracted the attention of practicing stereochemists. Readily available commercial instruments for experimental chiroptical spectroscopic measurements, and user-friendly quantum chemical programs for predicting chiroptical properties, have attracted new researchers to chiroptical spectroscopy. All of the recent technological developments have made it not only easier for the novice to enter this research area, but also secondary (or even unnecessary) for them to have a formal theoretical background. Examples illustrating the ability to determine the AC of chiral molecules for the first time, and to uncover the previously published incorrect AC assignments, have rendered an uncompromising role for chiroptical spectroscopy. Future investigations, with careful measurements avoiding spurious signals, accompanied by pragmatic, unbiased, and rigorous theoretical analyses are expected to propel the role of chiroptical spectroscopy to new heights in the coming years. Finally the current developments in chiroptical spectroscopy permit the introduction of some chiroptical spectroscopic experiments into the undergraduate curriculum.

References

Aamouche, A., F. J. Devlin, and P. J. Stephens. 1999. Determination of absolute configuration using circular dichroism: Troger's Base revisited using vibrational circular dichroism. *Chem. Commun.* (4):361–362.

Aamouche, A., F. J. Devlin, P. J. Stephens, J. Drabowicz, B. Bujnicki, and M. Mikołajczyk. 2000. Vibrational circular dichroism and absolute configuration of chiral sulfoxides: tert-butyl methyl sulfoxide. *Chem.Eur. J.* 6 (24):4479–4486.

Abbate, S., E. Castiglioni, F. Gangemi, R. Gangemi, G. Longhi, R. Ruzziconi, and S. Spizzichino. 2007. Harmonic and anharmonic features of IR and NIR absorption and VCD spectra of chiral 4-X-[2.2]paracyclophanes. *J. Phys. Chem. A* 111(30):7031–7040.

Abbate, S., R. Gangemi, F. Lebon, G. Longhi, M. Passarello, A. Ruggirello, and V. T. Liveri. 2012. Investigations of methyl lactate in the presence of reverse micelles by vibrational spectroscopy and circular dichroism. *Vib. Spectrosc* 60(0):54–62.

Abbate, S., F. Lebon, R. Gangemi, G. Longhi, S. Spizzichino, and R. Ruzziconi. 2009a. Electronic and vibrational circular dichroism spectra of chiral 4-X-[2.2]paracyclophanes with X containing fluorine atoms†. *J. Phys. Chem. A* 113(52):14851–14859.

Abbate, S., F. Lebon, S. Lepri, G. Longhi, R. Gangemi, S. Spizzichino, G. Bellachioma, and R. Ruzziconi. 2011. Vibrational circular dichroism: A valuable tool for conformational analysis and absolute configuration assignment of chiral 1-Aryl-2,2,2-trifluoroethanols. *ChemPhysChem.* 12(18):3519–3523.

Abbate, S., F. Lebon, G. Longhi, F. Fontana, T. Caronna, and D. A. Lightner. 2009b. Experimental and calculated vibrational and electronic circular dichroism spectra of 2-Br-hexahelicene. *Phys. Chem. Chem. Phys.* 11(40):9039–9043.

Abbate, S., G. Longhi, F. Lebon, E. Castiglioni, S. Superchi, L. Pisani, F. Fontana, F. Torricelli, T. Caronna, C. Villani, R. Sabia, M. Tommasini, A. Lucotti, D. Mendola, A. Mele, and D. A. Lightner. 2013. Helical sense-responsive and substituent-sensitive features in vibrational and electronic circular dichroism, in circularly polarized luminescence, and in Raman spectra of some simple optically active hexahelicenes. *J. Phys. Chem. C* 118(3):1682–1695.

Abbate, S., G. Mazzeo, S. Meneghini, G. Longhi, S. E. Boiadjiev, and D. A. Lightner. 2015. Bicamphor: A prototypic molecular system to investigate vibrational excitons. *J. Phys. Chem. A* 119:4261–4267.

Abbate, S., M. Passarello, F. Lebon, G. Longhi, A. Ruggirello, V. T. Liveri, F. Viani, F. Castiglione, D. Mendola, and A. Mele. 2014. Chiroptical phenomena in reverse micelles: The case of (1R,2S)-dodecyl (2-hydroxy-1-methyl-2-phenylethyl)dimethylammonium bromide (DMEB). *Chirality* 26(9):532–538.

Allenmark, S. G. 2000. Chiroptical methods in the stereochemical analysis of natural products. *Nat. Prod. Rep.* 17(2):145–155.

Andrushchenko, V., J. L. McCann, J. H. van de Sande, and H. Wieser. 2000. Determining structures of polymeric molecules by vibrational circular dichroism (VCD) spectroscopy. *Vib. Spectrosc.* 22(1–2):101–109.

Annamalai, A., K. J. Jalkanen, U. Narayanan, M. C. Tissot, T. A. Keiderling, and P. J. Stephens. 1990. Theoretical study of the vibrational circular dichroism of 1,3-dideuterioallene: Comparison of methods. *J. Phys. Chem.* 94(1):194–199.

Annamalai, A., T. A. Keiderling, and J. S. Chickos. 1985. Vibrational circular dichroism of trans-1,2-dideuteriocyclobutane. Experimental and calculational results in the mid-infrared. *J. Am. Chem. Soc.* 107(8):2285–2291.

Aparicio, F., B. Nieto-Ortega, F. Nájera, F. J. Ramírez, J. T. López Navarrete, J. Casado, and L. Sánchez. 2014. Inversion of supramolecular helicity in oligo-p-phenylene-based supramolecular polymers: Influence of molecular atropisomerism. *Angew. Chem. Int. Ed. Engl.* 53(5):1373–1377.

Armstrong, D. W., F. Albert Cotton, A. G. Petrovic, P. L. Polavarapu, and M. M. Warnke. 2007. Resolution of enantiomers in solution and determination of the chirality of extended metal atom chains. *Inorg. Chem.* 46 (5):1535–1537.

Asai, T., T. Taniguchi, T. Yamamoto, K. Monde, and Y. Oshima. 2013. Structures of spiroindicumides A and B, unprecedented carbon skeletal spirolactones, and determination of the absolute configuration by vibrational circular dichroism exciton approach. *Org. Lett.* 15:4320–4323.

Ashvar, C. S., P. J. Stephens, T. Eggimann, and H. Wieser. 1998. Vibrational circular dichroism spectroscopy of chiral pheromones: Frontalin (1,5-dimethyl-6,8-dioxabicyclo[3.2.1]octane). *Tetrahedron Asymmetry* 9(7):1107–1110.

Autschbach, J., L. Nitsch-Velasquez, and M. Rudolph. 2011. Time-dependent density functional response theory for electronic chiroptical properties of chiral molecules. In: R. Naaman, D. N. Beratan, and D. H. Waldeck (eds). *Electronic and Magnetic Properties of Chiral Molecules and Supramolecular Architectures.*Heidelberg, Germany: Springer.

Baciocchi, R., M. Juza, J. Classen, M. Mazzotti, and M. Morbidelli. 2004. Determination of the dimerization equilibrium constants of omeprazole and Pirkle's alcohol through optical-rotation measurements. *Helv. Chim. Acta* 87:1917–1926.

Baciocchi, R., G. Zenoni, M. Valentini, M. Mazzotti, and M. Morbidelli. 2002. Measurement of the dimerization equilibrium constants of enantiomers. *J. Phys. Chem. A* 106:10461–10469.

Back, D. M., and P. L. Polavarapu. 1984. Fourier-transform infrared, vibrational circular dichroism of sugars. A spectra-structure correlation. *Carbohydr. Res.* 133(1):163–167.

Bao, J., W. D. Wulff, J. B. Dominy, M. J. Fumo, E. B. Grant, A. C. Rob, M. C. Whitcomb, S.-M. Yeung, R. L. Ostrander, and A. L. Rheingold. 1996. Synthesis, resolution, and determination of absolute configuration of a vaulted 2,2′-binaphthol and a vaulted 3,3′-biphenanthrol (VAPOL). *J. Am. Chem. Soc.* 118(14):3392–3405.

Bara, R., I. Zerfass, A. H. Aly, H. Goldbach-Gecke, V. Raghavan, P. Sass, A. Mándi, V. Wray, P. L. Polavarapu, A. Pretsch, W. Lin, T. Kurtán, A. Debbab, H. Brötz-Oesterhelt, and P. Proksch. 2013. Atropisomeric dihydroanthracenones as inhibitors of multiresistant staphylococcus aureus. *J. Med. Chem.* 56(8):3257–3272.

Barron, L. 2013. Raman optical activity studies of structure and behavior of biomolecules. In: G. K. Roberts (ed). *Encyclopedia of Biophysics*. Berlin, Germany: Springer-Verlag.

Barron, L. D. 1977. Raman optical activity due to isotopic substitution: [[small alpha]-2H]benzyl alcohol. *J. Chem. Soc., Chem. Commun.* (9):305–306.

Barron, L. D. 2012. Structure and behavior of biomolecules from Raman optical activity. In: N. Berova, P. L. Polavarapu, K. Nakanishi, and R. W. Woody (eds). *Comprehensive Chiroptical Spectrosocpy*. New York, NY: Wiley.

Barron, L. D., M. P. Bogaard, and A. D. Buckingham. 1973. Raman scattering of circularly polarized light by optically active molecules. *J. Am. Chem. Soc.* 95(2):603–605.

Barron, L. D., and L. Hecht. 2012. Structure and behavior of biomolecules from Raman optical activity. In: N. Berova, P. L. Polavarapu, K. Nakanishi, and R. W. Woody (eds). *Comprehensive Chiroptical Spectroscopy: Applications in Stereochemical Analysis of Synthetic Compounds, Natural Products, and Biomolecules*. Vol. 2. New York, NY: John Wiley & Sons.

Barron, L. D., L. Hecht, E. W. Blanch, and A. F. Bell. 2000. Solution structure and dynamics of biomolecules from Raman optical activity. *Prog. Biophys. Mol. Biol.* 73(1):1–49.

Barron, L. D., H. Numan, and H. Wynberg. 1978. Raman optical activity due to isotopic substitution: (1S)-4,4-dideuterioadamantan-2-one. *J. Chem. Soc., Chem.Commun.* (6):259–260.

Barron, L. D., F. Zhu, L. Hecht, G. E. Tranter, and N. W. Isaacs. 2007. Raman optical activity: An incisive probe of molecular chirality and biomolecular structure. *J. Mol. Struct.* 834–836(0):7–16.

Barth, G., and C. Djerassi. 1981. Circular dichroism of molecules with isotopically engendered chirality. *Tetrahedron* 37(24):4123–4142.

Bartus, J., and O. Vogl. 1992. Solid state measurements of optical activity. *Polym. Bull.* 28(2):203–210.

Bas, D., T. Bürgi, J. Lacour, J. Vachon, and J. Weber. 2005. Vibrational and electronic circular dichroism of Δ-TRISPHAT [tris(tetrachlorobenzenediolato)phosphate(V)] anion. *Chirality* 17(S1):S143–S148.

Batista, J. M. 2013. Determination of absolute configuration using chiroptical methods. In: V. Andrushko and N. Andrushko (eds) *Stereoselective Synthesis of Drugs and Natural Products*. New York, NY: John Wiley & Sons, Inc.

Batista, J. M., A. N. L. Batista, J. S. Mota, Q. B. Cass, M. J. Kato, V. S. Bolzani, T. B. Freedman, S. N. Loóez, M. Furlan, and L. A. Nafie. 2011. Structure elucidation and absolute stereochemistry of isomeric monoterpene chromane esters. *J. Org. Chem.* 76(8):2603–2612.

Batista, J. M. Jr, A. N. L. Batista, D. Rinaldo, W. Vilegas, Q. B. Cass, V. S. Bolzani, M. J. Kato, S. N. López, M. Furlan, and L. A. Nafie. 2010. Absolute configuration reassignment of two chromanes from Peperomia obtusifolia (Piperaceae) using VCD and DFT calculations. *Tetrahedron Asymmetry* 21(19):2402–2407.

Batista, J. M., Jr, E. W. Blanch, and V. da S. Bolzani. 2015a. Recent advances in the use of vibrational chiroptical spectroscopic methods for stereochemical characterization of natural products. *Nat. Prod. Rep.* 32(9):1280–1302.

B Batista, J. M., Jr, and V. S. Bolzani. 2014. Determination of the absolute configuration of natural product molecules using vibrational circular dichroism. In: A.-U. Rahman (ed). *Studies in Natural Products Chemistry*. Amsterdam, the Netherlands: Elsevier.

Batista, J. M., S. N. López, J. da S. Mota, K. L. Vanzolini, Q. B. Cass, D. Rinaldo, W. Vilegas, V. da S. Bolzani, M. J. Kato, and M. Furlan. 2009. Resolution and absolute configuration assignment of a natural racemic chromane from Peperomia obtusifolia (Piperaceae). *Chirality* 21(9):799–801.

Batista, J. M., Jr, B. Wang, M. V. Castelli, E. W. Blanch, and S. N. López. 2015b. Absolute configuration assignment of an unusual homoisoflavanone from Polygonum ferrugineum using a combination of chiroptical methods. *Tetrahedron Lett.* 56(44):6142–6144.

Bell, A. F., L. D. Barron, and L. Hecht. 1994. Vibrational Raman optical activity study of d-glucose. *Carbohydr. Res.* 257(1):11–24.

Bell, A. F., L. Hecht, and L. D. Barron. 1993. Low-wavenumber vibrational Raman optical activity of carbohydrates. *J. Raman Spectrosc.* 24(9):633–635.

Bell, A. F., L. Hecht, and L. D. Barron. 1994. Disaccharide solution stereochemistry from vibrational Raman optical activity. *J. Am. Chem. Soc.* 116(12):5155–5161.

Bell, A. F., L. Hecht, and L. D. Barron. 1995. Polysaccharide vibrational Raman optical activity: Laminarin and pullulan. *J. Raman Spectrosc.* 26(12):1071–1074.

Ben-Moshe, A., B. M. Maoz, A. O. Govorov, and G. Markovich. 2013. Chirality and chiroptical effects in inorganic nanocrystal systems with plasmon and exciton resonances. *Chem. Soc. Rev.* 42(16):7028–7041.

Berthod, A., T. L. Xiao, Y. Liu, W. S. Jenks, and D. W. Armstrong. 2002. Separation of chiral sulfoxides by liquid chromatography using macrocyclic glycopeptide chiral stationary phases. *J. Chromatogr. A* 955(1):53–69; (see also *J. Chromatography A* 2004, 1047, 163).

Besse, P., G. Baziard-Mouysset, K. Boubekeur, P. Palvadeau, H. Veschambre, M. Payard, and G. Mousset. 1999. Microbiological reductions of chromen-4-one derivatives. *Tetrahedron Asymmetry* 10(24):4745–4754.

Bettelheim, F. A. 1971. *Experimental Physical Chemistry*. Philadelphia, PA: Saunders.

Bhattacharjya, S., J. Venkatraman, P. Balaram, and A. Kumar. 1999. Fluoroalcohols as structure modifiers in peptides and proteins: Hexafluoroacetone hydrate stabilizes a helical conformation of melittin at low pH. *J. Pep. Res.* 54(2):100–111.

Birke, S. S., I. Agbaje, and M. Diem. 1992. Experimental and computational infrared CD studies of prototypical peptide conformations. *Biochemistry* 31:450–455.

Blanch, E. W., D. J. Robinson, L. Hecht, C. D. Syme, K. Nielsen, and L. D. Barron. 2002. Solution structures of potato virus X and narcissus mosaic virus from Raman optical activity. *J. Gen. Virol.* 83(1):241–246.

Bose, P. K., and P. L. Polavarapu. 1999a. Acetate groups as probes of the stereochemistry of carbohydrates: A vibrational circular dichroism study. *Carbohydr. Res.* 322(1–2):135–141.

Bose, P. K., and P. L. Polavarapu. 1999b. Vibrational circular dichroism is a sensitive probe of the glycosidic linkage: Oligosaccharides of glucose. *J. Am. Chem. Soc.* 121(25):6094–6095.

Bose, P. K., and P. L. Polavarapu. 1999c. Vibrational circular dichroism of monosaccharides. *Carbohydr. Res.* 319(1–4):172–183.

Bour, P., and T. A. Keiderling. 1992. Computational evaluation of the coupled oscillator model in the vibrational circular dichroism of selected small molecules. *J. Am. Chem. Soc.* 114:9100–9105.

Bouř, P., and T. A. Keiderling. 2005. Vibrational spectral simulation for peptides of mixed secondary structure: Method comparisons with the trpzip model hairpin. *J. Phys. Chem. B* 109(49):23687–23697.

Bouř, P., J. Sopková, L. Bednárová, P. Maloň, and T. A. Keiderling. 1997. Transfer of molecular property tensors in cartesian coordinates: A new algorithm for simulation of vibrational spectra. *J. Comput. Chem.* 18(5):646–659.

Bouř, P., C. N. Tam, M. Shaharuzzaman, J. S. Chickos, and T. A. Keiderling. 1996. Vibrational optical activity study of trans-succinic-d2 anhydride. *J. Phys. Chem.* 100(37):15041–15048.

Bringmann, G., T. Bruhn, K. Maksimenka, and Y. Hemberger. 2009. The assignment of absolute stereostructures through quantum chemical circular dichroism calculations. *Eur. J. Org. Chem.* 2009(17):2717–2727.

Brizard, A., D. Berthier, C. Aimé, T. Buffeteau, D. Cavagnat, L. Ducasse, I. Huc, and R. Oda. 2009. Molecular and supramolecular chirality in gemini-tartrate amphiphiles studied by electronic and vibrational circular dichroisms. *Chirality* 21(1E):E153–E162.

Brotin, T., D. Cavagnat, and T. Buffeteau. 2008. Conformational changes in cryptophane having C1-symmetry studied by vibrational circular dichroism. *J. Phys. Chem. A* 112(36):8464–8470.

Brotin, T., D. Cavagnat, J.-P. Dutasta, and T. Buffeteau. 2006. Vibrational circular dichroism study of optically pure cryptophane-A. *J. Am. Chem. Soc.* 128(16):5533–5540.

Brotin, T., N. Daugey, N. Vanthuyne, E. Jeanneau, L. Ducasse, and T. Buffeteau. 2015. Chiroptical properties of cryptophane-223 and -233 investigated by ECD, VCD, and ROA spectroscopy. *J. Phys. Chem. B* 119(27):8631–8639.

Brotin, T., N. Vanthuyne, D. Cavagnat, L. Ducasse, and T. Buffeteau. 2014. Chiroptical properties of nona- and dodecamethoxy cryptophanes. *J. Org. Chem.* 79(13):6028–6036.

Bruhn, T., G. Pescitelli, S. Jurinovich, A. Schaumlöffel, F. Witterauf, J. Ahrens, M. Bröring, and G. Bringmann. 2014. Axially chiral BODIPY DYEmers: An apparent exception to the exciton chirality rule. *Angew. Chem. Int. Ed. Engl.* 53:14592–14595.

Brunsveld, L., J. A. J. M. Vekemans, J. H. K. K. Hirschberg, R. P. Sijbesma, and E. W. Meijer. 2002. Hierarchical formation of helical supramolecular polymers via stacking of hydrogen-bonded pairs in water. *Proc. Nat. Acad. Sci.* 99(8):4977–4982.

Buffeteau, T., L. Ducasse, L. Poniman, N. Delsuc, and I. Huc. 2006. Vibrational circular dichroism and ab initio structure elucidation of an aromatic foldamer. *Chem. Commun.* (25):2714–2716.

Bultinck, P., F. L. Cherblanc, M. J. Fuchter, W. A. Herrebout, Y.-P. Lo, H. S. Rzepa, G. Siligardi, and M. Weimar. 2015. Chiroptical studies on brevianamide B: Vibrational and electronic circular dichroism confronted. *J. Org. Chem.* 80(7):3359–3367.

Burgi, T., A. Urakawa, B. Behzadi, K.-H. Ernst, and A. Baiker. 2004. The absolute configuration of heptahelicene: A VCD spectroscopy study. *New J. Chem.* 28(3):332–334.

Canary, J. W. 2009. Redox-triggered chiroptical molecular switches. *Chem. Soc. Rev.* 38(3):747–756.

Canary, J. W., and Z. Dai. 2012. *Dynamic* stereochemistry and chiroptical spectroscopy of metallo-organic compounds. In: N. Berova, P. L. Polavarapu, K. Nakanishi, and R. W. Woody (eds). *Comprehensive Chiroptical Spectroscopy: Applications in Stereochemical Analysis of Synthetic Compounds, Natural Products and Biomolecules.* Vol. 2. New York, NY: John Wiley & Sons.

Canceill, J., A. Collet, and G. Gottarelli. 1984. Optical activity due to isotopic substitution. Synthesis, stereochemistry, and circular dichroism of (+)- and (−)-[2,7,12-2H3]cyclotribenzylene. *J. Am. Chem. Soc.* 106(20):5997–6003.

Canceill, J., A. Collet, G. Gottarelli, and P. Palmieri. 1987. Synthesis and exciton optical activity of D3-cryptophanes. *J. Am. Chem. Soc.* 109(21):6454–6464.

Cao, H., T. Ben, Z. Su, M. Zhang, Y. Kan, X. Yan, W. Zhang, and Y. Wei. 2005. Absolute configuration determination of a new chiral rigid bisetherketone macrocycle containing binaphthyl and thioether moieties by vibrational circular dichroism. *Macromol. Chem. Phys.* 206(11):1140–1145.

Caricato, M., P. H. Vaccaro, T. D. Crawford, K. B. Wiberg, and P. Lahiri. 2014. Insights on the origin of the unusually large specific rotation of (1S,4S)-norbornenone. *J. Phys. Chem. A* 118(26):4863–4871.

Carosati, E., R. Budriesi, P. Ioan, G. Cruciani, F. Fusi, M. Frosini, S. Saponara, F. Gasparrini, A. Ciogli, C. Villani, P. J. Stephens, F. J. Devlin, D. Spinelli, and A. Chiarini. 2009. Stereoselective behavior of the functional diltiazem analogue 1-[(4-chlorophenyl)sulfonyl]-2-(2-thienyl)pyrrolidine, a new L-type calcium channel blocker. *J. Med. Chem.* 52(21):6637–6648.

Carosati, E., G. Cruciani, A. Chiarini, R. Budriesi, P. Ioan, R. Spisani, D. Spinelli, B. Cosimelli, F. Fusi, M. Frosini, R. Matucci, F. Gasparrini, A. Ciogli, P. J. Stephens, and F. J. Devlin. 2006. Calcium channel antagonists discovered by a multidisciplinary approach. *J. Med. Chem.* 49(17):5206–5216.

Cavagnat, D., T. Buffeteau, and T. Brotin. 2007. Synthesis and chiroptical properties of cryptophanes having C1-symmetry. *J. Org. Chem.* 73(1):66–75.

Cheeseman, J. R., M. S. Shaik, P. L. A. Popelier, and E. W. Blanch. 2011. Calculation of Raman optical activity spectra of methyl-β-d-glucose incorporating a full molecular dynamics simulation of hydration effects. *J. Am. Chem. Soc.* 133(13):4991–4997.

Chen, G., and G. J. Mohamed. 2010. Complex protein patterns formation via salt-induced self-assembly and droplet evaporation. *Eur. Phys. J. E* 33(1):19–26.

Cherblanc, F., Y.-P. Lo, E. De Gussem, L. Alcazar-Fuoli, E. Bignell, Y. He, N. Chapman-Rothe, P. Bultinck, W. A. Herrebout, R. Brown, H. S. Rzepa, and M. J. Fuchter. 2011. On the determination of the stereochemistry of semisynthetic natural product analogues using chiroptical spectroscopy: Desulfurization of epidithiodioxopiperazine fungal metabolites. *Chem.—Eur. J.* 17(42):11868–11875.

Cherblanc, F. L., Y.-P. Lo, W. A. Herrebout, P. Bultinck, H. S. Rzepa, and M. J. Fuchter. 2013. Mechanistic and chiroptical studies on the desulfurization of epidithiodioxopiperazines reveal universal retention of configuration at the bridgehead carbon atoms. *J. Org. Chem.* 78(23):11646–11655.

Chi, H., W. R. W. Welch, J. Kubelka, and T. A. Keiderling. 2013. Insight into the packing pattern of β2 fibrils: A model study of glutamic acid rich oligomers with 13C isotopic edited vibrational spectroscopy. *Biomacromolecules* 14(11):3880–3891.

Choi, J.-H., J.-S. Kim, and M. Cho. 2005. Amide I vibrational circular dichroism of polypeptides: Generalized fragmentation approximation method. *J. Chem. Phys.* 122(17):174903.

Cianciosi, S. J., N. Ragunathan, T. B. Freedman, L. A. Nafie, and J. E. Baldwin. 1990. Racemization and geometrical isomerization of (−)-(R,R)-cyclopropane-1,2-2H2. *J. Am. Chem. Soc.* 112(22):8204–8206.

Cianciosi, S. J., N. Ragunathan, T. B. Freedman, L. A. Nafie, D. K. Lewis, D. A. Glenar, and J. E. Baldwin. 1991. Racemization and geometrical isomerization of (2S,3S)-cyclopropane-1-13C-1,2,3-d3 at 407 .degree.C: Kinetically competitive one-center and two-center thermal epimerizations in an isotopically substituted cyclopropane. *J. Am. Chem. Soc.* 113(5):1864–1866.

Cianciosi, S. J., K. M. Spencer, T. B. Freedman, L. A. Nafie, and J. E. Baldwin. 1989. Synthesis and gas-phase vibrational circular dichroism of (+)-(S,S)-cyclopropane-1,2-2H2. *J. Am. Chem. Soc.* 111(5):1913–1915.

Cichewicz, R. H., L. J. Clifford, P. R. Lassen, X. Cao, T. B. Freedman, L. A. Nafie, J. D. Deschamps, V. A. Kenyon, J. R. Flanary, T. R. Holman, and P. Crews. 2005. Stereochemical determination and bioactivity assessment of (S)-(+)-curcuphenol dimers isolated from the marine sponge Didiscus aceratus and synthesized through laccase biocatalysis. *Biorg. Med. Chem.* 13(19):5600–5612.

Cocinero, E. J., A. Lesarri, P. Écija, Á. Cimas, B. G. Davis, F. J. Basterretxea, J. A. Fernández, and F. Castaño. 2013. Free fructose is conformationally locked. *J. Am. Chem. Soc.* 135(7):2845–2852.

Collet, A. 1996. *Cryptophans.* In: J. L. Artwood, J. E. D. Davies, D. D. Macnicol, and F. Vogtle (eds). *Comprehensive Supramolecular Chemistry*. New York, NY: Elsevier.

Colombo, L. M., C. Nastruzzi, P. Luigi Luisi, and R. M. Thomas. 1991. Chiroptical properties of lecithin reverse micelles and organogels. *Chirality* 3(6):495–502.

Compton, R. N. and M. A. Duncan. 2016. *Laser Experiments for Chemistry and Physics.* Oxford, UK: Oxford University Press.

Corradini, R., T. Tedeschi, S. Sforza, and R. Marchelli. 2012. Electronic circular dichroism of peptide nucleic acids and their analogues. In: N. Berova, P. L. Polavarapu, K. Nakanishi, and R. W. Woody (eds). *Comprehensive Chiroptical Spectroscopy.* Vol. 2. New York, NY: Wiley.

Costante, J., L. Hecht, P. L. Polavarapu, A. Collet, and L. D. Barron. 1997. Absolute configuration of bromochlorofluoromethane from experimental and ab initio theoretical vibrational Raman optical activity. *Angew. Chem. Int. Ed. Engl.* 36(8):885–887.

Covington, C. L., V. P. Nicu, and P. L. Polavarapu. 2015. Determination of absolute configurations using exciton chirality method for vibrational circular dichroism: Right answers for the wrong reasons? *J. Phys. Chem.* 119:10589–10601.

Crassous, J., Z. Jiang, V. Schurig, and P. L. Polavarapu. 2004. Preparation of (+)-chlorofluoroiodomethane, determination of its enantiomeric excess and of its absolute configuration. *Tetrahedron Asymmetry* 15(13):1995–2001.

Crassous, J., J. Rivera, N. S. Fender, L. Shu, L. Echegoyen, C. Thilgen, A. Herrmann, and F. Diederich. 1999. Chemistry of C84: Separation of three constitutional isomers and optical resolution of D2-C84 by using the "Bingel–Retro-Bingel" strategy. *Angew. Chem. Int. Ed. Engl.* 38(11):1613–1617.

Crawford, T. D., L. S. Owens, M. C. Tam, P. R. Schreiner, and H. Koch. 2005. Ab initio calculation of optical rotation in (P)-(+)-[4]triangulane. *J. Am. Chem. Soc.* 127(5):1368–1369.

da Silva, C. O., B. Mennucci, and T. Vreven. 2004. Density functional study of the optical rotation of glucose in aqueous solution. *J. Org. Chem.* 69(23):8161–8164.

Dai, Z., J. Lee, and W. Zhang. 2012. Chiroptical switches: Applications in sensing and catalysis. *Molecules* 17(2):1247–1277.

Daugey, N., T. Brotin, N. Vanthuyne, D. Cavagnat, and T. Buffeteau. 2014. Raman optical activity of enantiopure cryptophanes. *J. Phys. Chem. B* 118(19):5211–5217.

de Meijere, A., A. F. Khlebnikov, R. R. Kostikov, S. I. Kozhushkov, P. R. Schreiner, A. Wittkopp, and D. S. Yufit. 1999. The first enantiomerically pure triangulane (M)-trispiro[2.0.0.2.1.1]nonane Is a σ-[4]helicene. *Angew. Chem. Int. Ed. Engl.* 38(23):3474–3477.

de Meijere, A., A. F. Khlebnikov, S. I. Kozhushkov, R. R. Kostikov, P. R. Schreiner, A. Wittkopp, C. Rinderspacher, H. Menzel, D. S. Yufit, and Judith A. K. Howard. 2002. The first enantiomerically pure [n]triangulanes and analogues: σ-[n] helicenes with remarkable features. *Chem.—Eur. J.* 8(4):828–842.

Debie, E., L. Jaspers, P. Bultinck, W. Herrebout, and B. Van Der Veken. 2008. Induced solvent chirality: A VCD study of camphor in CDCl3. *Chem. Phys. Lett.* 450(4–6):426–430.

Derewacz, D. K., C. R. McNees, G. Scalmani, C. L. Covington, G. Shanmugam, L. J. Marnett, P. L. Polavarapu, and B. O. Bachmann. 2014. Structure and stereochemical determination of hypogeamicins from a Cave-derived actinomycete. *J. Nat. Prod.* 77(8):1759–1763.

Devlin, F. J., and P. J. Stephens. 1999. Ab initio density functional theory study of the structure and vibrational spectra of cyclohexanone and its isotopomers. *J. Phys. Chem. A* 103(4):527–538.

Devlin, F. J., P. J. Stephens, and P. Besse. 2005a. Are the absolute configurations of 2-(1-hydroxyethyl)-chromen-4-one and its 6-bromo derivative determined by X-ray crystallography correct? A vibrational circular dichroism study of their acetate derivatives. *Tetrahedron Asymmetry* 16(8):1557–1566.

Devlin, F. J., P. J. Stephens, and P. Besse. 2005b. Conformational rigidification via derivatization facilitates the determination of absolute configuration using chiroptical spectroscopy: A case study of the chiral alcohol endo-Borneol. *J. Org. Chem.* 70(8):2980–2993.

Devlin, F. J., P. J. Stephens, and B. Figadère. 2009. Determination of the absolute configuration of the natural product klaivanolide via density functional calculations of vibrational circular dichroism (VCD). *Chirality* 21(1E):E48–E53.

Devlin, F. J., P. J. Stephens, and O. Bortolini. 2005c. Determination of absolute configuration using vibrational circular dichroism spectroscopy: Phenyl glycidic acid derivatives obtained via asymmetric epoxidation using oxone and a keto bile acid. *Tetrahedron Asymmetry* 16(15):2653–2663.

Devlin, F. J., P. J. Stephens, P. Scafato, S. Superchi, and C. Rosini. 2001. Determination of absolute configuration using vibrational circular dichroism spectroscopy: The chiral sulfoxide 1-thiochroman S-oxide. *Tetrahedron Asymmetry* 12(11):1551–1558.

Devlin, F. J., P. J. Stephens, P. Scafato, S. Superchi, and C. Rosini. 2002a. Conformational analysis using infrared and vibrational circular dichroism spectroscopies: The chiral cyclic sulfoxides 1-thiochroman-4-one S-oxide, 1-thiaindan S-oxide and 1-thiochroman S-oxide. *J. Phys. Chem. A* 106(44):10510–10524.

Devlin, F. J., P. J. Stephens, P. Scafato, S. Superchi, and C. Rosini. 2002b. Determination of absolute configuration using vibrational circular dichroism spectroscopy: The chiral sulfoxide 1-thiochromanone S-oxide. *Chirality* 14(5):400–406.

Dezhahang, Z., M. R. Poopari, F. E. Hernandez, C. Diaz, and Y. Xu. 2014. Diastereomeric preference of a triply axial chiral binaphthyl based molecule: A concentration dependent study by chiroptical spectroscopies. *Phys. Chem. Chem. Phys.* 16(25):12959–12967.

Di Meo, F., M. N. Pedersen, J. Rubio-Magnieto, M. Surin, M. Linares, and P. Norman. 2015. DNA electronic circular dichroism on the inter-base pair scale: An experimental–theoretical case study of the AT homo-oligonucleotide. *J. Phys. Chem. Lett.* 6(3):355–359.

Djerassi, C. 1960. *Optical Rotatory Dispersion; Applications to Organic Chemistry, McGraw-Hill Series in Advanced Chemistry*. New York, NY: McGraw-Hill.

Dolain, C., H. Jiang, J.-M. Léger, P. Guionneau, and I. Huc. 2005. Chiral induction in quinoline-derived oligoamide foldamers: Assignment of helical handedness and role of steric effects. *J. Am. Chem. Soc.* 127(37):12943–12951.

Domingos, S. R., A. Huerta-Viga, L. Baij, S. Amirjalayer, D. A. E. Dunnebier, A. J. C. Walters, M. Finger, L. A. Nafie, B. de Bruin, W. J. Buma, and S. Woutersen. 2014a. Amplified vibrational circular dichroism as a probe of local biomolecular structure. *J. Am. Chem. Soc.* 136(9):3530–3535.

Domingos, S. R., H. J. Sanders, F. Hartl, W. J. Buma, and S. Woutersen. 2014b. Switchable amplification of vibrational circular dichroism as a probe of local structure. *Angewandte Chemie Int. Ed.* 53:14042–14045.

Donnoli, M. I., E. Giorgio, S. Superchi, and C. Rosini. 2003. Circular dichroism spectra and absolute configuration of some aryl methyl sulfoxides. *Org. Biomol. Chem.* 1(19):3444–3449.

Doyle, D. A., and B. A. Wallace. 1998. Shifting the equilibrium mixture of gramicidin double helices toward a single conformation with multivalent cationic salts. *Biophys. J.* 75(2):635–640.

Drabowicz, J., B. Dudziński, M. Mikołajczyk, F. Wang, A. Dehlavi, J. Goring, M. Park, C. J. Rizzo, P. L. Polavarapu, P. Biscarini, M. W. Wieczorek, and W. R. Majzner. 2001. Absolute configuration, predominant conformations, and vibrational circular dichroism spectra of enantiomers of n-butyl tert-butyl sulfoxide. *J. Org. Chem.* 66(4):1122–1129.

Drabowicz, J., A. Zajac, P. Lyzwa, P. J. Stephens, J.-J. Pan, and F. J. Devlin. 2008. Determination of the absolute configurations of isotopically chiral molecules using vibrational circular dichroism (VCD) spectroscopy: The isotopically chiral sulfoxide, perdeuteriophenyl-phenyl-sulfoxide. *Tetrahedron Asymmetry* 19(3):288–294.

Ducasse, L., F. Castet, A. Fritsch, I. Huc, and T. Buffeteau. 2007. Density functional theory calculations and vibrational circular dichroism of aromatic foldamers. *J. Phys. Chem. A* 111(23):5092–5098.

Dukor, R. K., and T. A. Keiderling. 1991. Reassessment of the random coil conformation: Vibrational CD study of proline oligopeptides and related polypeptides. *Biopolymers* 31(14):1747–1761.

Dunmire, D., T. B. Freedman, L. A. Nafie, C. Aeschlimann, J. G. Gerber, and J. Gal. 2005. Determination of the absolute configuration and solution conformation of the antifungal agents ketoconazole, itraconazole, and miconazole with vibrational circular dichroism. *Chirality* 17(S1):S101–S108.

Dyatkin, A. B., T. B. Freedman, X. Cao, R. K. Dukor, B. E. Maryanoff, C. A. Maryanoff, J. M. Matthews, R. D. Shah, and L. A. Nafie. 2002. Determination of the absolute configuration of a key tricyclic component of a novel vasopressin receptor antagonist by use of vibrational circular dichroism. *Chirality* 14(2–3):215–219.

Dzwolak, W. 2014. Chirality and chiroptical properties of amyloid fibrils. *Chirality* 26(9):580–587.

Eliel, E. L., S. H. Wilen, and M. P. Doyle. 2001. *Basic Organic Stereochemistry*. New York, NY: Wiley.

Ferrié, L., S. Ferhi, G. Bernadat, and B. Figadère. 2014. Toward the total synthesis of klaivanolide: Complete reinterpretation of its originally assigned structure. *Eur. J. Org. Chem.* 2014(28):6183–6189.

Figadere, B., F. J. Devlin, J. G. Millar, and P. J. Stephens. 2008. Determination of the absolute configuration of the sex pheromone of the obscure mealybug by vibrational circular dichroism analysis. *Chem. Commun.* (9):1106–1108.

Freedman, T. B., X. Cao, A. Rajca, H. Wang, and L. A. Nafie. 2003c. Determination of absolute configuration in molecules with chiral axes by vibrational circular dichroism: A C2-symmetric annelated heptathiophene and a D2-symmetric dimer of 1,1′-binaphthyl. *J. Phys. Chem. A* 107(39):7692–7696.

Freedman, T. B., X. Cao, L. A. Nafie, M. Kalbermatter, A. Linden, and A. J. Rippert. 2003a. An unexpected atropisomerically stable 1,1-biphenyl at ambient temperature in solution, elucidated by vibrational circular dichroism (VCD). *Helv. Chim. Acta* 86(9):3141–3155.

Freedman, T. B., X. Cao, L. A. Nafie, M. Kalbermatter, A. Linden, and A. J. Rippert. 2005. Determination of the atropisomeric stability and solution conformation of asymmetrically substituted biphenyls by means of vibrational circular dichroism (VCD). *Helv. Chim. Acta* 88(8):2302–2314.

Freedman, T. B., X. Cao, R. V. Oliveira, Q. B. Cass, and L. A. Nafie. 2003b. Determination of the absolute configuration and solution conformation of gossypol by vibrational circular dichroism. *Chirality* 15(2):196–200.

Freedman, T. B., S. J. Cianciosi, N. Ragunathan, J. E. Baldwin, and L. A. Nafie. 1991a. Optical activity arising from carbon-13 substitution: Vibrational circular dichroism study of (2S,3S)-cyclopropane-1-13C,2H-2,3-2H2. *J. Am. Chem. Soc.* 113(22):8298–8305.

Freedman, T. B., R. K. Dukor, P. J. C. M. van Hoof, E. R. Kellenbach, and L. A. Nafie. 2002. Determination of the absolute configuration of (−)-mirtazapine by vibrational circular dichroism. *Helv. Chim. Acta* 85(4):1160–1165.

Freedman, T. B., F. Long, M. Citra, and L. A. Nafie. 1999. Hydrogen-stretching vibrational circular dichroism spectroscopy: Absolute configuration and solution conformation of selected pharmaceutical molecules. *Enantiomer* 4(2):103–119.

Freedman, T. B., M. G. Paterlini, N. S. Lee, L. A. Nafie, J. M. Schwab, and T. Ray. 1987. Vibrational circular dichroism in the carbon-hydrogen and carbon-deuterium stretching modes of (S,S)-[2,3-2H2]oxirane. *J. Am. Chem. Soc.* 109(15):4727–4728.

Freedman, T. B., N. Ragunathan, and S. Alexander. 1994. Vibrational circular dichroism in ephedra molecules. Experimental measurement and ab initio calculation. *Faraday Discuss.* 99(0):131–149.

Freedman, T. B., K. M. Spencer, N. Ragunathan, Laurence A. Nafie, Jeffrey A. Moore, and J. M. Schwab. 1991b. Vibrational circular dichroism of (S,S)-[2,3-2H2]oxirane in the gas phase and in solution. *Can. J. Chem.* 69(11):1619–1629.

Fugmann, B., S. Arnold, W. Steglich, J. Fleischhauer, C. Repges, A. Koslowski, and G. Raabe. 2001. Pigments from the puffball calvatia rubro-flava—Isolation, structural elucidation and synthesis. *Eur. J. Org. Chem.* 2001(16):3097–3104.

Fulara, A., A. Lakhani, S. Woójcik, H. Nieznańska, T. A. Keiderling, and W. Dzwolak. 2011. Spiral superstructures of amyloid-like fibrils of polyglutamic acid: An infrared absorption and vibrational circular dichroism study. *J. Phys. Chem. B* 115(37):11010–11016.

Furche, F., and R. Ahlrichs. 2002. Absolute configuration of D2-symmetric fullerene C84. *J. Am. Chem. Soc.* 124(15):3804–3805.

Furche, F., R. Ahlrichs, C. Wachsmann, E. Weber, A. Sobanski, F. Vögtle, and S. Grimme. 2000. Circular dichroism of helicenes investigated by time-dependent density functional theory. *J. Am. Chem. Soc.* 122(8):1717–1724.

Furo, T., T. Mori, Y. Origane, T. Wada, H. Izumi, and Y. Inoue. 2006. Absolute configuration determination of donor–acceptor [2.2]paracyclophanes by comparison of theoretical and experimental vibrational circular dichroism spectra. *Chirality* 18(3):205–211.

Furo, T., T. Mori, T. Wada, and Y. Inoue. 2005. Absolute configuration of chiral [2.2] paracyclophanes with intramolecular charge-transfer interaction. Failure of the exciton chirality method and use of the sector rule applied to the cotton effect of the CT transition. *J. Am. Chem. Soc.* 127(23):8242–8243.

Gaussian 09. 2013. Gaussian Inc., Willingford, CT.

Gautier, C., and T. Burgi. 2005. Vibrational circular dichroism of N-acetyl-l-cysteine protected gold nanoparticles. *Chem. Commun.* (43):5393–5395.

Gautier, C., and T. Bürgi. 2006. Chiral N-isobutyryl-cysteine protected gold nanoparticles: Preparation, size selection, and optical activity in the UV–vis and infrared. *J. Am. Chem. Soc.* 128(34):11079–11087.

Gautier, C., and T. Bürgi. 2009. Chiral gold nanoparticles. *ChemPhysChem* 10(3):483–492.

Gautier, C., and T. Bürgi. 2010. Vibrational circular dichroism of adsorbed molecules: BINAS on gold nanoparticles†. *J. Phys. Chem. C* 114(38):15897–15902.

George, J., and K. G. Thomas. 2010. Surface plasmon coupled circular dichroism of Au nanoparticles on peptide nanotubes. *J. Am. Chem. Soc.* 132(8):2502–2503.

Gobi, S., K. Knapp, E. Vass, Z. Majer, G. Magyarfalvi, M. Hollosi, and G. Tarczay. 2010. Is [small beta]-homo-proline a pseudo-[gamma]-turn forming element of [small beta]-peptides? An IR and VCD spectroscopic study on Ac-[small beta]-HPro-NHMe in cryogenic matrices and solutions. *Phys. Chem. Chem. Phys.* 12(41):13603–13615.

Gobi, S., E. Vass, G. Magyarfalvi, and G. Tarczay. 2011. Effects of strong and weak hydrogen bond formation on VCD spectra: A case study of 2-chloropropionic acid. *Phys. Chem. Chem. Phys.* 13(31):13972–13984.

Goldsmith, M.-R., N. Jayasuriya, D. N. Beratan, and P. Wipf. 2003. Optical rotation of noncovalent aggregates. *J. Am. Chem. Soc.* 125:15696–15697.

Goto, H., N. Harada, J. Crassous, and F. Diederich. 1998. Absolute configuration of chiral fullerenes and covalent derivatives from their calculated circular dichroism spectra. *J. Chem. Soc., Perkin Trans.* 2(8):1719–1724.

Gotoh, H., S. H. Lin, and H. Eyring. 1974. A statistical study of optical rotation for synthetic D,L-copolypeptide solutions. *Proc. Nat. Acad. Sci.* 71(12):4675–4678.

Gottarelli, G., G. Proni, G. P. Spada, D. Fabbri, S. Gladiali, and C. Rosini. 1996. Conformational and configurational analysis of 4,4′-biphenanthryl derivatives and related helicenes by circular dichroism spectroscopy and cholesteric induction in nematic mesophases. *J. Org. Chem.* 61(6):2013–2019.

Gouin, J., T. Bürgi, L. Guénée, and J. Lacour. 2014. Convergent synthesis, resolution, and chiroptical properties of dimethoxychromenoacridinium ions. *Org. Lett.* 16(14):3800–3803.

Gray, D. M. 2012. Circular dichroism of protein-nucleic acid interactions. In: N. Berova, P. L. Polavarapu, K. Nakanishi, and R. W. Woody (eds). *Comprehensive Chiroptical Spectroscopy*. Vol. 2. New York, NY: Wiley.

Greenstein, J. P., and M. Winitz (eds). 1961. *Chemistry of the Amino Acids*. Vol. 1. New York, NY: John Wiley & Sons.

Grell, E., and Th. Funck. 1973. Dynamic properties and membrane activity of ion specific antibiotics. *J. Supramol. Struct.* 1(4–5):307–335.

Grimme, S., and F. Neese. 2007. Double-hybrid density functional theory for excited electronic states of molecules. *J. Chem. Phys.* 127(15):154116.
Guerrero-Martínez, A., B. Auguié, J. L. Alonso-Gómez, Z. Džolić, S. Gómez-Graña, M. Žinić, M. M. Cid, and L. M. Liz-Marzán. 2011. Intense optical activity from three-dimensional chiral ordering of plasmonic nanoantennas. *Angew. Chem. Int. Ed. Engl.* 50(24):5499–5503.
Guo, C., R. D. Shah, R. K. Dukor, X. Cao, T. B. Freedman, and L. A. Nafie. 2004. Determination of enantiomeric excess in samples of chiral molecules using fourier transform vibrational circular dichroism spectroscopy: Simulation of real-time reaction monitoring. *Anal. Chem.* 76(23):6956–6966.
Haesler, J., I. Schindelholz, E. Riguet, C. G. Bochet, and W. Hug. 2007. Absolute configuration of chirally deuterated neopentane. *Nature* 446(7135):526–529.
Haleema, S., P. V. Sasi, I. Ibnusaud, P. L. Polavarapu, and H. B. Kagan. 2012. Enantiomerically pure compounds related to chiral hydroxy acids derived from renewable resources. *RSC Adv.* 2(25):9257–9285.
Hanabusa, K., M. Yamada, M. Kimura, and H. Shirai. 1996. Prominent gelation and chiral aggregation of alkylamides derived from trans-1,2-diaminocyclohexane. *Angew. Chem. Int. Ed. Engl.* 35(17):1949–1951.
Harada, N., T. Hattori, T. Suzuki, A. Okamura, H. Ono, S. Miyano, and H. Uda. 1993. Absolute stereochemistry of 1-(9-phenanthryl)-2-naphthoic acid as determined by CD and X-ray methods. *Tetrahedron Asymmetry* 4(8):1789–1792.
Harada, N., and K. Nakanishi. 1983. *Circular Dichroic Spectroscopy: Exciton Coupling in Organic Stereochemistry*. Mill Valley, CA: University Science Books.
Hase, Y., K. Nagai, H. Iida, K. Maeda, N. Ochi, K. Sawabe, K. Sakajiri, K. Okoshi, and E. Yashima. 2009. Mechanism of helix induction in poly(4-carboxyphenyl isocyanide) with chiral amines and memory of the macromolecular helicity and its helical structures. *J. Am. Chem. Soc.* 131(30):10719–10732.
Hattori, T., K. Sakurai, N. Koike, S. Miyano, H. Goto, F. Ishiya, and N. Harada. 1998. Is the CD exciton chirality method applicable to chiral 1,1′-biphenanthryl compounds? *J. Am. Chem. Soc.* 120(35):9086–9087.
He, Y., X. Cao, L. A. Nafie, and T. B. Freedman. 2001. Ab initio VCD calculation of a transition-metal containing molecule and a new intensity enhancement mechanism for VCD. *J. Am. Chem. Soc.* 123 (45):11320–11321.
He, J., and P. L. Polavarapu. 2005a. Determination of the absolute configuration of chiral α-aryloxypropanoic acids using vibrational circular dichroism studies: 2-(2-chlorophenoxy) propanoic acid and 2-(3-chlorophenoxy) propanoic acid. *Spectrochim. Acta A Mol Biomol Spectrosc.* 61(7):1327–1334.
He, J., and P. L. Polavarapu. 2005b. Determination of intermolecular hydrogen bonded conformers of α-aryloxypropanoic acids using density functional theory predictions of vibrational absorption and vibrational circular dichroism spectra. *J. Chem. Theory Comput.* 1(3):506–514.
He, P., X. Wang, X. Guo, Y. Ji, C. Zhou, S. Shen, D. Hu, X. Yang, D. Luo, R. Dukor, and H. Zhu. 2014. Vibrational circular dichroism study for natural bioactive schizandrin and reassignment of its absolute configuration. *Tetrahedron Lett.* 55(18):2965–2968.
He, J., F. Wang, and P. L. Polavarapu. 2005. Absolute configurations of chiral herbicides determined from vibrational circular dichroism. *Chirality* 17(S1):S1–S8.
Hecht, L., A. L. Phillips, and L. D. Barron. 1995. Determination of enantiomeric excess using Raman optical activity. *J. Raman Spectrosc.* 26(8–9):727–732.

Hembury, G. A., V. V. Borovkov, and Y. Inoue. 2007. Chirality-sensing supramolecular systems. *Chem. Rev.* 108(1):1–73.

Henderson, D. O., and P. L. Polavarapu. 1986. Fourier transform infrared vibrational circular dichroism of matrix-isolated molecules. *J. Am. Chem. Soc.* 108(22):7110–7111.

Herrmann, C., K. Ruud, and M. Reiher. 2006. Can Raman optical activity separate axial from local chirality? A theoretical study of helical deca-alanine. *ChemPhysChem* 7(10):2189–2196.

Herse, C., D. Bas, F. C. Krebs, T. Bürgi, J. Weber, T. Wesolowski, B. W. Laursen, and J. Lacour. 2003. A highly configurationally stable [4]heterohelicenium cation. *Angew. Chem. Int. Ed. Engl.* 42(27):3162–3166.

Heshmat, M., E. J. Baerends, P. L. Polavarapu, and V. P. Nicu. 2014. The importance of large-amplitude motions for the interpretation of mid-infrared vibrational absorption and circular dichroism spectra: 6,6′-dibromo-[1,1′-binaphthalene]-2,2′-diol in dimethyl sulfoxide. *J. Phys. Chem. A* 118:4766–4777.

Holmén, A., J. Oxelbark, and S. Allenmark. 2003. Direct determination of the absolute configuration of a cyclic thiolsulfinate by VCD spectroscopy. *Tetrahedron Asymmetry* 14(15):2267–2269.

Holzwarth, G., E. C. Hsu, H. S. Mosher, T. R. Faulkner, and A. Moscowitz. 1974. Infrared circular dichroism of carbon-hydrogen and carbon-deuterium stretching modes. Observations. *J. Am. Chem. Soc.* 96(1):251–252.

Hopmann, K. H., J. Šebestík, J. Novotná, W. Stensen, M. Urbanová, J. Svenson, J. S. Svendsen, P. Bouř, and K. Ruud. 2012. Determining the absolute configuration of two marine compounds using vibrational chiroptical spectroscopy. *J. Org. Chem.* 77(2):858–869.

Horeau, A. 1969. Interactions d'enantiomeres en solution ; influence sur le pouvoir rotatoire: Purete optique et purete enantiomerique. *Tetrahedron Lett.* 10:3121–3124.

Huang, X., B. Borhan, B. H. Rickman, K. Nakanishi, and N. Berova. 2000. Zinc porphyrin tweezer in host-guest complexation: Determination of absolute configurations of primary monoamines by circular dichroism. *Chem.—Eur. J.* 6(2):216–224.

Huang, X., B. H. Rickman, B. Borhan, N. Berova, and K. Nakanishi. 1998. Zinc porphyrin tweezer in host–guest complexation: Determination of absolute configurations of diamines, amino acids, and amino alcohols by circular dichroism. *J. Am. Chem. Soc.* 120(24):6185–6186.

Hug, W., G. Zuber, A. de Meijere, A. F. Khlebnikov, and H. J. Hansen. 2001. Raman optical activity of a purely σ-bonded helical chromophore: (−)-(M)-σ-[4]helicene. *Helv. Chim. Acta* 84(1):1–21.

Humbert-Droz, M., P. Oulevey, L. M. L. Daku, S. Luber, H. Hagemann, and T. Burgi. 2014. Where does the Raman optical activity of [Rh(en)3]3+ come from? Insight from a combined experimental and theoretical approach. *Phys. Chem. Chem. Phys.* 16(42):23260–23273.

Izumi, H., S. Futamura, N. Tokita, and Y. Hamada. 2006. Fliplike motion in the thalidomide dimer: Conformational analysis of (R)-thalidomide using vibrational circular dichroism spectroscopy. *J. Org. Chem.* 72(1):277–279.

Izumi, H., A. Ogata, L. A. Nafie, and R. K. Dukor. 2008. Vibrational circular dichroism analysis reveals a conformational change of the baccatin III ring of paclitaxel: Visualization of onformations using a new code for tructure–activity relationships. *J. Org. Chem.* 73(6):2367–2372.

Izumi, H., A. Ogata, L. A. Nafie, and R. K. Dukor. 2009a. A revised conformational code for the exhaustive analysis of conformers with one-to-one correspondence between conformation and code: Application to the VCD analysis of (S)-ibuprofen. *J. Org. Chem.* 74(3):1231–1236.

Izumi, H., A. Ogata, L. A. Nafie, and R. K. Dukor. 2009b. Structural determination of molecular stereochemistry using VCD spectroscopy and a conformational code: Absolute configuration and solution conformation of a chiral liquid pesticide, (R)-(+)-malathion. *Chirality* 21(1E):E172–E180.

Jacob, C. R., S. Luber, and M. Reiher. 2009. Understanding the signatures of secondary-structure elements in proteins with Raman optical activity spectroscopy. *Chem. Eur. J.* 15(48):13491–13508.

Jacob, C. R., and M. Reiher. 2009. Localizing normal modes in large molecules. *J. Chem. Phys.* 130(8):084106.

Jiang, N., R. X. Tan, and J. Ma. 2011. Simulations of solid-state vibrational circular dichroism spectroscopy of (S)-alternarlactam by using fragmentation quantum chemical calculations. *J. Phys. Chem. B* 115(12):2801–2813.

Johannessen, C., E. W. Blanch, C. Villani, S. Abbate, G. Longhi, N. R. Agarwal, M. Tommasini, and D. A. Lightner. 2013. Raman and ROA spectra of (−)- and (+)-2-Br-hexahelicene: Experimental and DFT studies of a π-conjugated chiral system. *J. Phys. Chem. B* 117(7):2221–2230.

Johannessen, C. and P. W. Thulstrup. 2007. Vibrational circular dichroism spectroscopy of a spin-triplet bis-(biuretato) cobaltate(iii) coordination compound with low-lying electronic transitions. *Dalton Transactions* (10):1028–1033.

Jose, K. V. J., D. Beckett, and K. Raghavachari. 2015. Vibrational circular dichroism spectra for large molecules through molecules-in-molecules fragment-based approach. *J. Chem. Theory Comput.* 11:4238–4247.

Jose, K. V. J., and K. Raghavachari. 2016. Raman optical activity spectra for large molecules through molecules-in-molecules fragment-based approach. *J. Chem. Theory Comput.* 12(2):585–594.

Joseph-Nathan, P., and B. Gordillo-Romain. 2015. Vibrational circular dichroism absolute configuration determination of natural products. In: D. A. Kinghorn, H. Falk, and J. I. Kobayashi (eds). *Progress in the Chemistry of Organic Natural Products*. Vol. 100. Berlin: Springer.

Julinek, O., M. Krupicka, W. Lindner, and M. Urbanova. 2010. Enantioselective interaction of carbamoylated quinine and (S)-3,5-dinitrobenzoyl alanine: Theoretical and experimental circular dichroism study. *Phys. Chem. Chem. Phys.* 12(37):11487–11497.

Jurinovich, S., C. A. Guido, T. Bruhn, G. Pescitelli, and B. Mennucci. 2015. The role of magnetic-electric coupling in exciton-coupled ECD spectra: The case of bis-phenanthrenes. *Chem. Commun.* 51(52):10498–10501.

Julínek, O., M. Urbanová, and W. Lindner. 2009. Enantioselective complexation of carbamoylated quinine and quinidine with N-blocked amino acids: Vibrational and electronic circular dichroism study. *Anal. Bioanal. Chem.* 393:303–312.

Kaizaki, S. 2012. Applications of electronic circular dichroism to inorganic stereochemistry. In: N. Berova, P. L. Polavarapu, K. Nakanishi, and R. W. Woody (eds). *Comprehensive Chiroptical Spectroscopy: Applications in Stereochemical Analysis of Synthetic Compounds, Natural Products and Biomolecules*. Vol. 2. New York, NY: John Wiley & Sons.

Kawauchi, T., A. Kitaura, J. Kumaki, H. Kusanagi, and E. Yashima. 2008a. Helix-sense-controlled synthesis of optically active poly(methyl methacrylate) stereocomplexes. *J. Am. Chem. Soc.* 130(36):11889–11891.

Kawauchi, T., J. Kumaki, A. Kitaura, K. Okoshi, H. Kusanagi, K. Kobayashi, T. Sugai, H. Shinohara, and E. Yashima. 2008b. Encapsulation of fullerenes in a helical PMMA cavity leading to a robust processable complex with a macromolecular helicity memory. *Angew. Chem. Int. Ed. Engl.* 47(3):515–519.

Keiderling, T. A. 1986. Vibrational CD of biopolymers. *Nature* 322(6082):851–852.

Keiderling, T. A. 2000. Peptide and protein conformational studies with vibrational circular dichroism and related spectroscopies. In: N. Berova, K. Nakanishi, and R. W. Woody (eds). *Circular Dichroism*. New York, NY: John Wiley & Sons.

Kessinger, R., J. Crassous, A. Herrmann, M. Rüttimann, L. Echegoyen, and F. Diederich. 1998. Preparation of enantiomerically pure C76 with a general electrochemical method for the removal of Di(alkoxycarbonyl)methano bridges from methanofullerenes: The retro-bingel reaction. *Angew. Chem. Int. Ed. Engl.* 37(13–14):1919–1922.

Keiderling, T. A., and A. Lakhani. 2012. Conformational studies of biopolymers, peptides, proteins, and nucleic acids: A role for vibrational circular dichroism. In: N. Berova, P. L. Polavarapu, K. Nakanishi, and R. W. Woody (eds). *Comprehensive Chiroptical Spectroscopy: Applications in Stereochemical Analysis of Synthetic Compounds, Natural Products, and Biomolecules*. Vol. 2. New York, NY: John Wiley & Sons.

Kessler, J., T. A. Keiderling, and P. Bouř. 2014. Arrangement of fibril side chains studied by molecular dynamics and simulated infrared and vibrational circular dichroism spectra. *J. Phys. Chem. B* 118(24):6937–6945.

Klevens, H. B., and J. R. Platt. 1949. Spectral resemblances of cata-condensed hydrocarbons. *J. Chem. Phys.* 17(5):470–481.

Kohno, K., K. Morimoto, N. Manabe, T. Yajima, A. Yamagishi, and H. Sato. 2012. Promotion effects of optical antipodes on the formation of helical fibrils: Chiral perfluorinated gelators. *Chem. Commun.* 48(32):3860–3862.

Komori, K., T. Taniguchi, S. Mizutani, K. Monde, K. Kuramochi, and K. Tsubaki. 2014. Short synthesis of berkeleyamide D and determination of the absolute configuration by the vibrational circular dichroism exciton chirality method. *Org. Lett.* 16:1386–1389.

Kondru, R. K., S. Lim, P. Wipf, and D. N. Beratan. 1997. Synthetic and model computational studies of optical angle additivity for interacting chiral centers: A reinvestigation of van't Hoff's principle. *Chirality* 9:468–477.

Kondru, R. K., P. Wipf, and D. N. Beratan. 1998. Theory assisted determination of absolute stereochemistry for complex natural products via computation of molar rotation angles. *J. Am. Chem. Soc.* 120:2204–2205.

Kondru, R. K., P. Wipf, and D. N. Beratan. 2000. Chiral action at a distance: Remote substituent effects on the optical activity of calyculins A and B. *Org. Lett.* 2:1509–1512.

Kraszewska, A., P. Rivera-Fuentes, G. Rapenne, J. Crassous, A. G. Petrovic, J. L. Alonso-Gómez, E. Huerta, F. Diederich, and C. Thilgen. 2010. Regioselectivity in tether-directed remote functionalization—The addition of a cyclotriveratrylene-based trimalonate to C60 revisited. *Eur. J. Org. Chem.* 2010(23):4402–4411.

Kudo, M., T. Hanashima, A. Muranaka, H. Sato, M. Uchiyama, I. Azumaya, T. Hirano, H. Kagechika, and A. Tanatani. 2009. Identification of absolute helical structures of aromatic multilayered oligo(m-phenylurea)s in solution. *J. Org. Chem.* 74(21):8154–8163.

Kundrat, M. D., and J. Autschbach. 2008. Computational modeling of the optical rotation of amino acids: Taking a new look at an old rule for the pH dependence of the optical rotation. *J. Am. Chem. Soc.* 130:4404–4414.

Kuppens, T., W. Herrebout, B. van der Veken, and P. Bultinck. 2006. Intermolecular association of tetrahydrofuran-2-carboxylic acid in solution: A vibrational circular dichroism study. *J. Phys. Chem. A* 110(34):10191–10200.

Kurouski, D., R. K. Dukor, X. Lu, L. A. Nafie, and I. K. Lednev. 2012. Spontaneous inter-conversion of insulin fibril chirality. *Chem. Commun.* 48(23):2837–2839.

Kurouski, D., K. Kar, R. Wetzel, R. K. Dukor, I. K. Lednev, and L. A. Nafie. 2013. Levels of supramolecular chirality of polyglutamine aggregates revealed by vibrational circular dichroism. *FEBS Lett.* 587(11):1638–1643.

Kurouski, D., J. D. Handen, R. K. Dukor, L. A. Nafie, and I. K. Lednev. 2015. Supramolecular chirality in peptide microcrystals. *Chem. Commun.* 51:89–92.

Kuwahara, S., K. Obata, K. Yoshida, T. Matsumoto, N. Harada, N. Yasuda, Y. Ozawa, and K. Toriumi. 2005. Conclusive determination of the absolute configuration of chiral C60-fullerene cis-3 bisadducts by X-ray crystallography and circular dichroism. *Angew. Chem. Int. Ed. Engl.* 44(15):2262–2265.

Kypr, J., I. Kejnovska, K. Bednarova, and M. Vorlickova. 2012. Circular dichroism spectroscopy of nucleic acids. In: N. Berova, P. L. Polavarapu, K. Nakanishi, and R. W. Woody (eds). *Comprehensive Chiroptical Spectrosocpy*. New York, NY: Wiley.

Lahiri, P., K. B. Wiberg, P. H. Vaccaro, M. Caricato, and T. D. Crawford. 2014. Large solvation effect in the optical rotatory dispersion of norbornenone. *Angew. Chem.* 126(5):1410–1413.

Lahoz, I. R., S. Castro-Fernández, A. Navarro-Vázquez, J. L. Alonso-Gómez, and M. M. Cid. 2014. Conformational stable alleno-acetylenic cyclophanes bearing chiral axes. *Chirality* 26(9):563–573.

Lamparska, E., V. Liégeois, O. Quinet, and B. Champagne. 2006. Theoretical determination of the vibrational Raman optical activity signatures of helical polypropylene chains. *ChemPhysChem* 7(11):2366–2376.

Lebon, F., G. Longhi, S. Abbate, M. Catellani, C. Zhao, and P. L. Polavarapu. 2001. Vibrational circular dichroism spectra of chirally substituted polythiophenes. *Synth. Met.* 119(1–3):75–76.

Lebon, F., G. Longhi, S. Abbate, M. Catellani, F. Wang, and P. L. Polavarapu. 2002. Circular dichroism spectra of regioregular poly{3[(S)-2-methylbutyl]-thiophene} and of Poly{3,4-di[(S)-2-methylbutyl]-thiophene}. *Enantiomer* 7(4–5):207–212.

Legault, C. Y. 2009. CYLview, 1.0b. Université de sherbrooke. http://www.cylview.org.

Lehtola, S., and H. Jónsson. 2013. Unitary optimization of localized molecular orbitals. *J. Chem. Theory Comput.* 9(12):5365–5372.

Li, X.-C., D. Ferreira, and Y. Ding. 2010. Determination of absolute configuration of natural products: Theoretical calculation of electronic circular dichroism as a tool. *Curr. Org. Chem.* 14:1678–1697.

Li, L., C. Li, Y.-K. Si, and D.-L. Yin. 2013. Absolute configuration of Buagafuran: An experimental and theoretical electronic circular dichroism study. *Chin. Chem. Lett.* 24(6):500–502.

Liégeois, V., and B. Champagne. 2009. Vibrational Raman optical activity of π-conjugated helical systems: Hexahelicene and heterohelicenes. *J. Comput. Chem.* 30(8):1261–1278.

Liégeois, V., C. R. Jacob, B. Champagne, and M. Reiher. 2010. Analysis of vibrational Raman optical activity signatures of the (TG)N and (GG)N conformations of isotactic polypropylene chains in terms of localized modes. *J. Phys. Chem. A* 114(26):7198–7212.

Liégeois, V., O. Quinet, and B. Champagne. 2005. Vibrational Raman optical activity as a mean for revealing the helicity of oligosilanes: A quantum chemical investigation. *J. Chem. Phys.* 122(21):214304.

Liégeois, V., O. Quinet, and B. Champagne. 2006. Investigation of polyethylene helical conformations: Theoretical study by vibrational Raman optical activity. *Int. J. Quantum Chem.* 106(15):3097–3107.

Lightner, D. A., and J. E. Gurst. 2000. *Organic Conformational Analysis and Stereochemistry from Circular Dichroism Spectroscopy*. Weinheim: Wiley-VCH.

Lightner, D. A., D. T. Hefelfinger, T. W. Powers, G. W. Frank, and K. N. Trueblood. 1972. Hexahelicene. Absolute configuration. *J. Am. Chem. Soc.* 94(10):3492–3497.

Lin, N., H. Solheim, K. Ruud, M. Nooijen, F. Santoro, X. Zhao, M. Kwit, and P. Skowronek. 2012. Vibrationally resolved circular dichroism spectra of a molecule with isotopically engendered chirality. *Phys. Chem. Chem. Phys.* 14(10):3669–3680.

Liquori, A. M., and B. Pispisa. 1967. Stereospecific polymerization of some optically active monomers. *J. Polym. Sci., Part B: Polym. Lett.* 5(5):375–385.

Losada, M., P. Nguyen, and Y. Xu. 2008a. Solvation of propylene oxide in water: Vibrational circular dichroism, optical rotation, and computer simulation studies. *J. Phys. Chem. A* 112(25):5621–5627.

Losada, M., H. Tran, and Y. Xu. 2008b. Lactic acid in solution: Investigations of lactic acid self-aggregation and hydrogen bonding interactions with water and methanol using vibrational absorption and vibrational circular dichroism spectroscopies. *J. Chem. Phys.* 128(1):014508.

Losada, M., and Y. Xu. 2007. Chirality transfer through hydrogen-bonding: Experimental and ab initio analyses of vibrational circular dichroism spectra of methyl lactate in water. *Phys. Chem. Chem. Phys.* 9(24):3127–3135.

Lovchik, M. A., G. Fráter, A. Goeke, and W. Hug. 2008. Total synthesis of junionone, a natural monoterpenoid from juniperus communis L., and determination of the absolute configuration of the naturally occurring enantiomer by ROA spectroscopy. *Chem. Biodiversity* 5(1):126–139.

Lowe, M. A., G. A. Segal, and P. J. Stephens. 1986. The theory of vibrational circular dichroism: Trans-1,2-dideuteriocyclopropane. *J. Am. Chem. Soc.* 108(2):248–256.

Lu, J., A. Xia, N. Zhou, W. Zhang, Z. Zhang, X. Pan, Y. Yang, Y. Wang, and X. Zhu. 2015. A versatile cyclic 2,2′-azobenzenophane with a functional handle and its polymers: Efficient synthesis and effect of topological structure on chiroptical properties. *Chem.—Eur. J.* 21(6):2324–2329.

Ma, S., C. A. Busacca, K. R. Fandrick, T. Bartholomeyzik, N. Haddad, S. Shen, H. Lee, A. Saha, N. Yee, C. Senanayake, and N. Grinberg. 2010. Directly probing the racemization of imidazolines by vibrational circular dichroism: Kinetics and mechanism. *Org. Lett.* 12(12):2782–2785.

Ma, S., X. Cao, M. Mak, A. Sadik, C. Walkner, T. B. Freedman, I. K. Lednev, R. K. Dukor, and L. A. Nafie. 2007. Vibrational circular dichroism shows unusual sensitivity to protein fibril formation and development in solution. *J. Am. Chem. Soc.* 129(41):12364–12365.

Ma, S., S. Shen, N. Haddad, W. Tang, J. Wang, H. Lee, N. Yee, C. Senanayake, and N. Grinberg. 2009a. Chromatographic and spectroscopic studies on the chiral recognition of sulfated β-cyclodextrin as chiral mobile phase additive: Enantiomeric separation of a chiral amine. *J. Chromatogr. A* 1216(8):1232–1240.

Ma, S., S. Shen, H. Lee, M. Eriksson, X. Zeng, J. Xu, K. Fandrick, N. Yee, C. Senanayake, and N. Grinberg. 2009b. Mechanistic studies on the chiral recognition of polysaccharide-based chiral stationary phases using liquid chromatography and vibrational circular dichroism: Reversal of elution order of N-substituted alpha-methyl phenylalanine esters. *J. Chromatogr. A* 1216(18):3784–3793.

Macleod, N. A., C. Johannessen, L. Hecht, L. D. Barron, and J. P. Simons. 2006. From the gas phase to aqueous solution: Vibrational spectroscopy, Raman optical activity and conformational structure of carbohydrates. *Int. J. Mass Spectrom.* 253(3):193–200.

Malon, P., T. A. Keiderling, J. Y. Uang, and J. C. Chickos. 1991. Vibrational circular dichroism study of [3R,4R]-dideuteriocyclobutane-1,2-dione. Preliminary comparison of experiment and calculations. *Chem. Phys. Lett.* 179(3):282–290.

Malon, P., L. J. Mickley, K. M. Sluis, C. N. Tam, T. A. Keiderling, S. Kamath, J. Uang, and J. S. Chickos. 1992. Vibrational circular dichroism study of (2S,3S)-dideuteriobutyrolactone. Synthesis, normal mode analysis, and comparison of experimental and calculated spectra. *J. Phys. Chem.* 96(25):10139–10149.

Mammana, A., G. T. Carroll, and B. L. Feringa. 2012. Cicrcular dichroism of dynamic systems: Switching molecular and supramolecular chirality. In: N. Berova, P. L. Polavarapu, K. Nakanishi, and R. W. Woody (eds). *Comprehensive Chiroptical Spectroscopy: Applications in Stereochemical Analysis of Synthetic Compounds, Natural Products and Biomolecules*. Vol. 2. New York, NY: John Wiley & Sons.

Marcott, C., H. A. Havel, J. Overend, and A. Moscowitz. 1978. Vibrational circular dichroism and individual chiral centers. An example from the sugars. *J. Am. Chem. Soc.* 100(22):7088–7089.

Maryanoff, B. E., D. F. McComsey, R. K. Dukor, L. A. Nafie, T. B. Freedman, X. Cao, and V. W. Day. 2003. Structural studies on McN-5652-X, a high-affinity ligand for the serotonin transporter in mammalian brain. *Biorg. Med. Chem.* 11(11):2463–2470.

Mason, S. F., G. W. Vane, K. Schofield, R. J. Wells, and J. S. Whitehurst. 1967. The circular dichroism and absolute configuration of Troger's base. *J. Chem. Soc. B* (0):553–556.

Mazzeo, G., E. Santoro, A. Andolfi, A. Cimmino, P. Troselj, A. G. Petrovic, S. Superchi, A. Evidente, and N. Berova. 2013. Absolute configurations of fungal and plant metabolites by chiroptical methods. ORD, ECD, and VCD studies on phyllostin, scytolide, and oxysporone. *J. Nat. Prod.* 76(4):588–599.

McCann, J. L., A. Rauk, and H. Wieser. 1997. Infrared absorption and vibrational circular dichroism spectra of poly(vinyl ether) containing diastereomeric menthols as pendants. *J. Mol. Struct.* 408–409(0):417–420.

McCann, J., D. Tsankov, N. Hu, G. Liu, and H. Wieser. 1995. VCD study of synthetic chiral polymers: Stereoregularity in poly-menthyl methacrylate. *J. Mol. Struct.* 349(0):309–312.

McColl, I. H., E. W. Blanch, L. Hecht, and L. D. Barron. 2004. A study of α-helix hydration in polypeptides, proteins, and viruses using vibrational Raman optical activity. *J. Am. Chem. Soc.* 126(26):8181–8188.

McConnell, O., A. Bach, C. Balibar, N. Byrne, Y. Cai, G. Carter, M. Chlenov, L. Di, K. Fan, I. Goljer, Y. He, D. Herold, M. Kagan, E. Kerns, F. Koehn, C. Kraml, V. Marathias, B. Marquez, L. McDonald, L. Nogle, C. Petucci, G. Schlingmann, G. Tawa, M. Tischler, R. T. Williamson, A. Sutherland, W. Watts, M. Young, M.-Y. Zhang, Y. Zhang, D. Zhou, and D. Ho. 2007. Enantiomeric separation and determination of absolute stereochemistry of asymmetric molecules in drug discovery—Building chiral technology toolboxes. *Chirality* 19(9):658–682.

Measey, T. J., and R. Schweitzer-Stenner. 2010. Vibrational circular dichroism as a probe of fibrillogenesis: The origin of the anomalous intensity enhancement of amyloid-like fibrils. *J. Am. Chem. Soc.* 133(1):1066–1076.

Merten, C., L. D. Barron, L. Hecht, and C. Johannessen. 2011. Determination of the helical screw sense and side-group chirality of a synthetic chiral polymer from Raman optical activity. *Angew. Chem. Int. Ed. Engl.* 50(42):9973–9976.

Merten, C., and A. Hartwig. 2010. Structural examination of dissolved and solid helical chiral poly(trityl methacrylate) by VCD spectroscopy. *Macromolecules* 43(20):8373–8378.

Merten, C., J. F. Reuther, J. D. DeSousa, and B. M. Novak. 2014. Identification of the specific, shutter-like conformational reorientation in a chiroptical switching polycarbodiimide by VCD spectroscopy. *Phys. Chem. Chem. Phys.* 16(23):11456–11460.

Merten, C., and Y. Xu. 2013. Matrix isolation-vibrational circular dichroism spectroscopy of 3-butyn-2-ol and its binary aggregates. *ChemPhysChem* 14(1):213–219.

Minick, D. J., R. C. B. Copley, J. R. Szewczyk, R. D. Rutkowske, and L. A. Miller. 2007. An investigation of the absolute configuration of the potent histamine H3 receptor antagonist GT-2331 using vibrational circular dichroism. *Chirality* 19(9):731–740.

Mislow, K., M. M. Green, P. Laur, J. T. Melillo, T. Simmons, and A. L. Ternay. 1965. Absolute configuration and optical rotatory power of sulfoxides and sulfinate esters1,2. *J. Am. Chem. Soc.* 87(9):1958–1976.

Mobian, P., C. Nicolas, E. Francotte, T. Bürgi, and J. Lacour. 2008. Synthesis, resolution, and VCD analysis of an enantiopure diazaoxatricornan derivative. *J. Am. Chem. Soc.* 130(20):6507–6514.

Monde, K., N. Miura, M. Hashimoto, T. Taniguchi, and T. Inabe. 2006. Conformational analysis of chiral helical perfluoroalkyl chains by VCD. *J. Am. Chem. Soc.* 128(18):6000–6001.

Monde, K., T. Taniguchi, N. Miura, and S.-I. Nishimura. 2004. Specific band observed in VCD predicts the anomeric configuration of carbohydrates. *J. Am. Chem. Soc.* 126(31):9496–9497.

Monde, K., T. Taniguchi, N. Miura, S.-I. Nishimura, N. Harada, R. K. Dukor, and L. A. Nafie. 2003. Preparation of cruciferous phytoalexin related metabolites, (−)-dioxibrassinin and (−)-3-cyanomethyl-3-hydroxyoxindole, and determination of their absolute configurations by vibrational circular dichroism (VCD). *Tetrahedron Lett.* 44(32):6017–6020.

Moore, B., and J. Autschbach. 2012. Density functional study of tetraphenylporphyrin long-range exciton coupling. *Chemistry Open* 1:184–194.

Moore, B., M. Srebro, and J. Autschbach. 2012. Analysis of optical activity in terms of bonds and lone-pairs: The exceptionally large optical rotation of norbornenone. *J. Chem. Theory Comput.* 8(11):4336–4346.

Mori, T., and Y. Inoue. 2011. Recent theoretical and experimental advances in the electronic circular dichroism of planar chiral cyclophanes. *Top. Curr. Chem.* 298:99–128

Mori, T., Y. Inoue, and S. Grimme. 2007a. Experimental and theoretical study of the CD spectra and conformational properties of axially chiral 2,2′-, 3,3′-, and 4,4′-biphenol ethers. *J. Phys. Chem. A* 111(20):4222–4234.

Mori, T., Y. Inoue, and S. Grimme. 2007b. Quantum chemical study on the circular dichroism spectra and specific rotation of donor–acceptor cyclophanes. *J. Phys. Chem. A* 111(32):7995–8006.

Mukhopadhyay, P., G. Zuber, P. Wipf, and D. N. Beratan. 2007. Contribution of a solute's chiral solvent imprint to optical rotation. *Angew. Chem. Int. Ed. Engl.* 46(34):6450–6452.

Muñoz, M. A., C. Chamy, M. A. Bucio, A. Hernández-Barragán, and P. Joseph-Nathan. 2014. Absolute configuration of scopadulane diterpenes from Calceolaria species. *Tetrahedron Lett.* 55(30):4274–4277.

Mutter, S. T., F. Zielinski, J. R. Cheeseman, C. Johannessen, P. L. A. Popelier, and E. W. Blanch. 2015a. Conformational dynamics of carbohydrates: Raman optical activity of d-glucuronic acid and N-acetyl-d-glucosamine using a combined molecular dynamics and quantum chemical approach. *Phys. Chem. Chem. Phys.* 17(8):6016–6027.

Mutter, S. T., F. Zielinski, P. L. A. Popelier, and E. W. Blanch. 2015b. Calculation of Raman optical activity spectra for vibrational analysis. *Analyst.* 140(9):2944–2956.

Nafie, L. A. 2008. Vibrational circular dichroism: A new tool for the solution-state determination of the strcuture and absolute configuration of chiral natural product molecules. *Nat. Prod. Commun.* 3:451–466.

Nakai, Y., T. Mori, and Y. Inoue. 2012a. Circular dichroism of (Di)methyl- and diaza[6]helicenes. A combined theoretical and experimental study. *J. Phys. Chem. A* 117(1):83–93.

Nakai, Y., T. Mori, and Y. Inoue. 2012b. Theoretical and experimental studies on circular dichroism of carbo[n]helicenes. *J. Phys. Chem. A* 116(27):7372–7385.

Nakai, Y., T. Mori, K. Sato, and Y. Inoue. 2013. Theoretical and experimental studies of circular dichroism of mono- and diazonia[6]helicenes. *J. Phys. Chem. A* 117(24):5082–5092.

Nakao, K., Y. Kyogoku, and H. Sugeta. 1994. Vibrational circular dichroism of the OH-stretching vibration in 2,2′-dihydroxy-1,1′-binaphthyl. *Faraday Discuss.* 99(0):77–85.

Nakashima, N., R. Ando, T. Muramatsu, and T. Kunitake. 1994. Unusually large induced circular dichroism of an aromatic compound bound to helical superstructures of chiral ammonium bilayers. *Langmuir* 10(1):232–234.

Naubron, J.-V., L. Giordano, F. Fotiadu, T. Bürgi, N. Vanthuyne, C. Roussel, and G. Buono. 2006. Chromatographic resolution, solution and crystal phase conformations, and absolute configuration of tert-Butyl(dimethylamino)phenylphosphine–borane complex. *J. Org. Chem.* 71(15):5586–5593.

Nicu, V. P., E. J. Baerends, and P. L. Polavarapu. 2012. Understanding solvent effects in vibrational circular dichroism spectra: [1,1′-Binaphthalene]-2,2′-diol in dichloromethane, acetonitrile, and dimethyl sulfoxide solvents. *J. Phys. Chem. A* 116(32):8366–8373.

Nicu, V. P., J. Neugebauer, and E. J. Baerends. 2008. Effects of complex formation on vibrational circular dichroism spectra. *J. Phys. Chem. A* 112(30):6978–6991.

Nieto-Ortega, B., J. Casado, E. W. Blanch, J. T. López Navarrete, A. R. Quesada, and F. J. Ramírez. 2011. Raman optical activity spectra and conformational elucidation of chiral drugs. The case of the antiangiogenic aeroplysinin-1. *J. Phys. Chem. A* 115(13):2752–2755.

Nitsch-Velasquez, L., and J. Autschbach. 2010. Toward a generalization of the Clough-Lutz-Jirgensons effect: Chiral organic acids with alkyl, hydroxyl, and halogen substituents. *Chirality* 22(Suppl 1):E81–E95.

Noguez, C., and F. Hidalgo. 2014. Ab initio electronic circular dichroism of fullerenes, single-walled carbon nanotubes, and ligand-protected metal nanoparticles. *Chirality* 26(9):553–562.

Norman, P., and M. Linares. 2014. On the interplay between chirality and exciton coupling: A DFT calculation of the circular dichroism in π-stacked ethylene. *Chirality* 26(9):483–489.

Norman, P., J. Parello, P. Polavarapu, and M. Linares. 2015. Predicting near-UV electronic circular dichroism in nucleosomal DNA by means of DFT response theory. *Phys. Chem. Chem. Phys.* 17(34):21866–21879.

Novotná, P., and M. Urbanová. 2012. Vibrational circular dichroism study of polypeptide model–membrane systems. *Anal. Biochem.* 427(2):211–218.

Nugroho, A., and H. Morita. 2014. Circular dichroism calculation for natural products. *J. Nat. Med.* 68(1):1–10.

Parchansky, V., J. Kapitan, and P. Bour. 2014. Inspecting chiral molecules by Raman optical activity spectroscopy. *RSC Adv.* 4(100):57125–57136.

Paterlini, M. G., T. B. Freedman, and L. A. Nafie. 1986. Ring current enhanced vibrational circular dichroism in the carbon-hydrogen bond stretching motions of sugars. *J. Am. Chem. Soc.* 108(7):1389–1397.

Perera, A. S., J. Thomas, M. R. Poopari, and Y. Xu. 2016. The clusters-in-a liquid approach for solvation: New insights from the conformer specific gas phase spectroscopy and vibrational optical activity spectroscopy. *Frontiers in Chemistry* 4 (9):1–17.

Pescitelli, G., L. D. Bari, and N. Berova. 2014. Application of electronic circular dichroism in the study of supramolecular systems. *Chem. Soc. Rev.* 43(15):5211–5233.

Petrovic, A. G., P. K. Bose, and P. L. Polavarapu. 2004. Vibrational circular dichroism of carbohydrate films formed from aqueous solutions. *Carbohydr. Res.* 339(16):2713–2720.

Petrovic, A. G., J. He, P. L. Polavarapu, L. S. Xiao, and D. W. Armstrong. 2005a. Absolute configuration and predominant conformations of 1,1-dimethyl-2-phenylethyl phenyl sulfoxide. *Org. Biomol. Chem.* 3(10):1977–1981.

Petrovic, A. G., and P. L. Polavarapu. 2007. Chiroptical spectroscopic determination of molecular structures of chiral sulfinamides: t-Butanesulfinamide. *J. Phys. Chem. A* 111(43):10938–10943.

Petrovic, A. G., P. L. Polavarapu, J. Drabowicz, P. Łyżwa, M. Mikołajczyk, W. Wieczorek, and A. Balińska. 2008a. Diastereomers of N-α-phenylethyl-t-butylsulfinamide: Absolute configurations and predominant conformations. *J. Org. Chem.* 73(8):3120–3129.

Petrovic, A. G., P. L. Polavarapu, J. Drabowicz, Y. Zhang, O. J. McConnell, and H. Duddeck. 2005b. Absolute configuration of C2-symmetric spiroselenurane: 3,3,3′,3′-tetramethyl-1,1′-spirobi[3 H,2,1]benzoxaselenole. *Chem.—Eur. J.* 11(14):4257–4262.

Petrovic, A. G., S. E. Vick, and P. L. Polavarapu. 2008b. Determination of the absolute stereochemistry of chiral biphenanthryls in solution phase using chiroptical spectroscopic methods: 2,2′-Diphenyl-[3,3′-biphenanthrene]-4,4′-diol. *Chirality* 20(3–4):501–510.

Pivonka, D. E., and S. S. Wesolowski. 2013. Vibrational circular dichroism (VCD) chiral assignment of atropisomers: Application to γ-aminobutyric acid (GABA) modulators designed as potential anxiolytic drugs. *Appl. Spectrosc.* 67(4):365–371.

Platt, J. R. 1949. Classification of spectra of cata-condensed hydrocarbons. *J. Chem. Phys.* 17(5):484–495.

Pohl, G., A. Perczel, E. Vass, G. Magyarfalvi, and G. Tarczay. 2007. A matrix isolation study on Ac-Gly-NHMe and Ac-l-Ala-NHMe, the simplest chiral and achiral building blocks of peptides and proteins. *Phys. Chem. Chem. Phys.* 9(33):4698–4708.

Pohl, G., A. Perczel, E. Vass, G. Magyarfalvi, and G. Tarczay. 2008. A matrix isolation study on Ac–l-Pro–NH2: A frequent structural element of β- and γ-turns of peptides and proteins. *Tetrahedron* 64(9):2126–2133.

Polavarapu, P. L. 1990. Ab initio vibrational Raman and Raman optical activity spectra. *J. Phys. Chem.* 94(21):8106–8112.

Polavarapu, P. L. 1997a. Ab initio molecular optical rotations and absolute configurations. *Mol. Phys.* 91(3):551–554.

Polavarapu, P. L. 1997b. Molecular optical rotations and structures. *Tetrahedron Asymmetry* 8(20):3397–3401.

Polavarapu, P. L. 1998. *Vibrational Spectra: Principles and Applications with Emphasis on Optical Activity*. New York: Elsevier.

Polavarapu, P. L. 2002. The absolute configuration of bromochlorofluoromethane. *Angew. Chem. Int. Ed. Engl.* 41(23):4544–4546.

Polavarapu, P. L. 2005. Kramers–Kronig transformation for optical rotatory dispersion studies. *J. Phys. Chem. A* 109(32):7013–7023.

Polavarapu, P. L. 2008. Why is it important to simultaneously use more than one chiroptical spectroscopic method for determining the structures of chiral molecules? *Chirality* 20(5):664–672.

Polavarapu, P. L. 2011. Determination of molecular stereochemistry using optical rotatory dispersion, vibrational circular dichroism and vibrational Raman optical activity. In: K. W. Busch and M. A. Busch (eds). *Chiral Analysis*. New York, NY: Elsevier Science.

Polavarapu, P. L. 2012. Determination of the structures of chiral natural products using vibrational circular dichroism. In: N. Berova, P. L. Polavarapu, K. Nakanishi, and R. W. Woody (eds). *Comprehensive Chiroptical Spectroscopy*. Hoboken, NJ: John Wiley & Sons.

Polavarapu, P. L., A. L. Cholli, and G. Vernice. 1992. Absolute configuration of isoflurane. *J. Am. Chem. Soc.* 114(27):10953–10955.

Polavarapu, P. L., A. L. Cholli, and G. Vernice. 1993. Determination of absolute configurations and predominant conformations of general inhalation anesthetics: Desflurane. *J. Pharm. Sci.* 82(8):791–793.

Polavarapu, P. L., A. L. Cholli, and G. Vernice. 1997. Determination of absolute configurations and predominant conformations of general inhalation anesthetics: Desflurane. *J. Pharm. Sci.* 86:267.

Polavarapu, P. L., and C. L. Covington. 2015. Wavelength resolved specific optical rotations and homochiral equilibria. *Phys. Chem. Chem. Phys.* 17:21630–21633.

Polavarapu, P. L., J. He, J. Crassous, and K. Ruud. 2005. Absolute configuration of C76 from optical rotatory dispersion. *ChemPhysChem* 6(12):2535–2540.

Polavarapu, P. L., L. A. Nafie, S. A. Benner, and T. H. Morton. 1981. Optical activity due to isotopic substitution. Vibrational circular dichroism and the absolute configurations of .alpha.-deuterated cyclohexanones. *J. Am. Chem. Soc.* 103(18):5349–5354.

Polavarapu, P. L., N. Jeirath, T. Kurtán, G. Pescitelli, and K. Krohn. 2009a. Determination of the absolute configurations at stereogenic centers in the presence of axial chirality. *Chirality* 21(1E):E202–E207.

Polavarapu, P. L., N. Jeirath, and S. Walia. 2009b. Conformational sensitivity of chiroptical spectroscopic methods: 6,6′-dibromo-1,1′-bi-2-naphthol. *J. Phys. Chem. A* 113(18):5423–5431.

Polavarapu, P. L., A. G. Petrovic, S. E. Vick, W. D. Wulff, H. Ren, Z. Ding, and R. J. Staples. 2009c. Absolute configuration of 3,3′-diphenyl-[2,2′-binaphthalene]-1,1′-diol revisited. *J. Org. Chem.* 74(15):5451–5457.

Polavarapu, P. L., and R. Vijay. 2012. Chiroptical spectroscopy of surfactants. *J. Phys. Chem. A* 116(21):5112–5118.

Polavarapu, P. L., and C. Zhao. 1998. A comparison of ab initio optical rotations obtained with static and dynamic methods. *Chem. Phys. Lett.* 296(1–2):105–110.

Polavarapu, P. L., and C. Zhao. 2000. Vibrational circular dichroism: A new spectroscopic tool for biomolecular structural determination. *Fresenius J. Anal. Chem.* 366(6–7):727–734.

Polavarapu, P. L., C. Zhao, A. L. Cholli, and G. G. Vernice. 1999a. Vibrational circular dichroism, absolute configuration, and predominant conformations of volatile anesthetics: Desflurane. *J. Phys. Chem. B* 103(29):6127–6132.

Polavarapu, P. L., C. Zhao, and K. Ramig. 1999b. Vibrational circular dichroism, absolute configuration and predominant conformations of volatile anesthetics: 1,2,2,2-tetrafluoroethyl methyl ether. *Tetrahedron Asymmetry* 10(6):1099–1106.

Poopari, M. R., Z. Dezhahang, K. Shen, L. Wang, T. L. Lowary, and Y. Xu. 2015. Absolute configuration and conformation of two Fráter–Seebach Alkylation reaction products by film VCD and ECD spectroscopic analyses. *J. Org. Chem.* 80(1):428–437.

Poopari, M. R., Z. Dezhahang, G. Yang, and Y. Xu. 2012a. Conformational distributions of N-acetyl-L-cysteine in aqueous solutions: A combined implicit and explicit solvation treatment of VA and VCD spectra. *ChemPhysChem* 13(9):2310–2321.

Poopari, M. R., Z. Dezhahang, and Y. Xu. 2013. A comparative VCD study of methyl mandelate in methanol, dimethyl sulfoxide, and chloroform: Explicit and implicit solvation models. *Phys. Chem. Chem. Phys.* 15(5):1655–1665.

Poopari, M. R., P. Zhu, Z. Dezhahang, and Y. Xu. 2012b. Vibrational absorption and vibrational circular dichroism spectra of leucine in water under different pH conditions: Hydrogen-bonding interactions with water. *J. Chem. Phys.* 137(19):194308.

Qiu, S., G. Li, S. Lu, B. Huang, Z. Feng, and C. Li. 2012. Chiral sulfur compounds studied by Raman optical activity: tert-Butanesulfinamide and its precursor tert-Butyl tert-Butanethiosulfinate. *Chirality* 24(9):731–740.

Quesada-Moreno, M. M., J. R. Avilés-Moreno, J. J. López-González, R. M. Claramunt, C. López, I. Alkorta, and J. Elguero. 2014. Chiral self-assembly of enantiomerically pure (4S,7R)-campho[2,3-c]pyrazole in the solid state: A vibrational circular dichroism (VCD) and computational study. *Tetrahedron Asymmetry* 25(6–7):507–515.

Rajca, A., M. Pink, S. Xiao, M. Miyasaka, S. Rajca, K. Das, and K. Plessel. 2009. Functionalized thiophene-based [7]helicene: Chirooptical properties versus electron delocalization. *J. Org. Chem.* 74(19):7504–7513.

Ren, J., G.-Y. Li, L. Shen, G.-L. Zhang, L. A. Nafie, and H.-J. Zhu. 2013. Challenges in the assignment of relative and absolute configurations of complex molecules: Computation can resolve conflicts between theory and experiment. *Tetrahedron* 69(48):10351–10356.

Romaine, I. M., J. E. Hempel, G. Shanmugam, H. Hori, Y. Igarashi, P. L. Polavarapu, and G. A. Sulikowski. 2011. Assignment and stereocontrol of hibarimicin atropoisomers. *Org. Lett.* 13(17):4538–4541.

Rosini, C., M. I. Donnoli, and S. Superchi. 2001. Towards a correlation of absolute configuration and chiroptical properties of alkyl aryl sulfoxides: A coupled-oscillator foundation of the empirical mislow rule? *Chem.—Eur. J.* 7(1):72–79.

Rüther, A., M. Pfeifer, V. A. Lórenz-Fonfría, and S. Lüdeke. 2014. Reaction monitoring using mid-infrared laser-based vibrational circular dichroism. *Chirality* 26(9):490–496.

Sadlej, J., J. C. Dobrowolski, and J. E. Rode. 2010. VCD spectroscopy as a novel probe for chirality transfer in molecular interactions. *Chem. Soc. Rev.* 39(5):1478–1488.

Sakamoto, A., N. Ohya, T. Hasegawa, H. Izumi, N. Tokita, and Y. Hamada. 2012. Determination of the absolute stereochemistry of limonene and alpha-santalol by Raman optical activity spectroscopy. *Nat Prod Commun.* 7:419–421.

Santoro, E., G. Mazzeo, A. G. Petrovic, A. Cimmino, J. Koshoubu, A. Evidente, N. Berova, and S. Superchi. 2015. Absolute configurations of phytotoxins seiricardine A and inuloxin A obtained by chiroptical studies. *Phytochemistry* 116:359–366.

Sato, H., K. Hori, T. Sakurai, and A. Yamagishi. 2008. Long distance chiral transfer in a gel: Experimental and ab initio analyses of vibrational circular dichroism spectra of R- and S-12-hydroxyoctadecanoic acid gels. *Chem. Phys. Lett.* 467(1–3):140–143.

Sato, H., T. Nakae, K. Morimoto, and K. Tamura. 2012. Critical effects of alkyl chain length on fibril structures in benzene-trans(RR)- or (SS)-N,N[prime or minute]-alkanoyl-1,2-diaminocyclohexane gels. *Org. Biomol. Chem.* 10(8):1581–1586.

Sato, H., E. Nogami, T. Yajima, and A. Yamagishi. 2014a. Terminal effects on gelation by low molecular weight chiral gelators. *RSC Adv.* 4(4):1659–1665.

Sato, H., T. Sakurai, and A. Yamagishi. 2011a. Comparison of vibrational circular dichroism between the Langmuir–Blodgett films and gels of 12-hydroxyoctadecanoic acid. *Chem. Lett.* 40(1):25–27.

Sato, H., T. Yajima, and A. Yamagishi. 2011b. Molecular origin for helical winding of fibrils formed by perfluorinated gelators. *Chem. Commun.* 47(13):3736–3738.

Sato, H., T. Yajima, and A. Yamagishi. 2014b. An intermediate state in gelation as revealed by vibrational circular dichroism spectroscopy. *RSC Adv.* 4(49):25867–25870.

Schaaff, T. G., G. Knight, M. N. Shafigullin, R. F. Borkman, and R. L. Whetten. 1998. Isolation and selected properties of a 10.4 kDa gold: Glutathione cluster compound. *J. Phys. Chem. B* 102(52):10643–10646.

Schlosser, D. W., F. Devlin, K. Jalkanen, and P. J. Stephens. 1982. Vibrational circular dichroism of matrix-isolated molecules. *Chem. Phys. Lett.* 88(3):286–291.

Schreiner, P. R., A. A. Fokin, H. P. Reisenauer, B. A. Tkachenko, E. Vass, M. M. Olmstead, D. Blaeser, R. Boese, J. E. P. Dahl, and R. M. K. Carlson. 2009. [123]Tetramantane: Parent of a new family of σ-helicenes. *J. Am. Chem. Soc.* 131(32):11292–11293.

Schurig, V., M. Juza, B. S. Green, J. Horakh, and A. Simon. 1996. Absolute configurations of the inhalation anesthetics isoflurane and desflurane. *Angew. Chem. Int. Ed. Engl.* 35(15):1680–1682.

Schwartz, E., S. R. Domingos, A. Vdovin, M. Koepf, W. Jan Buma, J. J. L. M. Cornelissen, A. E. Rowan, R. J. M. Nolte, and S. Woutersen. 2010. Direct access to polyisocyanide screw sense using vibrational circular dichroism. *Macromolecules* 43(19):7931–7935.

Schweitzer-Stenner, R. 2004. Secondary structure analysis of polypeptides based on an excitonic coupling model to describe the band profile of Amide I′ of IR, Raman, and vibrational circular dichroism spectra. *J. Phys. Chem. B* 108(43):16965–16975.

Sen, A. C., and T. A. Keiderling. 1984. Vibrational circular dichroism of polypeptides. III. Film studies of several α-helical and β-sheet polypeptides. *Biopolymers* 23(8):1533–1545.

Setnička, V., M. Urbanová, P. Bouř, V. Král, and K. Volka. 2001. Vibrational circular dichroism of 1,1′-binaphthyl derivatives: Experimental and theoretical study. *J. Phys. Chem. A* 105(39):8931–8938.

Setnička, V., M. Urbanová, S. Pataridis, V. Král, and K. Volka. 2002. Sol-gel phase transition of brucine-appended porphyrin gelator: A study by vibrational circular dichroism spectroscopy. *Tetrahedron Asymmetry* 13(24):2661–2666.

Sett, A., S. Bag, S. Dasgupta, and S. DasGupta. 2015. Interfacial force driven pattern formation during drying of Aβ (25-35) fibrils. *Int. J. Biol. Macromol.* 79(0):344–352.

Shallenberger, R. S. 1982. *Advanced Sugar Chemistry: Principles of Sugar Stereochemistry*. Westport, CT: Avi Publishing Co.

Shanmugam, G., N. Phambu, and P. L. Polavarapu. 2011. Unusual structural transition of antimicrobial VP1 peptide. *Biophys. Chem.* 155(2–3):104–108.

Shanmugam, G., and P. L. Polavarapu. 2004. Vibrational circular dichroism of protein films. *J. Am. Chem. Soc.* 126(33):10292–10295.

Shanmugam, G., and P. L. Polavarapu. 2011. Isotope-assisted vibrational circular dichroism investigations of amyloid β peptide fragment, Aβ(16–22). *J. Struct. Biol.* 176(2):212–219.

Shanmugam, G., and P. L. Polavarapu. 2013. Site-specific structure of Aβ(25–35) peptide: Isotope-assisted vibrational circular dichroism study. *Biochim. Biophys. Acta* 1834(1):308–316.

Shanmugam, G., P. L. Polavarapu, D. Gopinath, and R. Jayakumar. 2005a. The structure of antimicrobial pexiganan peptide in solution probed by Fourier transform infrared absorption, vibrational circular dichroism, and electronic circular dichroism spectroscopy. *Pept. Sci.* 80(5):636–642.

Shanmugam, G., P. L. Polavarapu, A. Kendall, and G. Stubbs. 2005b. Structures of plant viruses from vibrational circular dichroism. *J. Gen. Virol.* 86(8):2371–2377.

Shen, S., S. Ma, H. Lee, C. Manolescu, M. Grinberg, N. Yee, C. Senanayake, and N. Grinberg. 2009. HPLC enantiomeric separation of aromatic amines using crown ether tetracarboxylic acid. *J. Liq. Chromatogr. Related Technol.* 33(2):153–166.

Shen, J., J. Yang, W. Heyse, H. Schweitzer, N. Nagel, D. Andert, C. Zhu, V. Morrison, G. A. Nemeth, T.-M. Chen, Z. Zhao, T. A. Ayers, and Y.-M. Choi. 2014. Enantiomeric characterization and structure elucidation of Otamixaban. *J. Pharm. Anal.* 4(3):197–204.

Sherer, E. C., J. R. Cheeseman, and R. T. Williamson. 2015. Absolute configuration of remisporines A & B. *Org. Biomol. Chem.* 13(14):4169–4173.

Sherer, E. C., C. H. Lee, J. Shpungin, J. F. Cuff, C. Da, R. Ball, R. Bach, A. Crespo, X. Gong, and C. J. Welch. 2014. Systematic approach to conformational sampling for assigning absolute configuration using vibrational circular dichroism. *J. Med. Chem.* 57(2):477–494.

Shi, Z., R. W. Woody, and N. R. Kallenbach. 2002. Is polyproline II a major backbone conformation in unfolded proteins? *Adv. Protein Chem.* 62:163–240.

Shinitzky, M., and R. Haimovitz. 1993. Chiral surfaces in micelles of enantiomeric N-palmitoyl- and N-stearoylserine. *J. Am. Chem. Soc.* 115(26):12545–12549.

Silva, R. A. G. D., J. Kubelka, P. Bour, S. M. Decatur, and T. A. Keiderling. 2000. Site-specific conformational determination in thermal unfolding studies of helical peptides using vibrational circular dichroism with isotopic substitution. *Proc. Nat. Acad. Sci.* 97(15):8318–8323.

Simpson, S., J. Autschbach, and E. Zurek. 2013. Computational modeling of the optical rotation of amino acids: An in-silico experiment for physical chemistry. *J Chem. Ed.* 90, 650-656.

Slocik, J. M., A. O. Govorov, and R. R. Naik. 2011. Plasmonic circular dichroism of peptide-functionalized gold nanoparticles. *Nano Lett.* 11(2):701–705.

Smith, A. M., R. F. Collins, R. V. Ulijn, and E. Blanch. 2009. Raman optical activity of an achiral element in a chiral environment. *J. Raman Spectrosc.* 40(9):1093–1095.

Smulders, M. M. J., T. Buffeteau, D. Cavagnat, M. Wolffs, A. P. H. J. Schenning, and E. W. Meijer. 2008. C3-symmetrical self-assembled structures investigated by vibrational circular dichroism. *Chirality* 20(9):1016–1022.

Solladié-Cavallo, A., C. Marsol, G. Pescitelli, L. Di Bari, P. Salvadori, X. Huang, N. Fujioka, N. Berova, X. Cao, T. B. Freedman, and L. A Nafié. 2002. (R)-(+)- and (S)-(−)-1-(9-Phenanthryl)ethylamine: Assignment of absolute configuration by CD tweezer and VCD methods, and difficulties encountered with the CD exciton chirality method. *Eur. J. Org. Chem.* 2002(11):1788–1796.

Spencer, K. M., S. J. Cianciosi, J. E. Baldwin, T. B. Freedman, and L. A. Nafie. 1990. Determination of enantiomeric excess in deuterated chiral hydrocarbons by vibrational circular dichroism spectroscopy. *Appl. Spectrosc.* 44(2):235–238.

Spencer, K. M., R. B. Edmonds, and R. David Rauh. 1996. Analytical chiral purity verification using Raman optical activity. *Appl. Spectrosc.* 50(5):681–685.

Spencer, K. M., R. B. Edmonds, R. David Rauh, and Michael M. Carrabba. 1994. Analytical determination of enantiomeric purity using Raman optical activity. *Anal. Chem.* 66(8):1269–1273.

Srebro, M., N. Govind, W. A. de Jong, and J. Autschbach. 2011. Optical rotation calculated with time-dependent density functional theory: The OR45 benchmark. *J. Phys. Chem. A* 115(40):10930–10949.

Sreerama, N., and R. W. Woody. 2000. Estimation of protein secondary structure from circular dichroism spectra: Comparison of CONTIN, SELCON, and CDSSTR methods with an expanded reference set. *Anal. Biochem.* 287(2):252–260.

Steiner, L. A., and S. Lowey. 1966. Optical rotatory dispersion studies of rabbit γG-immunoglobulin and its papain fragments. *J. Biol. Chem.* 241(1):231–240.

Stephens, P. J., A. Aamouche, F. J. Devlin, S. Superchi, M. I. Donnoli, and C. Rosini. 2001. Determination of absolute configuration using vibrational circular dichroism spectroscopy: The chiral sulfoxide 1-(2-methylnaphthyl) methyl sulfoxide. *J. Org. Chem.* 66(11):3671–3677.

Stephens, P. J., F. J. Devlin, J. R. Cheeseman, M. J. Frisch, and C. Rosini. 2002. Determination of absolute configuration using optical rotation calculated using density functional theory. *Org. Lett.* 4(26):4595–4598.

Stephens, P. J., F. J. Devlin, F. Gasparrini, A. Ciogli, D. Spinelli, and B. Cosimelli. 2007. Determination of the absolute configuration of a chiral oxadiazol-3-one calcium channel blocker, resolved using chiral chromatography, via concerted density functional theory calculations of its vibrational circular dichroism, electronic circular dichroism, and optical rotation. *J. Org. Chem.* 72(13):4707–4715.

Stephens, P. J., K. J. Jalkanen, F. J. Devlin, and C. F. Chabalowski. 1993. Ab initio calculation of vibrational circular dichroism spectra using accurate post-self-consistent-field force fields: Trans-2,3-dideuteriooxirane. *J. Phys. Chem.* 97(23):6107–6110.

Su, C. N., and T. A. Keiderling. 1980. Conformation of dimethyl tartrate in solution Vibrational circular dichroism results. *J. Am. Chem. Soc.* 102:511–515.

Tachibana, T., T. Mori, and K. Hori. 1979. New type of twisted mesophase in jellies and solid films of chiral 12-hydroxyoctadecanoic acid. *Nature* 278 (5704):578–579.

Takaishi, K., M. Kawamoto, K. Tsubaki, and T. Wada. 2009. Photoswitching of dextro/levo rotation with axially chiral binaphthyls linked to an azobenzene. *J. Org. Chem.* 74(15):5723–5726.

Takaishi, K., A. Muranaka, M. Kawamoto, and M. Uchiyama. 2012. Photoinversion of cisoid/transoid binaphthyls. *Org. Lett.* 14(1):276–279.

Tamburro, A. M., V. Guantieri, L. Pandolfo, and A. Scopa. 1990. Synthetic fragments and analogues of elastin. II. Conformational studies. *Biopolymers* 29(4–5):855–870.

Tamoto, R., N. Daugey, T. Buffeteau, B. Kauffmann, M. Takafuji, H. Ihara, and R. Oda. 2015. *In situ* helicity inversion of self-assembled nano-helices. *Chem. Commun.* 51(17):3518–3521.

Tang, H.-Z., E. R. Garland, B. M. Novak, J. He, P. L. Polavarapu, F. Chen Sun, and S. S. Sheiko. 2007. Helical polyguanidines prepared by helix-sense-selective polymerizations of achiral carbodiimides using enantiopure binaphthol-based titanium catalysttts. *Macromolecules* 40(10):3575–3580.

Tang, H.-Z., B. M. Novak, J. He, and P. L. Polavarapu. 2005. A thermal and solvocontrollable cylindrical nanoshutter based on a single screw-sense helical polyguanidine. *Angew. Chem. Int. Ed. Engl.* 44(44):7298–7301.

Taniguchi, T., D. Manai, M. Shibata, Y. Itabashi, and K. Monde. 2015. Stereochemical analysis of glycerophospholipids by vibrational circular dichroism. *J. Am. Chem. Soc.* 137(38):12191–12194.

Taniguchi, T., C. L. Martin, K. Monde, K. Nakanishi, N. Berova, and L. E. Overman. 2009. Absolute configuration of actinophyllic acid as determined through chiroptical Data⊥. *J. Nat. Prod.* 72(3):430–432.

Taniguchi, T., and K. Monde. 2012a. Exciton chirality method in vibrational circular dichroism. *J. Am. Chem. Soc.* 134:3695–3698.

Taniguchi, T., and K. Monde. 2012b. Optical rotation, electronic circular dichroism, and vibrational circular dichroism of carbohydrates and glycoconjugates. In: N. Berova, P. L. Polavarapu, K. Nakanishi and R. W. Woody (eds). *Comprehensive Chiroptical Spectroscopy: Applications in Stereochemical Analysis of Synthetic Compounds, Natural Products, and Biomolecules*. Vol. 2. New York, NY: John Wiley & Sons.

Taniguchi, T., K. Monde, N. Miura, and S.-I. Nishimura. 2004. A characteristic CH band in VCD of methyl glycosidic carbohydrates. *Tetrahedron Lett.* 45(46):8451–8453.

Taniguchi, T., K. Monde, K. Nakanishi, and N. Berova. 2008. Chiral sulfinates studied by optical rotation, ECD and VCD: The absolute configuration of a cruciferous phytoalexin brassicanal C. *Org. Biomol. Chem.* 6(23):4399–4405.

Tarczay, G., S. Góbi, E. Vass, and G. Magyarfalvi. 2009. Model peptide–water complexes in Ar matrix: Complexation induced conformation change and chirality transfer. *Vib. Spectrosc* 50(1):21–28.

Tarczay, G., G. Magyarfalvi, and E. Vass. 2006. Towards the determination of the absolute configuration of complex molecular systems: Matrix isolation vibrational circular dichroism study of (R)-2-amino-1-propanol. *Angew. Chem. Int. Ed. Engl.* 45(11):1775–1777.

Teodorescu, F., S. Nica, C. Uncuta, E. Bartha, P. Ivan Filip, N. Vanthuyne, C. Roussel, A. Mándi, L. Tóth, T. Kurtán, J.-V. Naubron, and I.-C. Man. 2015. Vibrational and electronic circular dichroism studies on the axially chiral pyridine-N-oxide: Trans-2,6-di-ortho-tolyl-3,4,5-trimethylpyridine-N-oxide. *Tetrahedron Asymmetry* 26(18–19):1043–1049.

Terzi, E., G. Hoelzemann, and J. Seelig. 1994. Alzheimer .beta.-amyloid peptide 25-35: Electrostatic interactions with phospholipid membranes. *Biochemistry* 33(23):7434–7441.

Thilgen, C., A. Herrmann, and F. Diederich. 1997. Configurational description of chiral fullerenes and fullerene derivatives with a chiral functionalization pattern. *Helv. Chim. Acta* 80(1):183–199.

Toniolo, C., F. Formaggio, and R. W. Woody. 2012. Electronic circular dichroism of peptides. In: N. Berova, P. L. Polavarapu, K. Nakanishi, and R. W. Woody (eds). *Comprehensive Chiroptical Spectroscopy*. Vol. 2. New York, NY: Wiley.

Tinoco, I, Jr., and C. R. Cantor. 1970. Application of optical rotatory dispersion and circular dichroism in biochemical analysis. In: D. Glick (ed). *Methods of Biochemical Analysis*. Vol. 18. New York, NY: John Wiley & Sons.

Tummalapalli, C. M., D. M. Back, and P. L. Polavarapu. 1988. Fourier-transform infrared vibrational circular dichroism of simple carbohydrates. *J. Chem. Soc., Faraday Trans. 1* 84(8):2585–2594.

Urbanová, M., V. Setnička, F. J. Devlin, and P. J. Stephens. 2005. Determination of molecular structure in solution using vibrational circular dichroism spectroscopy: The supramolecular tetramer of S-2,2′-dimethyl-biphenyl-6,6′-dicarboxylic Acid. *J. Am. Chem. Soc.* 127(18):6700–6711.

Urbanová, M., V. Setnička, and K. Volka. 2000. Measurements of concentration dependence and enantiomeric purity of terpene solutions as a test of a new commercial VCD spectrometer. *Chirality* 12(4):199–203.

Urray, D. W., M. Mednikes, and E. Bejnarowicz. 1967. Optical rotation of mitochondrial membranes. *Pro. Natl. Acad. Sci.* 57:1043–1049.

van Delden, R. A., T. Mecca, C. Rosini, and B. L. Feringa. 2004. A chiroptical molecular switch with distinct chiral and photochromic entities and its application in optical switching of a cholesteric liquid crystal. *Chem.—Eur. J.* 10(1):61–70.

van't Hoff, G. H. 1875. Sur les formules de structure dans l'espace. *Bull. Soc. Chim. Fr.* 23:295–301.

Vanthuyne, N., C. Roussel, J.-V. Naubron, N. Jagerovic, P. M. Lázaro, I. Alkorta, and J. Elguero. 2011. Determination of the absolute configuration of 1,3,5-triphenyl-4,5-dihydropyrazole enantiomers by a combination of VCD, ECD measurements, and theoretical calculations. *Tetrahedron Asymmetry* 22(10):1120–1124.

Vijay, R., G. Baskar, A. B. Mandal, and P. L. Polavarapu. 2013. Unprecedented relationship between the size of spherical chiral micellar aggregates and their specific optical rotations. *J. Phys. Chem. A* 117(18):3791–3797.

Vijay, R., and P. L. Polavarapu. 2012. FMOC-amino acid surfactants: Discovery, characterization and chiroptical spectroscopy. *J. Phys. Chem. A* 116(44):10759–10769.

Vijay, R., and P. L. Polavarapu. 2013. Molecular structural transformations induced by spatial confinement in barium fluoride cells. *J. Phys. Chem. A* 117(51):14086–14094.

Wang, F., and P. L. Polavarapu. 2003. Conformational analysis of melittin in solution phase: Vibrational circular dichroism study. *Biopolymers* 70(4):614–619.

Wang, F., P. L. Polavarapu, J. Drabowicz, P. Kiełbasinski, M. J. Potrzebowski, M. Mikołajczyk, M. W. Wieczorek, W. W. Majzner, and I. Łażewska. 2004. Solution and crystal structures of chiral molecules can be significantly different: tert-Butylphenylphosphinoselenoic acid. *J. Phys. Chem. A* 108(11):2072–2079.

Wang, F., P. L. Polavarapu, J. Drabowicz, and M. Mikołajczyk. 2000. Absolute configurations, predominant conformations and tautomeric structures of enantiomeric tert-Butylphenylphosphine oxides. *J. Org. Chem.* 65(22):7561–7565.

Wang, F., P. L. Polavarapu, J. Drabowicz, M. Mikołajczyk, and P. Łyżwa. 2001. Absolute configurations, predominant conformations, and tautomeric structures of enantiomeric tert-Butylphenylphosphinothioic acid. *J. Org. Chem.* 66(26):9015–9019.

Wang, F., P. L. Polavarapu, F. Lebon, G. Longhi, S. Abbate, and M. Catellani. 2002a. Absolute configuration and conformational stability of (S)-(+)-3-(2-methylbutyl) thiophene and (+)-3,4-Di[(S)-2-methylbutyl)]thiophene and their polymers. *J. Phys. Chem. A* 106(24):5918–5923.

Wang, F., P. L. Polavarapu, V. Schurig, and R. Schmidt. 2002b. Absolute configuration and conformational analysis of a degradation product of inhalation anesthetic Sevoflurane: A vibrational circular dichroism study. *Chirality* 14(8):618–624.

Wang, Y., G. Raabe, C. Repges, and J. Fleischhauer. 2003. Time-dependent density functional theory calculations on the chiroptical properties of rubroflavin: Determination of its absolute configuration by comparison of measured and calculated CD spectra. *Int. J. Quantum Chem.* 93(4):265–270.

Wang, F., Y. Wang, P. L. Polavarapu, T. Li, J. Drabowicz, K. M. Pietrusiewicz, and K. Zygo. 2002c. Absolute configuration of tert-Butyl-1-(2-methylnaphthyl)phosphine oxide. *J. Org. Chem.* 67(18):6539–6541.

Wang, F., C. Zhao, and P. L. Polavarapu. 2004. A study of the conformations of valinomycin in solution phase. *Biopolymers* 75(1):85–93.

Wen, Z. Q., L. D. Barron, and L. Hecht. 1993. Vibrational Raman optical activity of monosaccharides. *J. Am. Chem. Soc.* 115(1):285–292.

Weymuth, T., C. R. Jacob, and M. Reiher. 2011. Identifying protein β-turns with vibrational Raman optical activity. *ChemPhysChem.* 12(6):1165–1175.

Weymuth, T., and M. Reiher. 2013. Characteristic Raman optical activity signatures of protein β-sheets. *J. Phys. Chem. B* 117(40):11943–11953.

Wilen, S. H., K. A. Bunding, C. M. Kasheres, and M. J. Wieder. 1985. On the optical activity of bromochlorofluoromethane. *J. Am. Chem. Soc.* 107(24):6997–6998.

Wilson, E. B., J. C. Decius, and P. C. Cross. 1980. *Molecular Vibrations: The Theory of Infrared and Raman Vibrational Spectra*. New York, NY: Dover Publications.

Wilen, S. H., J. Z. Qi, and P. G. Williard. 1991. Resolution, asymmetric transformation, and configuration of Troeger's base. Application of Troeger's base as a chiral solvating agent. *J. Org. Chem.* 56(2):485–487.

Woody, R. W. 1992. Circular dichroism and conformation of unordered polypeptides. In: C. A. Bush (ed). *Advances in Biophysical Chemistry*. Greenwich: JAI Press.

Woody, R. W. 2012. Electronic circular dichroism of proteins. N. Berova, P. L. Polavarapu, K. Nakanishi and R. W. Woody (eds). *Comprehensive Chiroptical Spectroscopy*. Vol. 2. New York, NY: Wiley.

Wu, T., and X. You. 2012. Exciton coupling analysis and enolization monitoring by vibrational circular dichroism spectra of camphor diketones. *J. Phys. Chem. A* 116:8959–8964.

Wu, T., X.-Z. You, and P. Bouř. 2015. Applications of chiroptical spectroscopy to coordination compounds. *Coord. Chem. Rev.* 284:1–18.

Xiao, D., W. Yang, J. Yao, L. Xi, X. Yang, and Z. Shuai. 2004. Size-dependent exciton chirality in (R)-(+)-1,1′-Bi-2-naphthol dimethyl ether nanoparticles. *J. Am. Chem. Soc.* 126(47):15439–15444.

Yamabe, S., W. Guan, and S. Sakaki. 2013. Three competitive transition states at the glycosidic bond of sucrose in its acid-catalyzed hydrolysis. *J. Org. Chem.* 78(6):2527–2533.

Yamamoto, S., T. Furukawa, P. Bouř, and Y. Ozaki. 2014. Solvated states of poly-l-alanine α-helix explored by Raman optical activity. *J. Phys. Chem. A* 118(20):3655–3662.

Yamamoto, S., X. Li, K. Ruud, and P. Bouř. 2012. Transferability of various molecular property tensors in vibrational spectroscopy. *J. Chem. Theory Comput.* 8(3):977–985.

Yamamoto, S., M. Straka, H. Watarai, and P. Bour. 2010. Formation and structure of the potassium complex of valinomycin in solution studied by Raman optical activity spectroscopy. *Phys. Chem. Chem. Phys.* 12(36):11021–11032.

Yang, J. T., and P. Doty. 1957. The optical rotatory dispersion of polypeptides and proteins in relation to configuration. *J. Am. Chem. Soc.* 79(7):761–775.

Yang, G., H. Tran, E. Fan, W. Shi, T. L. Lowary, and Y. Xu. 2010. Determination of the absolute configurations of synthetic daunorubicin analogues using vibrational circular dichroism spectroscopy and density functional theory. *Chirality* 22(8):734–743.

Yang, G., and Y. Xu. 2008. The effects of self-aggregation on the vibrational circular dichroism and optical rotation measurements of glycidol. *Phys. Chem. Chem. Phys.* 10(45):6787–6795.

Yang, G., and Y. Xu. 2011. Vibrational circular dichroism spectroscopy of chiral molecules. In: R. Naaman, D. N. Beratan, and D. Waldeck (eds). *Electronic and Magnetic Properties of Chiral Molecules and Supramolecular Architectures*. Springer Berlin Heidelberg.

Yang, G., Y. Xu, J. Hou, H. Zhang, and Y. Zhao. 2010. Determination of the absolute configuration of pentacoordinate chiral phosphorus compounds in solution by using vibrational circular dichroism spectroscopy and density functional theory. *Chem.—Eur. J.* 16(8):2518–2527.

Yao, H., N. Nishida, and K. Kimura. 2010. Conformational study of chiral penicillamine ligand on optically active silver nanoclusters with IR and VCD spectroscopy. *Chem. Phys.* 368(1–2):28–37.

Yu, G.-S., D. Che, T. B. Freedman, and L. A. Nafie. 1993. Backscattering dual circular polarization raman optical activity in ephedrine molecules. *Tetrahedron Asymmetry* 4(3):511–516.

Zhang, P., A. Mándi, X.-M. Li, F.-Y. Du, J.-N. Wang, X. Li, T. Kurtán, and B.-G. Wang. 2014. Varioxepine A, a 3H-oxepine-containing alkaloid with a new oxa-cage from the marine algal-derived endophytic fungus paecilomyces variotii. *Org. Lett.* 16(18):4834–4837.

Zhao, C., and P. L. Polavarapu. 1999. Vibrational circular dichroism is an incisive structural probe: Ion-induced structural changes in gramicidin D. *J. Am. Chem. Soc.* 121(48):11259–11260.

Zhao, C., and P. L. Polavarapu. 2001a. Comparative evaluation of vibrational circular dichroism and optical rotation for determination of enantiomeric purity. *Appl. Spectrosc.* 55(7):913–918.

Zhao, C., and P. L. Polavarapu. 2001b. Vibrational circular dichroism of gramicidin D in vesicles and micelles. *Biopolymers* 62(6):336–340.

Zhao, C., and P. L. Polavarapu. 2002. Conformational study of gramicidin D in organic solvents in the presence of cations using vibrational circular dichroism. In: J. M. Hicks (ed). *Chirality: Physical Chemistry*. Washington, DC: American Chemical Society.

Zhao, C., P. L. Polavarapu, H. Grosenick, and V. Schurig. 2000. Vibrational circular dichroism, absolute configuration and predominant conformations of volatile anesthetics: Enflurane. *J. Mol. Struct.* 550–551(0):105–115.

Zhang, P., and P. L. Polavarapu. 2007. Spectroscopic investigation of the structures of dialkyl tartrates and their cyclodextrin complexes. *J. Phys. Chem. A* 111(5):858–871.

Zhu, F., N. W. Isaacs, L. Hecht, G. E. Tranter, and L. D. Barron. 2006a. Raman optical activity of proteins, carbohydrates and glycoproteins. *Chirality* 18(2):103–115.

Zhu, F., G. E. Tranter, N. W. Isaacs, L. Hecht, and L. D. Barron. 2006b. Delineation of protein structure classes from multivariate analysis of protein Raman optical activity data. *J. Mol. Biol.* 363(1):19–26.

Zielinski, F., S. T. Mutter, C. Johannessen, E. W. Blanch, and P. L. A. Popelier. 2015. The Raman optical activity of [small beta]-d-xylose: Where experiment and computation meet. *Phys. Chem. Chem. Phys.* 17:21799–21809.

Appendix 1: Polarization States and Stokes Parameters

The details of Maxwell's equations associated with electromagnetic radiation will not be given here, but it is to be noted that a time-dependent plane electromagnetic wave propagating along the z-axis satisfies the second-order differential equation known as the wave equation:

$$\frac{\partial^2 F}{\partial z^2} - \frac{1}{c^2}\frac{\partial^2 F}{\partial t^2} = 0 \tag{A1.1}$$

A1.1 Real-Form Expressions for Polarization States

A solution to this equation is $\boldsymbol{F} = \boldsymbol{F}_0 \cos \Theta$, where $\Theta = 2\pi(\nu t - z/\lambda)$. Since there is no component of $\boldsymbol{F}$ along the z-axis, $\boldsymbol{F}_0$ can have components along the x- and y-axes only. Then $\boldsymbol{F}$ can be written as $\boldsymbol{F} = \left(\mathbf{u}F_{x0} + \mathbf{v}F_{y0}\right) \cos\Theta$, where $\boldsymbol{u}$ and $\boldsymbol{v}$ are unit vectors along the x- and y-axes, respectively. It is a common practice to fix the value of z at zero in this function and view the propagation of the wave in time. In that case, Θ is replaced by $\Theta_t = 2\pi\nu t$. For changing the polarization states, both x- and y-component waves have to be coherent, and these coherent components are derived (see Chapter 1) from a linearly polarized wave by taking its projection at +45° and −45° to the plane of polarization. Thus, $F_{x0} = F_{y0} = \frac{F_0}{\sqrt{2}}$, where F_0 is the amplitude of plane-polarized parent wave. Thus a linearly polarized light wave, generated from two coherent wave components, is written as

$$\boldsymbol{F}_{\mathrm{LP}} = \frac{F_0}{\sqrt{2}}(\mathbf{u}\cos\Theta_t \pm \mathbf{v}\cos\Theta_t) \tag{A1.2}$$

As demonstrated in Chapter 1, adding $\delta = +\pi/2$ to the phase of the y-polarized wave results in a right circularly polarized wave and $\delta = -\pi/2$ results in a left circularly polarized wave. Then, following the discussion in Chapter 1, circularly polarized waves can be written as

$$\boldsymbol{F}_{\mathrm{RCP}} = \frac{F_0}{\sqrt{2}}(\mathbf{u}\cos\Theta_t - \mathbf{v}\sin\Theta_t) \tag{A1.3}$$

$$\boldsymbol{F}_{\mathrm{LCP}} = \frac{F_0}{\sqrt{2}}(\mathbf{u}\cos\Theta_t + \mathbf{v}\sin\Theta_t) \tag{A1.4}$$

Equations A1.2 through A1.4 represent the real-form expressions for polarization states.

A1.2 Complex-Form Expressions for Polarization States

Analogous expressions in complex form can also be obtained by noting that a complex function $F = F_0\, e^{-i\Theta}$, where $\Theta = 2\pi(\nu t - z/\lambda)$ is also a solution to the second-order differential equation A1.1. Then using the x- and y-polarized coherent wave components, the resulting electric vector is written as $\boldsymbol{F} = \mathbf{u}F_x + \mathbf{v}F_y = \left(\mathbf{u}F_{x0} + \mathbf{v}F_{y0}\right) e^{-i\Theta}$. Different polarization states can now be generated by multiplying the y-polarized component, $F_{y0}\, e^{-i\Theta}$ with a complex phase difference quantity, $e^{-i\delta}$. For $\delta = 0$, $e^{-i\delta} = 1$; for $\delta = \pi/2$, $e^{-i\delta} = -i$; for $\delta = \pm\pi$, $e^{-i\delta} = -1$; and for $\delta = -\pi/2$, $e^{-i\delta} = +i$. For a linearly polarized light wave, $F_{x0} = F_{y0} = \frac{F_0}{\sqrt{2}}$ and $\delta = 0$ or $\pm\pi$. Therefore, electric vector for linearly polarized light wave is

$$\boldsymbol{F}_{\mathrm{LP}} = \frac{F_0}{\sqrt{2}}(\mathbf{u} \pm \mathbf{v})\, e^{-i\Theta} \tag{A1.5}$$

For a right circularly polarized wave, $F_{x0} = F_{y0} = \frac{F_0}{\sqrt{2}}$ and $\delta = \pi/2$. Therefore, electric vector for right circularly polarized wave is

$$\boldsymbol{F}_{\mathrm{RCP}} = \frac{F_0}{\sqrt{2}}\,(\mathbf{u} - i\mathbf{v})\, e^{-i\Theta} \tag{A1.6}$$

For a left circularly polarized wave, $F_{x0} = F_{y0} = \frac{F_0}{\sqrt{2}}$ and $\delta = -\pi/2$. Therefore, electric vector for left circularly polarized wave becomes

$$\boldsymbol{F}_{\mathrm{LCP}} \frac{F_0}{\sqrt{2}}(\mathbf{u} + i\mathbf{v})\, e^{-i\Theta} \tag{A1.7}$$

Equations A1.5 through A1.7 represent the complex-form expressions for polarization states.

A1.3 Stokes Parameters

For an arbitrary phase difference between x- and y-polarized coherent light waves with unequal amplitudes, the projection of the resulting electric vector in the xy plane takes the shape of an ellipse. The angle between the major axis of ellipse and $+x$-axis is called the azimuth. The length of the major axis of ellipse is designated as "a" and that of the minor axis as "b." The inverse tangent of (b/a) is called ellipticity, η; that is, $\eta = \tan^{-1}(b/a)$. The state of the resulting electric vector can be completely specified by its intensity, I, azimuth θ, and ellipticity η. These three quantities can be derived from four Stokes parameters, designated as S_0, S_1, S_2, and S_3 (Barron 2004). S_0 determines the total intensity of x- and y-polarized waves, and S_1 determines the difference in the intensities of x- and y-polarized waves. This difference can be measured by placing a linear polarization analyzer in the light beam and orienting this analyzer such that it transmits only the x-polarized or y-polarized waves. The Stokes parameter S_2 determines the intensity difference when the above-mentioned analyzer is placed at +45° (between the $+x$- and $+y$-axes) and at −45° (between $-x$- and $+y$-axes). Finally, S_3 determines the intensity difference between right and left circularly polarized waves.

To see the connection between Stokes parameters and the projected shape of an electric vector in the xy plane, consider a general ellipse depicted in Figure A1.1. The major axis of the ellipse, which is at an angle θ from the x-axis, is aligned with the x' axis and has a length "a." The minor axis of ellipse is aligned with the y' axis and has a length "b." The resulting complex electric vector, $\boldsymbol{P}$ of the ellipse, oriented at an angle η from the x' axis, is given as

$$\boldsymbol{P} = F_0\left(\mathbf{u}P_x + \mathbf{v}P_y\right) e^{-i\Theta} \tag{A1.8}$$

Upon taking the components of this complex vector along the x- and y-axes, one obtains

$$P_x = (\cos\eta\cos\theta - i\sin\eta\sin\theta) \tag{A1.9}$$

$$P_y = (\cos\eta\sin\theta + i\sin\eta\cos\theta) \tag{A1.10}$$

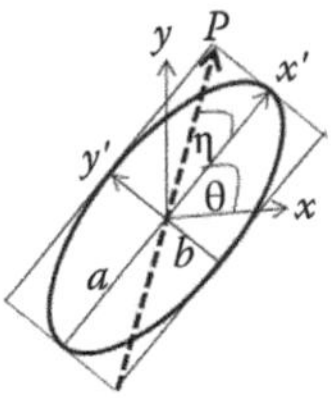

FIGURE A1.1
Projection of an elliptically polarized electric vector in the xy plane and depiction of ellipticity η, azimuth θ, and resultant complex vector P associated with the ellipse. Axes x' and y' are aligned with major and minor axes of the ellipse. The length of major axis is designated as "a" and that of the minor axis as "b."

Stokes parameters S_0, S_1, S_2, and S_3 are written as follows:

$$S_0 = F_0^2\left(P_xP_x^* + P_yP_y^*\right) = F_0^2 \tag{A1.11}$$

$$S_1 = F_0^2\left(P_xP_x^* - P_yP_y^*\right) = F_0^2\cos 2\eta\cos 2\theta \tag{A1.12}$$

$$S_2 = F_0^2\left(P_xP_y^* + P_yP_x^*\right) = F_0^2\cos 2\eta\sin 2\theta \tag{A1.13}$$

$$S_3 = iF_0^2\left(P_xP_y^* - P_yP_x^*\right) = F_0^2\sin 2\eta \tag{A1.14}$$

Note that regardless of the differences in the orientation of the ellipse used, one will find that $S_0^2 = S_1^2 + S_2^2 + S_3^2$, so only three of the four Stokes parameters are independent. By measuring these parameters, one can determine the azimuth and ellipticity of a light wave from the following equations:

$$\theta = \frac{1}{2}\tan^{-1}\left(\frac{S_2}{S_1}\right) \tag{A1.15}$$

$$\eta = \frac{1}{2}\tan^{-1}\left[\frac{S_3}{\left(S_1^2 + S_2^2\right)^{\frac{1}{2}}}\right] = \tan^{-1}\left(\frac{b}{a}\right) \tag{A1.16}$$

For linear polarization states generated from two orthogonal plane-polarized coherent waves with $\delta = 0$ or $\pm\pi$ and equal amplitude, S_1 should be zero because $I_x = I_y$ or $F_{x0}^2 = F_{y0}^2$. For a linear polarization state, $\eta = 0$ because the projected shape of the resulting electric vector collapses into a straight line. Then from Equation A1.12, we can infer that θ should be 45° or 135°, as seen in Figure 1.3 for $\delta = 0°$ or 180°. By setting $\eta = 0$ and $\theta = 45°$ or 135° in Equations A1.12 through A.1.14, one obtains $S_1 = S_3 = 0$ and $S_2 = F_0^2$.

For linear polarization states generated from two orthogonal plane-polarized coherent waves with $\delta = 0$ or $\pm\pi$ and unequal amplitudes $S_1 \neq 0$ because $I_x \neq I_y$ or $F_{x0}^2 \neq F_{y0}^2$. If $S_1 \neq 0$, then Equation A1.12 suggests that θ cannot be 45° or 135°, which explains why the projection of the resulting polarization in Figure 1.8 of Chapter 1 was not at 45° to the x-axis. Substituting $\theta \neq 45°$ or $\theta \neq 135°$ and $\eta = 0$ in Equations A1.12 through A1.14, one would obtain $\left(S_1^2 + S_2^2\right) = F_0^2$ and $S_3 = 0$.

For circularly polarized waves generated from two orthogonal plane-polarized coherent waves with $\delta = \pm\pi/2$ or $\pm 3\pi/2$ and equal amplitudes (see Figure 1.3 of Chapter 1 with $\delta = 90°$ or 270°), the major and minor axes have equal lengths and therefore $b = a$. The condition of $b = a$ is satisfied by Equation A1.16 when $\eta = 45°$. Substituting $\eta = 45°$ in Equations

A1.12 through A1.14, one finds that $S_1 = S_2 = 0$ and $S_3 = F_0^2$. For circularly polarized light, the intensity transmitted by a linear polarization analyzer will be independent of the analyzer orientation; hence $S_1 = 0$ and $S_2 = 0$.

For elliptical polarization states generated from two orthogonal plane-polarized coherent waves of equal amplitudes, S_1 should equal zero because $I_x = F_{x0}^2 = I_y = F_{y0}^2$. This can be verified by setting $\theta = 45°$ or 135° in Equation A1.12. However, for an ellipse (see Figure A.1.1), $\eta \neq 0$ and η cannot be 45° (because 45° angle applies only for circularly polarized waves). Substituting $\theta = 45°$ or 135° and $\eta \neq 0$ in Equations A1.12 through A.1.14 gives $S_1 = 0$ and $\left(S_2^2 + S_3^2\right)^{1/2} = F_0^2$.

For elliptical polarization states generated from two orthogonal plane-polarized coherent waves with a phase difference of $\pi/2$, and unequal amplitudes, S_1 should not be 0 (because $F_{x0}^2 \neq F_{y0}^2$; $I_x \neq I_y$) and $\eta \neq 0$. Then Equation A1.12 suggests that θ cannot be ±45° or ±135°. Also η cannot be 45° for an ellipse. Among the remaining possible values for θ, special situations occur when $\theta = 0°$ or 90°. When two orthogonal plane-polarized coherent waves combine with a phase difference of $\pi/2$, the resultant wave would be circularly polarized, and the projected shape is circular, if the amplitudes of participating waves are equal. However, if the amplitudes are not equal, the circular shape distorts into an ellipse with its major axis aligned with that of a participating component wave whose amplitude is larger. For such cases, $\theta = 0°$ or 90° will apply as can be verified in Figures 1.10 and 1.11. Substitution of $\theta = 0°$ or 90° in Equations A1.12 through A1.14 results in the Stokes parameters as $S_2 = 0$ and $\left(S_1^2 + S_3^2\right)^{1/2} = F_0^2$.

In the previous paragraphs, at least one Stokes parameter was seen to be zero, which is not a general situation. For a general case of elliptical polarization, as depicted in Figure 1.9, all three Stokes parameters, S_1, S_2, and S_3, are required to describe the polarization state of a light wave.

More information on the material covered in this Appendix 1 can be found in the references Barron (2004) and Fitts (1974).

References

Barron, L. D. 2004. *Molecular Light Scattering and Optical Activity*, 2nd ed. Cambridge, UK: Cambridge University Press.

Fitts, D. D. 1974. *Vector Analysis in Chemistry*. New York City, NY: McGraw Hill.

Appendix 2: Kramers–Kronig Transformation

Electronic circular dichroism (ECD) and optical rotatory dispersion (ORD) are related via the Kramers–Kronig (KK) transform (Moscowitz 1962; Tinoco and Cantor 1970). Thus, if one of these two properties is measured as a function of wavelength, then the second can be obtained, at least in principle, via the KK transform. This transform requires the measured data in an infinitely wide spectral region (see later), but practical limitations restrict the region of measurement. Therefore, the result obtained from the KK transform is usually approximate. In the older literature (Tinoco and Cantor 1970) when ECD instruments were not widely developed, the KK transform of experimental ORD into ECD was practiced. However, in the current times, measurements of ECD at a desired wavelength resolution are more common than measuring ORD at that resolution (most commercially available polarimeters currently measure ORD at discrete wavelengths, not at the wavelength resolution needed for the KK transform; see Chapter 7). For this reason, the presentation here is limited to the KK transform of ECD to ORD, but the transform from ORD to ECD can be carried out in a similar way.

A2.1 KK Transformation of Molar Ellipticity into Molar Rotation

ECD and ORD are expressed, respectively, as molar ellipticity and molar rotation, in the same units of deg·cm^2 ·dmol^{-1} (see Chapter 3, Equations 3.24 and 3.44) before subjecting them to the KK transform (note: dmol = decimol = 0.1 mol). The KK transformation from molar ellipticity $[\theta(\mu)]$ (as a function of wavelength μ) to the molar rotation $[\phi(\lambda)]$ at wavelength λ is given as

$$[\phi(\lambda)] = \frac{2}{\pi}\int_0^{\infty}[\theta(\mu)]\frac{\mu}{(\lambda^2-\mu^2)}d\mu \quad \text{(A2.1)}$$

The integration in Equation A2.1 will be truncated to a limited region, because it is not practical to measure ECD in the entire wavelength region indicated by the integral limits. The integral on the right-hand side of Equation A2.1 has singularity at $\lambda = \mu$, which requires the use of numerical methods for evaluating this integral.

A2.2 Numerical Integration Methods for KK Transform

Four different numerical methods have been suggested in the literature to overcome the problem in evaluating the integral in Equation A2.1 (Moscowitz 1962; Emeis et al. 1967; Parris and Van der Walt 1975; Ohta and Ishida 1988). Three of these proposed methods (Moscowitz 1962; Emeis et al. 1967; Parris and Van der Walt 1975) deal with transformation between ECD and ORD. The fourth method, originally proposed for the KK transform between absorption and refractive index (Ohta and Ishida 1988), has been adopted in the author's laboratory (Polavarapu 2005) for the KK transform between ECD and ORD. Computer programs for using three of these methods (Moscowitz 1962; Emeis et al. 1967; Ohta and Ishida 1988) have been developed, and the results evaluated (Polavarapu 2005). The adoption of the method of Ohta and Ishida (Ohta and Ishida 1988) is most convenient, from a practical point of view, so the presentation here is limited to this method.

Assuming that the ECD spectrum is available at constant wavelength intervals of "h" and that the wavelengths and spectral intensities at these intervals are labeled, respectively, as μ_j and $\left[\theta(\mu_j)\right]$ with $j = 1,2 \ldots N$, Equation A2.1 can be approximated (Ohta and Ishida 1988) as

$$[\phi(\lambda)] = \frac{2}{\pi}\int_0^\infty [\theta(\mu)]\frac{\mu}{(\lambda^2-\mu^2)}d\mu \approx \left(\frac{2}{\pi}\right)(2h)\left(\frac{1}{2}\right)\sum_j{}^{\#}\left[\frac{[\theta(\mu_j)]}{\lambda-\mu_j}-\frac{[\theta(\mu_j)]}{\lambda+\mu_j}\right] \quad \text{(A2.2)}$$

In this equation, the summation $\sum^{\#}$ signifies that the summation uses alternate data points to avoid singularity at $\lambda = \mu$. If the wavelength λ, where molar rotation is to be calculated, corresponds to an odd data number, then the summation is carried over even data numbers. In contrast, if the wavelength where molar rotation is to be calculated corresponds to an even data number, then the summation is carried over odd data numbers.

The KK transformation of ECD to ORD using Equation A2.2 can be undertaken using the freely available CDSpecTech program (Covington and Polavarapu 2015).

Recent applications of the KK transform have been in the quantum theoretical predictions of ECD and ORD (Krykunov et al. 2006; Rudolph and Autschbach 2008; Rudolph and Autschbach 2011).

References

Covington, C. L., and P. L. Polavarapu. 2015. CDSpecTech: Computer programs for calculating similarity measures for experimental and calculated dissymmetry factors and circular intensity differentials. https://sites.google.com/site/cdspectech1/ (accessed May 4, 2016).

Emeis, C. A., L. J. Oosterhoff, and G. De Vries. 1967. Numerical evaluation of Kramers-Kronig relations. *Proc. Roy. Soc. London* A 297:54–65.

Krykunov, M., M. D. Kundrat, and J. Autschbach. 2006. Calculation of circular dichroism spectra from optical rotatory dispersion, and vice versa, as complementary tools for theoretical studies of optical activity using time-dependent density functional theory. *J. Chem. Phys.* 125(19):194110–194113.

Moscowitz, A. 1962. Theoretical aspects of optical activity part one: Small molecules. *Adv. Chem. Phys.* 4:67–112.

Ohta, K., and H. Ishida. 1988. Comparison among several numerical integration methods for Kramers-Kronig transformation. *Appl. Spectrosc.* 42(6):952–957.

Parris, D., and S. J. Van der Walt. 1975. A new numerical method for evaluating the Kramers-Kronig transformation. *Anal. Biochem.* 68:321–327.

Polavarapu, P. L. 2005. Kramers–Kronig transformation for optical rotatory dispersion studies. *J. Phys. Chem. A* 109(32):7013–7023.

Rudolph, M., and J. Autschbach. 2008. Fast generation of nonresonant and resonant optical rotatory dispersion curves with the help of circular dichroism calculations and Kramers-Kronig transformations. *Chirality* 20:995–1008.

Rudolph, M., and J. Autschbach. 2011. Calculation of optical rotatory dispersion and electronic circular dichroism for tris-bidentate groups 8 and 9 metal complexes, with emphasis on exciton coupling. *J. Phys. Chem. A* 115(12):2635–2649.

Tinoco, I., Jr., and C. R. Cantor. 1970. Application of optical rotatory dispersion and circular dichroism in biochemical analysis. In: D. Glick. *Methods Biochemistry Analysis*. Vol. 18. New York, NY: John Wiley & Sons.

Appendix 3: Spectral Simulations

Quantum chemical calculations of absorption, circular dichroism (CD), vibrational Raman, and vibrational Raman optical activity (VROA) spectra predict integrated band intensity, along with transition energy (reported as wavenumber/wavelength for absorption spectra or as Raman shift in the case of Raman spectra) for each of the transitions. Using this information one can only obtain a stick-line spectrum, where integrated intensity is represented as a line at the transition energy. The experimentally observed spectra, however, do not exhibit line spectra, but instead contain bands, each with a certain width. The width associated with experimentally observed spectral bands originates (Bernath 1995) from lifetimes associated with excited states, collisions between molecules, and an inhomogeneous environment. The excited state lifetimes and molecular collisions result in Lorentzian band shapes, while the inhomogeneous environment results in a Gaussian band shape. For this reason, Lorentzian bandwidths are generally referred to as homogeneous line widths and Gaussian bandwidths as inhomogeneous line widths.

To convert the theoretically predicted stick-line spectrum into the one that resembles the experimental spectra, one has to simulate the spectral intensity distribution by assuming band shapes and bandwidths.

In Chapter 5, we have noted that quantum chemical predictions of electronic absorption (EA) and electronic CD (ECD) spectra can be achieved in two different ways: (1) The predicted transition wavelengths, dipole strengths (or oscillator strengths), and rotational strengths are used to simulate EA and ECD spectra using a Gaussian or Lorentzian band shape function with a selected bandwidth. A Gaussian band shape function is most commonly used for simulating the theoretical EA/ECD spectra using this procedure. (2) EA and ECD spectra are calculated at a chosen spectral resolution using a complex propagator approach where lifetimes of excited electronic states are assumed (Jiemchooroj and Norman 2007; Krykunov et al. 2006). These two procedures yield equivalent results only when a Lorentzian distribution function is used in the spectral simulation procedure of (1).

Quantum chemical predictions of vibrational absorption (VA) spectra provide transition wavenumbers (in cm^{-1}) and dipole strengths, while those of vibrational CD (VCD) spectra provide transition wavenumbers and rotational strengths (see Chapter 5). Using these data, predicted VA and VCD spectra can be simulated using a Gaussian or Lorentzian band shape function with a selected bandwidth. A Lorentzian band shape function is most commonly used here.

Quantum chemical predictions of vibrational Raman and VROA spectra provide vibrational Raman shifts as wavenumbers (in cm^{-1}). In addition, normal coordinate derivatives of electric dipole–electric dipole polarizability, electric dipole-magnetic dipole polarizability, and electric dipole–electric quadrupole polarizability tensors are provided (see Chapter 5). For the appropriate scattering geometry, vibrational Raman and VROA activities are assembled from these tensors and Raman and VROA spectra can be simulated using a Gaussian or Lorentzian band shape function with a selected bandwidth. A Lorentzian band shape function is most commonly used here.

Thus the simulations of predicted EA/ECD/VA/VCD/vibrational Raman/VROA spectra utilize a Gaussian or Lorentzian intensity distribution function. The working details involved for these spectral simulations are provided in the following sections.

A3.1 Gaussian Distribution

A commonly used function to represent the band shapes is the Gaussian function, given as

$$y = y_{ko} e^{\frac{-1}{2}\left(\frac{x - x_{ko}}{\sigma_k}\right)^2} \tag{A3.1}$$

In Equation A3.1, x_{ko} is the center position of band k, y_{ko} is the y-value at x_{ko}, x is the running x-axis value and y is the corresponding value at x; σ_k is the width parameter for the kth band (a larger width results in a broader band; a smaller width results in a narrower/sharper band). The integrated area of the kth band, I_k, is obtained by integrating this Gaussian function (Taylor 1982).

$$I_k = y_{ko} \int_{-\infty}^{\infty} e^{\frac{-1}{2}\left(\frac{x - x_{ko}}{\sigma_k}\right)^2} dx = y_{ko} \sigma_k \sqrt{2\pi} \tag{A3.2}$$

In spectral simulations using the Gaussian band profile, the first task is to assume a width parameter σ_k for each band and convert the integrated band intensity, I_k, into peak intensity, y_{ko}, at its band center using Equation A3.2. Then calculate the spectral distribution using Equation A3.1. A spectrum normally consists of several transitions, and therefore it is necessary to sum the distributions over all bands. Then overall spectral distribution as a function of running x-value, is calculated as

$$y = \sum_k y_{ko} e^{\frac{-1}{2}\left(\frac{x - x_{ko}}{\sigma_k}\right)^2} \tag{A3.3}$$

where the summation index k runs over all available transitions.

In reporting the spectral simulations, the width parameter is reported using varying terminologies (Stephens and Harada 2010), such as full-width at half-maximum (FWHM), half-width at half-maximum (HWHM), standard deviation, and so on. The meaning and connection between these terms is as follows. For Gaussian function given by Equation A3.1, σ_k is called standard deviation (Taylor 1982) of Gaussian distribution function. At half-maximum (HM)

$$y = \frac{y_{ko}}{2} = y_{ko} e^{\frac{-1}{2}\left(\frac{x_{HM} - x_{ko}}{\sigma_k}\right)^2} \tag{A3.4}$$

where x_{HM} is the x-value corresponding to one-half of the peak band intensity. Taking logarithm on both sides, one obtains

$$\ln\frac{1}{2} = -\ln 2 = -\frac{1}{2}\left(\frac{x_{HM} - x_{ko}}{\sigma_k}\right)^2 \tag{A3.5}$$

$$\left(\frac{x_{HM} - x_{ko}}{\sigma_k}\right)^2 = 2\ln 2 \tag{A3.6}$$

$$x_{HM} - x_{ko} = \pm\sigma_k\sqrt{2\ln 2} \tag{A3.7}$$

The difference $x_{HM} - x_{k0}$ is one-half of the bandwidth at HM (see Figure A3.1) and therefore, the absolute value of HWHM is $\sigma_k\sqrt{2\ln 2} = 1.1774\sigma_k$. Accordingly FWHM is $2\sigma_k\sqrt{2\ln 2} = 2.3548\sigma_k$. Following the preceding procedure for HWHM, one can find that the full-width at 0.6065 of y_{ko} is $2\sigma_k$. Similarly, the full-width at 1/*e* of band maximum (FW1OeM) is $2\sqrt{2}\,\sigma_k = 2.8284\sigma_k$ and the half-width at 1/*e* of band maximum (HW1OeM) is $\sqrt{2}\,\sigma_k = 1.4142\sigma_k$. These widths are depicted in Figure A3.1 for a band centered at 1650 cm^{-1} with $\sigma = 20$ cm^{-1}.

It is convenient to rewrite Equation A3.1 by substituting Δ_k for $\sqrt{2}\,\sigma_k$ as follows:

$$y = y_{ko} e^{-\left(\frac{x - x_{ko}}{\Delta_k}\right)^2} \tag{A3.8}$$

If this equation is used, then the integrated band area becomes

$$I_k = y_{ko}\int_{-\infty}^{\infty} e^{-\left(\frac{x - x_{ko}}{\Delta_k}\right)^2} dx = y_{ko}\Delta_k\sqrt{\pi} \tag{A3.9}$$

In this equation, Δ_k represents HW1OeM.

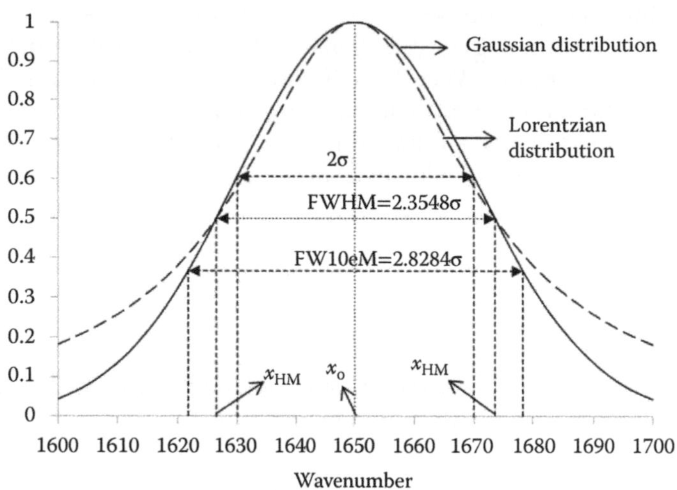

FIGURE A3.1
Gaussian (solid line) and Lorentzian (dashed line) distributions with band center at 1650 cm^{-1}. The width parameters, σ, full-width at HM (FWHM) and full-width at $1/e$ of maximum (FW1OeM) shown are for the Gaussian distribution. Peak intensity of 1 **(AU)** and $\sigma = 20$ cm^{-1} are assumed.

A3.2 Lorentzian Distribution

Another commonly used function to represent the band shapes is the Lorentzian function, given as

$$y = y_{ko}\left(\frac{\Delta_k^2}{(x - x_{ko})^2 + \Delta_k^2}\right) \tag{A3.10}$$

Note that even though we are using the same symbol Δ_k in Equations A3.8 and A3.10, it has different meanings for Gaussian and Lorentzian functions: Δ_k for the kth band represents HWHM for Lorentzian functions, and HW1OeM for Gaussian functions. The integrated band area, I_k, is obtained by integrating this Lorentzian function, noting that

$$\int_{-\infty}^{\infty} \frac{1}{x^2 + a^2}\, dx = \frac{\pi}{a} \tag{A3.11}$$

$$I_k = y_{ko}\int_{-\infty}^{\infty} \frac{\Delta_k^2}{(x - x_{ko})^2 + \Delta_k^2}\, dx = y_{ko}\Delta_k^2\int_{-\infty}^{\infty} \frac{1}{(x - x_{ko})^2 + \Delta_k^2}\, dx = y_{ko}\Delta_k\pi \tag{A3.12}$$

In spectral simulations using Lorentzian band profiles, the first task is to assume a width parameter Δ_k for each band and convert the integrated band intensity, I_k, into peak intensity, y_{ko}, at its band center using Equation A3.12. Then calculate the spectral distribution using Equation A3.10.

When multiple transitions are involved, the sum over spectral distributions of all bands should be calculated as a function of x, using

$$y = \sum_k y_{ko}\left(\frac{\Delta_k^2}{(x - x_{ko})^2 + \Delta_k^2}\right) \tag{A3.13}$$

where the summation index k runs over all available transitions.

For a Gaussian function with $\sigma = 20\ \text{cm}^{-1}$, as demonstrated earlier, HWHM = $1.1774\ \sigma = 23.548\ \text{cm}^{-1}$. For comparison, a Lorentzian function with the same HWHM (i.e., $\Delta = 23.548\ \text{cm}^{-1}$) is also shown in Figure A3.1. It can be seen from this figure that, for the same width parameter, the Lorentzian function yields a wider distribution in the wings of a band, than the Gaussian function.

A3.3 Absorption Spectral Simulations

Modern spectrometers present infrared spectra with wavenumber as the x-axis, and electronic spectra with wavelength as the x-axis. There are some subtle differences in using wavenumber versus wavelength for the x-axis in simulating the theoretical spectra, so we will present both procedures for spectral simulations.

A3.3.1 Vibrational Absorption Spectral Simulation as a Function of Wavenumber

The integrated molar absorption coefficient for the kth band is related to the dipole strength, D_k (see Chapter 5, Equation 5.26) as follows:

$$A_k = \int \alpha(\bar{\nu})d\bar{\nu} = \frac{8\pi^3 N_A \bar{\nu}_k^0}{3ch} D_k \tag{A3.14}$$

In Equation A3.14, N_A, h, and c are, respectively, Avogadro's number, Planck's constant, and speed of light, D_k is the dipole strength of the kth transition and $\bar{\nu}_k^0$ is the band center in cm^{-1}. Substituting the constants in Equation A3.14, and converting the units (see Appendix 8, Section A8.7) one obtains the expression for integrated absorption intensities in units of km mol^{-1}.

$$A_k = 2.5066 \times 10^{-4} \times \bar{\nu}_k^0 \times D_k \tag{A3.15}$$

In Equation A3.15, dipole strength D_k is in units of 10^{-40} $\text{esu}^2\ \text{cm}^2$ and $\bar{\nu}_k^0$ in cm^{-1} units.

The molar absorption coefficient, $\alpha(\bar{\nu})$, and the decadic molar absorption coefficient, $\varepsilon(\bar{\nu})$, are related as follows (see Chapter 4, Equations 4.143 and 4.144):

$$\alpha(\bar{\nu}) = \frac{1}{Cl}\log_e \frac{I_0}{I} \quad \text{(A3.16)}$$

$$\varepsilon(\bar{\nu}) = \frac{1}{Cl}\log_{10} \frac{I_0}{I} \quad \text{(A3.17)}$$

$$\alpha(\bar{\nu}) = 2.303\varepsilon(\bar{\nu}) \quad \text{(A3.18)}$$

Here C represents the molar concentration of the sample (in mol L^{-1} units) and l represents the path length of the cell used to hold the sample. I_0 and I are, respectively, the intensities of light incident on and transmitted through the sample.

Infrared absorption spectral bands are normally associated with Lorentzian distribution, in which case, the peak intensity of the kth band, in terms of the decadic molar absorption coefficient, ε_{k0}, is determined by substituting the Lorentzian function for $\varepsilon(\bar{\nu})$.

$$A_k = \int 2.303\varepsilon(\bar{\nu})\, d\bar{\nu} = \int 2.303\varepsilon_{k0} \frac{\Delta_k^2}{\Delta_k^2 + \left(\bar{\nu} - \nu_k^0\right)^2} d\bar{\nu} = 2.303\varepsilon_{k0}\Delta_k\pi \quad \text{(A3.19)}$$

Expressing the bandwidth parameter Δ_k in cm^{-1} and using A_k from Equation A3.15 in units of km mol^{-1}, Equation A3.19 can be rearranged as

$$\begin{aligned} \varepsilon_{k0} &= \frac{A_k}{2.303\Delta_k\pi} \text{km mol}^{-1} \text{ cm} \times \frac{10^5 \text{cm}}{\text{km}} \times \frac{L}{1000 \text{cm}^3} \\ &= \frac{100 \times A_k}{2.303\Delta_k\pi} L \text{ mol}^{-1}\text{cm}^{-1} \end{aligned} \quad \text{(A3.20)}$$

As stated earlier, Δ_k is the HWHM of the Lorentzian band. Using Equation A3.20, the quantum chemically predicted Lorentzian spectral distribution for absorption intensity can be calculated with Equation A3.10.

If a Gaussian spectral distribution is desired, π in Equation A3.20 is to be replaced with $\sqrt{\pi}$, and Δ_k will then represent HW1OeM of the Gaussian band. With the resulting peak intensity, the Gaussian spectral distribution can be calculated with Equation A3.8.

A3.3.2 Electronic Absorption Spectral Simulation as a Function of Wavelength

It should be remembered at the outset that the simulation of intensity distribution as a function of the running x-values in wavenumber units is not equivalent to that in wavelength units, because energy is proportional to the wavenumber, but inversely proportional to the wavelength. For a given bandwidth in wavelength units, longer wavelength transitions will have smaller width in energy units and vice versa. Therefore, the simulation of intensity distribution as a function of running *x*-value in wavelength units is appropriate when this implication is not an issue. On the other hand, if the goal is to have the same energy width for all transitions, then different transitions should be given different widths in wavelength units or simulations are to be carried out with running *x*-value in wavenumber units.

The quantum chemically predicted EA spectral intensities are often reported as oscillator strength (a dimensionless quantity). The oscillator strength of kth transition is related to its dipole strength (see Chapter 4, Equation 4.130 and Appendix 8, Section A8.9)

$$D_k = \frac{3e^2h}{8\pi^2 m_e c}\lambda_k^0 f_k \tag{A3.21}$$

Substituting the constants (see Appendix 8, Section A8.9), one obtains

$$D_k = 2127\lambda_k f_k \times 10^{-40}\text{esu}^2\text{cm}^2 \tag{A3.22}$$

where wavelength λ_k is in units of nm. The integrated molar absorption coefficient for the kth electronic band, presented as a function of wavelength as *x*-axis, in analogy to A3.14, is given as

$$A_k = \int \alpha(\lambda)d\lambda = \frac{8\pi^3 N_A}{3hc}\lambda_k^0 D_k \tag{A3.23}$$

Expressing D_k in 10^{-40} esu^{2}cm^2, and wavelength λ_k in nm, the integrated molar absorption coefficient can be expressed in cm^3 mol^{-1} units (see Appendix 8, Section A8.10) as

$$A_k = 25.066 \times 10^{-7}\lambda_k^0 D_k \tag{A3.24}$$

The EA bands are often associated with Gaussian spectral distribution, in which case, the peak intensity of the kth band, in terms of decadic molar absorption coefficient, ε_{k0}, is determined by substituting the Gaussian function for $\varepsilon(\lambda)$:

$$A_k = \int 2.303\varepsilon(\lambda)d\lambda = \int 2.303\varepsilon_{k0} e^{-\left(\frac{\lambda-\lambda_{k0}}{\Delta_k}\right)^2} d\lambda = 2.303\varepsilon_{k0}\Delta_k\sqrt{\pi} \tag{A3.25}$$

Expressing Δ_k in nm and substituting A3.24 for A_k, the relation between peak intensity and dipole strength is obtained as follows:

$$\begin{aligned}\varepsilon_{k0} &= \frac{A_k}{2.303\Delta_k\sqrt{\pi}} = \frac{25.066 \times \lambda_k^0 D_k}{2.303\Delta_k\sqrt{\pi}}\text{cm}^2\,\text{mol}^{-1} \\ &= \frac{25.066 \times \lambda_k^0 D_k}{2.303\Delta_k\sqrt{\pi}} \times 10^{-3}L\,\text{mol}^{-1}\,\text{cm}^{-1}\end{aligned} \tag{A3.26}$$

The relation between peak intensity and oscillator strength is obtained by substituting Equation A3.22 for D_k in Equation A3.26:

$$\varepsilon_{k0} = \frac{25.066\times 10^{-3}\times 2127\times\left(\lambda_k^0\right)^2 f_k}{2.303\Delta_k\sqrt{\pi}} = 13.06\times\frac{\left(\lambda_k^0\right)^2 f_k}{\Delta_k} L\ \text{mol}^{-1}\ \text{cm}^{-1} \qquad \text{(A3.27)}$$

From the peak intensity in Equation A3.27, the quantum chemically predicted Gaussian spectral distribution for absorption intensity can be calculated using Equation A3.8.

When Lorentzian spectral distribution is used, $\pi^{½}$ is to be replaced with π in Equation A3.27. Then,

$$\varepsilon_{k0} = \frac{25.066\times 10^{-3}\times 2127\times\left(\lambda_k^0\right)^2 f_k}{2.303\Delta_k\pi} = 7.369\times\frac{\left(\lambda_k^0\right)^2 f_k}{\Delta_k} L\ \text{mol}^{-1}\text{cm}^{-1} \qquad \text{(A3.28)}$$

From the peak intensity in Equation A3.28, the quantum chemically predicted Lorentzian spectral distribution for absorption intensity can be calculated using Equation A3.10.

A3.4 Circular Dichroism Spectral Simulations

VCD spectra are commonly presented using wavenumber, as the *x*-axis, while ECD spectra are presented using wavelength as the *x*-axis. Therefore, these two cases are presented separately.

A3.4.1 VCD Spectral Simulation as a Function of Wavenumber

The integrated difference molar absorption coefficient, ΔA_k, for the kth CD band is often represented (see Chapter 5, Equation 5.31) by its rotational strength, R_k, as follows:

$$\Delta A_k = \int \Delta\alpha(\bar{\nu})d\nu = \int 2.303\Delta\varepsilon(\bar{\nu})d\bar{\nu} = 4\times\frac{8\pi^3 N_A}{3ch}\bar{\nu}_k^0 R_k \qquad \text{(A3.29)}$$

The rotational strengths for vibrational transitions are normally reported as 10^{-44} esu^2 cm^2, and in such cases, following the procedure given earlier for absorption bands (see Equation A3.15), one obtains the integrated difference molar absorption coefficient in cm mol^{-1} units:

$$\begin{aligned}\Delta A_k &= 4\times 2.5066\times 10^{-8}\bar{\nu}_k^0 R_k\ \text{km mol}^{-1}\times\frac{10^5\,\text{cm}}{\text{km}} \\ &= 4\times 25.066\times 10^{-4}\bar{\nu}_k^0 R_k\ \text{cm mol}^{-1}\end{aligned} \qquad \text{(A3.30)}$$

For Lorentzian spectral distribution, with bandwidth, Δ_k, expressed in cm^{-1} units, the peak value of difference molar absorption coefficient, $\Delta\varepsilon_{k0}$, is obtained as follows:

$$\Delta A_k = \int 2.303\ \Delta\varepsilon(\bar{\nu})\, d\bar{\nu} = \int 2.303\ \Delta\varepsilon_{k0} \frac{\Delta_k^2}{\Delta_k^2 + \left(\bar{\nu} - \bar{\nu}_k^0\right)^2} d\bar{\nu} = 2.303\ \Delta\varepsilon_{k0}\Delta_k\pi \quad \text{(A3.31)}$$

$$\Delta\varepsilon_{k0} = \frac{\Delta A_k}{2.303\ \Delta_k\ \pi} = \frac{4\times 25.066\,\bar{\nu}_k^0\ R_k \times 10^{-4}}{2.303\ \Delta_k\pi} \text{cm}^2\text{mol}^{-1}$$
$$= \frac{\bar{\nu}_k^0\ R_k \times 10^{-4}}{22.97\ \Delta_k\pi} L\,\text{mol}^{-1}\ \text{cm}^{-1} \quad \text{(A3.32)}$$

Using the peak intensity in Equation A3.32, the quantum chemically predicted Lorentzian spectral distribution for VCD intensity can be calculated with Equation A3.10.

The analogous expression for $\Delta\varepsilon_{k0}$ with Gaussian spectral distribution is obtained by replacing π in Equation 3.32 with $\sqrt{\pi}$ and recognizing that Δ_k in that case represents HW1OeM of the Gaussian band. With the resulting peak intensity, the Gaussian intensity distribution of VCD bands can be calculated with Equations A3.8.

A3.4.2 ECD Spectral Simulation as a Function of Wavelength

ECD spectra are normally presented using wavelength as the *x*-axis, in which case

$$\Delta A_k = \int \Delta\alpha(\lambda) d\lambda = 4\times \frac{8\pi^3 N_A}{3ch} \lambda_k^0 R_k \quad \text{(A3.33)}$$

Note that rotational strengths, R_k, for electronic transitions are reported in 10^{-40} esu^2 cm^2 (unlike for vibrational transitions where these are reported in 10^{-44} esu^2 cm^2) and wavelength in nm. Then the integrated difference molar extinction coefficient, in units of cm^3 mol^{-1}, becomes (see Appendix 8, Section A8.10)

$$\Delta A_k = 4\times 25.066\times 10^{-7}\times \lambda_k^0 R_k \quad \text{(A3.34)}$$

Band simulations for electronic transitions are associated with Gaussian spectral distribution. Then the peak value of difference molar absorption coefficient can be determined from

$$\Delta A_k = \int 2.303\Delta\varepsilon(\lambda) d\lambda = \int 2.303\,\Delta\varepsilon_{k0}\ e^{-\left(\frac{\lambda-\lambda_{k0}}{\Delta_k}\right)^2} d\lambda = 2.303\times \Delta\varepsilon_{k0}\times \Delta_k\sqrt{\pi} \quad \text{(A3.35)}$$

Expressing the bandwidth Δ_k in nm, and using Equation A3.34, one obtains

$$\Delta\varepsilon_{k0} = \frac{4\times 25.066\times 10^{-7}\lambda_k^0 R_k}{2.303\,\Delta_k\sqrt{\pi}}\frac{\text{cm}^3\text{mol}^{-1}}{10^{-7}\,\text{cm}}\times\frac{L}{1000\text{cm}^3}$$
$$= \frac{\lambda_k^0 R_k}{22.97\,\Delta_k\sqrt{\pi}} L\,\text{mol}^{-1}\,\text{cm}^{-1} \tag{A3.36}$$

Using the peak intensity in Equation A3.36, the quantum chemically predicted Gaussian spectral distribution for the kth ECD band can be calculated with Equation A3.8.

For Lorentzian distributions, $\sqrt{\pi}$ in Equation A3.36 is replaced with π and Equation A3.10 is used in place of Equation A3.8.

Note the restrictions pointed in the first paragraph in Section A3.3.2 when wavelength is used for the x-axis.

A3.5 Vibrational Raman and Raman Optical Activity Spectral Simulations

Unlike the case of absorption and CD spectra, where the measured intensities should in principle be independent of the instrument used, vibrational Raman scattering intensities depend on the incident laser power, solid angle of collection, and so on. Therefore, vibrational Raman and VROA spectral intensities are not generally reported on an absolute scale, and are reported as photon counts in arbitrary units (AU). When vibrational Raman and VROA spectra are obtained simultaneously for a given measurement, their ratio will, however, be independent of the instrumental factors. The spectral simulations procedure presented here follows that reported for the first quantum chemical predicted VROA spectra (Polavarapu, 1990).

For vibrational Raman spectral simulations, the intensity expression given in Chapter 3, by Equation 3.50

$$\left(I_\alpha^\gamma\right)_k = K\frac{(\bar{\nu}\mp\bar{\nu}_k)^4}{\bar{\nu}_k\left(\pm 1\mp e^{\frac{\mp hc\bar{\nu}_k}{bT}}\right)}S_k \tag{A3.37}$$

is taken as the integrated intensity and the peak intensity, $\left(I_\alpha^\gamma\right)_k^0$ is obtained as $\left(I_\alpha^\gamma\right)_k^0 = \left(I_\alpha^\gamma\right)_k / \Delta_k\pi$ (see Equation A3.12) for Lorentzian intensity distribution. In Equation A3.37, S_k is the scattering geometry-dependent Raman activity (see Chapter 3, Equations 3.51 through 3.53) that depends on the scattering geometry employed for Raman measurements. The Raman spectral intensity distribution is obtained from Equation A3.38.

$$I_\alpha^\gamma(\bar{\nu}) = \sum_k \left(I_\alpha^\gamma\right)_k^0 f(\bar{\nu}) \tag{A3.38}$$

where $f(\bar{\nu})$ is the desired distribution function (A3.8 or A3.10).

For VROA spectral simulations, the intensity expression given in Chapter 3 (see Equation 3.57)

$$\left(I_\alpha^\gamma - I_\beta^\delta\right)_k = K \frac{(\bar{\nu} \mp \bar{\nu}_k)^4}{\bar{\nu}_k \left(\pm 1 \mp e^{\frac{\mp hc\bar{\nu}_k}{bT}}\right)} P_k \tag{A3.39}$$

is taken as the integrated intensity and the peak intensity, $\left(I_\alpha^\gamma - I_\beta^\delta\right)_k^0$ is obtained as $\left(I_\alpha^\gamma - I_\beta^\delta\right)_k^0 = \left(I_\alpha^\gamma - I_\beta^\delta\right)_k / \Delta_k \pi$ (see Equation A3.12) for Lorentzian intensity distribution. In Equation A3.39, P_k is the scattering geometry-dependent Raman circular intensity difference activity (see Chapter 3, Equations 3.58 through 3.61), that depends on the scattering geometry employed for VROA measurements. The VROA spectral intensity distribution is obtained from Equation A3.40.

$$I_\alpha^\gamma(\bar{\nu}) - I_\beta^\delta(\bar{\nu}) = \sum_k \left(I_\alpha^\gamma - I_\beta^\delta\right)_k^0 f(\bar{\nu}) \tag{A3.40}$$

where $f(\bar{\nu})$ is the desired distribution function (A3.8 or A3.10)

Since the experimental Raman and VROA intensities are in AU, the experimental, or corresponding predicted, spectra have to be multiplied with a *y*-axis scale factor to compare the experimental and predicted spectra on the same scale. This *y*-axis scale factor should be the same for vibrational Raman and VROA spectra, if the experimental Raman and VROA spectra are obtained simultaneously in the same measurement. Also the corresponding predicted Raman and VROA spectra should have been obtained at the same theoretical level. This arbitrary *y*-axis scale factor can be avoided when circular intensity differential (CID), the ratio of VROA to vibrational Raman spectra

$$\Delta(\bar{\nu}) = \frac{I_\alpha^\gamma(\bar{\nu}) - I_\beta^\delta(\bar{\nu})}{I_\alpha^\gamma(\bar{\nu}) + I_\beta^\delta(\bar{\nu})} \tag{A3.41}$$

is compared. When this ratio spectrum is taken, the use of baseline tolerance (Polavarapu and Covington 2014) is necessary to avoid division by small numbers (see Chapter 8).

A3.6 Spectral Simulation Software

All of the spectral simulations discussed in this appendix have been implemented in the freely available computer program (Covington and Polavarapu 2015).

References

Bernath, P. F. 1995. *Spectra of Atoms and Molecules*. Oxford: Oxford University Press.

Covington, C. L., and P. L. Polavarapu. 2015. CDSpecTech: Computer programs for calculating similarity measures for experimental and calculated dissymmetry factors and circular intensity differentials. https://sites.google.com/site/cdspectech1/.

Jiemchooroj, A., and P. Norman. 2007. Electronic circular dichroism spectra from the complex polarization propagator. *J. Chem. Phys.* 126(13):134102.

Krykunov, M., Kundrat, M. D., and Jochen, A. 2006. Calculation of circular dichroism spectra from optical rotatory dispersion, and vice versa, as complementary tools for theoretical studies of optical activity using time-dependent density functional theory. *J. Chem. Phys.* 125(19):194110–13.

Polavarapu, P. L. 1990. Ab initio vibrational Raman and Raman optical activity spectra. *J. Phys. Chem.* 94(21):8106–8112.

Polavarapu, P. L., and Covington, C. L. 2014. Comparison of experimental and calculated chiroptical spectra for chiral molecular structure determination. *Chirality* 26:179–192.

Stephens, P. J., and Harada, N. 2010. ECD cotton effect approximated by the Gaussian curve and other methods. *Chirality* 22 (2):229–233.

Taylor, J. R. 1982. *An Introduction Error Analysis: The Study of Uncertainties in Physical Measurements*. Herndon, VA: University Science Books.

Appendix 4: Exciton Coupling, Exciton Chirality, or Coupled Oscillator Model

The exciton coupling (EC) model was developed for electronic circular dichroism (ECD) spectral interpretations (Tinoco 1963; Tinoco and Cantor 1970). This model, when adopted for vibrational circular dichroism (VCD) spectral interpretations (Holzwarth and Chabay 1972), was referred to as the coupled oscillator (CO) model. The interaction between two electric dipole transition moments is considered in both cases, but the nature of transitions is different in the interpretations of ECD and VCD. For ECD spectral interpretations, the coupling between electric dipole transition moments of electronic transitions from two chromophores is considered to arise from dipolar interaction. For VCD spectral interpretations, on the other hand, the coupling between electric dipole transition moments of vibrational transitions of two (harmonic) oscillators is considered. Although the literature articles adopted the dipolar interaction mechanism (as in the case of ECD), the mechanism of interaction was not specified in the original article (Holzwarth and Chabay 1972). Except for these differences, the conceptual details involved in arriving at the final result are identical for EC and CO models. For this reason the procedure given in the following is cast in general terms. The chromophores used for electronic transitions, or the oscillators used for vibrational transitions, are referred to as the groups. The wave functions of the groups are to be viewed as those associated with electronic states when considering ECD, and vibrational states when considering VCD. Similarly, the associated transitions are to be viewed as electronic transitions when considering ECD and vibrational transitions when considering VCD. Except for the aforementioned contextual differences, the EC and CO models are identical. These models are also referred to as the exciton chirality (also abbreviated as EC) model in the recent literature.

As EC/CO model has been widely used in the literature, it is described in greater detail here. But the reason for presenting this material as an appendix, instead of as a main chapter, is that the predictions obtained with this method are of approximate nature and are not quantitatively accurate. The explosive use of quantum chemical (QC) calculations has led to a diminishing use of the EC/CO model. Yet the EC/CO model may find some use in situations where QC calculations are not feasible.

A4.1 Two Noninteracting Groups

If there is no interaction between the groups, then the total Hamiltonian is written as the sum of Hamiltonian operators, $\hat{H}_A$ and $\hat{H}_B$, for the individual groups A and B. The caret (inverted V) symbol on top of H identifies the operator property. For both groups in their ground states, the ground state wave function of the two-group system, ψ_{00} is given as the product of orthonormal wave functions of individual groups:

$$\psi_{00} = \varphi_A^0 \varphi_B^0 \tag{A4.1}$$

Here, φ_A^0 is a ground state wave function of group A and φ_B^0 is that of group B. The energy of ground state is

$$\begin{aligned}\left\langle \psi_{00} \middle| \hat{H}_A + \hat{H}_B \middle| \psi_{00} \right\rangle &= \left\langle \varphi_A^0\, \varphi_B^0 \middle| \hat{H}_A + \hat{H}_B \middle| \left\langle \varphi_A^0\, \varphi_B^0 \right\rangle \right. \\ &= \left\langle \varphi_A^0 \middle| \hat{H}_A \middle| \varphi_A^0 \right\rangle \left\langle \varphi_B^0 \middle| \varphi_B^0 \right\rangle + \left\langle \varphi_B^0 \middle| \hat{H}_B \middle| \varphi_B^0 \right\rangle \left\langle \varphi_A^0 \middle| \varphi_A^0 \right\rangle = E_A^0 + E_B^0\end{aligned} \tag{A4.2}$$

Note that $\left\langle \varphi_A^0 \middle| \varphi_A^0 \right\rangle = \left\langle \varphi_B^0 \middle| \varphi_B^0 \right\rangle = 1$ because the individual wave functions are normalized. For two identical groups, $E_A^0 = E_B^0 = E^0$.

There are two possible excited states for this system.

(1) Group A in its first excited state represented by the wave function φ_A^1 and group B in its ground state. The total wave function and energy of this excited state are, respectively, $\psi_{10} = \varphi_A^1 \varphi_B^0$ and

$$\begin{aligned}E_{10} &= \left\langle \psi_{10} \middle| \hat{H}_A + \hat{H}_B \middle| \psi_{10} \right\rangle = \left\langle \varphi_A^1 \varphi_B^0 \middle| \hat{H}_A + \hat{H}_B \middle| \varphi_A^1 \varphi_B^0 \right\rangle \\ &= \left\langle \varphi_A^1 \middle| \hat{H}_A \middle| \varphi_A^1 \right\rangle \left\langle \varphi_B^0 \middle| \varphi_B^0 \right\rangle + \left\langle \varphi_B^0 \middle| \hat{H}_B \middle| \varphi_B^0 \right\rangle \left\langle \varphi_A^1 \middle| \varphi_A^1 \right\rangle = E_A^1 + E_B^0\end{aligned} \tag{A4.3}$$

Note that $\left\langle \varphi_A^1 \middle| \varphi_A^1 \right\rangle = 1$, because the individual wave functions are normalized.

(2) Group B in its first excited state represented by the wave function φ_B^1 and group A in its ground state. The total wave function and energy of this excited state are, respectively,

$$\psi_{01} = \varphi_A^0 \varphi_B^1$$

and

$$\begin{aligned}E_{01} &= \left\langle \psi_{01} \middle| \hat{H}_A + \hat{H}_B \middle| \psi_{01} \right\rangle = \left\langle \varphi_A^0 \varphi_B^1 \middle| \hat{H}_A + \hat{H}_B \middle| \varphi_A^0 \varphi_B^1 \right\rangle \\ &= \left\langle \varphi_A^0 \middle| \hat{H}_A \middle| \varphi_A^0 \right\rangle \left\langle \varphi_B^1 \middle| \varphi_B^1 \right\rangle + \left\langle \varphi_B^1 \middle| \hat{H}_B \middle| \varphi_B^1 \right\rangle \left\langle \varphi_A^0 \middle| \varphi_A^0 \right\rangle = E_A^0 + E_B^1\end{aligned} \tag{A4.4}$$

There is no interaction between the excited states ψ_{10} and ψ_{01} because, the off-diagonal energy integral is zero. that is,

$$\begin{aligned}\left\langle \psi_{01}\middle|\hat{H}_A+\hat{H}_B\middle|\psi_{10}\right\rangle &= \left\langle \varphi_A^0\varphi_B^1\middle|\hat{H}_A+\hat{H}_B\middle|\varphi_A^1\varphi_B^o\right\rangle \\ &= \left\langle \varphi_A^0\middle|\hat{H}_A\middle|\varphi_A^1\right\rangle\left\langle \varphi_B^1\middle|\varphi_B^o\right\rangle+\left\langle \varphi_B^1\middle|\hat{H}_B\middle|\varphi_B^0\right\rangle\left\langle \varphi_A^0\middle|\varphi_A^1\right\rangle \\ &= 0+0=0\end{aligned} \tag{A4.5}$$

In the preceding equation, $\left\langle \varphi_B^1\middle|\varphi_B^o\right\rangle$ and $\left\langle \varphi_A^0\middle|\varphi_A^1\right\rangle$ are zero because individual group wave functions are considered to be orthogonal. Figures A4.1a through A4.1c depict the energy levels for this noninteracting system. Transition wavenumbers for the two transitions, $\psi_{00}\rightarrow\psi_{10}$ and $\psi_{00}\rightarrow\psi_{01}$, are

$$\bar{\nu}_A = \frac{E_A^1+E_B^0-\left(E_A^0+E_B^0\right)}{hc} = \frac{E_A^1-E_A^0}{hc} \tag{A4.6}$$

$$\bar{\nu}_B = \frac{E_A^0+E_B^1-\left(E_A^0+E_B^0\right)}{hc} = \frac{E_B^1-E_B^0}{hc} \tag{A4.7}$$

The energies of transitions in individual groups to their respective excited states will be the same for identical groups; that is, $\bar{\nu}_A=\bar{\nu}_B=\bar{\nu}_0$. The integrated absorption intensity associated with a transition from 0 tol state is determined by the dipole strength (see Chapter 4, Equations 4.129 and 4.140)

$$D=\langle 0|\hat{\mu}|1\rangle^2 \tag{A4.8}$$

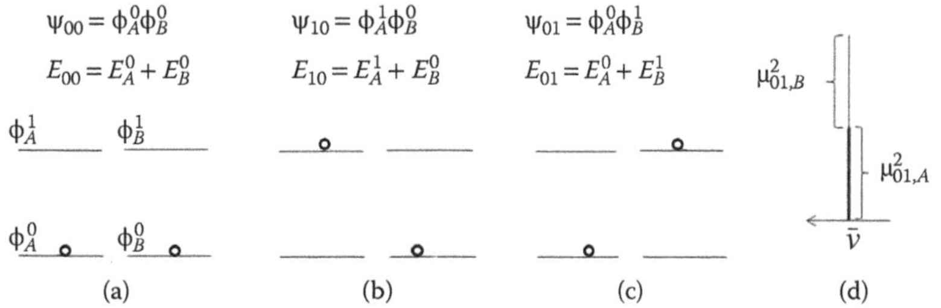

FIGURE A4.1
Representation of wave functions and energies of two groups without any interaction between them: (a) Groups *A* and *B* in their ground states. (b) Group *A* in its first excited state and Group *B* in its ground state. (c) Group *B* in its first excited state and Group *A* in its ground state. (d) Predicted absorption spectrum with both transitions resulting in an equal amount of absorption and appearing at the same wavelength/wavenumber.

where $\langle 0|\hat{\mu}|1\rangle$ is the electric dipole transition moment and $\hat{\mu}$ is the electric dipole moment operator. The two transitions of identical groups will be located at the same wavenumber, or wavelength, with the same amount of absorption intensity. To see this point, let us designate the electric dipole moment operator as, $\hat{\mu} = \hat{\mu}_A + \hat{\mu}_B$ and write the transition moment integral for each of the two transitions as follows:

$$\begin{aligned} \langle \psi_{00}|\hat{\mu}|\psi_{10}\rangle &= \langle \varphi_A^0\varphi_B^0|\hat{\mu}_A + \hat{\mu}_B|\varphi_A^1\varphi_B^0\rangle \\ &= \langle \varphi_A^0|\hat{\mu}_A|\varphi_A^1\rangle\langle \varphi_B^0|\varphi_B^0\rangle + \langle \varphi_B^0|\hat{\mu}_B|\varphi_B^0\rangle\langle \varphi_A^0|\varphi_A^1\rangle \\ &= \vec{\mu}_{01,A}\times 1 + \vec{\mu}_B^0 \times 0 = \vec{\mu}_{01,A} \end{aligned} \tag{A4.9}$$

$$\begin{aligned} \langle \psi_{00}|\hat{\mu}|\psi_{01}\rangle &= \langle \varphi_A^0\varphi_B^0|\hat{\mu}_A + \hat{\mu}_B|\varphi_A^0\varphi_B^1\rangle \\ &= \langle \varphi_A^0|\hat{\mu}_A|\varphi_A^0\rangle\langle \varphi_B^0|\varphi_B^1\rangle + \langle \varphi_B^0|\hat{\mu}_B|\varphi_B^1\rangle\langle \varphi_A^0|\varphi_A^0\rangle \\ &= \vec{\mu}_A^0\times 0 + \vec{\mu}_{01,B} \times 1 = \vec{\mu}_{01,B} \end{aligned} \tag{A4.10}$$

In the preceding equations, $\vec{\mu}_{01,A} = \langle \varphi_A^0|\hat{\mu}_A|\varphi_A^1\rangle$ is the electric dipole transition moment vector of group A and $\vec{\mu}_{01,B} = \langle \varphi_B^0|\hat{\mu}_B|\varphi_B^1\rangle$ is the electric dipole transition moment vector of group B; $\vec{\mu}_A^0 = \langle \varphi_A^0|\hat{\mu}_A|\varphi_A^0\rangle$ is the electric dipole moment vector of group A; $\vec{\mu}_B^0 = \langle \varphi_B^0|\hat{\mu}_B|\varphi_B^0\rangle$ is the electric dipole moment vector of group B. Each of the two transitions will be associated with a dipole strength of $D_0 = \mu_{01,A}^2 = \mu_{01,B}^2$. In this case, the transition frequencies are also equal, so the corresponding absorption bands will appear at the same wavelength/wavenumber, as shown in Figure A4.1d.

The model considered thus far, with two identical groups without any interaction between them, is also referred to as the dimer model.

A4.2 Two Degenerate Interacting Groups

The two groups under consideration with identical transition energies are referred to as the degenerate groups. If there is interaction between these groups, then the total Hamiltonian is written with an additional interaction operator, $\hat{V}$. Here it is assumed that the interaction takes place between the excited states (referred to as excitons and hence the name EC). The ground state is assumed to be unperturbed. The interaction is represented by the dipolar coupling term. The corresponding interaction operator is given as

$$\hat{V} = \frac{\hat{\mu}_A \bullet \hat{\mu}_B}{R_{AB}^3} - \frac{3\left(\hat{\mu}_A \bullet \vec{R}_{AB}\right)\left(\vec{R}_{AB} \bullet \hat{\mu}_B\right)}{R_{AB}^5} \tag{A4.11}$$

In this equation, $\vec{R}_{AB}$ is the distance vector between the two groups A and B. Equation A4.11 has been expressed in terms of the angles between $\boldsymbol{\mu}$ and $\mathbf{R}_{AB}$ vectors (Abbate et al. 2015; Moore and Autschbach 2012), but that is not done here because Cartesian coordinates, often available these days, can be used to compute the dot products and cross products between appropriate vectors.

In the presence of this dipolar interaction, the energies and wave functions of the excited states are altered from those of unperturbed excited states ψ_{10} and ψ_{01}. The energies and wave functions can be determined by solving the 2 × 2 secular determinant

$$\begin{vmatrix} E_A^1 - E & V_{AB} \\ V_{AB} & E_B^1 - E \end{vmatrix} = 0 \tag{A4.12}$$

where

$$V_{AB} = \left\langle \varphi_A^1 \varphi_B^0 \middle| \hat{V} \middle| \varphi_A^0 \varphi_B^1 \right\rangle = \left\langle \varphi_A^1 \varphi_B^0 \middle| \frac{\hat{\mu}_A \bullet \hat{\mu}_B}{R_{AB}^3} - \frac{3\left(\hat{\mu}_A \bullet R_{AB}\right)\left(R_{AB} \bullet \hat{\mu}_B\right)}{R_{AB}^5} \middle| \varphi_A^0 \varphi_B^1 \right\rangle$$

$$= \frac{\left\langle \varphi_A^1 \middle| \hat{\mu}_A \middle| \varphi_A^0 \right\rangle \bullet \left\langle \varphi_B^0 \middle| \hat{\mu}_B \middle| \varphi_B^1 \right\rangle}{R_{AB}^3} - \tag{A4.13}$$

$$\frac{3\left(\left\langle \varphi_A^1 \middle| \hat{\mu}_A \middle| \varphi_A^0 \right\rangle \bullet R_{AB}\right)\left(R_{AB} \bullet \left\langle \varphi_B^0 \middle| \hat{\mu}_B \middle| \varphi_B^1 \right\rangle\right)}{R_{AB}^5}$$

Note that for electric dipole transition moment integrals, $\left\langle \varphi_A^0 \middle| \hat{\mu}_A \middle| \varphi_A^1 \right\rangle = \left\langle \varphi_A^1 \middle| \hat{\mu}_A \middle| \varphi_A^0 \right\rangle$ and $\left\langle \varphi_B^0 \middle| \hat{\mu}_B \middle| \varphi_B^1 \right\rangle = \left\langle \varphi_B^1 \middle| \hat{\mu}_B \middle| \varphi_B^0 \right\rangle$. Equation A4.13 indicates that the coupling is between electric dipole transition moments of groups A and B. In Equation A4.12, $E_A^1 = E_B^1$ because the interaction considered here is between electric dipole transition moments of two degenerate groups. This model is therefore referred to as the degenerate EC (DEC) or degenerate CO (DCO) model.

If desired, the excited state energies and interaction energy V_{AB}, in Equations A4.12 and A4.13, can be converted to the wavenumber units by dividing them with hc, where h is Planck's constant and c is the speed of light, that is, $\frac{E_i^1}{hc} = \bar{E}_i^1$ and $\frac{V_{AB}}{hc} = \bar{V}_{AB}$.

The secular determinant, Equation A4.12, can be solved from the quadratic equation:

$$\left(E_A^1 - E\right)\left(E_B^1 - E\right) - V_{AB}^2 = 0 \tag{A4.14}$$

In the DEC/DCO model, $E_A^1 = E_B^1 = E^1$. Then one obtains,

$$E_{\pm} = E^1 \pm V_{AB}. \tag{A4.15}$$

The wave functions of the perturbed excited states can be determined by solving the simultaneous equations:

$$\begin{pmatrix} E^1 - E & V_{AB} \\ V_{AB} & E^1 - E \end{pmatrix} \begin{pmatrix} c_1 \\ c_2 \end{pmatrix} = 0 \tag{A4.16}$$

Substituting E_+ from Equation A4.15 for E in Equation 4.16, the coefficients and wave function are obtained after normalization, as

$$c_1 = c_2 = \frac{1}{\sqrt{2}} \tag{A4.17}$$

$$\psi_+ = \frac{\psi_{10} + \psi_{01}}{\sqrt{2}} \tag{A4.18}$$

Substituting $E-$ from Equation A4.15 for E in Equation 4.16, the wave function is obtained after normalization, as

$$\psi_- = \frac{\psi_{10} - \psi_{01}}{\sqrt{2}} \tag{A4.19}$$

The energy levels and wave functions for the perturbed excited states of two-group system are shown in Figure A4.2. Transitions to the two perturbed excited states now have different energies, as revealed by Equation A4.15. Therefore, unlike in the noninteracting case, the two transitions are expected to appear at separated x-axis (wavelength or wavenumber) positions (Figure A4.3).

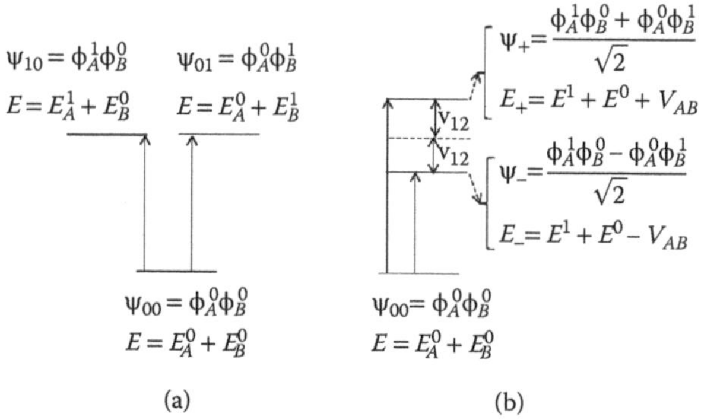

FIGURE A4.2
Representation of wave functions and energies of two identical groups with dipolar interaction between them: (a) Before interaction. (b) After interaction, assuming $V_{AB} > 0$.

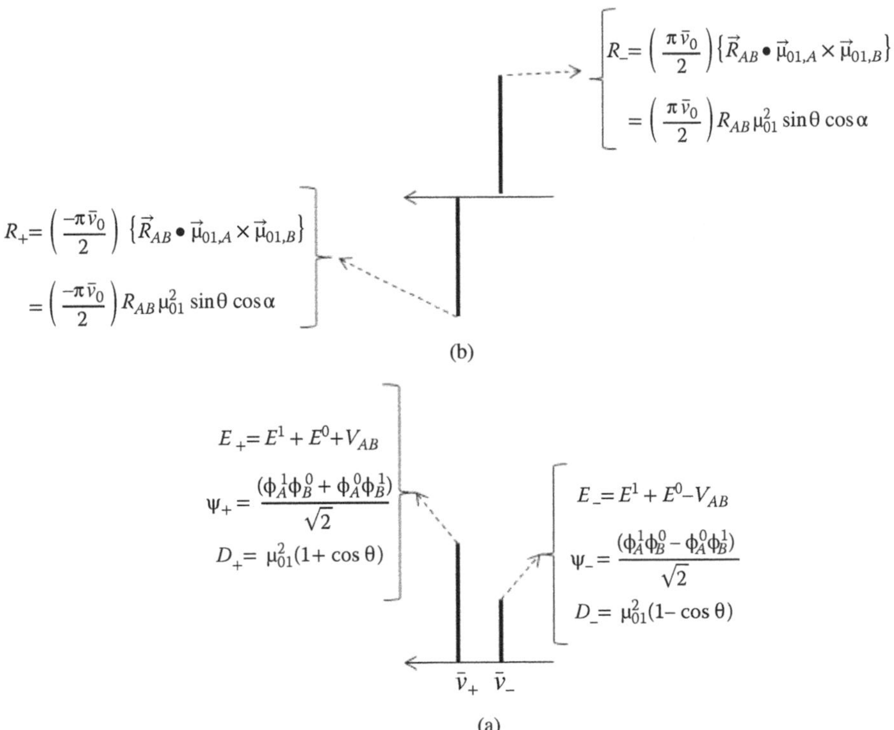

FIGURE A4.3
Depiction of spectral intensities for two identical groups with dipolar interaction between them, assuming $V_{AB} > 0$; θ is the angle between transition moment vectors; α is the angle between intergroup distance vector and the resultant of cross product between transition moment vectors. The angles θ and α can take values in the range $0° < \theta < 180°$, and $0° \leq \alpha \leq 180°$, respectively. (a) Dipole strengths for $0° < \theta < 90°$. (b) Rotational strengths for $0° < \theta < 180°$ and $90° > \alpha \geq 0°$. In the case of vibrational transitions, $\bar{\nu}_+$ and $\bar{\nu}_-$ correspond, respectively, to symmetric and antisymmetric vibrational frequencies.

Now let us consider the electric dipole transition moments and dipole strengths associated with transitions to the perturbed excited states. The electric dipole moment vector of the two-group system is given as

$$\vec{\mu}=\sum_i \xi_i\vec{r}_i=\sum_{i,A}\xi_{i,A}\vec{r}_{i,A}+\sum_{i,B}\xi_{i,B}\vec{r}_{i,B} \tag{A4.20}$$

where the summation index i,A represents the atoms/particles with charges $\xi_{i,A}$ in group A and the summation index i,B represents those in group B. It is convenient to refer the position vector of particles, r_i from the center of mass of individual groups. If $\vec{l}_{i,A}$ is the position vector of particle i in group A from the center of mass of group A and $\vec{R}_{AB}$ is the distance vector from the center of mass of group A to that of group B, then the position vectors of particles in group A become $\vec{r}_{i,A}=\vec{l}_{i,A}-\frac{\vec{R}_{AB}}{2}$ (Figure A4.4)

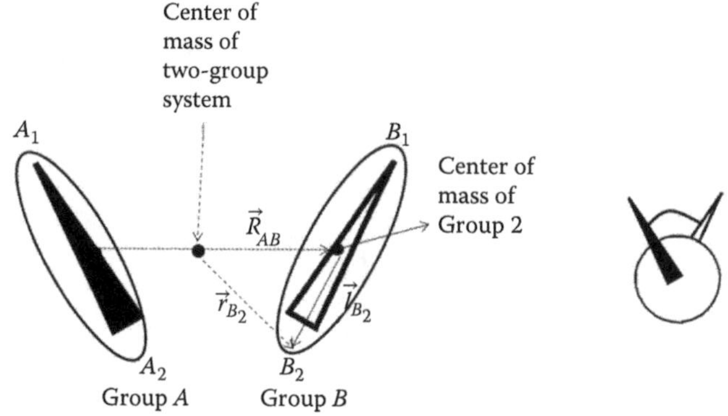

FIGURE A4.4
Two-group system considered for EC/CO model. $\vec{R}_{AB}$ is the distance vector from center of mass of group A to that of group B; $\vec{l}_{B_2}$ is the distance vector from center of mass of group B to atom B_2; $\vec{r}_{B_2}$ is the distance vector from center of mass of the system to atom B_2. Newman projection of the groups is shown on the right.

and those in group B become $\vec{r}_{i,B} = \vec{l}_{i,B} + \frac{\vec{R}_{AB}}{2}$. Then electric dipole moment vector is rewritten as

$$\vec{\mu} = \sum_{i,A} \xi_{i,A}\left(\vec{l}_{i,A} - \frac{\vec{R}_{AB}}{2}\right) + \sum_{i,2} \xi_{i,B}\left(\vec{l}_{i,B} + \frac{\vec{R}_{AB}}{2}\right)$$
$$= \sum_{i,1} \xi_{i,A}\vec{l}_{i,A} + \sum_{i,B} \xi_{i,B}\vec{l}_{i,B} - \frac{\vec{R}_{AB}}{2}\sum_{i,1} \xi_{i,A} + \frac{\vec{R}_{AB}}{2}\sum_{i,2} \xi_{i,B} \tag{A4.21}$$

For neutral groups, $\sum_{i,A} \xi_{i,A} = 0; \sum_{i,B} \xi_{i,B} = 0$. For identical groups, $\sum_{i,A} \xi_{i,A} = \sum_{i,B} \xi_{i,B}$. Thus the last two terms on the right side of preceding equation drop out, and the electric dipole moment vector becomes

$$\vec{\mu} = \sum_{i,A} \xi_{i,A}\vec{l}_{i,A} + \sum_{i,B} \xi_{i,B}\vec{l}_{i,B} = \vec{\mu}_A + \vec{\mu}_B \tag{A4.22}$$

Using Equation A4.22, the electric dipole transition moments and dipole strengths are obtained as follows:

$$\langle \psi_{00} | \hat{\mu} | \psi_+ \rangle = \left\langle \varphi_A^0 \varphi_B^0 \middle| \hat{\mu}_A + \hat{\mu}_B \middle| \frac{(\varphi_A^1 \varphi_B^0 + \varphi_A^0 \varphi_B^1)}{\sqrt{2}} \right\rangle$$
$$= \frac{1}{\sqrt{2}}\left\{ \langle \varphi_A^0 | \hat{\mu}_A | \varphi_A^1 \rangle \langle \varphi_B^0 | \varphi_B^0 \rangle + \langle \varphi_B^0 | \hat{\mu}_B | \varphi_B^1 \rangle \langle \varphi_A^0 | \varphi_A^0 \rangle \right\} \tag{A4.23}$$
$$= \frac{\vec{\mu}_{01,A} \times 1 + \vec{\mu}_{01,B} \times 1}{\sqrt{2}} = \frac{\vec{\mu}_{01,A} + \vec{\mu}_{01,B}}{\sqrt{2}}$$

$$D_+ = \langle \psi_{00}|\hat{\mu}|\psi_+\rangle^2 = \frac{\mu_{01,A}^2 + \mu_{01,B}^2 + 2\mu_{01,A}\mu_{01,B}\cos\theta}{2}$$

$$= \mu_{01}^2(1+\cos\theta) = D_0(1+\cos\theta) \tag{A4.24}$$

$$\langle \psi_{00}|\hat{\mu}|\psi_-\rangle = \left\langle \varphi_A^0\varphi_B^0 \middle| \hat{\mu}_A + \hat{\mu}_B \middle| \frac{(\varphi_A^1\varphi_B^0 - \varphi_A^0\varphi_B^1)}{\sqrt{2}} \right\rangle$$

$$= \frac{1}{\sqrt{2}}\left\{\langle \varphi_A^0|\hat{\mu}_A|\varphi_A^1\rangle\langle \varphi_B^0|\varphi_B^0\rangle - \langle \varphi_B^0|\hat{\mu}_B|\varphi_B^1\rangle\langle \varphi_A^0|\varphi_A^0\rangle\right\} \tag{A4.25}$$

$$= \frac{\vec{\mu}_{01,A}\times 1 - \vec{\mu}_{01,B}\times 1}{\sqrt{2}} = \frac{\vec{\mu}_{01,A} - \vec{\mu}_{01,B}}{\sqrt{2}}$$

$$D_- = \frac{\mu_{01,A}^2 + \mu_{01,B}^2 - 2\mu_{01,A}\mu_{01,B}\cos\theta}{2}$$

$$= \mu_{01}^2(1-\cos\theta) = D_0(1-\cos\theta) \tag{A4.26}$$

In the preceding equations, $0^0 \leq \theta \leq 180°$ is the angle between electric dipole transition moment vectors; $D_0 = \mu_{01,A}^2 = \mu_{01,B}^2$ is the dipole strength associated with transitions in individual groups before interaction. Note that θ, the angle between electric dipole transition moment vectors, is not same as the dihedral angle (which is the angle between the planes containing the transition moments).

Equations A4.24 and A4.26 have been expressed in terms of the angles between $\boldsymbol{\mu}$ and $\mathbf{R}_{AB}$ vectors (Abbate et al. 2015; Moore and Autschbach 2012), but that is not done here because Cartesian coordinates, often available these days, can be used to compute the dot products and cross products between appropriate vectors.

If V_{AB} is positive, then the transition that appears at higher wavenumber (or energy) gains absorption intensity, while the transition that appears at lower wavenumber loses absorption intensity by the same amount (Figure A4.3a) for $0° \leq \theta < 90°$. These relative intensities reverse for $90° < \theta \leq 180°$. The two transitions will have the same absorption intensity when $\theta = 90°$.

Just as dipole strength represents the integrated absorption intensity, rotational strength represents the integrated CD intensity (see Chapter 4, Equations 4.127 and 4.141). For transition from state 0 to state 1, rotational strength is given as

$$R = \text{Im}\langle 0|\hat{\mu}|1\rangle\langle 1|\hat{m}|0\rangle \tag{A4.27}$$

where $\hat{m}$ is the magnetic dipole moment operator and "Im" stands for "imaginary part of." The electric dipole transition moment integrals, $\langle 0|\hat{\mu}|1\rangle$, have already been discussed earlier, so let us now consider the magnetic dipole transition moment integral, $\langle 0|\hat{m}|1\rangle$. The magnetic dipole moment vector (in esu.cm units) is given as

$$\vec{m} = \frac{e}{2c}\sum_i \vec{r}_i \times \frac{\vec{p}_i}{m_i} \tag{A4.28}$$

where the summation is over all atoms/particles in the two-group system, $\vec{r}$ is the position vector, and $\vec{p}$ is the linear momentum vector. Using the coordinate system shown in Figure A4.4, the magnetic dipole moment vector for the two-group system becomes

$$\vec{m} = \frac{e}{2c}\left(\sum_{i,A} \frac{\vec{l}_{i,A} \times \vec{p}_{i,A}}{m_{i,A}} - \vec{R}_{AB} \times \sum_{i,1} \frac{\vec{p}_{i,A}}{2m_{i,A}} + \sum_{i,B} \frac{\vec{l}_{i,B} \times \vec{p}_{i,B}}{m_{i,B}} + \vec{R}_{AB} \times \sum_{i,2} \frac{\vec{p}_{i,B}}{2m_{i,B}}\right) \tag{A4.29}$$

In this equation, the terms

$$\vec{m}_A = \frac{e}{2c}\sum_{i,A} \frac{\vec{l}_{i,A} \times \vec{p}_{i,A}}{m_{i,A}} \tag{A4.30}$$

and

$$\vec{m}_B = \frac{e}{2c}\sum_{i,B} \frac{\vec{l}_{i,B} \times \vec{p}_{i,B}}{m_{i,B}} \tag{A4.31}$$

represent, respectively, the intrinsic magnetic moments from atoms/particles in groups *A* and *B*. It is a common practice to ignore the intrinsic magnetic moment operators, $\hat{m}_A$ and $\hat{m}_B$, in developing the expression for EC/CO model. However, there are cases where the neglect of the intrinsic magnetic dipole moment operators of groups *A* and *B* is not a good approximation (Bruhn et al. 2014; Jurinovich et al. 2015). In such cases, the use of the EC/CO model can result in erroneous conclusions.

A4.2.1 Two Degenerate Interacting Groups without Intrinsic Magnetic Dipole Transition Moments

In the commonly adopted approximation that the intrinsic magnetic moment operators, $\hat{m}_A$ and $\hat{m}_B$, can be ignored, the magnetic dipole moment operator becomes

$$\hat{m} = \frac{e}{2c}\left(-\vec{R}_{AB} \times \sum_{i,A} \frac{\hat{p}_{i,A}}{2m_{i,A}} + \vec{R}_{AB} \times \sum_{i,2} \frac{\hat{p}_{i,B}}{2m_{i,B}}\right) \tag{A4.32}$$

The momentum integrals can be converted to electric dipole moment integrals, for transition from state 0 to state 1, using the velocity-dipole relation (see Appendix 18 of Atkins 1983)

$$\langle 0 \left| \frac{e\hat{p}}{2mc} \right| 1 \rangle = -i\pi\bar{\nu}\langle 0|\hat{\mu}|1\rangle \tag{A4.33}$$

where $\bar{\nu}$ is the transition wavenumber and "i" is the imaginary term (so that $i^2 = -1$). Then, the magnetic dipole transition moments associated with transitions to perturbed excited states become

$$\langle \psi_{00} | \hat{m} | \psi_+ \rangle = \left\langle \varphi_A^0 \varphi_B^0 | \hat{m} | \frac{(\varphi_A^1 \varphi_B^0 + \varphi_A^0 \varphi_B^1)}{\sqrt{2}} \right\rangle$$

$$= \frac{-i\pi}{2\sqrt{2}} \left\{ -\bar{\nu}_A \vec{R}_{AB} \times \vec{\mu}_{01,A} + \bar{\nu}_B \vec{R}_{AB} \times \vec{\mu}_{01,B} \right\} \tag{A4.34}$$

$$= \frac{i\pi\bar{\nu}_0}{2\sqrt{2}} \vec{R}_{AB} \times (\vec{\mu}_{01,A} - \vec{\mu}_{01,B})$$

$$\langle \psi_{00} | \hat{m} | \psi_- \rangle = \frac{-i\pi}{2\sqrt{2}} \left\{ -\bar{\nu}_A \vec{R}_{AB} \times \vec{\mu}_{01,A} - \bar{\nu}_B \vec{R}_{AB} \times \vec{\mu}_{01,B} \right\}$$

$$= \frac{i\pi\bar{\nu}_0}{2\sqrt{2}} \vec{R}_{AB} \times (\vec{\mu}_{01,A} + \vec{\mu}_{01,B}) \tag{A4.35}$$

where $\bar{\nu}_A$ and $\bar{\nu}_B$ are transition wavenumbers, respectively, of $\varphi_A^0\varphi_B^0 \rightarrow \varphi_A^1\varphi_B^0$ and $\varphi_A^0\varphi_B^0 \rightarrow \varphi_A^0\varphi_B^1$ transitions and for the degenerate groups $\bar{\nu}_A = \bar{\nu}_B = \bar{\nu}_0$. Note that for magnetic dipole transition moment integrals, $\langle \psi_{00} | \hat{m} | \psi_+ \rangle = -\langle \psi_+ | \hat{m} | \psi_{00} \rangle$ and $\langle \psi_{00} | \hat{m} | \psi_- \rangle = -\langle \psi_- | \hat{m} | \psi_{00} \rangle$. Using these relations, rotational strength for transition from ψ_{00} state to ψ_+ state becomes

$$R_+ = \mathrm{Im} \langle \psi_{00} | \hat{\mu} | \psi_+ \rangle \langle \psi_+ | \hat{m} | \psi_{00} \rangle$$

$$= \mathrm{Im} \left\{ \frac{(\vec{\mu}_{01,A} + \vec{\mu}_{01,B})}{\sqrt{2}} \bullet \left(\frac{-i\pi\bar{\nu}_0}{2\sqrt{2}} \right) \left\{ \vec{R}_{AB} \times (\vec{\mu}_{01,A} - \vec{\mu}_{01,B}) \right\} \right\}$$

$$= \left(\frac{-\pi\bar{\nu}_0}{4} \right) \left\{ \vec{\mu}_{01,A} \bullet \vec{R}_{AB} \times \vec{\mu}_{01,A} - \vec{\mu}_{01,A} \bullet \vec{R}_{AB} \times \vec{\mu}_{01,B} \right\} \tag{A4.36}$$

$$+ \left(\frac{-\pi\bar{\nu}_0}{4} \right) \left\{ \vec{\mu}_{01,B} \bullet \vec{R}_{AB} \times \vec{\mu}_{01,A} - \vec{\mu}_{01,B} \bullet \vec{R}_{AB} \times \vec{\mu}_{01,B} \right\}$$

Since $\vec{R}_{AB} \times \vec{\mu}_{01,A}$ will yield a vector that is perpendicular to both $\vec{R}_{AB}$ and $\vec{\mu}_{01,A}$, the dot product of $\vec{\mu}_{01,A}$ with the resultant of cross product will be zero; that is, $\vec{\mu}_{01,A} \bullet (\vec{R}_{AB} \times \vec{\mu}_{01,A}) = 0$. Similarly, $\vec{\mu}_{01,B} \bullet (\vec{R}_{AB} \times \vec{\mu}_{01,B}) = 0$. Then, using the relation among triple products

$$\vec{a} \bullet (\vec{b} \times \vec{c}) = \vec{b} \bullet (\vec{c} \times \vec{a}) = \vec{c} \bullet (\vec{a} \times \vec{b}) \tag{A4.37}$$

the preceding equation can be simplified as follows:

$$R_{+} = \left(\frac{-\pi\bar{\nu}_0}{4}\right)\left\{-\vec{\mu}_{01,A} \bullet \vec{R}_{AB} \times \vec{\mu}_{01,B} + \vec{\mu}_{01,B} \bullet \vec{R}_{AB} \times \vec{\mu}_{01,A}\right\}$$

$$= \left(\frac{-\pi\bar{\nu}_0}{4}\right)\left\{\vec{R}_{AB} \bullet \vec{\mu}_{01,A} \times \vec{\mu}_{01,B} + \vec{R}_{AB} \bullet \vec{\mu}_{01,A} \times \vec{\mu}_{01,B}\right\} \quad \text{(A4.38)}$$

$$= \left(\frac{-\pi\bar{\nu}_0}{2}\right)\vec{R}_{AB} \bullet \left(\vec{\mu}_{01,A} \times \vec{\mu}_{01,B}\right)$$

Similarly, rotational strength for transition from ψ_{00} state to ψ_{-} state becomes

$$R_{-} = \mathrm{Im}\langle\psi_{00}|\hat{\mu}|\psi_{-}\rangle\langle\psi_{-}|\hat{m}|\psi_{00}\rangle$$

$$= \mathrm{Im}\left\{\frac{\left(\vec{\mu}_{01,A} - \vec{\mu}_{01,B}\right)}{\sqrt{2}} \bullet \frac{-i\pi\bar{\nu}_0}{2\sqrt{2}}\vec{R}_{AB} \times \left(\vec{\mu}_{01,A} + \vec{\mu}_{01,B}\right)\right\}$$

$$= \left(\frac{-\pi\bar{\nu}_0}{4}\right)\left\{\vec{\mu}_{01,A} \bullet \left(\vec{R}_{AB} \times \vec{\mu}_{01,A}\right) + \vec{\mu}_{01,A} \bullet \left(\vec{R}_{AB} \times \vec{\mu}_{01,B}\right)\right\}$$

$$+ \left(\frac{-\pi\bar{\nu}_0}{4}\right)\left\{-\vec{\mu}_{01,B} \bullet \left(\vec{R}_{AB} \times \vec{\mu}_{01,A}\right) - \vec{\mu}_{01,B} \bullet \left(\vec{R}_{AB} \times \vec{\mu}_{01,B}\right)\right\} \quad \text{(A4.39)}$$

$$= \left(\frac{-\pi\bar{\nu}_0}{4}\right)\left\{-\vec{R}_{AB} \bullet \left(\vec{\mu}_{01,A} \times \vec{\mu}_{01,B}\right) - \vec{R}_{AB} \bullet \left(\vec{\mu}_{01,A} \times \vec{\mu}_{01,B}\right)\right\}$$

$$= \left(\frac{\pi\bar{\nu}_0}{2}\right)\vec{R}_{AB} \bullet \left(\vec{\mu}_{01,A} \times \vec{\mu}_{01,B}\right)$$

Combining the two preceding expressions

$$R_{\pm} = \left(\frac{\mp\pi\bar{\nu}_0}{2}\right)\vec{R}_{AB} \bullet \left(\vec{\mu}_{01,A} \times \vec{\mu}_{01,B}\right) \quad \text{(A4.40)}$$

Equation A4.40 has been expressed in terms of angles between $\boldsymbol{\mu}$ and $\boldsymbol{R}_{\mathrm{AB}}$ vectors (Abbate et al. 2015; Moore and Autschbach 2012), but that is not done here because, Cartesian coordinates, often available these days, can be used to compute the dot products and cross products between appropriate vectors.

In the DEC/DCO model, electric dipole transition moments, $\mu_{01,A} = \mu_{01,B}$. Designating α as the angle between the vector $\vec{R}_{AB}$ and the resultant vector of the cross product, $\vec{\mu}_{01,A} \times \vec{\mu}_{01,B}$, the rotational strengths simplify as

$$R_{\pm} = \left(\frac{\mp\pi\bar{\nu}_0}{2}\right)R_{AB}\mu_{01}^2 \sin\theta\cos\alpha \quad \text{(A4.41)}$$

In Equation A4.41, $0° \leq \theta \leq 180°$ is the angle between the two electric transition dipole moment vectors (which is not same as the dihedral angle as mentioned earlier).

The literature references on the EC/CO model did not express Equation A4.40 in terms of α as is done here in Equation A4.41. Therefore the analysis of Equation A4.40 takes a different approach.

A special situation can be noted when the two electric dipole transition moment vectors are at 90° to the axis connecting them (see angle ϕ in Figure A4.5). In that special situation, the resultant vector of $\vec{\mu}_{01,A} \times \vec{\mu}_{01,B}$ will be along $\vec{R}_{AB}$ for clockwise (CW) or *P* chiral orientation, yielding $\alpha =$ 0°; for counterclockwise (CCW) or *M* chiral orientation, the resultant vector of $\vec{\mu}_{01,A} \times \vec{\mu}_{01,B}$ will be in the opposite direction to $\vec{R}_{AB}$, yielding $\alpha = 180°$. Thus the $\cos\alpha$ term will cause a change in the sign of rotational strengths from *P* chiral orientation to *M* chiral orientation. As a result, it is the angle α that provides the connection between the handedness and the signs of rotational strengths. Note that $\sin\theta$ is always positive because as noted earlier, $0° \leq \theta \leq 180°$, regardless of CW or CCW orientations (Figure A4.5).

If the two electric dipole transition moment vectors are not at 90° to the axis connecting them, then $\cos\alpha$ will be between + 1 and −1 and the resultant vector of $\vec{\mu}_{01,A} \times \vec{\mu}_{01,B}$ need not align parallel to $\vec{R}_{AB}$. For CW or *P* chiral orientation, $0° < \alpha < 90°$ and $0 < \cos\alpha < 1$; for CCW or *M* chiral orientation, $90° < \alpha < 180°$ and $-1 < \cos\alpha < 0$.

For $\alpha = 0°$ or 180° (which happens for $\phi = 90°$, see Figure A4.5), rotational strengths depend on the angle between electric dipole transition moment vectors ($0° \leq \theta \leq 180°$). For CW or *P* chiral orientation with $\alpha = 0°$,

$$R_{\pm} = \left(\frac{\mp\pi\bar{\nu}_0}{2}\right) R_{AB}\mu_{01}^2 \sin\theta; \quad (0° \leq \theta \leq 180°) \tag{A4.42}$$

and for CCW or *M* chiral orientation with $\alpha = 180°$,

$$R_{\pm} = \left(\frac{\pm\pi\bar{\nu}_0}{2}\right) R_{AB}\mu_{01}^2 \sin\theta; \quad (0° \leq \theta \leq 180°) \tag{A4.43}$$

Rotational strengths for the two transitions, as predicted by the preceding equation for *P* chiral orientation, and depicted in Figure A4.3b, are of opposite signs and of equal magnitude. For $V_{AB} > 0$, the higher wavenumber (shorter wavelength) transition has negative rotational strength, and the lower wavenumber (longer wavelength) transition has positive rotational strength. This type of bisignate CD band pairs, with positive CD on the longer wavelength side and negative CD on the shorter wavelength side, are called positive CD couplets. The opposite pattern is referred to as the negative CD couplet. Thus *P* chiral arrangement of electric dipole transition moment vectors yields a positive CD couplet and *M* chiral arrangement yields a negative CD couplet, both for $V_{AB} > 0$. This observation is the basis for DEC/DCO model

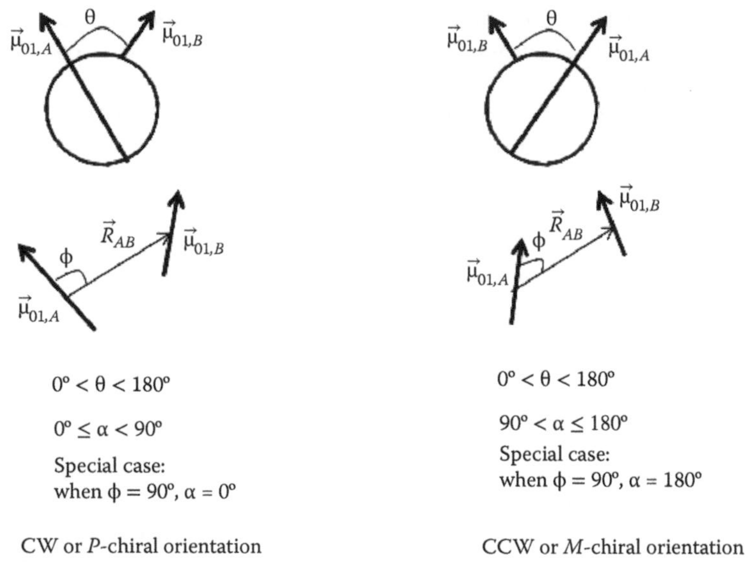

$0° < \theta < 180°$

$0° \leq \alpha < 90°$

Special case:
when $\phi = 90°$, $\alpha = 0°$

CW or *P*-chiral orientation

$0° < \theta < 180°$

$90° < \alpha \leq 180°$

Special case:
when $\phi = 90°$, $\alpha = 180°$

CCW or *M*-chiral orientation

FIGURE A4.5
Depiction of the orientation of electric dipole transition moment vectors $\vec{\mu}_{01,A}$ and $\vec{\mu}_{01,B}$, and identification of the associated angles θ, ϕ, and α in two opposite chiral orientations. θ is the angle between two transition moment vectors; ϕ is the angle between transition moment vector and the distance vector, $\vec{R}_{AB}$; α is the angle between the resultant vector of cross product, $\vec{\mu}_{01,A} \times \vec{\mu}_{01,B}$ and the distance vector, $\vec{R}_{AB}$. Rotational strengths for the two transitions are given, for a general situation, by the relation, $R_{\pm} = \left(\frac{\mp\pi\bar{\nu}_0}{2}\right) R_{AB}\mu_{01}^2 \sin\theta\cos\alpha$. Newmann projections of transition dipole moment vectors, when looking along $\vec{R}_{AB}$, are shown at the top.

predictions. The visual representations of these predictions for absorption and CD are shown in Figure A4.6.

Most articles use the expressions (Equation A4.42 and A4.43) without emphasizing the approximations behind them. There are some important points one should remember in interpreting the experimentally observed spectra using the EC/CO model: (1) First, one needs to know the orientations of electric dipole transition moment vectors in each of the groups. For common groups, experience/literature can provide some guidance. For C=O groups, for example, the electric dipole transition moment vector associated with its vibration is assumed to be along the C=O bond axis (although deviations from this assumptions are well known). For other groups, this information may not be obvious and is assumed. (2) From Equation A4.13, it can be estimated that V_{AB} can have both positive and negative signs and the CD sign pattern expected for a given chiral arrangement will reverse for $V_{AB} > 0$ and $V_{AB} < 0$. When V_{AB} is positive the transition represented by the positive (or symmetric) combination of excited state wave functions will appear at a higher energy than that

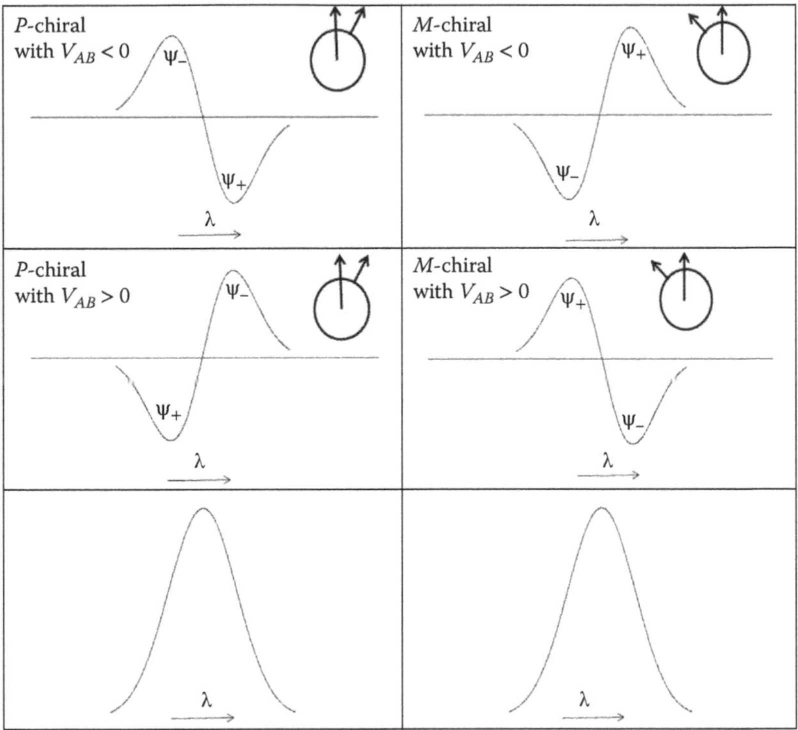

FIGURE A4.6
CD sign patterns as a function of wavelength λ and their relation to *P*-chiral and *M*-chiral arrangement of electric dipole transition moment vectors, as expected for the degenerate exciton chirality, or CO, model. The corresponding absorption (bottom panels) is shown as an unresolved broad band.

represented by a negative (or antisymmetric) combination of excited state wave functions. The opposite is true when V_{AB} is negative. The following information can be gleaned on V_{AB} from Equation A4.13: When the transition dipole moment vectors are at 90° to the axis connecting them (angle ϕ in Figure A4.5), the second term in Equation A4.13 becomes zero and V_{AB} becomes $\frac{\mu_{01}^2 \cos\theta}{R_{AB}^3}$, and is positive for $0° < \theta < 90°$ and negative for $90° < \theta < 180°$ (θ is the angle between transition moment vectors). The second term in Equation A4.13 can be positive or negative depending on how transition moment vectors are oriented. If one transition moment vector is at an angle $0° < \phi < 90°$ and the other at $90° < \phi < 180°$ to the R_{AB} vector, then the second term is negative and the minus sign in front of it makes the contribution positive. However, if both transition moment vectors are at either $0° < \phi < 90°$ or $90° < \phi < 180°$ angles to the R_{AB} vector, then the second term becomes positive and the minus sign in front of it makes the contribution negative. Thus the net sign of V_{AB} depends on the signs and relative magnitudes of the two terms in Equation A4.13. (3) Dipolar coupling is unlikely to be the dominant interaction in all situations. For vibrational transitions, the wavenumber order of the appearance of symmetric and antisymmetric vibrations can be different from that predicted with V_{AB} (Abbate et al. 2015; Covington et al. 2015), because the sign of interaction force constant is important for predicting the order of vibrational frequencies (Polavarapu et al. 1987). As mentioned earlier, when the wavenumber positions of two transitions interchange, a corresponding change takes place in the visual pattern of the CD spectra.

A freely available computer program for DEC/DCO model predictions of circular dichroism, which can also take the input from QC calculations of VCD, was developed (Covington and Polavarapu 2015).

The pattern of dipole strengths of two transitions changes when θ is above and below 90°. As a result, the absorption spectrum should also be evaluated along with CD spectrum. For this purpose, the analysis should focus on the dissymmetry factor

$$g_{\pm} = \frac{4R_{\pm}}{D_{\pm}} = \frac{\mp 2\pi\bar{\nu}_0 R_{AB} \sin\theta \cos\alpha}{(1 \pm \cos\theta)} \tag{A4.44}$$

but this aspect is rarely practiced, probably because separate absorption bands corresponding to the two transitions may not be resolved in the experimental spectra.

The DEC/DCO model has been characterized in the literature as a non-empirical method, due to the theoretical basis available for this model as described earlier. However, such categorization should be qualified by indicating that the underlining theory of the EC/CO model is one of conceptual nature and approximate. Hence the EC/CO model is indeed an approximate one that can fail in certain circumstances (see Chapter 10).

Note that the predicted rotational strength expressions (Equation A4.40) have also been used subsequently for predicting the optical rotation (Tinoco 1963).

A model based on Eckart–Sayvetz conditions (Wilson et al. 1980) that molecular vibrations satisfy was formulated for interpreting the VCD associated with A-B stretching vibrations (Polavarapu 1986) and A-B-B bending vibrations of A_2B_2 type molecules (Polavarapu 1987). This model emphasizes charge flow during vibrations and interaction force constants (Polavarapu et al. 1987). In the absence of charge flow and negligible interaction force constants, A_2B_2 model predictions are similar to those of DEC/DCO models.

A4.2.2 Two Degenerate Interacting Groups with Intrinsic Magnetic Dipole Transition Moments

In this case, the intrinsic magnetic moments, $\vec{m}_A$ and $\vec{m}_B$ are carried over from Equation A4.29 to Equation A4.32, as

$$\hat{m} = \frac{e}{2c}\left(-\vec{R}_{AB} \times \sum_{i,1} \frac{\hat{p}_{i,A}}{2m_{i,A}} + \vec{R}_{AB} \times \sum_{i,2} \frac{\hat{p}_{i,B}}{2m_{i,B}}\right) + \hat{m}_A + \hat{m}_B \tag{A4.45}$$

and Equations A4.33 through A4.40 are modified accordingly. Then Equation A4.40 changes to

$$R_{\pm} = \left(\frac{\mp\pi\bar{\nu}_0}{2}\right)\left\{\vec{R}_{AB} \bullet \vec{\mu}_{01,A} \times \vec{\mu}_{01,B}\right\} + \frac{1}{2}\mathrm{Im}\left[\left(\vec{\mu}_{01,A} \pm \vec{\mu}_{01,B}\right) \bullet \left(\vec{m}_{01,A} \pm \vec{m}_{01,B}\right)\right] \tag{A4.46}$$

The presence of nonzero intrinsic magnetic transition dipole moments, $\vec{m}_{01,A}$ and $\vec{m}_{01,B}$, respectively, of groups *A* and *B*, in Equation A 4.46 can reverse the signs of rotational strengths predicted by Equation A 4.40, thereby rendering the predictions of widely used EC/CO model invalid (Bruhn et al. 2014; Jurinovich et al. 2015).

A4.3 Two Nondegenerate Interacting Groups

The expressions for the rotational strength obtained for nonidentical interacting groups using nonorthogonal wave functions (Harada and Nakanishi 1983), and wave functions corrected to first order (see Chapter 4, Equation 4.33) (Rodger and Norden 1997), have been reported. However the wave functions given by Equations A4.18 and A4.19 can be rewritten transparently in a general form (Covington et al. 2015). For nonidentical interacting groups, the orthonormal wave functions can be given, in a general form, as

$$\psi_+ = c_1\psi_{10} + c_2\psi_{01} \tag{A4.47}$$

$$\psi_- = -c_2\psi_{10} + c_1\psi_{01} \tag{A4.48}$$

The coefficients c_1 and c_2 in the preceding two equations are positive and satisfy the normalization condition, $c_1^2 + c_2^2 = 1$. They can be obtained as the eigenvectors of 2 × 2 matrix

$$\begin{pmatrix} E_A^1 - E & V_{AB} \\ V_{AB} & E_B^1 - E \end{pmatrix}\begin{pmatrix} c_1 \\ c_2 \end{pmatrix} = 0 \tag{A4.49}$$

where E_i^1 is the unperturbed excited state energy of the ith group and V_{AB} is the interaction energy.

For vibrational transitions, Equation A4.49 can be written in terms of wavenumbers, as

$$\begin{pmatrix} \bar{\nu}_A - \bar{\nu} & \bar{V}_{AB} \\ \bar{V}_{AB} & \bar{\nu}_B - \bar{\nu} \end{pmatrix}\begin{pmatrix} c_1 \\ c_2 \end{pmatrix} = 0 \tag{A4.50}$$

where, $\bar{\nu}_A$ and $\bar{\nu}_B$ represent the transition wavenumbers of nondegenerate groups *A* and *B*. In the case of each group representing a chemical bond, and vibrational transitions associated with stretching of chemical bonds, the unperturbed vibrational wavenumbers of nonidentical groups can be determined from the respective chemical bond force constants and reduced mass associated with each of the chemical bonds. For chemical bond *A*

$$\bar{\nu}_A = \frac{1}{2\pi c}\sqrt{\frac{f_{AA}}{M_A}} \tag{A4.51}$$

where f_{AA} is the diagonal force constant and M_A is the reduced mass. The interaction energy for vibrational transitions can be derived from interaction force constants, instead of dipolar interaction energy through Equation A4.11. From the chemical bond force constants, f_{AA}, f_{BB}, and interaction force constants f_{AB}, the interaction term $\bar{V}_{AB}$ in Equation A4.50 can be determined (Polavarapu et al. 1987). QC calculations can be used for determining the force constants mentioned earlier.

Using the wave functions given by Equations A4.47 and A4.48, electric dipole transition moments can be obtained as

$$\begin{aligned} \langle \psi_{00} | \hat{\mu} | \psi_+ \rangle &= \langle \varphi_A^0 \varphi_B^0 | \hat{\mu}_A + \hat{\mu}_B | (c_1 \varphi_A^1 \varphi_B^0 + c_2 \varphi_A^0 \varphi_B^1) \rangle \\ &= c_1 \vec{\mu}_{01,A} + c_2 \vec{\mu}_{01,B} \end{aligned} \tag{A4.52}$$

$$\begin{aligned} \langle \psi_{00} | \hat{\mu} | \psi_- \rangle &= \langle \varphi_A^0 \varphi_B^0 | \hat{\mu}_A + \hat{\mu}_B | (-c_2 \varphi_A^1 \varphi_B^0 + c_1 \varphi_A^0 \varphi_B^1) \rangle \\ &= -c_2 \vec{\mu}_{01,A} + c_1 \vec{\mu}_{01,B} \end{aligned} \tag{A4.53}$$

The dipole strengths associated with the two transitions become

$$D_+ = c_1^2\mu_{01,A}^2 + c_2^2\mu_{01,B}^2 + 2c_1c_2\mu_{01,A}\mu_{01,B}\cos\theta \tag{A4.54}$$

$$D_- = c_2^2\mu_{01,A}^2 + c_1^2\mu_{01,B}^2 - 2c_1c_2\mu_{01,A}\mu_{01,B}\cos\theta \tag{A4.55}$$

Similarly the magnetic dipole transition moments can be obtained, without intrinsic magnetic dipole moment operators following Equations A4.32 through A4.35, as

$$\begin{aligned}
&\langle\psi_{00}|\hat{m}|\psi_+\rangle = \langle\psi_{00}|\hat{m}|(c_1\psi_{10}+c_2\psi_{01})\rangle = \\
&= \left\langle\psi_{00}\left|\frac{e}{2c}\left(-\vec{R}_{AB}\times\sum_{i,A}\frac{\hat{p}_{i,A}}{2m_{i,A}}+\vec{R}_{AB}\times\sum_{i,B}\frac{\hat{p}_{i,B}}{2m_{i,B}}\right)\right|(c_1\psi_{10})\right\rangle \\
&+ \left\langle\psi_{00}\left|\frac{e}{2c}\left(-\vec{R}_{AB}\times\sum_{i,A}\frac{\hat{p}_{i,A}}{2m_{i,A}}+\vec{R}_{AB}\times\sum_{i,2}\frac{\hat{p}_{i,B}}{2m_{i,B}}\right)\right|(c_2\psi_{01})\right\rangle \\
&= \left\langle\psi_{00}\left|\frac{e}{2c}\left(-\vec{R}_{AB}\times\sum_{i,A}\frac{\hat{p}_{i,A}}{2m_{i,A}}\right)\right|(c_1\psi_{10})\right\rangle \\
&+ \left\langle\psi_{00}\left|\frac{e}{2c}\left(\vec{R}_{AB}\times\sum_{i,B}\frac{\hat{p}_{i,B}}{2m_{i,B}}\right)\right|(c_1\psi_{10})\right\rangle \\
&+ \left\langle\psi_{00}\left|\frac{e}{2c}\left(-\vec{R}_{AB}\times\sum_{i,A}\frac{\hat{p}_{i,A}}{2m_{i,A}}\right)\right|(c_2\psi_{01})\right\rangle \\
&+ \left\langle\psi_{00}\left|\frac{e}{2c}\left(\vec{R}_{AB}\times\sum_{i,2}\frac{\hat{p}_{i,B}}{2m_{i,B}}\right)\right|(c_2\psi_{01})\right\rangle \\
&= \frac{-i\pi}{2}\left\{-c_1\bar{\nu}_A\vec{R}_{AB}\times\vec{\mu}_{01,A}+c_2\bar{\nu}_B\vec{R}_{AB}\times\vec{\mu}_{01,B}\right\} \\
&= \frac{i\pi}{2}\vec{R}_{AB}\times\left(c_1\bar{\nu}_A\vec{\mu}_{01,A}-c_2\bar{\nu}_B\vec{\mu}_{01,B}\right)
\end{aligned} \tag{A4.56}$$

$$\langle \psi_{00} | \hat{m} | \psi_- \rangle = \langle \psi_{00} | \hat{m} | -c_2\psi_{10} + c_1\psi_{01} \rangle =$$

$$= \left\langle \psi_{00} \left| \frac{e}{2c} \left(-\vec{R}_{AB} \times \sum_{i,A} \frac{\hat{p}_{i,A}}{2m_{i,A}} + \vec{R}_{AB} \times \sum_{i,B} \frac{\hat{p}_{i,B}}{2m_{i,B}} \right) \right| -c_2\psi_{10} \right\rangle$$

$$+ \left\langle \psi_{00} \left| \frac{e}{2c} \left(-\vec{R}_{AB} \times \sum_{i,A} \frac{\hat{p}_{i,A}}{2m_{i,A}} + \vec{R}_{AB} \times \sum_{i,B} \frac{\hat{p}_{i,B}}{2m_{i,B}} \right) \right| c_1\psi_{01} \right\rangle$$

$$= \left\langle \psi_{00} \left| \frac{e}{2c} \left(-\vec{R}_{AB} \times \sum_{i,A} \frac{\hat{p}_{i,A}}{2m_{i,A}} \right) \right| -c_2\psi_{10} \right\rangle$$

$$+ \left\langle \psi_{00} \left| \frac{e}{2c} \left(\vec{R}_{AB} \times \sum_{i,B} \frac{\hat{p}_{i,B}}{2m_{i,B}} \right) \right| -c_2\psi_{10} \right\rangle \quad \text{(A4.57)}$$

$$+ \left\langle \psi_{00} \left| \frac{e}{2c} \left(-\vec{R}_{AB} \times \sum_{i,1} \frac{\hat{p}_{i,A}}{2m_{i,A}} \right) \right| c_1\psi_{01} \right\rangle$$

$$+ \left\langle \psi_{00} \left| \frac{e}{2c} \left(\vec{R}_{AB} \times \sum_{i,2} \frac{\hat{p}_{i,B}}{2m_{i,B}} \right) \right| c_1\psi_{01} \right\rangle$$

$$= \frac{-i\pi}{2} \left\{ -(-c_2\bar{\nu}_A)\vec{R}_{AB} \times \vec{\mu}_{01,A} + (c_1\bar{\nu}_B)\vec{R}_{AB} \times \vec{\mu}_{01,B} \right\}$$

$$= \frac{-i\pi}{2} \vec{R}_{AB} \times (c_2\bar{\nu}_A\vec{\mu}_{01,A} + c_1\bar{\nu}_B\vec{\mu}_{01,B})$$

The rotational strength for the ψ_{00} to ψ_+ transition can be obtained for neutral groups, as in Section 4.2.1

$$R_+ = \text{Im} \langle \psi_{00} | \hat{\mu} | \psi_+ \rangle \langle \psi_+ | \hat{m} | \psi_{00} \rangle$$

$$= \text{Im} \left\{ (c_1\vec{\mu}_{01,A} + c_2\vec{\mu}_{01,B}) \bullet \left(\frac{-i\pi}{2} \right) \left\{ \vec{R}_{AB} \times (c_1\bar{\nu}_A\vec{\mu}_{01,A} - c_2\bar{\nu}_B\vec{\mu}_{01,B}) \right\} \right\}$$

$$= \left(\frac{-\pi}{2} \right) \left\{ c_1^2\bar{\nu}_A\vec{\mu}_{01,A} \bullet \vec{R}_{AB} \times \vec{\mu}_{01,A} - c_1c_2\bar{\nu}_B\vec{\mu}_{01,A} \bullet \vec{R}_{AB} \times \vec{\mu}_{01,B} \right\} \quad \text{(A4.58)}$$

$$+ \left(\frac{-\pi}{2} \right) \left\{ c_1c_2\bar{\nu}_A\vec{\mu}_{01,B} \bullet \vec{R}_{AB} \times \vec{\mu}_{01,A} - c_2^2\bar{\nu}_B\vec{\mu}_{01,B} \bullet \vec{R}_{AB} \times \vec{\mu}_{01,B} \right\}$$

Noting as mentioned earlier (see the text following Equation A4.36) that $\vec{\mu}_{01,A} \bullet \left(\vec{R}_{AB} \times \vec{\mu}_{01,A}\right) = 0$, and $\vec{\mu}_{01,B} \bullet \left(\vec{R}_{AB} \times \vec{\mu}_{01,B}\right) = 0$, the preceding equation simplifies to

$$R_+ = \left(\frac{-\pi}{2}\right)\left\{-c_1c_2\bar{\nu}_B\vec{\mu}_{01,A} \bullet \vec{R}_{AB} \times \vec{\mu}_{01,B} + c_1c_2\bar{\nu}_A\vec{\mu}_{01,B} \bullet \vec{R}_{AB} \times \vec{\mu}_{01,A}\right\} \quad \text{(A4.59)}$$

Making use of the relation among triple products (see Equation A4.37), this expression can be rewritten as

$$\begin{aligned} R_+ &= \left(\frac{-\pi}{2}c_1c_2\right)\left\{\bar{\nu}_B\vec{R}_{AB} \bullet \vec{\mu}_{01,A} \times \vec{\mu}_{01,B} + \bar{\nu}_A\vec{R}_{AB} \bullet \vec{\mu}_{01,A} \times \vec{\mu}_{01,B}\right\} \\ &= \left(-\pi\bar{\nu}_{0,nd}c_1c_2\right)\left\{\vec{R}_{AB} \bullet \vec{\mu}_{01,A} \times \vec{\mu}_{01,B}\right\} \end{aligned} \quad \text{(A4.60)}$$

where,

$$\bar{\nu}_{0,nd} = \left(\frac{\bar{\nu}_A + \bar{\nu}_B}{2}\right) \quad \text{(A4.61)}$$

Similarly, the rotational strength for ψ_{00} to $\psi-$ transition can be written as follows:

$$\begin{aligned} R_- &= \operatorname{Im}\langle\psi_{00}|\hat{\mu}|\psi_-\rangle\langle\psi_-|\hat{m}|\psi_{00}\rangle \\ &= \operatorname{Im}\left\{\left(-c_2\vec{\mu}_{01,A} + c_1\vec{\mu}_{01,B}\right) \bullet \frac{i\pi}{2}\vec{R}_{AB} \times \left(c_2\bar{\nu}_A\vec{\mu}_{01,A} + c_1\bar{\nu}_B\vec{\mu}_{01,B}\right)\right\} \\ &= \left(\frac{\pi}{2}\right)\left\{-c_2^2\bar{\nu}_A\vec{\mu}_{01,A} \bullet \left(\vec{R}_{AB} \times \vec{\mu}_{01,A}\right) - c_2c_1\bar{\nu}_B\vec{\mu}_{01,A} \bullet \left(\vec{R}_{AB} \times \vec{\mu}_{01,B}\right)\right\} \\ &\quad + \left(\frac{\pi}{2}\right)\left\{c_1c_2\bar{\nu}_A\vec{\mu}_{01,B} \bullet \left(\vec{R}_{AB} \times \vec{\mu}_{01,A}\right) + c_1^2\bar{\nu}_B\vec{\mu}_{01,B} \bullet \left(\vec{R}_{AB} \times \vec{\mu}_{01,B}\right)\right\} \\ &= \left(\frac{\pi}{2}\right)\left\{c_2c_1\bar{\nu}_B\vec{R}_{AB} \bullet \left(\vec{\mu}_{01,A} \times \vec{\mu}_{01,B}\right) + c_2c_1\bar{\nu}_A\vec{R}_{AB} \bullet \left(\vec{\mu}_{01,A} \times \vec{\mu}_{01,B}\right)\right\} \\ &= \left(\pi\bar{\nu}_{0,nd}c_1c_2\right)\left\{\vec{R}_{AB} \bullet \left(\vec{\mu}_{01,A} \times \vec{\mu}_{01,B}\right)\right\} \end{aligned} \quad \text{(A4.62)}$$

The rotational strengths in the aforementioned two equations are written together as

$$R_\pm = \left(\mp c_1c_2\pi\bar{\nu}_{0,nd}\right)\vec{R}_{AB} \bullet \left(\vec{\mu}_{01,A} \times \vec{\mu}_{01,B}\right) \quad \text{(A4.63)}$$

Comparing Equation A4.63 with A4.40, it can be seen that the magnitudes of rotational strengths in the non-degenerate exciton coupling (NDEC) model will be reduced from those in the DEC model. This is because the product c_1c_2 in Equation A4.63 will always be less than 0.5 for nonidentical groups. The percent reduction can be estimated from the difference between Equations A4.40 and A4.63, with the same geometrical parameters as

$$\frac{|R_{\pm,\mathrm{DEC}}|-|R_{\pm,\mathrm{NDEC}}|}{|R_{\pm,\mathrm{DEC}}|}\times 100 \approx \frac{\left(0.5\bar{\nu}_0 D_0 - c_1c_2\bar{\nu}_{0,nd}\sqrt{D_A D_B}\right)}{0.5\bar{\nu}_0 D_0}\times 100$$
$$\approx \frac{(0.5-c_1c_2)}{0.5}\times 100 \tag{A4.64}$$

For larger separation between the transition energies of nonidentical groups, and for weaker interaction energy between them, the significance of EC will be diminished. A stronger interaction between the two nondegenerate groups would be required to obtain any significant EC CD contribution in the NDEC model.

Indiscriminate use of the EC/CO model for interpreting the CD associated with vibrational transitions may lead to the right answers for the wrong reasons or wrong predictions altogether (Covington et al. 2015).

A4.4 Multiple Interacting Transitions

The original formulation (Tinoco 1963) was presented for multiple interacting transitions (i.e., for a polymer), with dimer as a special case. The equations given earlier for the two interacting transitions can be generalized for such cases. Such generalized expressions (Harada and Nakanishi 1983; Harada et al. 1975), and a matrix method (Moore and Autschbach 2012) have been used in analyzing multiple transitions in the ECD spectra. The generalized expressions have also been applied for analyzing the VCD spectra (Gulotta et al. 1989). More recently, Norman and Linares have presented density functional theory calculations on stacked ethylene molecules, using achiral ethylene as well as chiral (twisted) ethylene molecules and found that excitonic contributions are intertwined with those from chiral units, thereby making EC predictions unpredictable (Norman and Linares 2014).

An alternate approach for multiple interacting transitions, originally proposed by DeVoe, is also widely used (DeVoe 1964, 1965). In this approach, the electric dipole transition moment associated with each transition is represented by a point dipole. Each of these point dipoles is considered to be influenced by the electric field induced at other dipoles by the incident electric field. This method is often referred to as Devoe's polarizability model and, sometimes, also as the CO model (which is not same as the one discussed earlier). A review on the use of Devoe's polarizability model has been reported (Superchi et al. 2004).

References

Abbate, S., G. Mazzeo, S. Meneghini, G. Longhi, S. E. Boiadjiev, and D. A. Lightner. 2015. Bicamphor: A prototypic molecular system to investigate vibrational excitons. *J. Phys. Chem. A* 119:4261–4267.

Atkins, P. W. 1983. *Molecular Quantum Mechanics*. Oxford, UK: Oxford University Press.

Bruhn, T., G. Pescitelli, S. Jurinovich, A. Schaumlöffel, F. Witterauf, J. Ahrens, M. Bröring, and G. Bringmann. 2014. Axially chiral BODIPY DYEmers: An apparent exception to the exciton chirality rule. *Angew. Chem. Int. Ed. Engl.* 53:14592–14595.

Covington, C. L., and P. L. Polavarapu. 2015. CDSpecTech: Computer programs for calculating similarity measures for experimental and calculated dissymmetry factors and circular intensity differentials. https://sites.google.com/site/cdspectech1/.

Covington, C. L., V. P. Nicu, and P. L. Polavarapu. 2015. Determination of absolute configurations using exciton chirality method for vibrational circular dichroism: Right answers for the wrong reasons? *J. Phys. Chem.* 119:10589–10601.

DeVoe, H. 1964. Optical properties of molecular aggregates. I. Classical model of electronic absorption and refraction. *J. Chem. Phys.* 41(2):393–400.

DeVoe, H. 1965. Optical properties of molecular aggregates. II. Classical theory of the refraction, absorption, and optical activity of solutions and crystals. *J. Chem. Phys.* 43(9):3199–3208.

Gulotta, M., D. J. Goss, and M. Diem. 1989. Ir vibrational CD in model deoxyoligonucleotides: Observation of the B → Z phase transition and extended coupled oscillator intensity calcuations. *Biopolymers* 28(12):2047–2058.

Harada, N., S.-M. L. Chen, and K. Nakanishi. 1975. Quantitative definition of exciton chirality and the distant effect in the exciton chirality method. *J. Am. Chem. Soc.* 97(19):5345–5352.

Harada, N., and K. Nakanishi. 1983. *Circular Dichroic Spectroscopy : Exciton Coupling in Organic Stereochemistry*. Mill Valley, CA: University Science Books.

Holzwarth, G., and I. Chabay. 1972. Optical activity of vibrational transitions: A coupled oscillator model. *J. Chem. Phys.* 57:1632–1635.

Jurinovich, S., C. A. Guido, T. Bruhn, G. Pescitelli, and B. Mennucci. 2015. The role of magnetic-electric coupling in exciton-coupled ECD spectra: The case of bis-phenanthrenes. *Chem. Commun.* 51(52):10498–10501.

Moore, B., and J. Autschbach. 2012. Density functional study of tetraphenylporphyrin long-range exciton coupling. *Chem. Open* 1:184–194.

Norman, P., and M. Linares. 2014. On the interplay between chirality and exciton coupling: A DFT calculation of the circular dichroism in π-stacked ethylene. *Chirality* 26(9):483–489.

Polavarapu, P. L. 1986. Vibrational circular dichroism of A_2B_2 molecules. *J. Chem. Phys.* 85:6245–6246.

Polavarapu, P. L. 1987. Absorption and circular dichroism due to bending vibrations of A_2B_2 molecules with C_2 symmetry. *J. Chem. Phys.* 87:4419–4422.

Polavarapu, P. L., C. S. Ewig, and T. Chandramouly. 1987. Conformations of tartaric acid and its esters. *J. Am. Chem. Soc.* 109:7382–7386.

Rodger, A., and B. Norden. 1997. *Circular Dichroism and Linear Dichroism*. Oxford, UK: Oxford University Press.

Superchi, S., E. Giorgio, and C. Rosini. 2004. Structural determinations by circular dichroism spectra analysis using coupled oscillator methods: An update of the applications of the DeVoe polarizability model. *Chirality* 16(7):422–451.

Tinoco, I., Jr. 1963. The exciton contribution to the optical rotation of polymers. *Radiat. Res.* 20:133–139.

Tinoco, I., Jr., and C. R. Cantor. 1970. Application of optical rotatory dispersion and circular dichroism in biochemical analysis. In: D. Glick (ed). *Methods Biochemistry Analysis.* Vol. 18. New York, NY: John Wiley & Sons.

Wilson, E. B., J. C. Decius, and P. C. Cross. 1980. *Molecular Vibrations: The Theory of Infrared and Raman Vibrational Spectra*. New York, NY: Dover Publications.

Appendix 5: Averaging Properties Using Boltzmann Populations

A5.1 Conformational Averaging

In most cases, the desired property needs to be averaged over all conformers, for example, as

$$P_{av} = \sum_i f_i P_i \quad \text{(A5.1)}$$

where f_i is the fractional population of conformer i, P_i is the property of conformer i, and P_{av} is the property averaged over all conformers. For this purpose, one needs to know the populations of different conformers. The populations of different conformers can be predicted using their quantum chemical calculated energies. Although both electronic energies and Gibbs energies are used for the purpose of determining the population of conformers, the preference for one over the other involves some debate.

Nevertheless, in calculating the populations of different conformers, fractional populations, f_i, are calculated using the Boltzmann distribution law:

$$f_i = \frac{N_i}{N_{tot}} = \frac{e^{-E_i/RT}}{\sum_a e^{-E_a/RT}} \quad \text{(A5.2)}$$

In Equation A5.2, N_i is the number of molecules with conformation i and energy E_i; N_{tot} is the total number of molecules and the summation in the denominator extends over all possible conformations. The ratio N_i/N_{tot} represents the fraction of molecules in conformation i, which is also called the fractional population of the ith conformation. Note that the sum of fractional populations of all conformers becomes 1, that is,

$$\sum_i f_i = 1 \quad \text{(A5.3)}$$

The use of Equation A5.3 is one way to visualize the fractional populations.

It is often customary to arrange the conformers in the increasing order of energy and to define their fractional populations in relation to the energy of lowest energy conformer. For this purpose, let us designate conformer i

as the lowest energy conformer, *j* as the next higher energy conformer, and so on. Then, the ratios of fractional populations of conformer *j*, *k*, and so on, with respect to that of *i* can be defined as

$$\frac{f_\alpha}{f_i} = \frac{\left(\dfrac{N_\alpha}{N_{tot}}\right)}{\left(\dfrac{N_i}{N_{tot}}\right)} = \frac{\left(\dfrac{e^{-E_\alpha/RT}}{\sum_a e^{-E_a/RT}}\right)}{\left(\dfrac{e^{-E_i/RT}}{\sum_a e^{-E_a/RT}}\right)} = \frac{e^{-E_\alpha/RT}}{e^{-E_i/RT}}; \quad \alpha = j, k\ldots \tag{A5.4}$$

Since the sum of fractional populations, $f_i + f_j + f_k + \ldots = 1$, f_i can be determined as follows:

$$1 = \sum_a f_a = f_i + f_j + f_k + \ldots \tag{A5.5}$$

$$1 = f_i\left[1 + \frac{f_j}{f_i} + \frac{f_k}{f_i} + \ldots\right] = f_i\left[1 + \frac{e^{-E_j/RT}}{e^{-E_i/RT}} + \frac{e^{-E_k/RT}}{e^{-E_i/RT}} + \ldots\right]$$
$$= f_i \frac{\left[e^{-E_i/RT} + e^{-E_j/RT} + e^{-E_k/RT} + \ldots\right]}{e^{-E_i/RT}} \tag{A5.6}$$

Rearranging this equation leads to

$$f_i = \frac{e^{-E_i/RT}}{\left[e^{-E_i/RT} + e^{-E_j/RT} + e^{-E_k/RT} + \ldots\right]} = \frac{e^{-E_i/RT}}{\sum_a e^{-E_a/RT}} \tag{A5.7}$$

which is same as Equation A5.2.

One may find the determination of populations using Equation A5.4 more convenient, since fractional populations in relation to lowest energy conformers can be visualized. However, this procedure is no different from using Equation A5.2.

A5.2 Vibrational Averaging

The experimentally measured electronic properties are averaged over nuclear motions. Therefore, even in the absence of vibronic coupling, the measured electronic properties, such as specific optical rotation, are to be averaged over vibrational motions. For such vibrational averaging, the averaged property is written as (Crawford and Allen 2009)

$$P_{av} = \frac{\sum_i e^{-E_i/kT} \sum_{v_i} g_{v_i} e^{-E_{v_i}/kT} P_{v_i}}{\sum_i e^{-E_i/kT} \sum_{v_i} g_{v_i} e^{-E_{v_i}/kT}} \tag{A5.8}$$

In this equation, there are two different summations: one is over conformers i (as in the previous section), and the second summation is over vibrational states ν_i, each with degeneracy g_{v_i}, of conformer i; P_{v_i} is the property of interest in vibrational state ν of conformer i; E_{v_i} is the energy in vibrational state ν of conformer i. Equation A5.8 reduces to Equation A5.1 under two approximations:

1. The property P_{v_i} in each vibrational state of conformer i is the same, that is, $P_{v_i} = P_i$.
2. The vibrational partition functions are the same for all conformers. Under this approximation,

$$\sum_{v_i} g_{v_i} e^{-E_{v_i}/kT} = \sum_{v} g_v e^{-E_v/kT} \tag{A5.9}$$

Substituting these two approximations in Equation A5.8, one obtains

$$P_{av} = \frac{\sum_i e^{-E_i/kT} P_i}{\sum_i e^{-E_i/kT}} \tag{A5.10}$$

which is same as Equation A5.1 (see Section 5.2.1.1 of Chapter 5 for more information on vibrational averaging of specific optical rotation).

Reference

Crawford, T. D., and W. D. Allen. 2009. Optical activity in conformationally flexible molecules: A theoretical study of large-amplitude vibrational averaging in (R)-3-chloro-1-butene. *Mol. Phys.* 107(8–12):1041–1057.

Appendix 6: Monochromators and Interferometers

A6.1 Monochromators

A monochromator uses the principle of wavelength dispersion to spatially separate the individual wavelength components of a polychromatic light source (LS). Although both prisms and gratings can be used for dispersion, a majority of the instruments use gratings. Visible or infrared light from an appropriate LS is focused onto the entrance slit of a monochromator equipped with a diffraction grating. Reflection gratings are made either by ruling fine groves with a diamond point on a polished metallic mirror or using holography. The light entering the monochromator (see Figure A6.1) is directed by mirrors such that collimated light falls on the grating and the light diffracted from the grating is focused at the exit slit of the monochromator. The interference of light waves reflected at the grooves results in a diffraction pattern that is dependent on groove density. The grating equation (Jenkins and White 1976) is given as

$$m\lambda = d(\sin i + \sin\theta) \tag{A6.1}$$

where d is the groove spacing (distance between grooves), i is the angle of incidence on grating, θ is the diffraction angle, and m is an integer referred to as the order. First-order diffraction ($m = 1$) is normally used. The above equation indicates that for a given angle of incidence, different wavelength components of light reflect at different diffraction angles. Such spatial separation of light components of different wavelengths is called dispersion. The angular dispersion is given by the following relation:

$$\frac{\Delta\theta}{\Delta\lambda} = \frac{m}{d\cos\theta} \tag{A6.2}$$

These spatially dispersed light components of different wavelengths are passed through the exit slit of the monochromator. The groove density and exit slit width determine the wavelength resolution that can be realized with the monochromator used. As the grating is rotated, the light component of the desired wavelength is brought into the exit slit. This procedure is called scanning, and for this reason, the word "scanning monochromator" is often

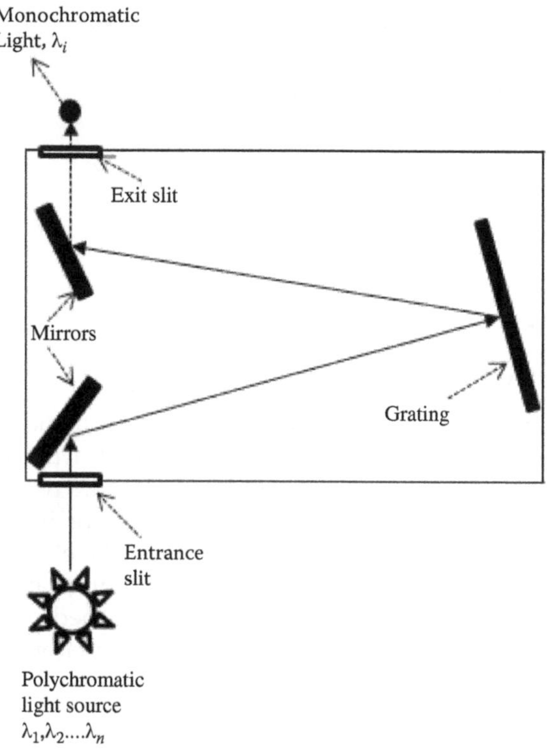

FIGURE A6.1
Schematic diagram of a monochromator.

used. Monochromators are equipped with a servomotor to electronically control the rotation of the grating. The scanning of light components of different wavelengths that transmit through the exit slit is thus achieved using this electronically controlled servomotor.

A6.2 Interferometers

There are different kinds of interferometers, operating on different principles (Gåsvik 2003). The Michelson interferometer, one of the most widely used, works on the principle of amplitude division while the Martin–Puplett interferometer (Martin and Puplett 1970; Dignam and Baker 1981) works on the principle of polarization division. The former type is referred to as an amplitude division interferometer and the latter as a polarization division interferometer (PDI).

A6.2.1 Amplitude Division Interferometers

In a Michelson interferometer (see Figure A6.2), the light coming from a polychromatic source (LS) is directed to a beamsplitter (BS), which is oriented at 45° to the direction of propagation of incoming light. This BS splits the incident light into two components, one transmitting (T_b) through the BS and another reflected (R_b) at 90° to the direction of propagation of the incident beam. These two split components, T_b and R_b, are brought back to the beamspiltter by reflecting mirrors. One of these mirrors is fixed in its position while the other is placed on a translating stage. Upon reflection at the fixed mirror, the resulting light beam, R_bR, is split into two components: one transmitting (R_bRT_b) through the BS and another reflected (R_bRR_b) at the BS toward LS. Upon reflection at the movable mirror, the resulting light beam, T_bR, is split into two components: one transmitting (T_bRT_b) through the BS toward LS and another reflected (T_bRR_b) at the BS. As a result, one-half of the light leaves through the exit port of the interferometer and the other half returns toward the LS. If the two mirrors are equidistant from the BS, the two light components, R_bRT_b and T_bRR_b, exiting the interferometer would have traveled the same distance. This is referred to as the zero path difference (ZPD) position. At the ZPD position, the electric vectors of light components exiting the interferometer would have the same phase at all wavelengths. At each wavelength, the electric vectors from the two arms of the interferometer would have undergone constructive interference. When the moving mirror is displaced away from the equidistance position, the electric vectors of light coming from the two arms of the interferometer

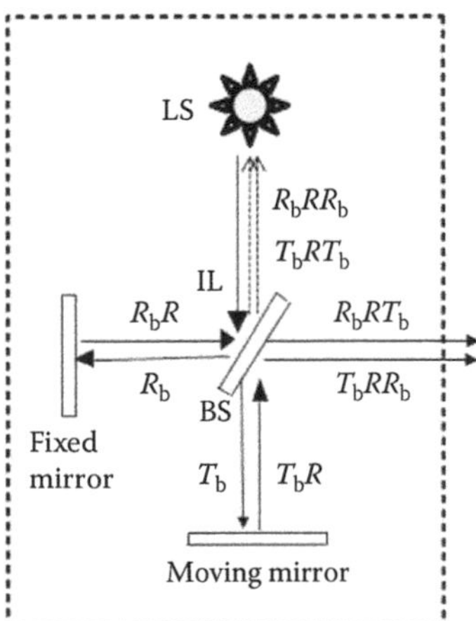

FIGURE A6.2
Schematic diagram of a Michelson interferometer. LS, light source; IL, incident light; BS, beamsplitter; T_b, transmitted beam; T_bR, T_b reflected by the mirror; T_bRR_b, T_bR reflected by the BS; T_bRT_b, T_bR transmitted by the BS; R_b, reflected beam; R_bR, R_b reflected by the mirror; R_bRT_b, R_bR transmitted by the BS; R_bRR_b, R_bR reflected by the BS.

would have traversed different distances. If δ is the path difference (twice the distance traversed by the moving mirror) in the two arms of the interferometer, then they would have acquired a phase difference, $\theta(\lambda)=\frac{2\pi\delta}{\lambda}$. Since this phase difference depends on the wavelength, electric vectors of different wavelengths from the two arms of the interferometer would have acquired different phase differences. As a result, the Michelson interferometer introduces a wavelength-dependent phase difference beyond the ZPD position. Constructive interference occurs at some wavelengths, and destructive interference occurs at others. The intensity signal recorded at the exit port, by a suitable detector, as a function of path difference introduced by the moving mirror is given by the following expression:

$$I(\delta)=\int_{-\infty}^{\infty}\frac{I_{LS}^{o}(\lambda)}{2}\left(1+\cos\theta(\lambda)\right)d\lambda \tag{A6.3}$$

The portion of this equation that is independent of δ is called the direct current/voltage (DC) signal and is eliminated by the electronic filters. The part of Equation A.6.3 that is dependent on δ is called the interferogram, and the Fourier transform of this interferogram yields the intensity distribution of the LS as a function of wavelength, $I_{LS}^{o}(\lambda)$. If the mirror is continuously scanned back and forth at velocity V, the path difference varies as $\delta=2Vt$, where t is the time. The spectral resolution depends on the distance traversed by the mirror: the longer the distance traveled the higher will be the spectral resolution. Additional details on Michelson interferometers, which are the central components of Fourier transform infrared instruments, can be found in standard books on the subject (Griffiths and De Haseth 2007).

A6.2.2 Polarization Division Interferometers

A schematic diagram of a polarization division interferometer (PDI) is shown in Figure A6.3. Light from a polychromatic source (LS) is linearly polarized using an input polarizer (P). This linearly polarized light falls on a polarizing wiregrid BS, which is constructed from parallel wires. The BS is oriented such that the incoming linear polarization axis is at 45° to the wires of BS. Then the polarization of the incoming light can be resolved into two components, one parallel and the other perpendicular to the wires of BS. The parallel component (p_1) is reflected to a fixed roof–top mirror (M_1), while the perpendicular component (s_2) is transmitted toward another rooftop mirror (M_2) that is on a translational stage. Upon reflection by M_1 the polarization component p_1 is rotated 90° and becomes the perpendicular polarization component (s_1), and this component is transmitted through the BS. Similarly, the mirror M_2 rotates the polarization component s_2 to become parallel polarization component (p_2), and this component is reflected by the BS. The components s_1 and p_2 exit at 90° to the direction of the input beam. When the

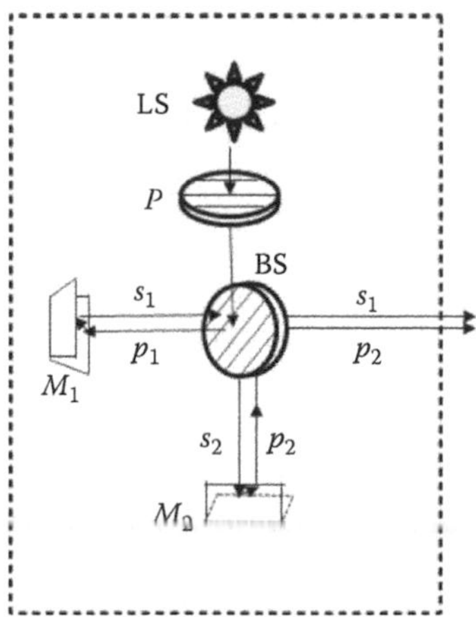

FIGURE A6.3
Schematic diagram of a polarization division interferometer. LS, polychromatic light source; P, linear polarizer; BS, polarization division beamsplitter; p_1, p_2, polarization components parallel to the BS grids; s_1, s_2, polarization components perpendicular to the BS grids; M_1, M_2, roof-top mirrors.

two mirrors are equidistant from BS, that is, at the ZPD point of the interferometer, the polarization of the output beam is identical to that of the beam coming through polarizer P. As the mirror M_2 travels away from the ZPD point, the polarization of the output beam changes, depending on the length of the travel by the mirror M_2. For monochromatic input light of wavelength λ, the mirror travel introduces varying phase differences between s_1 and p_2. In other words, the polarization state changes among left-circular, vertical, right-circular, and horizontal polarizations. This modulation cycle is repeated as the mirror travels to longer distances. For polychromatic light, different wavelengths go through the above-mentioned polarization states at different mirror travel positions, as correspondingly longer mirror travel is needed to complete one cycle of modulation for a longer wavelength. As a consequence, PDI serves the purpose of polarization modulation in the entire wavelength region (appropriate for a given BS).

For randomly polarized light, the intensity of light exiting the polarizer P is one-half of the intensity of LS. The relative phase difference between the polarization components s_1 and p_2 exiting the interferometer is $\theta(\lambda)=\frac{2\pi\delta}{\lambda}$ with δ representing the optical path difference. If no other components are placed in the light path, and the light exiting the interferometer is directly passed to the detector, then there is no intensity modulation and light exiting the interferometer is a constant equal to the total intensity of light transmitted by the input polarizer P. Then the detector signal should contain no interferogram. However, if a chiral sample is placed in the beam, the chiral sample can absorb right and circular polarized light differently, so intensity modulation appears at the detector, which appears as an interferogram at the detector. The sine Fourier transform of this interferogram yields the circular dichroism (CD) signal (Dignam and Baker 1981; Ragunathan et al. 1990; Polavarapu et al. 1994, 2002).

PDI can also be used for normal spectroscopic measurements, by placing an analyzer in the path of the light exiting the interferometer. Since the analyzer blocks and transmits polarization components selectively, intensity modulation, and an interferogram, would be registered at the detector. The cosine Fourier transform of this interferogram yields the intensity distribution of light coming through the polarizer P.

A large number of novel applications are possible with PDI, including CD measurements. More details on the additional applications can be found in Polavarapu (1997).

References

Dignam, M. J., and M. D. Baker. 1981. Analysis of a polarizing Michelson interferometer for dual beam Fourier-transform infrared, circular dichroism infrared, and reflectance ellipsometric infrared spectroscopies *Appl. Spectrosc.* 35(2):186–193.

Gåsvik, K. J. 2003. *Optical Metrology*. New York, NY: John Wiley & Sons.

Griffiths, P., and J. A. De Haseth. 2007. *Fourier Transform Infrared Spectrometry*. Hoboken, NJ: Wiley.

Jenkins, F. A., and H. E. White. 1976. *Fundamentals of Optics*. New York, NY: McGraw Hill Book Co.

Martin, D. H., and E. Puplett. 1970. Polarised interferometric spectrometry for the millimetre and submillimetre spectrum. *Infrared Phys.* 10(2):105–109.

Polavarapu, P. 1997. *Principles and Applications of Polarization-Division Interferometry*. New York, NY: John Wiley & Sons.

Polavarapu, P. L., P. K. Bose, A. J. Rilling, H. Buijs, and J. R. Roy. 2002. Development and evaluation of a polarization-division interferometer with cube corner mirrors. *Appl. Spectrosc.* 56(12):1626–1632.

Polavarapu, P. L., G. C. Chen, and S. Weibel. 1994. Development, justification, and applications of a midinfrared polarization-division interferometer. *Appl. Spectrosc.* 48(10):1224–1235.

Ragunathan, N., N. S. Lee, T. B. Freedman, L. A. Nafie, C. Tripp, and H. Buijs. 1990. Measurement of vibrational circular dichroism using a polarizing Michelson interferometer. *Appl. Spectrosc.* 44(1):5–7.

Appendix 7: Polarization Modulation

The central component of modern circular dichroism (CD) instruments is the one that modulates the polarization of light between left and right circular polarization states. Modern instruments use photoelastic modulators (PEMs) to achieve this modulation. The optical component of PEM is an isotropic crystal, with good transmission for the appropriate wavelength region of interest. Fused silica is used for the visible region while ZnSe is used for the mid-infrared region. CaF_2 is also suitable from the vacuum ultraviolet (UV)-visible region up to a portion of the mid-infrared region. This isotropic crystal is subjected to compression and extension along one direction using voltage driven transducers glued to the crystal (see Figure A7.1). The axis along which this extension and compression take place is called the stress axis. A linearly polarized monochromatic light of wavelength λ, with its electric vector oriented at 45° to the stress axis of PEM, can be resolved into two components: one parallel to the stress axis and another perpendicular to it. Because of the stress applied, the electric vector parallel to the stress axis will propagate at a different speed from that perpendicular to the stress axis. That means there will be a birefringence and consequential phase difference between two mutually perpendicular electric vectors of light waves exiting the PEM. When the voltage applied along the stress axis results in a phase difference that corresponds to quarter wave, light exiting the PEM optical element would have been circularly polarized (see Chapter 1 for the dependence of polarization state of light as a function of the phase difference between combining coherent waves). If the voltage applied to the transducers is modulated at a certain frequency, causing compression and extension of the crystal at that frequency, the polarization state of light exiting the PEM optical element would have synchronously changed between left and right circular polarizations at the same frequency. This modulation frequency by PEM, a characteristic of the optical element (usually ~37 kHz for ZnSe crystal and ~50 kHz for CaF_2 crystal) is achieved using sinusoidal variation of the voltage that necessitates a complex signal processing (vide infra). In the reminder of this section, the necessary mathematical details needed to understand the signal processing are presented. For more details on the subject, see the cited references (Cheng et al. 1975; Osborne et al. 1973; Polavarapu 1985).

The phase difference between the two electric vectors (parallel and perpendicular to the stress axis) changes synchronously with time-dependent variation in the birefringence of the crystal, that is, with the sinusoidal variation of voltage applied to the transducers. The time-dependent phase difference is given by the expression

$$\delta_t = \delta_\lambda^0 \sin 2\pi \nu_m t \tag{A7.1}$$

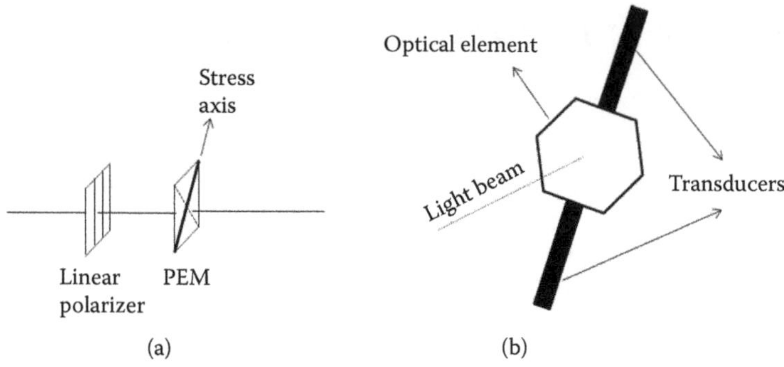

FIGURE A7.1
(a) Depiction of relative orientation of the axes of linear polarizer and photoelastic modulator (PEM) for generating circular polarizations. Vertical lines embedded in the linear polarizer depiction represent the direction of the electric vector emerging from the linear polarizer. The thick line embedded in the PEM depiction represents the stress axis. (b) Expanded frontal view of the PEM optical element.

where ν_m is the frequency of modulation in Hz, and δ_λ^0 is the maximum phase shift introduced between the electric vectors of wavelength λ. For the operation of PEM, the user selects a voltage that corresponds to quarter wave retardation at wavelength, say λ_q. For a given voltage setting, the retardation at other wavelengths is governed by the relation

$$\delta_\lambda^0 = \frac{\pi}{2}\frac{\lambda_q}{\lambda} \tag{A7.2}$$

which indicates that light components of all wavelengths do not experience the same quarter wave retardation at one voltage setting. Therefore, for a light component of selected wavelength, the PEM controller has to be adjusted to impart quarter wave retardation at that wavelength. In dispersive scanning CD instruments, the servocontroller of monochromator (see Appendix 6) can be synchronized with the voltage applied by the PEM controller, such that quarter wave retardation is imparted for each of the light components exiting the monochromator. However, this procedure is not applicable for Fourier transform (FT) CD instruments (vide infra).

A7.1 Signal Processing in Dispersive CD Spectrometers

In Chapter 1 (and also Appendix 1), we have seen that linearly polarized light can be written as a combination of two orthogonal electric vectors. For the present case, the incoming linear polarization, at 45° to the stress axis of the PEM optical element, can be resolved into two orthogonal components (one along stress axis and the other perpendicular to it). Using unit vectors

u and **v** along these axes, electric vector for linearly polarized light wave is (see Equation A1.5)

$$\mathbf{F}_{\mathrm{LP}} = \frac{F_0}{\sqrt{2}}(\mathbf{u}+\mathbf{v})\, e^{-i2\pi(\nu t - z/\lambda)} \tag{A7.3}$$

The time-varying phase difference, δ_t, introduced by the PEM can be incorporated into this equation in complex representation so the resultant electric vector is given as

$$\mathbf{F} = \frac{F_0}{\sqrt{2}}\left(\mathbf{u}+\mathbf{v}e^{-i\delta_t}\right) e^{-i2\pi(\nu t - z/\lambda)} \tag{A7.4}$$

This phase-modulated electric vector can be rewritten as

$$\mathbf{F} = \frac{F_0}{\sqrt{8}}\left[(\mathbf{u}+i\mathbf{v})\left(1 - ie^{-i\delta_t}\right)+(\mathbf{u}-i\mathbf{v})\left(1 + ie^{-i\delta_t}\right)e^{i2\pi(\nu t - z/\lambda)}\right] \tag{A7.5}$$

From Appendix 1, we see that $(\mathbf{u} - i\mathbf{v})\, e^{-i2\pi\,(vt - z\lambda)}$ and $(\mathbf{u} + i\mathbf{v})\, e^{-i2\pi\,(vt - z\lambda)}$ represent, respectively, right and left circular polarization components (see Equations A1.6 and A1.7). In the optical configuration to measure sample CD (see Figure A7.2), the amplitudes of these circular polarization components are attenuated differently by the chiral sample due to differential absorption. Thus, the above equation is modified to account for absorption as

$$\mathbf{F} = \frac{F_0}{\sqrt{8}}\left[(\mathbf{u}+i\mathbf{v})e^{\frac{-2.303A_L}{2}}\left(1-ie^{-i\delta_t}\right)+(\mathbf{u}-i\mathbf{v})e^{\frac{-2.303A_R}{2}}\left(1+ie^{-i\delta_t}\right)\right]e^{-i2\pi(\nu t - z/\lambda)} \tag{A7.6}$$

where A_L and A_R are decadic absorbances (i.e., $A = \log_{10}(I^0/I)$, with I^0 and I representing, respectively, the intensity of light before entering and after exiting the sample) for left and right circular polarizations of wavelength λ. Note that A_L and A_R are wavelength dependent, although not explicitly designated here for the sake of brevity. Following differential absorption by the sample, the intensity of light reaching the detector is given by the product of this electric vector with its complex conjugate. Noting that complex

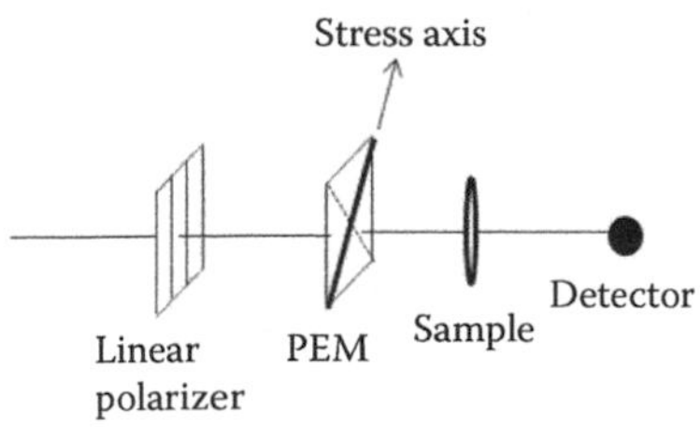

FIGURE A7.2
Optical layout for measuring sample circular dichroism.

conjugate is obtained by replacing "i" with "$-i$," $(\mathbf{u}+i\mathbf{v})(\mathbf{u}-i\mathbf{v})=2$, $(\mathbf{u}+i\mathbf{v})(\mathbf{u}+i\mathbf{v})=0$, $(1-ie^{-i\delta_t})(1+ie^{i\delta_t})=(2\text{–}2\sin\delta_t)$, and $(1+ie^{-i\delta_t})(1-ie^{i\delta_t})=(2+2\sin\delta_t)$, the intensity at the detector becomes

$$I=\frac{I_0}{2}\Big[\big(e^{-2.303A_R}+e^{-2.303A_L}\big)+\big(e^{-2.303A_R}-e^{-2.303A_L}\big)\sin\delta_t\Big] \quad \text{(A7.7)}$$

Since δ_t itself is a time-varying function (see Equation A7.1), $\sin\delta_t$ becomes a complicated function, expressed as a Fourier series,

$$\sin\delta_t=\sin\big(\delta_\lambda^0\sin(2\pi\nu_m t)\big)=2\sum_{n=1}^{\infty}J_{2n-1}\big(\delta_\lambda^0\big)\sin\big((2n-1)2\pi\nu_m t\big) \quad \text{(A7.8)}$$

In this equation, n is an integer and $J_n\big(\delta_\lambda^0\big)$ are Bessel functions. From Equations A7.7 and A7.8, it can be noticed that the signal at the detector contains a time-independent signal, referred to as the direct current/voltage (DC) signal,

$$I_{DC}=\frac{I_0}{2}\big(e^{-2.303A_R}+e^{-2.303A_L}\big)\times G_f \quad \text{(A7.9)}$$

and time-dependent signals oscillating at frequencies, ν_m, $3\nu_m$, and so on. The signal oscillating at ν_m,

$$I_{\nu_m}=I_{AC}=\frac{I_0}{2}\big(e^{-2.303A_R}-e^{-2.303A_L}\big)\times 2J_1\big(\delta_\lambda^0\big)\times G_l \quad \text{(A7.10)}$$

is referred to as the alternating current/voltage (AC) signal. In Equations A7.9 and A7.10, G_f and G_l are the gain factors that depend on the electronics (filters, lock-in amplifiers, etc.) associated with signal processing and vary from one instrument to another. The ratio of the above two equations leads to the CD signal as follows:

$$\frac{I_{\nu_m}}{I_{DC}}=2J_1\big(\delta_\lambda^0\big)\times\frac{G_l}{G_f}\times\frac{\big(e^{-2.303A_R}-e^{-2.303A_L}\big)}{\big(e^{-2.303A_R}+e^{-2.303A_L}\big)} \quad \text{(A7.11)}$$

Multiplying the numerator and denominator of this equation with $e^{2.303\frac{(A_R+A_L)}{2}}$,

$$\frac{I_{\nu_m}}{I_{DC}}=2J_1\big(\delta_\lambda^0\big)\times\frac{G_l}{G_f}\times\frac{\left(e^{2.303\frac{(A_L-A_R)}{2}}-e^{-2.303\frac{(A_L-A_R)}{2}}\right)}{\left(e^{2.303\frac{(A_L-A_R)}{2}}+e^{-2.303\frac{(A_L-A_R)}{2}}\right)} \quad \text{(A7.12)}$$

$$=2J_1\big(\delta_\lambda^0\big)\times\frac{G_l}{G_f}\times\tanh\left(2.303\frac{(A_L-A_R)}{2}\right)\sim J_1\big(\delta_\lambda^0\big)\times\frac{G_l}{G_f}\times 2.303\times\Delta A$$

where $\Delta A = (A_L - A_R)$ is the desired CD. One needs to know the value of $J_1(\delta_\lambda^0) \times \frac{G_l}{G_f}$ to extract the CD quantity, and that is the purpose of instrumental calibration.

A7.2 Calibration of Dispersive CD Spectrometers

In the calibration experiment, the chiral sample is replaced with a birefringent plate and linear polarization analyzer together. There are four different ways for orienting the axes of the birefringent plate and analyzer with respect to those of the polarizer and PEM (see Figure A7.3).

(1) The fast axis of the birefringent plate is parallel to that of the PEM optical element and the polarization axes of the polarizer and analyzer are parallel:

In this arrangement (see Figure A7.3a), the birefringent plate introduces additional retardation, δ_B, in the same direction as the PEM optical element, so the resulting electric vector is given as

$$\mathbf{F} = \frac{F_0}{2}\left(\mathbf{u} + \mathbf{v}e^{-i(\delta_t+\delta_B)}\right)(\mathbf{u}+\mathbf{v})e^{-i2\pi(\nu t - z/\lambda)} \tag{A7.13}$$

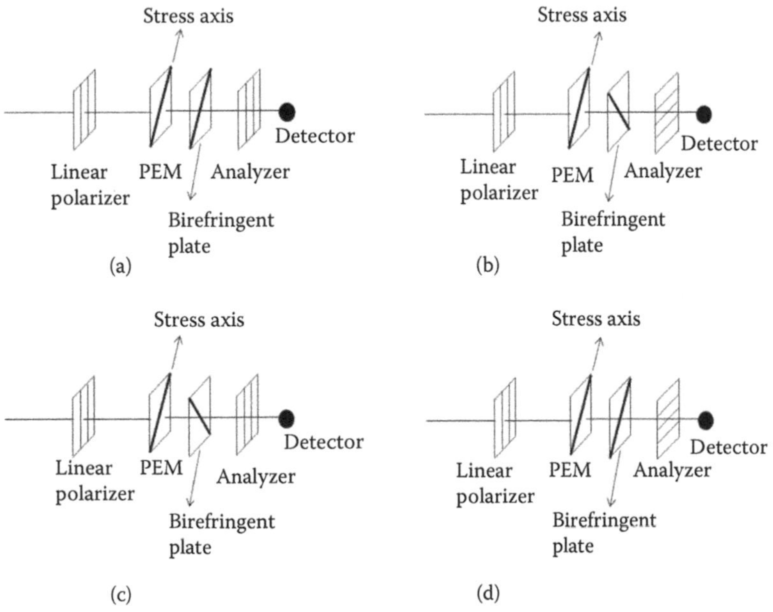

FIGURE A7.3
Four different optical arrangements (a)–(b) for a calibration experiment.

The intensity reaching the detector is obtained by multiplying Equation A7.13 with its complex conjugate:

$$\begin{aligned} I &= \frac{I_0}{4}\left(1+e^{-i(\delta_t+\delta_B)}\right)\left(1+e^{+i(\delta_t+\delta_B)}\right) = \frac{I_0}{2}\left(1+\cos\left(\delta_t+\delta_B\right)\right) \\ &= \frac{I_0}{2}\left(1+\cos\delta_t\cos\delta_B - \sin\delta_t\sin\delta_B\right) \end{aligned} \tag{A7.14}$$

In this equation, $\cos\delta_t$ can be expanded, similar to $\sin\delta_t$ in Equation A7.8, as

$$\cos\delta_t = \cos\left(\delta_\lambda^0\sin(2\pi\nu_m t)\right) = J_0\left(\delta_\lambda^0\right) + 2\sum_{n=1}^{\infty} J_{2n}\left(\delta_\lambda^0\right)\cos(4n\pi\nu_m t) \tag{A7.15}$$

In Equation A7.15, $J_0\left(\delta_\lambda^0\right)$, $J_2\left(\delta_\lambda^0\right)$, and so on, are referred to as zero-order, second-order, and so on, Bessel functions. Substituting the expansions of sin δ_t and cos δ_t (Equations A7.8 and A7.15) into Equation A7.14, and separating the time-varying signal at ν_m (the AC term) and the DC term, the ratio of AC to DC terms becomes

$$\frac{I_{\nu_m}}{I_{DC}} = \frac{-2J_1\left(\delta_\lambda^0\right)\sin\delta_B}{1+J_0\left(\delta_\lambda^0\right)\cos\delta_B} \times \frac{G_l}{G_f} \tag{A7.16}$$

(2) The fast axis of the birefringent plate is perpendicular to that of the PEM optical element and the polarization axes of the polarizer are perpendicular to those of the analyzer:

In this arrangement (see Figure A7.3b), the retardation of the birefringent plate works in the opposite direction to that in arrangement (A), and the axes of the linear polarizer and analyzer are orthogonal, so the resulting electric vector is given as

$$\mathbf{F} = \frac{F_0}{2}\left(\mathbf{u}+\mathbf{v}e^{-i(\delta_t-\delta_B)}\right)(\mathbf{u}-\mathbf{v})e^{-i2\pi(\nu t - z/\lambda)} \tag{A7.17}$$

Following the procedure used in obtaining Equation A7.14, the intensity reaching the detector becomes

$$\begin{aligned} I &= \frac{I_0}{4}\left(1-e^{-i(\delta_t-\delta_B)}\right)\left(1-e^{+i(\delta_t-\delta_B)}\right) = \frac{I_0}{2}\left(1-\cos\left(\delta_t-\delta_B\right)\right) \\ &= \frac{I_0}{2}\left(1-\cos\delta_t\cos\delta_B - \sin\delta_t\sin\delta_B\right) \end{aligned} \tag{A7.18}$$

Substituting the expansions of sin δ_t and cos δ_t (Equations A7.8 and A7.15) into Equation A7.18, and separating the DC and AC terms, the ratio of AC to DC signals becomes

$$\frac{I_{\nu_m}}{I_{DC}} = \frac{-2J_1\left(\delta_\lambda^0\right)\sin\delta_B}{1-J_0\left(\delta_\lambda^0\right)\cos\delta_B} \times \frac{G_l}{G_f} \tag{A7.19}$$

This equation is not same as that for arrangement shown in Figure A7.3a, because the second term in the denominator changed its sign.

(3) The fast axis of the birefringent plate is perpendicular to that of the PEM optical element, and the polarization axes of the polarizer and analyzer are parallel:

In this arrangement (see Figure A7.3c), the retardation of birefringent plate works in the opposite direction to that for the arrangement shown in Figure A7.3a and the resulting electric vector becomes

$$\mathbf{F} = \frac{F_0}{2}\left(\mathbf{u} + \mathbf{v}e^{-i(\delta_t - \delta_B)}\right)(\mathbf{u} + \mathbf{v})e^{-i2\pi(\nu t - z/\lambda)} \tag{A7.20}$$

Following the procedure used in obtaining Equations A7.14 and A7.18, the intensity reaching the detector becomes

$$\begin{aligned} I &= \frac{I_0}{4}\left(1 + e^{-i(\delta_t - \delta_B)}\right)\left(1 + e^{+i(\delta_t - \delta_B)}\right) = \frac{I_0}{2}\left(1 + \cos(\delta_t - \delta_B)\right) \\ &= \frac{I_0}{2}(1 + \cos\delta_t \cos\delta_B + \sin\delta_t \sin\delta_B) \end{aligned} \tag{A7.21}$$

Substituting the expansions of sin δ_t and cos δ_t (Equations A7.8 and A7.15) into Equation A7.21, and separating the DC and AC terms, the ratio of AC to DC signals becomes

$$\frac{I_{\nu_m}}{I_{DC}} = \frac{2J_1\left(\delta_\lambda^0\right)\sin\delta_B}{1 + J_0\left(\delta_\lambda^0\right)\cos\delta_B} \times \frac{G_l}{G_f} \tag{A7.22}$$

This equation is the negative of Equation A7.16.

(4) The fast axis of the birefringent plate is parallel to that of the PEM optical element, but the polarization axis of the analyzer is perpendicular to that of the polarizer:

In this arrangement (see Figure A7.3d), the retardation of the birefringent plate remains the same as that for the arrangement shown in Figure 7.3a, but the resulting electric vector changes to

$$\mathbf{F} = \frac{F_0}{2}\left(\mathbf{u} + \mathbf{v}e^{-i(\delta_t + \delta_B)}\right)(\mathbf{u} - \mathbf{v})e^{-i2\pi(\nu t - z/\lambda)} \tag{A7.23}$$

Following the procedure used in obtaining Equations A7.14, A7.18, and A7.21, the intensity reaching the detector becomes

$$\begin{aligned} I &= \frac{I_0}{4}\left(1 - e^{-i(\delta_t + \delta_B)}\right)\left(1 - e^{+i(\delta_t + \delta_B)}\right) = \frac{I_0}{2}\left(1 - \cos(\delta_t + \delta_B)\right) \\ &= \frac{I_0}{2}(1 - \cos\delta_t \cos\delta_B + \sin\delta_t \sin\delta_B) \end{aligned} \tag{A7.24}$$

Substituting the expansions of sin δ_t and cos δ_t (Equations A7.8 and A7.15) into Equation A7.24, and separating the DC and AC terms, the ratio of AC to DC signals becomes

$$\frac{I_{V_m}}{I_{DC}} = \frac{2J_1\left(\delta_\lambda^0\right)\sin\delta_B}{1 - J_0\left(\delta_\lambda^0\right)\cos\delta_B} \times \frac{G_1}{G_f} \tag{A7.25}$$

This equation is the negative of Equation A7.19.

Although there are four arrangements possible for this calibration experiment, the combination of (1) and (2) or of (3) and (4) is sufficient to determine the $J_1\left(\delta_\lambda^0\right) \times \frac{G_1}{G_f}$ term. This is because equations in each of these pairs of equations are equal when sin $\delta_B = 1$ and cos $\delta_B = 0$. This happens when

$$\delta_B = (2n+1)\frac{\pi}{2} = \frac{\lambda_B}{\lambda}\frac{\pi}{2} \tag{A7.26}$$

that is, when $\lambda = \frac{\lambda_B}{2n+1}$. Note that λ_B is the wavelength where the birefringent plate imparts 90° phase difference (i.e., one-quarter wave retardation). Moreover, all four equations become zero when sin $\delta_B = 0$, which occurs when

$$\delta_B = n\pi = \frac{\lambda_B}{\lambda}\frac{\pi}{2} \tag{A7.27}$$

that is, when $\lambda = \frac{\lambda_B}{2n}$.

In other words, the four curves represented by Equations A7.16, A7.19, A7.22, and A7.25 go through zero at $\lambda = \frac{\lambda_B}{2n}$, while the first two and latter two curves cross each other at $\lambda = \frac{\lambda_B}{2n+1}$. Thus, the nonzero crossings of Equations A7.16 and A7.19, or of Equations A7.22 and A7.25, which occur at $\lambda = \frac{\lambda_B}{2n+1}$, yield the values of $J_1\left(\delta_\lambda^0\right) \times \frac{G_1}{G_f}$ at those wavelengths. An interpolation can be used to determine the values of $J_1\left(\delta_\lambda^0\right) \times \frac{G_1}{G_f}$ for intermediate wavelengths.

A7.3 Signal Processing in FTCD Spectrometers

Unlike in dispersive spectrometers, all wavelength components originating from the light source will pass through the sample in FT spectrometers. In an FT spectrometer based on a Michelson interferometer, light exiting the interferometer, before falling on subsequent optical elements, is given as (see Appendix 6, Equation A6.3)

$$I(\delta) = \int_{-\infty}^{\infty} \frac{I_{LS}^0(\lambda)}{2}\left(1+\cos\theta(\lambda)\right)d\lambda \tag{A7.28}$$

where δ is the path difference (twice the distance traversed by the moving mirror) in the interferometer, $I_{LS}^0(\lambda)$ is the intensity of light source at wavelength λ, and $\theta(\lambda) = \frac{2\pi\delta}{\lambda}$.

For a given voltage setting on the PEM to yield quarter wave retardation, only one wavelength component would experience true quarter wave retardation, while the others would experience differing retardations as given by Equation A7.2. For this reason, one can never achieve true quarter wave retardation at all wavelengths with a single PEM setting in FTCD spectrometers. A limited wavelength range around λ_q can, however, be used for CD measurements with appropriate calibration. Two different configurations, as for dispersive CD spectrometers, are considered.

1. *Sample configuration* (see Figure A7.2): In this configuration, light exiting the Michelson interferometer is passed through a linear polarizer, PEM, sample, and detector in that order.

   The expression for light reaching the detector is obtained by substituting Equations A7.7 and A7.8 in Equation A7.28:

$$I(\delta) = \int_{-\infty}^{\infty} \frac{I_{LS}^0(\lambda)}{4}\left(1+\cos\theta(\lambda)\right)\Big[\left(e^{-2.303A_R}+e^{-2.303A_L}\right)+ \\ \left(e^{-2.303A_R}-e^{-2.303A_L}\right)2\sum_{n=1}^{\infty} J_{2n-1}\left(\delta_\lambda^0\right)\sin\left((2n-1)2\pi\nu_m t\right)\Big]d\lambda \tag{A7.29}$$

   Note that although not shown explicitly here, the absorptions A_R and A_L are wavelength dependent. This equation contains four components, but the only two components that depend on the mirror movement in the interferometer and PEM modulation frequency are relevant. When the detector signal is passed through an electronic filter that will only transmit the signals that depend on $\theta(\lambda)$ and block the signals varying at PEM frequency, one obtains the transmission interferogram, $I_t(\delta)$:

$$I_t(\delta) = \int_{-\infty}^{\infty} \frac{I_{LS}^0(\lambda)}{4}\left(e^{-2.303A_R}+e^{-2.303A_L}\right)G_f(\lambda)\cos\theta(\lambda)d\lambda \tag{A7.30}$$

   where $G_f(\lambda)$ is the wavelength-dependent gain factor introduced by the electronic filter. Similarly, when the detector signal is passed through a lock-in-amplifier tuned to the PEM frequency, followed by an electronic filter that transmits signals that depend on θ(λ), one obtains the CD interferogram, $I_{v_m}(\delta)$:

$$I_{v_m}(\delta)=\int^{\infty}\frac{I^0_{LS}(\lambda)}{4}\left(e^{-2.303A_R}-e^{-2.303A_L}\right)2J_1\left(\delta^0_\lambda\right)G_l(\lambda)G'_f(\lambda)\cos\theta(\lambda)d\lambda \quad \text{(A7.31)}$$

where $G_l(\lambda)$ and $G'_f(\lambda)$ are, respectively, the wavelength-dependent gain factors introduced by the lock-in amplifier and electronic filter. FT of $I_{v_m}(\delta)$, divided by the FT of $I_t(\delta)$, yields the wavelength-dependent CD signal, ΔA. That is (see Equation A7.12),

$$\begin{aligned}\frac{\mathrm{FT}\left[I_{v_m}(\delta)\right]}{\mathrm{FT}\left[I_t(\delta)\right]}&=\frac{G_l(\lambda)\times G'_f(\lambda)}{G_f(\lambda)}\times 2J_1\left(\delta^0_\lambda\right)\times\frac{\left(e^{-2.303A_R}-e^{-2.303A_L}\right)}{\left(e^{-2.303A_R}+e^{-2.303A_L}\right)}\\&=\frac{G_l(\lambda)\times G'_f(\lambda)}{G_f(\lambda)}\times J_1\left(\delta^0_\lambda\right)\times 2.303\times\Delta A\end{aligned} \quad \text{(A7.32)}$$

where FT[..] represents the Fourier transform of the function in parenthesis. The same modifications, as those described for going from Equations A7.11 and A7.12, are also used here. The value of $\frac{G_l(\lambda)\times G'_f(\lambda)}{G_f(\lambda)}\times J_1\left(\delta^0_\lambda\right)$ is determined from the calibration curves as described below.

2. *Calibration configuration* (see Figure A7.3): In this configuration, light exiting the Michelson interferometer is passed through a linear polarizer, PEM, birefringent plate, polarizer, and detector in that order. There are four possible arrangements (1)–(4) for calibration measurements (as discussed earlier for dispersive spectrometers). Proceeding in the same manner as for sample configuration, but by substituting Equations A7.14, A7.18, A7.21, or A7.24 in Equation 7.28, one obtains the expression for $I_t(\delta)$ and $I_{v_m}(\delta)$. The shapes of $I_{v_m}(\delta)$ interferograms, different from normal transmission interferograms, have been evaluated and discussed in relation to the observed interferograms (Polavarapu 1985). Fourier transforming the interferograms and taking the ratio, one obtains four calibration curves

$$\frac{\mathrm{FT}\left[I_{v_m}(\delta)\right]}{\mathrm{FT}\left[I_t(\delta)\right]}=\frac{\pm 2J_1\left(\delta^0_\lambda\right)\sin\delta_B}{1\pm J_0\left(\delta^0_\lambda\right)\cos\delta_B}\times\frac{G_lG'_f}{G_f} \quad \text{(A7.33)}$$

where the sign in "±" depends on the configuration (see Equations A.7.16, A7.19, A7.22, and A7.25). The value of $\frac{G_l(\lambda)\times G_f'(\lambda)}{G_f(\lambda)}\times 2J_1(\delta_\lambda^0)$ is determined from the nonzero crossings of the calibration curves as discussed earlier (following Equation A7.27):

A7.4 Polarization Scrambling

Due to the small magnitudes of vibrational circular dichroism (VCD) signals, VCD measurements are prone to artifacts and spurious signals can dominate the VCD signals. The equations given in the previous sections for sample CD measurements (see Figure A7.2) do not take artifacts arising from birefringence and linear dichroism from the components in the optical train after PEM. The artifacts that appear as a product of linear birefringence and the Bessel function J_0 can be eliminated in dispersive spectrometers (Cheng et al. 1975) by introducing a second PEM2 between the sample and detector (see Figure A7.4). The modulation frequency of PEM2 is chosen to be slightly different from that of PEM1. The voltage applied to the first PEM (PEM1) corresponds to that required for achieving quarter wave retardation at the wavelength of light probing the sample. The voltage applied to the PEM2 is chosen to give a retardation that makes the zero-order Bessel function, J_0, at that probing wavelength, equal to zero. The voltages applied to PEM1 and PEM2 can be synchronized to the wavelength of light that is probing the sample. This approach is referred to as polarization scrambling (PS).

The same approach when adopted for CD measurements with FT spectrometers does not work that transparently, because, in an FTCD instrument, all wavelengths go through the sample simultaneously. As a result, the quarter wave retardation selected by PEM1 is set in the middle of the wavelength region that is being investigated. For example, quarter wave retardation is set at 1400 cm^{-1} for studying the 2000–800 cm^{-1} region. The second PEM cannot be set to a retardation level that would make J_0 equal to zero for all of the

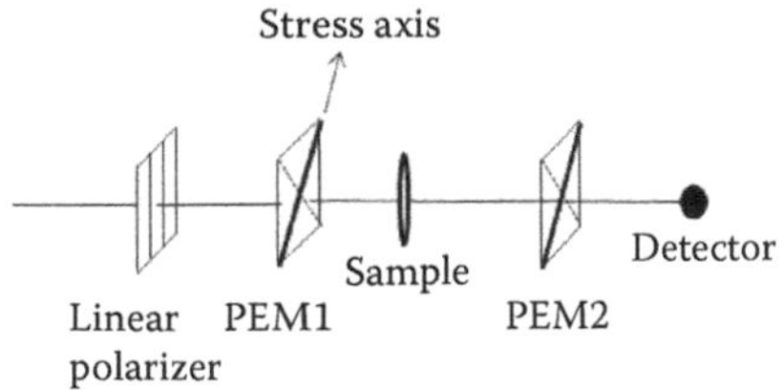

FIGURE A7.4
Optical layout for polarization scrambling in circular dichroism measurements.

wavelengths that are being probed in FTCD instrument. As a result, signal processing in FTCD instruments in the presence of postsample PEM is done differently and is referred to as the dual polarization modulation method (Nafie 2000). Additional details can be found in Chapter 7.

References

Cheng, J. C., L. A. Nafie, and P. J. Stephens. 1975. Polarization scrambling using a photoelastic modulator: Application to circular dichroism measurement. *J. Opt. Soc. Am.* 65:1031–1035.

Nafie, L. A. 2000. Dual polarization modulation: A real-time, spectral-multiplex separation of circular dichroism from linear birefringence spectral intensities. *Appl. Spectrosc.* 54(11):1634–1645.

Osborne, G. A., J. C. Cheng, and P. J. Stephens. 1973. Near-infrared circular dichroism and magnetic circular dichroism instrument. *Rev. Sci. Instrum.* 44:10–15.

Polavarapu, P. L. 1985. Fourier transform infrared vibrational circular dichroism. In: J. R. Ferraro, and L. J. Basile (eds). *Fourier Transform Infrared Spectroscopy. Applications to Chemical Systems.* Vol. 4. New York, NY: Academic Press.

Appendix 8: Units and Conversions

A8.1 Constants

Planck's constant: $h = 6.62606896 \times 10^{-34}$ J·s
Avogadro's number: $N_A = 6.02214179 \times 10^{23}$ mol^{-1}
Speed of light: $c = 2.99792458 \times 10^{10}$ cm·s^{-1}
Electron charge: $e = 4.803204 \times 10^{-10}$ esu $= 1.602176487 \times 10^{-19}$ C
Mass of electron: $m_e = 9.10938 \times 10^{-31}$ kg

A8.2 Different Units of Energy

Joule, J = C·V
Calorie = 4.184 J
erg = esu^2·cm^{-1}
eV = $1.602176487 \times 10^{-19}$ C·V = $1.602176487 \times 10^{-19}$ J
eV = 8065.5 cm^{-1}
Hartree = 27.2114 eV = 4.35974×10^{-18}J = 2.6255×10^{6} J·mol^{-1} = 627.51 kcal·mol^{-1}

A8.3 Unit Conversions

Joule, J = N·m = kg·m^2·s^{-2} = 10^7 erg
N·m^{-1} = kg·s^{-2}
dyn·cm^{-1} = g·s^{-2}
esu^2 = g·cm^3 s^{-2}
Debye, D = 10^{-18} esu·cm
Bohr radius, $a_0 = 0.529177 \times 10^{-8}$ cm
Atomic mass unit, amu = 1.6605×10^{-27} kg
Angstrom, Å = 10^{-8} cm
Forces in atomic units; $\left(\frac{e^2}{a_0^2}\right) = 8.2387 \times 10^{-3}$ dynes
Force constants in atomic units; $\left(\frac{e^2}{a_0^3}\right) = 15.5689 \times 10^{-3}$ dyn·Å^{-1}

$$\text{Wavelength (in nm)} = \frac{1239.84}{\text{Energy in eV}}$$

A8.4 Hartree

$$1\text{Hartree} = \frac{e^2}{a_0}$$

$$\text{Hartree} = \frac{\left(4.8032\times10^{-10}\text{esu}\right)^2}{0.529177\times10^{-8}\text{cm}} \times \frac{\text{erg}}{\text{esu}^2\text{cm}^{-1}} \times \frac{\text{J}}{10^{-7}\text{erg}} = 4.35974\times10^{-18}\text{J}$$

A8.5 Bohr Radius, a_0

$$a_0 = \frac{h^2}{4\pi^2 m_e e^2}$$

$$a_0 = \frac{\left(6.62606896\times10^{-27}\text{g cm}^2\text{s}^{-2}\text{s}\right)^2}{4\times\pi^2\times9.10938\times10^{-28}\text{g}\times\left(4.8032\times10^{-10}\text{esu}\right)^2}$$

$$= 5.29177\times10^{-6}\frac{\text{g cm}^4\text{s}^{-2}}{\text{esu}^2\times\dfrac{\text{erg cm}}{\text{esu}^2}\times\dfrac{\text{g cm}^2\text{s}^{-2}}{\text{erg}}} = 5.29177\times10^{-9}\text{cm}$$

$$= 5.29177\times10^{-11}\text{m}$$

$$= 0.529177\,\text{Å}$$

A8.6 Conversion between Dimensionless Normal Coordinate q_k to Normal Coordinate Q_k

$$q_k = \left(\frac{4\pi^2\nu_k}{h}\right)^{1/2} Q_k = \left(\frac{4\times(3.14159)^2\times\nu_k\left\{\text{s}^{-1}\right\}}{6.62606896\times10^{-27}\text{g cm}^2\text{s}^{-1}}\right)^{1/2} \times Q_k\left\{10^{-8}\text{cm}\times\left(1.6605\times10^{-24}\text{g}\right)^{1/2}\right\}$$

$$q_k = 6.4087\times10^{-12}\nu_k Q_k \quad (\nu_k \text{ in Hz and } Q_k \text{ in Å amu}^{1/2})$$

A8.7 Conversion of Dipole Strength to Absorption Intensity in km·mol^{-1}

Starting from Equation 5.26:

$$A_k = \int \alpha(\bar{\nu})d\bar{\nu} = \frac{8\pi^3 N_A \bar{\nu}_k^0}{3ch} D_k$$

$$A_k = \frac{8\times(3.14159)^3 \times 6.02214179\times 10^{23}\text{mol}^{-1}}{3\times 2.99792458\times 10^{10}\text{ cm s}^{-1}\times 6.62606896\times 10^{-34}\text{J s}} \times \bar{\nu}_k^0\{\text{cm}^{-1}\}\times D_k\{10^{-40}\text{esu}^2\text{cm}^2\}$$

$$A_k = 2.5066\times 10^8 \bar{\nu}_k^0 D_k \frac{\text{mol}^{-1}\times\text{esu}^2}{\text{J}}\times\frac{10^{-7}\text{J}}{\text{erg}}\times\frac{\text{erg}}{\text{esu}^2\text{cm}^{-1}}\times\frac{\text{m}}{100\text{ cm}}\times\frac{\text{km}}{1000\text{ m}}$$

$$A_k = 2.5066\times 10^{-4}\times\bar{\nu}_k^0\times D_k \text{ km mol}^{-1};\quad (\bar{\nu}_k^0 \text{ in cm}^{-1} \text{ and } D_k \text{ in } 10^{-40}\text{esu}^2\text{cm}^2)$$

A8.8 Conversion between Absorption Intensity and Electric Dipole Moment Derivatives

$A_k = \frac{N_A \pi}{3c^2}\left(\frac{\partial\mu}{\partial Q_k}\right)^2$. In this equation, express μ in Debye ($D = 10^{-18}$ esu cm) and normal coordinates in Å amu$^{1/2}$ (amu = 1.6605×10^{-24} g) and use the conversion esu^2 = g·cm^3·s^{-2}:

$$A_k = \frac{6.02214179\times 10^{23}\text{mol}^{-1}\times 3.14159}{3\left(2.99792458\times 10^{10}\text{cm s}^{-1}\right)^2}\times\left\{\frac{\partial\mu}{\partial Q}\left(\frac{10^{-10}\text{esu}}{\left(1.6605\times 10^{-24}\text{g}\right)^{1/2}}\right)\right\}^2$$

$$= 4.2256\times 10^6 \frac{\text{cm}}{\text{mol}}\times\frac{\text{m}}{100\text{ cm}}\times\frac{\text{km}}{1000\text{ m}}$$

$$= 42.256\left(\frac{\partial\mu}{\partial Q}\right)^2 \text{ km mol}^{-1}$$

A8.9 Conversion between Dipole Strength D_i and Oscillator Strength f_i

For the kth transition, starting from Equation 4.130,

$$f_k = \frac{8\pi^2 m_e \nu_k}{3e^2 h} D_k$$

and using the relation $c = \lambda\nu$

$$D_k = \frac{3e^2 h \lambda_k f_k}{8\pi^2 m_e c}$$

$$= \frac{3\left(4.80324\times10^{-10}\,\text{esu}\right)^2\left(6.62606896\times10^{-34}\,\dfrac{\text{g cm}^2}{\text{s}^2}\,\text{s}\right)}{8(3.1415926)^2\left(9.10938\times10^{-28}\,\text{g}\right)\left(2.99792458\times10^{10}\,\dfrac{\text{cm}}{\text{s}}\right)}\lambda_k\text{nm}\times\frac{10^{-7}\,\text{cm}}{\text{nm}}\times f_k$$

$$D_k = 2127\lambda_k f_k \times 10^{-40}\,\text{esu}^2\text{cm}^2 \quad (\lambda_k \text{ in nm})$$

A8.10 Conversion of Dipole Strength to Absorption Intensity in cm³·mol⁻¹

The experimental electronic spectra are presented with wavelength as the x-axis. To simulate the spectra on wavelength scale, the integrated intensities are written as

$$A_k = \int \alpha(\lambda)d\lambda = \frac{8\pi^3 N_A}{3\text{hc}}\lambda_k^0 D_k$$

$$A_k = \frac{8\times(3.14159)^3\times6.02214179\times10^{23}\,\text{mol}^{-1}}{3\times2.99792458\times10^{10}\,\text{cm s}^{-1}\times6.62606896\times10^{-34}\,\text{J s}}\times\lambda_k^0\{\text{nm}\}\times D_k\left\{10^{-40}\,\text{esu}^2\text{cm}^2\right\}$$

$$A_k = 2.5066\times10^8\lambda_k^0 D_k\,\frac{\text{mol}^{-1}\times\text{nm}\times\text{esu}^2\text{cm}^2}{\text{J}\times\text{cm}}\times\frac{10^{-7}\,\text{J}}{\text{erg}}\times\frac{\text{erg}}{\text{esu}^2\text{cm}^{-1}}\times\frac{10^{-7}\,\text{cm}}{\text{nm}}$$

$$A_k = 25.066\times10^{-7}\times\lambda_k^0\times D_k\,\text{cm}^3\text{mol}^{-1} \quad \left(\lambda_k^0 \text{ in nm and } D_k \text{ in } 10^{-40}\,\text{esu}^2\text{cm}^2\right)$$

A8.11 Conversion of Experimental Integrated Absorption Intensities to Dipole Strengths

Starting from Equation 5.26, dipole strength can be obtained from the integrated decadic molar absorption coefficient using the following relation:

$$A_k = 2.303\int \varepsilon(\bar{\nu})d\bar{\nu} = \frac{8\pi^3 N_A \bar{\nu}_k^0}{3\text{ch}} D_k$$

$$D_k = \frac{3ch \times 2.303}{8\pi^3 N_A \bar{\nu}_k^0}\int \varepsilon(\bar{\nu})d\bar{\nu}$$

$$D_k = \frac{3\times 2.99792458\times 10^{10}\,cms^{-1}\times 6.62606896\times 10^{-27}\,\text{erg s}\times\dfrac{\text{esu}^2\text{cm}^{-1}}{\text{erg}}\times 2.303}{8\times(3.141592654)^3\times 6.02214179\times 10^{23}\,\text{mol}^{-1}}\times\frac{\int \varepsilon(\bar{\nu})d\bar{\nu}}{\bar{\nu}_k^0}$$

$$D_k = 0.9188\times 10^{-38}\,\text{esu}^2\text{mol}\times\frac{\int \varepsilon(\bar{\nu})\left\{\dfrac{L}{\text{mol cm}}\times\dfrac{1000\text{ cm}^3}{L}\right\}d\bar{\nu}\{\text{cm}^{-1}\}}{\bar{\nu}_k^0\{\text{cm}^{-1}\}}$$

$$D_k = 0.9188\times 10^{-38}\times\frac{\int \varepsilon(\bar{\nu})d\bar{\nu}}{\bar{\nu}_k^0}\text{esu}^2\text{cm}^2 \quad (\varepsilon \text{ in L mol}^{-1}\text{cm}^{-1}; \bar{\nu} \text{ in cm}^{-1})$$

A8.12 Conversion of Experimental Integrated Circular Dichroism Intensities to Rotational Strengths

Staring from Equation 5.31, rotational strength can be obtained from integrated decadic difference molar absorption coefficient using the following relation:

$$\Delta A_k = 2.303\int \Delta\varepsilon(\bar{\nu})d\bar{\nu} \sim 4\times\left(\frac{8\pi^3 N_A \bar{\nu}_k^0}{3ch}\right)R_k$$

$$R_k = \frac{3ch\times 2.303}{4\times 8\pi^3 N_A \bar{\nu}_k^0}\int \Delta\varepsilon(\bar{\nu})d\bar{\nu}$$

Following the procedure in Section A8.11,

$$R_k = 0.2297\times 10^{-38}\times\frac{\int \Delta\varepsilon(\bar{\nu})d\bar{\nu}}{\bar{\nu}_k^0}\text{esu}^2\text{cm}^2 \quad (\Delta\varepsilon \text{ in L mol}^{-1}\text{cm}^{-1}; \bar{\nu} \text{ in cm}^{-1})$$

A8.13 Units for Raman and Raman Circular Intensity Difference Activities

1. *Electric dipole polarizability tensor* $\alpha_{\alpha\beta}$ (see Equation 4.39 or 4.68):

   $\alpha_{\alpha\beta}$ elements have the units of volume (in cm³, Å³, or Bohr³):

$$\alpha_{\alpha\beta} = \sum_{m \neq n} \frac{2\mu_{\alpha,nm}\{\text{esu cm}\}\mu_{\beta,mn}\{\text{esu cm}\}}{\left(E_m^o - E_n^o\right)\{\text{esu}^2\,\text{cm}^{-1}\}} = \sum_{m \neq n} \frac{2\mu_{\alpha,nm}\mu_{\beta,mn}}{\left(E_m^o - E_n^o\right)} \text{cm}^3$$

2. *The electric dipole–magnetic dipole polarizability tensor* $\omega^{-1}G'_{\alpha\beta}$ (see Equation 4.77):

   $\omega^{-1}G'_{\alpha\beta}$ elements have the units of cm⁴, Å⁴, or Bohr⁴:

$$\omega^{-1}G'_{\alpha\beta}(\omega) = -\sum_{m \neq n} \frac{2\,\text{Im}\left\{\mu_{\alpha,nm}\{\text{esu cm}\}m_{\beta,mn}\{\text{esucm}^2\text{s}^{-1}\}\right\}}{\hbar\left(\omega_{mn}^2 - \omega^2\right)\{\text{esu}^2\text{cm}^{-1}\text{s}^{-1}\}}$$

$$= -\sum_{m \neq n} \frac{2\,\text{Im}\left\{\mu_{\alpha,nm}m_{\beta,mn}\right\}}{\hbar\left(\omega_{mn}^2 - \omega^2\right)} \text{cm}^4$$

3. *The electric dipole–electric quadrupole polarizability tensor,* $\mathbf{A}_{\alpha\beta\gamma}$ (see Equation 5.38):

   $A_{\alpha\beta\gamma}$ elements have the units of cm⁴, Å⁴, or Bohr⁴:

$$A_{\alpha\beta\gamma} = 4\pi \sum_{n \neq m} \frac{\omega_{mn}}{h\left(\omega_{mn}^2 - \omega^2\right)} \{\text{esu}^{-2}\text{cm}\}\,\text{Re}\left\{\mu_{\alpha,nm}\{\text{esu cm}\}\Theta_{\beta\gamma,mn}\{\text{esu cm}^2\}\right\}$$

$$= \frac{4\pi}{h} \sum_{n \neq m} \frac{\omega_{mn}}{\left(\omega_{mn}^2 - \omega^2\right)} \text{Re}\left\{\mu_{\alpha,nm}\Theta_{\beta\gamma,mn}\right\}\text{cm}^4$$

4. *Normal coordinate derivatives:*

   Normal coordinates have units of cm·amu$^{1/2}$. $\frac{\partial\alpha_{\alpha\beta}}{\partial Q_k}, \frac{\partial\omega^{-1}G'_{\alpha\beta}}{\partial Q_k}$, and $\frac{\partial A_{\alpha\beta\gamma}}{\partial Q_k}$, therefore, have units of cm²·amu$^{-1/2}$, cm³·amu$^{-1/2}$, and cm³·amu$^{-1/2}$, respectively. It is common practice to use Å (or Bohr) instead of cm, so $\frac{\partial\alpha_{\alpha\beta}}{\partial Q_k}, \frac{\partial\omega^{-1}G'_{\alpha\beta}}{\partial Q_k}$, and $\frac{\partial A_{\alpha\beta\gamma}}{\partial Q_k}$ are expressed, respectively, in Å²·amu$^{-1/2}$ (or Bohr²·amu$^{-1/2}$), Å³·amu$^{-1/2}$ (or Bohr³·amu$^{-1/2}$), and Å³·amu$^{-1/2}$ (or Bohr³·amu$^{-1/2}$).

5. *Raman activities* (see Equation 3.51):

$$S_k = \left(a_1 \bar{\alpha}_k^2 \left\{ \frac{\mathring{A}^4}{\text{amu}} \right\} + a_2 \beta_k^2 \left\{ \frac{\mathring{A}^4}{\text{amu}} \right\} \right) = \left(a_1 \bar{\alpha}_k^2 + a_2 \beta_k^2 \right) \frac{\mathring{A}^4}{\text{amu}}$$

In the above equation, a_1 and a_2 are constants, and $\bar{\alpha}_k^2$ and β_k^2 are, respectively, the square of mean and the anisotropy (see Equations 3.52 and 3.53) of the $\frac{\partial \alpha_{\alpha\beta}}{\partial Q_k}$ tensor. The Raman activities, S_k, are thus expressed in units of $\mathring{A}^4 \cdot \text{amu}^{-1}$ or $\text{Bohr}^4 \cdot \text{amu}^{-1}$.

6. *Raman circular intensity difference activities* (see Equation 3.62):

$$P_k = \frac{2\pi}{\lambda \left\{ \mathring{A} \right\}} \left(a_3 \omega^{-1} \bar{\alpha}_k \bar{G}_k' \left\{ \frac{\mathring{A}^5}{\text{amu}} \right\} + a_4 \omega^{-1} \gamma_k^2 \left\{ \frac{\mathring{A}^5}{\text{amu}} \right\} + a_5 \omega^{-1} \delta_k^2 \left\{ \frac{\mathring{A}^5}{\text{amu}} \right\} \right)$$

$$= \frac{2\pi}{\lambda} \left(a_3 \omega^{-1} \bar{\alpha}_k \bar{G}_k' + a_4 \omega^{-1} \gamma_k^2 + a_5 \omega^{-1} \delta_k^2 \right) \frac{\mathring{A}^4}{\text{amu}}$$

In the above equation, a_3, a_4, and a_5 are constants, λ is the wavelength of exciting laser radiation in Å units, and $\omega^{-1} \bar{\alpha}_k \bar{G}_k'$, $\omega^{-1} \gamma_k^2$, and $\omega^{-1} \delta_k^2$ are appropriate products of $\frac{\partial \alpha_{\alpha\beta}}{\partial Q_k}$ tensor elements with $\frac{\partial \omega^{-1} G_{\alpha\beta}'}{\partial Q_k}$ and $\frac{\partial A_{\alpha\beta\gamma}}{\partial Q_k}$ tensor elements (see Equations 3.59 through 3.61). Due to the presence of λ in the denominator, the magnitudes of P_k are smaller than S_k by about 10,000. For this reason, the Raman circular intensity difference activities, P_k, are multiplied with 10^4 and expressed in units of $10^{-4}\ \mathring{A}^4 \cdot \text{amu}^{-1}$ or $10^{-4}\ \text{Bohr}^4 \cdot \text{amu}^{-1}$.

Appendix 9: References for Chiroptical Spectroscopic Studies on Transition Metal Complexes

TABLE A9.1

Chiroptical Spectroscopic Applications to Transition Metal Complexes

Year	Subject	Title	Authors	Reference
1971	ECD	Electric quadrupole contributions to the optical activity of transition metal complexes	L. D. Barron	*Mol. Phys.* 21 (1971) 241–246
1980	VCD	Vibrational-electronic interaction in the infrared circular dichroism spectra of transition-metal complexes	C. J. Barnett, A. F. Drake, R. Kuroda, S. F. Mason, and S. Savage	*Chem. Phys. Lett.* 70 (1980) 8–10
1985	VCD	Vibrational circular dichroism in amino acids and peptides. 9. Carbon-hydrogen stretching spectra of the amino acids and selected transition-metal complexes	M. R. Oboodi, B. B. Lal, D. A, Young, T. B. Freedman, and L. A. Nafie	*J. Am. Chem. Soc.* 107 (1985) 1547–1556
1985	VCD	Vibrational circular dichroism in bis(acetylacetonato)(L-alaninato) cobalt(III). Isolated occurrences of the coupled oscillator and ring current intensity mechanisms	D. A, Young, E. D. Lipp, and L. A. Nafie	*J. Am. Chem. Soc.* 107 (1985) 6205–6213
1985	ECD	Transient circular dichroism of the luminescent state of $Ru(bpy)_3^{2+}$	J. S. Gold, S. J. Milder, J. W. Lewis, and D. S. Kliger	*J. Am. Chem. Soc.* 107 (1985) 8285–8286
1986	VCD	Vibrational circular dichroism in transition-metal complexes. 2. Ion association, ring conformation, and ring currents of ethylenediamine ligands	D. A, Young, T. B. Freedman, E. D. Lipp, and L. A. Nafie	*J. Am. Chem. Soc.* 108 (1986) 7255–7263.
1986	ECD	Circular dichroism of a sub nanosecond state: (Δ)-$Fe(bpy)_3^{2+}$	S. J. Milder, J. S. Gold, and D. S. Kliger	*J Am. Chem. Soc.* 108 (1986) 8295–8296
1988	ECD	Time resolved circular dichroism of the lowest excited state of (Δ)-$Ru(bpy)_3^{2+}$	S. J. Milder, J. S. Gold, and D. S. Kliger	*Chem. Phys. Lett.* 144 (1988) 269–272

(*Continued*)

TABLE A9.1 ***(Continued)***

Chiroptical Spectroscopic Applications to Transition Metal Complexes

Year	Subject	Title	Authors	Reference
1990	ECD	Assignments of ground and excited state spectra from time resolved absorption and circular dichroism measurements of the ^2E state of (Δ)-$Cr(bpy)_3^{3+}$	S. J. Milder, J. S. Gold, and D. S. Kliger	*Inorg. Chem.* 29 (1990) 2506–2511
1998	ECD	Charge-transfer excited state properties of chiral transition metal coordination compounds studied by chiroptical spectroscopy	M. Ziegler and A. von Zelewsky	*Coord. Chem. Rev.* 177 (1998) 257–300
1998	ECD	Zinc porphyrin tweezer in host–guest complexation: Determination of absolute configurations of diamines, amino acids, and amino alcohols by circular dichroism	X. Huang, B. H. Rickman, B. Borhan, N. Berova, and K. Nakanishi	*J. Am. Chem. Soc.* 120 (1998) 6185–6186
2001	VCD	Ab initio VCD calculation of a transition-metal containing molecule and a new intensity enhancement mechanism for VCD	Y. He, X. Cao, L. A. Nafie, and T. B. Freedman	*J. Am. Chem. Soc.* 123 (2001) 11320–11321
2002	VCD	Density functional theory calculations of vibrational circular dichroism in transition metal complexes: Identification of solution conformations and mode of chloride ion association for (+)-tris(ethylenediaminato) cobalt(III)	T. B. Freedman, X. Cao, D. A. Young, and L. A. Nafie	*J. Phys. Chem. A* 106 (2002) 3560–3565
2003	ECD	Density functional calculations on electronic circular dichroism spectra of chiral transition metal complexes	J. Autschbach, F. E. Jorge, and T. Ziegler	*Inorg. Chem.* 42 (2003) 2867–2877
2003	ECD	Systematic investigation of modern quantum chemical methods to predict electronic circular dichroism spectra	C. Diedrich and S. Grimme	*J. Phys. Chem. A* 107 (2003) 2524–2539
2003	ECD	On the origin of the optical activity in the d–d transition region of tris-bidentate Co(III) and Rh(III) complexes	F. E. Jorge, J. Autschbach, and T. Ziegler	*Inorg. Chem.* 42 (2003) 8902–8910
2004	ECD	Synthesis and stereochemical properties of chiral square complexes of iron(II)	T. Bark, A. von Zelewsky, D. Rappoport, M. Neuburger, S. Schaffner, J. Lacour, and J. Jodry	*Chem. Eur. J.* 10 (2004) 4839–4845

(Continued)

TABLE A9.1 ***(Continued)***

Chiroptical Spectroscopic Applications to Transition Metal Complexes

Year	Subject	Title	Authors	Reference
2005	VCD	A chiral layered Co(II) coordination polymer with helical chains from achiral materials	G. Tian, G. S. Zhu, X. Y. Yang, Q. R. Fang, M. Xue, J. Y. Sun, Y. Wei, and S. L. Qiu	*Chem. Commun.* 41 (2005) 1396–1398
2005	ECD	On the origin of optical activity in tris-diamine complexes of Co(III) and Rh(III): A simple model based on time-dependent density function theory	F. E. Jorge, J. Autschbach, and T. Ziegler	*J. Am. Chem. Soc.* 127 (2005) 975–985
2005	ECD	Density functional calculation of the electronic circular dichroism spectra of the transition metal complexes [M(phen)3]2+ (M = Fe, Ru, Os)	B. L. Guennic, W. Hieringer, A. Gorling, and J. Autschbach	*J. Phys. Chem. A* 109 (2005) 4836–4846
2006	VCD	Dramatic effects of d-electron configurations on vibrational circular dichroism spectra of tris(acetylacetonato)metal(III)	H. Sato, T. Taniguchi, K. Monde, S. I. Nishimura, and A. Yamagishi	*Chem. Lett.* 35 (2006) 364–365
2006	VCD	Synthesis and vibrational circular dichroism of enantiopure chiral oxorhenium(V) complexes containing the hydrotris(1-pyrazolyl) borate ligand	P. R. Lassen, L. Guy, I. Karame, T. Roisnel, N. Vanthuyne, C. Roussel, X. Cao, R. Lombardi, J. Crassous, T. B. Freedman, and L. A. Nafie	*Inorg. Chem.* 45 (2006) 10230–10239
2006	ECD	The electronic structure and spectra of spin-triplet ground state bis(biuretato)cobalt(III) coordination compounds	P. W. Thulstrup and E. Larsen	*Dalton Trans.* 35 (2006) 1784–1789
2006	ECD	Circular dichroism spectrum of [Co(en)3]3+ in water: A discrete solvent reaction field study	L. Jensen, M. Swart, P. Th. van Duijnen, and J. Autschbach	*Int. J. Quant. Chem.* 106 (2006) 2479–2488
2006	ECD	Density-functional theory investigation of the geometric, energetic, and optical properties of the cobalt(II)tris(2,2′-bipyridine) complex in the high-spin and the Jahn–Teller active low-spin states	A. Vargas, M. Zerara, E. Krausz, A. Hauser, and L. M. L. Daku	*J. Chem. Theory Comput.* 2 (2006) 1342–1359

(Continued)

TABLE A9.1 (*Continued*)

Chiroptical Spectroscopic Applications to Transition Metal Complexes

Year	Subject	Title	Authors	Reference
2007	VCD	Effects of central metal ions on vibrational circular dichroism spectra of Tris-(β-diketonato) metal(III) complexes	H. Sato, T. Taniguchi, A. Nakahashi, K. Monde, and A. Yamagishi	*Inorg. Chem.* 46 (2007) 6755–6766
2007	VCD	Vibrational circular dichroism spectroscopy of a spin-triplet bis-(biuretato) cobaltate (III) coordination compound with low-lying electronic transitions	C. Johannessen and P. W. Thulstrup	*Dalton Trans.* 36 (2007) 1028–1033
2007	VCD	Resolution of enantiomers in solution and determination of the chirality of extended metal atom chains	D. W. Armstrong, F. A. Cotton, A. G. Petrovic, P. L. Polavarapu, and M. M. Warnke	*Inorg. Chem.* 46 (2007) 1535–1537
2007	ECD	Density functional theory applied to calculating optical and spectroscopic properties of metal complexes: NMR and optical activity	J. Autschbach	*Coord. Chem. Rev.* 251 (2007) 1796–1821
2007	ECD	Synthetic molecular motors and mechanical machines	E. R. Kay, D. A. Leigh, and F. Zerbetto	*Angew. Chem., Int.* Ed. 46 (2007) 72–191
2007	OR	Computation of optical rotation using time–dependent density functional theory	J. Autschbach	*Comp. Lett.* 3 (2007) 131–150
2008	VCD	Determination of the absolute configurations of chiral organometallic complexes via density functional theory calculations of their vibrational circular dichroism spectra: The chiral chromium tricarbonyl complex of N-pivaloyl-tetrahydroquinoline	P. J. Stephens, F. J. Devlin, C. Villani, F. Gasparrini, and S. L. Mortera	*Inorg. Chim. Acta* 361 (2008) 987–999
2008	ROA	Raman optical activity spectra of chiral transition metal complexes	S. Luber and M. Reiher	*Chem. Phys.* 346 (2008) 212–223
2008	ECD	On the origin of circular dichroism in trigonal dihedral cobalt (III) complexes of unsaturated bidentate ligands	J. Fan and T. Ziegler	*Inorg. Chem.* 47 (2008) 4762–4773

(*Continued*)

TABLE A9.1 (*Continued*)

Chiroptical Spectroscopic Applications to Transition Metal Complexes

Year	Subject	Title	Authors	Reference
2008	ECD	On the origin of circular dichroism in trigonal dihedral d6 complexes of bidentate ligands containing only σ-orbitals: A qualitative model based on a density functional theory study of Λ-[Co(en)3]3+	J. Fan and T. Ziegler	*Chirality* 20 (2008) 938–950
2008	ECD	Circular dichroism of trigonal dihedral chromium(iii) complexes: A theoretical study based on open-shell time-dependent density functional theory	J. Fan, M. Seth, J. Autschbach, and T. Ziegler	*Inorg. Chem.* 47 (2008) 11656–11668
2008	ECD	Determination of absolute configuration of chiral hemicage metal complexes using time-dependent density functional theory	F. J. Coughlin, K. D. Oyler, R. A. Jr. Pascal, and S. Bernhard	*Inorg. Chem.* 47 (2008) 974–979
2009	VCD	Syntheses and vibrational circular dichroism spectra of the complete series of [Ru((−)- or (+)-tfac) n(acac)3−n] (n = 0 3, tfac = 3-trifluoroacetylcamphorato and acac = Acetylacetonato)	H. Sato, Y. Mori, Y. Fukuda, and A. Yamagishi	*Inorg. Chem.* 48 (2009) 4353–4361
2009	VCD	Enhancement of IR and VCD intensities due to charge transfer	V. P. Nicu, J. Autschbach, and E. J. Baerends	*Phys. Chem. Chem. Phys.* 11 (2009) 1526–1538
2009	VCD	Subtle chirality in oxo- and sulfidorhenium(V) complexes	F. De Montigny, L. Guy, G. Pilet, N. Vanthuyne, C. Roussel, R. Lombardi, T. B. Freedman, L. A. Nafie, and J. Crassous	*Chem. Comm.* 46 (2009) 4841–4843
2009	VCD	Two 3D chiral coordination polymers with 4-connected 6^6 topological net: Synthesis, structure and magnetic properties	Z. B. Han, J. W. Ji, H. Y. An, W. Zhang, G. X. Han, G. X. Zhang, and L. G. Yang	*Dalton Trans.* 38 (2009) 9807–9811
2009	ECD	Theoretical analysis of the individual contributions of chiral arrays to the chiroptical properties of tris-diamine ruthenium chelates	Y. Wang, Y. K. Wang, J. M. Wang, Y. Liu, and Y. T. Yang	*J. Am. Chem. Soc.* 131 (2009) 8839–8847

(*Continued*)

TABLE A9.1 (***Continued***)

Chiroptical Spectroscopic Applications to Transition Metal Complexes

Year	Subject	Title	Authors	Reference
2009	ECD	Metal–Bis(helicene) assemblies incorporating π-conjugated phosphole-azahelicene ligands: Impacting chiroptical properties by metal variation	S. Graule, M. Rudolph, N. Vanthuyne, J. Autschbach, C. Roussel, J. Crassous, and R. Réau	*J. Am. Chem. Soc.* 131 (2009) 3183–3185
2009	ECD	Computing chiroptical properties with first-principles theoretical methods: Background and illustrative examples	J. Autschbach	*Chirality* 21 (2009) E116–E152
2009	ECD	Optically active oxo(phthalocyaninato) vanadium(iv) with geometric asymmetry: Synthesis and correlation between the circular dichroism sign and conformation	N. Kobayashi, F. Narita, K. Ishii, and A. Muranaka	*Chem. A Eur. J.* 15 (2009) 10173–10181
2009	ECD	Metal-bis(helicene) assemblies incorporating pi-conjugated phosphole-azahelicene ligands: Impacting chiroptical properties by metal variation	S. Graule, M. Rudolph, N. Vanthuyne, J. Autschbach, C. Roussel, J. Crassous, and R. Réau	*J Am Chem Soc.* 131(9) (2009) 3183–31855
2009	ECD	Probing molecular chirality by CD-sensitive dimeric metalloporphyrin hosts	N. Berova, G. Pescitelli, G. Ana, and G. Proni	*Chem. Commun.* (2009) 5958–5980
2010	VCD	Conformational analysis and vibrational circular dichroism study of a chiral metallocene catalyst	C. Merten, M. Amkreutz, and A. Hartwig	*J. Mol. Struct.* 970 (2010) 101–105
2010	*VCD*	Vibrational circular dichroism of Δ-SAPR-8-tetrakis[(+)-heptafluorobutyrylcamphorato] lanthanide(III) complexes with an encapsulated alkali metal ion	H. Sato, D. Shirotani, K. Yamanari, and S. Kaizaki	*Inorg. Chem.* 49 (2010) 356–358
2010	VCD	Synthesis, structure and chiroptical study of chiral macrocyclic imine nickel(II) coordination compounds derived from camphor	T. Wu, C. H. Li, Y. Z. Li, Z. G. Zhang, and X. Z. You	*Dalton Trans.* 39 (2010) 3227–3232
2010	ROA	Prediction of Raman optical activity spectra of chiral 3-acetylcamphorato-cobalt complexes	S. Luber, and M. Reiher	*Chem. Phys. Chem.* 11 (2010) 1876–1887
2010	ROA	Observation of resonance electronic and non-resonance-enhanced vibrational natural Raman optical activity	C. Merten, H. Li, X. Lu, A. Hartwig, and L. A. Nafie	*J. Raman Spectrosc.* 41 (2010) 1563–1565

(*Continued*)

TABLE A9.1 (*Continued*)
Chiroptical Spectroscopic Applications to Transition Metal Complexes

Year	Subject	Title	Authors	Reference
2010	ECD	Electronic structure and circular dichroism of tris bipyridyl metal complexes within density functional theory	J. Fan, J. Autschbach, and T. Ziegler	*Inorg. Chem.* 49 (2010) 1355–1362
2010	ECD	Transition metal-based chiroptical switches for nanoscale electronics and sensors	J. W. Canary, S. Mortezaei, and J. Liang	*Coord. Chem. Rev.* 254 (2010) 2249–2266
2010	ECD	Metallahelicenes: Easily accessible helicene derivatives with large and tunable chiroptical properties	L. Norel, M. Rudolph, N. Vanthuyne, J. A. G. Williams, C. Lescop, C. Roussel, J. Autschbach, J. Crassous, and R. Réau	*Angew. Chem. Int. Ed.* 49 (2010) 99–102
2010	ECD	Metallahelicenes: easily accessible helicene derivatives with large and tunable chiroptical properties	L. Norel, M. Rudolph, N. Vanthuyne, J. A. Williams, C. Lescop, C. Roussel, J. Autschbach, J. Crassous, and R. Réau	*Angew. Chem. Int. Ed. Engl.* 49(1) (2010) 99–102
2010	ECD	Assembly of pi-conjugated phosphole azahelicene derivatives into chiral coordination complexes: An experimental and theoretical study	S. Graule, M. Rudolph, W. Shen, J. A. Williams, C. Lescop, J. Autschbach, J. Crassous, and R. Réau	*Chemistry* 16(20) (2010) 5976–6005
2010	ECD	CD-sensitive Zn-porphyrin tweezer host-guest complexes. Part 1: MC/OPLS-2005 computational approach for predicting preferred intraporphyrin helicity	A. G. Petrovic, Y. Chen, G. Pescitelli, N. Berova, and G. Proni	*Chirality* 22(1) (2010) 129–139
2010	ECD	CD-sensitive Zn-porphyrin tweezer host-guest complexes. Part 2: Cis and trans 3-Hydroxy-4-aryl/alkyl-Lactams. A case study	Y. Chen, A. G. Petrovic, M. Roje, G. Pescitelli, M. M. Kayser, Y. Yang, N. Berova, and G. Proni	*Chirality* 22(1) (2010) 140–152

(*Continued*)

TABLE A9.1 (*Continued*)

Chiroptical Spectroscopic Applications to Transition Metal Complexes

Year	Subject	Title	Authors	Reference
2011	VCD	Theoretical study on vibrational circular dichroism spectra of tris(acetylacetonato)metal(III) complexes: Anharmonic effects and low-lying excited states	H. Mori, A. Yamagishi, and H. Sato	*J. Chem. Phys.* 135 (2011) 084506
2011	VCD	Conformational studies on chiral rhodium complexes by ECD and VCD spectroscopy	G. Szilvágyi, Z. Majer, E. Vass, and M. Hollósi	*Chirality* 23 (2011) 294–299
2011	VCD	Chirality and diastereoselection of δ/λ-configured tetrahedral zinc complexes through enantiopure Schiff base complexes: Combined vibrational circular dichroism, density functional theory, 1H NMR, and X-ray structural studies	A. C. Chamayou, S. Lüdeke, V. Brecht, T. B. Freedman, L. A. Nafie, and C. Janiak	*Inorg. Chem.* 50 (2011) 11363–11374
2011	VCD	Optically active tripodal dendritic polyoxometalates: Synthesis, characterization and their use in asymmetric sulfide oxidation with hydrogen peroxide	C. Jahier, M. F. Coustou, M. Cantuel, N. D. McClenaghan, T. Buffeteau, D. Cavagnat, M. Carraro, and S. Nlate	*Eur. J. Inorg. Chem.* 2011 (2011) 727–738
2011	ROA	Comparative study of measured and computed Raman optical activity of a chiral transition metal complex	C. Johannessen, L. Hecht, and C. Merten	*Chem. Phys. Chem.* 12 (2011) 1419–1421
2011	ECD	Time-dependent density functional response theory for electronic chiroptical properties of chiral molecules	J. Autschbach, L. Nitsch-Velasquez, and M. Rudolph	*Top. Curr. Chem.* 298 (2011) 1–98
2011	ECD	A theoretical study on the exciton circular dichroism of propeller-like metal complexes of bipyridine and tripodal tris(2-pyridylmethyl)amine derivatives	J. Fan and T. Ziegler	*Chirality* 23 (2011) 155–166
2011	ECD	Exciton coupling in coordination compounds	S. G. Telfer, T. M. McLean, and M. R. Waterland	*Dalton Trans.* 40 (2011) 3097–3108

(*Continued*)

TABLE A9.1 (*Continued*)

Chiroptical Spectroscopic Applications to Transition Metal Complexes

Year	Subject	Title	Authors	Reference
2011	ECD	From hetero- to homochiral bis(metallahelicene)s based on a PtIII–PtIII bonded scaffold: isomerization, structure, and chiroptical properties	E. Anger, M. Rudolph, C. Shen, N. Vanthuyne, L. Toupet, C. Roussel, J. Autschbach, J. Crassous, and R. Réau	*J. Am. Chem. Soc.* 133 (2011) 3800–3803
2011	ECD	Multifunctional and reactive enantiopure organometallic helicenes: Tuning chiroptical properties by structural variations of mono- and bis(platinahelicene)s	E. Anger, M. Rudolph, L. Norel, S. Zrig, C. Shen, N. Vanthuyne, L. Toupet, J. A. Williams, C. Roussel, J. Autschbach, J. Crassous, and R. Réau	*Chemistry* 17(50) (2011) 14178–14198
2011	ECD	Bulky melamine based Zn-porphyrin tweezer as a CD probe of molecular chirality	A. G. Petrovic, G. Vantomme, Y. L. Negrón-Abril, E. Lubian, G. Saielli, I. Menegazzo, R. Cordero, G. Proni, K. Nakanishi, T. Carofiglio, and N. Berova	*Chirality* 23 (2011) 808–819
2011	ECD,ORD	Calculation of optical rotatory dispersion and electronic circular dichroism for tris-bidentate groups 8 and 9 metal complexes, with emphasis on exciton coupling	M. Rudolph,and J. Autschbach	*J. Phys. Chem. A* 115(12) (2011) 2635–2649
2011	ECD	Role of morphology in the enhanced optical activity of ligand-protected metal nanoparticles	C. Noguez, A. Sánchez-Castillo, and F. Hidalgo	*J. Phys. Chem. Lett.* 2 (2011) 1038–1044

(*Continued*)

TABLE A9.1 (*Continued*)

Chiroptical Spectroscopic Applications to Transition Metal Complexes

Year	Subject	Title	Authors	Reference
2012	VCD	Shape-conserving enhancement of vibrational circular dichroism in lanthanide complexes	S. Lo Piano, S. Di Pietrob, and L. Di Bari	*Chem. Commun.* 48 (2012) 11996–11998
2012	VCD	Effects of electron configuration and coordination number on the vibrational circular dichroism spectra of metal complexes of trans-1,2-diaminocyclohexane	C. Merten, K. Hiller, and Y. J. Xu	*Phys. Chem. Chem. Phys.* 14 (2012) 12884–12891
2012	VCD	Chirality effects on core-periphery connection in a star-burst type tetranuclear Ru(III) complex: Application of vibrational circular dichroism spectroscopy	H. Sato, F. Sato, M. Taniguchi, and A. Yamagishi	*Dalton Trans.* 41 (2012) 1709–1712
2012	VCD	VCD spectroscopy probing of weak intermolecular interactions between copper coordination compounds and N-blocked amino acids	T. Wu, C. H. Li, and X. Z. You	*Vib. Spectrosc.* 63 (2012) 451–459
2012	ROA	Detection of molecular chirality by induced resonance Raman optical activity in europium complexes	S. Yamamoto, and P. Bouř	*Angew. Chem. Int. Ed.* 51 (2012) 11058–11061
2012	ECD	Ab initio electronic circular dichroism and optical rotatory dispersion: From organic molecules to transition metal complexes	J. Autschbach	in *Comprehensive Chiroptical Spectroscopy: Instrumentation, Methodologies, and Theoretical Simulations,* Volume 1, John Wiley & Sons, Inc., NJ (2012) p. 593–642

(*Continued*)

TABLE A9.1 (*Continued*)
Chiroptical Spectroscopic Applications to Transition Metal Complexes

Year	Subject	Title	Authors	Reference
2012	ECD	Application of electronic circular dichroism to inorganic stereochemistry	S. Kaizaki	in *Comprehensive Chiroptical Spectroscopy: Applications in Stereochemical Analysis of Synthetic Compounds, Natural Products, and Biomolecules,* Volume 2, John Wiley & Sons, Inc., NJ (2012) p. 451–471
2012	ECD	Vibrational and electronic circular dichroism monitoring of copper(II) coordination with a chiral ligand	T. Wu, X. P. Zhang, C. H. Li, P. Bouř, Y. Z. Li, and X. Z. You	*Chirality* 24 (2012) 451–458
2012	ECD	An exciton-coupled circular dichroism protocol for the determination of identity, chirality, and enantiomeric excess of chiral secondary alcohols	L. You, G. Pescitelli, E. V. Anslyn, and L. Di Bari	*J. Am. Chem. Soc.* 134 (2012) 7117–7125
2012	ECD	Ruthenium-vinylhelicenes: Remote metal-based enhancement and redox switching of the chiroptical properties of a helicene core	E. Anger, M. Srebro, N. Vanthuyne, L. Toupet, S. Rigaut, C. Roussel, J. Autschbach, J Crassous, and R. Réau	*J. Am. Chem. Soc.* 134 (2012), 15628–15631
2012	ECD	Chiroptical spectra of tetrakis (+)-3-heptafluorobutylrylcamphorate Ln(III) complexes with an encapsulated alkali metal ion: Solution structures as revealed by chiroptical spectra	D. Shirotani, K. Yamanari, R. Kuroda, T. Harada, J. L. Lunkley, G. Muller, H. Sato, and S. Kaizaki	*Chirality* 24(12) (2012) 1055–1062
2012	VCD	Chirality in copper nanoalloy clusters	H. Elgavi, C. Krekeler, R. Berger, and D. Avnir	*J. Phys. Chem. C* 116(1) (2012) 330–335

(*Continued*)

TABLE A9.1 (*Continued*)

Chiroptical Spectroscopic Applications to Transition Metal Complexes

Year	Subject	Title	Authors	Reference
2012	ECD	Conformations of [(R,R)-1,5-diaza-cis-declain] copper(II)complex and its hydrogen bonding interaction with the crystal water: A combined experimental VA, UV-Vis and ECD spectroscopic and DFT study	Z. Dezhahang, M. R. Poopari, and Y. Xu	*J. Mol. Struct.* 1024 (2012) 123–131
2012	ROA	Simultaneous resonance Raman optical activity involving two electronic states	C. Merten C, H. Li, and L. A. Nafie	*J. Phys. Chem. A* 116 (2012) 7329–7336
2013	VCD	A chiral rhenium complex with predicted high parity violation effects: Synthesis, stereochemical characterization by VCD spectroscopy and quantum chemical calculations	N. Saleh, S. Zrig, T. Roisnel, L. Guy, R. Bast, T. Saue, B. Darquie, and J. Crassous	*Phys. Chem. Chem. Phys.* 15 (2013) 10952–10959
2013	VCD	Chelate structure of a dirhodium–amino acid complex identified by chiroptical and NMR spectroscopy	Z. Majer, G. Szilvágyi, L. Benedek, A. Csámpai, M. Hollósi, and E. Vass	*Eur. J. Inorg. Chem.* 2013 (2013) 3020–3027
2013	VCD	Dirhodium complexes: Determination of absolute configuration by the exciton chirality method using VCD spectroscopy	G. Szilvágyi, B. Brém, G. Báti, L. Tölgyesi, M. Hollósi, and E. Vass	*Dalton Trans.* 42 (2013) 13137–13144
2013	VCD	A comparative vibrational CD study of homo- and heteroleptic complexes of the type [Cu(trans-1,2-diaminocyclohexane)2L](ClO4)2	C. Merten and Y. J. Xu	*Dalton Trans.* 42 (2013) 10572–10578
2013	VCD	A novel correlation of vibrational circular dichroism spectra with the electronic ground state for D-SAPR-8-cesium-tetrakis((+)-heptafluorobutyryl-camphorato) lanthanide(III) complexes	S. Kaizaki, D. Shirotani, and H. Sato	*Phys. Chem. Chem. Phys.* 15 (2013) 9513–9515
2013	VCD	VCD studies on chiral characters of metal complex oligomers	H. Sato, and A. Yamagishi	*Int. J. Mol. Sci.* 14 (2013) 964–978
2013	VCD	Stereo-chemical analysis of racemization of a chiral bipyridine	T. Wu, X. P. Zhang, and X. Z. You	*RSC Adv.* 3 (2013) 26047–26051
2013	VCD	Circular dichroism spectroscopy study of crystalline-to-amorphous transformation in chiral platinum(II) complexes	X. P. Zhang, T. Wu, J. Liu, J. C. Zhao, C. H. Li, and X. Z. You	*Chirality* 25 (2013) 384–392

(*Continued*)

TABLE A9.1 (*Continued*)

Chiroptical Spectroscopic Applications to Transition Metal Complexes

Year	Subject	Title	Authors	Reference
2013	ECD	Dimolybdenum tetracarboxylates as auxiliary chromophores in chiroptical studies of vic-diols	M. Jawiczuk, M. Górecki, A. Suszczyńska, M. Karchier, J. Jaźwiński, and J. Frelek	*Inorg. Chem.* 52 (2013) 8250–8263
2013	ECD	TDDFT studies on the determination of the absolute configurations and chiroptical properties of Strandberg-type polyoxometalates	Y. -M. Sang, L. -K. Yan, N. -N. Ma, J. -P. Wang, and Z. -M. Su	*J. Phys. Chem. A* 117 (2013) 2492–2498
2013	ECD	Assembly of helicene-capped N,P,N,P,N-helicands within Cu(I) helicates: Impacting chiroptical properties by ligand-ligand charge transfer	V. Vreshch, M. El Sayed Moussa, B. Nohra, M. Srebro, N. Vanthuyne, C. Roussel, J. Autschbach, J. Crassous, C. Lescop, and R. Réau	*Angew. Chem. Int. Ed. Engl.* 52(7) (2013) 1968–1972
2013	ECD	Chiroptical properties of carbo[6] helicene derivatives bearing extended π-conjugated cyano substituents	S. Moussa Mel, M. Srebro, E. Anger, N. Vanthuyne, C. Roussel, C. Lescop, J. Autschbach, and J. Crassous	*Chirality* 25(8) (2013) 455–65. doi: 10.1002/chir.22201
2013	ECD	Diastereo- and enantioselective synthesis of organometallic bis(helicene)s by a combination of C-H activation and dynamic isomerization	C. Shen, E. Anger, M. Srebro, N. Vanthuyne, L. Toupet, C. Roussel, J. Autschbach, R. Réau, and J. Crassous	*Chemistry* 19(49) (2013) 16722–16728
2014	VCD	Amplified vibrational circular dichroism as a probe of local biomolecular structure	S. R. Domingos, A. Huerta-Viga, L. Baij, S. Amirjalayer, D. A. E. Dunnebier, A. J. C. Walters, M. Finger, L. A. Nafie, B. de Bruin, W. J. Buma, and S. Woutersen	*J. Am. Chem. Soc.* 136 (2014) 3530–3535

(*Continued*)

TABLE A9.1 ***(Continued)***

Chiroptical Spectroscopic Applications to Transition Metal Complexes

Year	Subject	Title	Authors	Reference
2014	VCD	Strong solvent-dependent preference of Δ and Λ stereoisomers of a tris(diamine)nickel(ii) complex revealed by vibrational circular dichroism spectroscopy	C. Merten, R. McDonald, and Y. J. Xu	*Inorg. Chem.* 53 (2014) 3177–3182
2014	VCD	Chirality transfer in magnetic coordination complexes monitored by vibrational and electronic circular dichroism	T. Wu, X. P. Zhang, X. Z. You, Y. Z. Li, and P. Bouř	*Chem. Plus. Chem.* 79 (2014) 698–707
2014	VCD	Vapor-induced chiroptical switching in chiral cyclometalated platinum(II) complexes with pinene functionalized C^N^N ligands	X. P. Zhang, T. Wu, J. Liu, J. X. Zhang, C. H. Li, and X. Z. You	*J. Mater. Chem. C* 2 (2014) 184–194
2014	ECD	Induced chirality-at-metal and diastereoselectivity at Δ/Λ-configured distorted square-planar copper complexes by enantiopure Schiff base ligands: Combined circular dichroism, DFT and X-ray structural studies	M. Enamullah, A. K. M. Royhan Uddin, G. Pescitelli, R. Berardozzi, G. Makhloufi, V. Vasylyeva, A. Chamayoud, and C. Janiak	*Dalton Trans.* 43 (2014) 3313–3329
2014	ECD	Porphyrin tweezer receptors: Binding studies, conformational properties and applications	V. Valderreya, G. Aragaya, and P. Ballester	*Coord. Chem. Rev.* 258–259 (2014) 137–156
2014	ECD	Straightforward access to mono- and bis-cycloplatinated helicenes that display circularly polarized phosphorescence using crystallization resolution methods	C. Shen, E. Anger, M. Srebro, N. Vanthuyne, K. K. Deol, T. D., Jr., Jefferson, G. Muller, J. A. Gareth Williams, L. Toupet, C. Roussel, J. Autschbach, R. Réau, and J. Crassous	*Chem. Sci.* 5 (2014) 1915–1927
2014	ROA	Where does the Raman optical activity of [Rh(en)3]3+ come from? Insight from a combined experimental and theoretical approach	M. Humbert-Droz, P. Oulevey, L. M. L. Daku, S. Luber, H. Hagemann, and T. Bürgi	*Phys. Chem. Chem. Phys.* 16 (2014) 23260–23273

Index

B

F

M

N

Q

R

S

U

V